AF529807

Direct Taxes Glossary

(1860-2012)

Compilation
Rājendra. IRS

Published by:

Scientific Publishers (India)
5 A, New Pali Road, P.O. Box 91
Jodhpur 342 001 (India)

Branch Office
Scientific Publishers (India)
4806/24, Ansari Road, Daryaganj
New Delhi - 110 002 (India)

E-mail: info@scientificpub.com
Website: www.scientificpub.com

ISBN: 978-81-7233-834-3

Typeset by Rajesh Ojha
Printed in India

R.V. Easwar
Judge
HIGH COURT OF DELHI

AB-85, Shahjahan Road
New Delhi - 110 013
Res. : 011-23389300
Off. : 011-23382628

FOREWORD

This compilation fulfils the long felt need of the tax practitioners for a handy digest of case-law. I have read the draft of this digest and have found its content to be pragmatic, informative and useful. It is a first initiative of its kind in recent times taken by Mr. Rajendra to compile and index the important words and phrases used in the field of tax-law tagged along with the relevant provisions of law and precedants. He has put in his hard work and rich experience both as an officer of the IRS and as a representative of the income-tax department before the Income Tax Appellate Tribunal into good use which is indeed an appreciable effort.

This digest sums up all the relevant judgments on the crucial words and phrases used in the tax jurisprudence. It does not limit itself only to the fiscal laws but also contains important words and phrases from other pieces of legislation like the Constitution of India and the Companies Act, 1956. Importance has been given from all the courts in the country and the terms used in the old taxation laws prevalent in the years 1860 and 1922 respectively. The names of the courts and the parties to the cases have also been provided in the contents section for easy reference. This shall help the users to locate the respective words, phrases and judgments

R.V. Easwar
Judge
HIGH COURT OF DELHI

AB-85, Shahjahan Road
New Delhi - 110 013
Res. : 011-23389300
Off. : 011-23382628

with least trouble. Not only the department officers but also the legal practitioners can take huge advantage from this case-law glossary.

I appreciate the sincere efforts made by Mr. Rajendra in compiling this digest and congratulate him on its publication.

Date: 02.11.2012

(R.V. EASWAR)

From the desk of the Compiler

Words *(Sabd)* are the most effective tools used by the human beings in their day to day life. Emotions make a living-being human as well as humane and the best vehicle to emote one's feelings is word. Discovery of fire, invention of wheel and development of language (word collection) are acknowledged as the foundation-stones of human civilization by historians and social scientists. Perhaps that is the reason that Indian tradition equates word with the Almighty *(Brahm)*.[1] Not only this, a word is considered to be a representative of *Aakas* (sky)-one of the five basic elements (*Panchmahabhoot*) of Universe.

Like the sky, words also have different colours, shades and textures and they change according to their context. This phenomenon is a result of journey of human-race from cave-age to chip-age. Transition of agrarian-rural-society into semi-urban cottage-industry based society was a major change when knowledge started replacing physical strength as a power-center. Systematic segregation of knowledge resulted in evolving various branches of science and humanities. It led to collection of words of a particular branch of knowledge. Indexes of words, dictionaries and glossaries were result of such specialisation. *Nighantu and Amarkosh* two of the most referred word-collections for Vedic and Sanskrit literature, are the living examples of efforts made by the scholars to arrange words in a systemic and scientific manner for the benefit of common users. Now a days we have dictionaries/glossaries for almost all subjects read or taught under the sun-from Aeronautics to Zoology.

Many words have more than one meanings (*Arth*). A word as simple as plant can convey different meanings to different people e.g. to a farmer, to a botanist, to an engineer, to an Assessing Officer, to a businessman. But, for taxation matters it has a definite meaning and courts expect that while deciding the issue regarding a plant definite, proper and legal meaning should be adhered to. In literary world importance of using an appropriate word after understanding its meaning is considered to be using *Kamdhenu.*[2] In the field of law same rule applies but it applies more strictly-especially in matters of taxation. At the time of framing statues it is not possible for legislation to define all the words used in a particular Act. Courts while deciding matters related to a particular Act interpret the words used in the statues. Taxation legislations are not exceptions to this rule. Hon'ble Supreme Court and High Courts have defined, explained and analysed many words of Indian Income tax, 1922 and Income tax Act, 1961 during the course tax-litigations. It is expected that professionals and tax administrators should be aware of such meanings.

So, on joining ITAT, Mumbai as Departmental Representative I searched for an exclusive glossary of Income-tax and I was amazed to find that such a compilation was not readily available. Income tax is being collected since 1860 and after independence Acts like Estate Duty, Wealth Tax Act and Gift Tax Act also became applicable. Thousands of words used in the taxation statutes have been defined by the courts. I tried to compile such words in the form of a glossary. As the compilation was first of its kind and I had no format or precedent available with me, so I tried presenting in a manner that would be useful for the users.As per the advice of some of the seniors it was submitted to the CBDT for publishing.A committee was constituted for evaluating the usefulness and relevance of the compilation and to my pleasant surprise it was decided that it will see the light of day. When the professionals came to know about the 'Legal Glossary' they asked for the copies of it. As the glossary was for **Departmental Use only**, I supplied them the soft copies. I am firm believer of the principle that knowledge should always be shared. Feedback and inputs from the colleagues of the department and professionals were quite encouraging and invaluable. Their unbiased and independent opinions helped me improve it and make it more user-friendly. On their positive suggestion sections of the words compiled have also been included in the present glossary and a section wise index has been prepared. Though the relation between words and their meanings have been considered as the bond between *Uma and Mahesvar*,[3] but I feel such a bond happens when meaning of a word is understood and used in a correct perspective. I hope my humble and small effort will help users to get desired bond of words and their meanings.

Glossary in your hands contains words not only of Income tax, Gift Tax and Wealth Tax but also a few words of Estate Duty Act, Companies Act, Interest Act, other allied Acts and Constitution of India. I have also picked up a few words from the Acts of 1860 to 1886 for historical references.Words which were not falling in any category have been categorised as miscellaneous.

I am indeed grateful to Shri R.V. Easver, Hon'ble justice of the High Court of Delhi, for penning the foreword of this compilation. I had the privilege of appearing before him as a DR and was highly impressed by his knowledge, hardwork, sincerity and fair approach. The prompt encouraging message sent by him on receiving a copy of the Legal Glossary is still fresh in my mind. I am very grateful to Shri Pawan Kumar Sharma of Scientific Publishers (India), Jodhpur and members of his team for designing and publishing the Direct-taxes Glossary. I specially thank those IRS officers and Chartered Accountants who gave their valuable suggestions to improve the quality of this compilation. I am also thankful to Dr. B.R. Bishnoi, a reliable source for Sanskrit literature, for his guidance and generous help.

I know I have selected only a few words in the compilation, but there is always scope of improvement. I have tried my best to compile the Glossary, rest is left to the learned users. Suggestions, opinions, comments and views are welcome at my following e-mail address- rajendra82010@yahoo.co.in

1. शब्द इति ब्रह्म. *Sabd iti Braham*
2. एकः शब्दः सम्यगधीतः सम्यक् प्रयुक्तः स्वर्गे लोके कामधुग्भवति
   Ekh sabdh samyagadheeth samyak prayukth swarge loke kaamdhugbhavati
3. वागर्थाविव सम्पृक्तौ . . . . पार्वतीपरमेश्वरौ.
   Vaagaarthaaviv sampriktau.......Parvateeparmeshvarau.

GLOSSARY

CONTENTS

		Khem	
Month	Kar.	B.V. Aswathaiah and Brothers - ITO	384
Month	Guj.	CIT - S.L.M. Maneklal Industries Ltd.	384
More and less	Bom.	Gilbs Computer Ltd.	384
More reliable data	Bom.	CIT -Great Eastern Life Insurance Co. Ltd.	384
More reliable data	PC	CIT -Great Eastern Life Assurance Co. Ltd.	384
Mutatis-mutandis	Pun.	Khushi Ram Raghunath Sahai, - CIT	385
Mutual association	Ori.	Cuttack Club Pvt. Ltd. - CIT	385
Mutual concern	Bom.	Ismailia Grain Merchants Association - CIT	385
Mutual concern	Guj.	CIT - Adarsh Co- op. Housing Society Ltd.	385
Mutuality	Guj.	Sports Club of Gujarat Ltd. - CIT	386
Mutuality	Bom.	CIT - Bombay Oilseeds and Oil Exchange Ltd.	386
Mutuality	SC	Chelmsford Club - CIT	386
Mutuality	Bom.		387
Mutuality principle	Mad.	CIT- South Indian Film Chamber of Commerce	387
Necessarily	Bom.	Vodafone International Holdings B.V.- U O I	389
Necessarily	Bom.	3i Infotech Ltd. - ACIT	389
Net value of the assets of the	Guj.	CWT- Raipur Manufacturing Co. Ltd.	389
Net wealth	Bom.	Seth Ramnath K. Daga - CWT	389
Net wealth tax	SC	CWT - P.C.Oswal	389
New	Ker	CIT - Cochin Company	390
New machinery	SC	Cochin Company - CIT	390
No appeal shall lie	SC	CIT - Filmistan Ltd.	390
No income liable to tax	Bom.	CIT - Allied Silk Mills	390
No profits or gains chargeable for	SC	CIT - Virmani Industries Pvt. Ltd.	391
Non recovery receipt	Bom.	Mehboob Productions Pvt. Ltd. - CIT	391
Non recurring nature	All.	Rani Amrit Kunwar - CIT	391
Non-resident	Guj.	Chimanbhai K.Patel - CWT	392
Non-resident	Guj.	Rambhai L. Patel - CIT	392
Not a business loss	Cal.	Raghubir Prasad Gupta - CIT	392
Not being a charge created by	Bom.	CIT - Tarachand Kalyanji	392

A

A controlling interest **Bom.** **Vodafone International Holdings B.V.- UOI**

A controlling interest does not for the purpose of the Income-tax Act, 1961 constitute a distinct capital asset. That is simply because the assumption of control is a right which emanates from the acquisition of a sufficient number of shares in the company as would enable the holder of the shares to exercise a voting power of a degree and nature as would result in a control of the management. A controlling interest is an incident of the ownership of the shares in a company; something which flows out of the holding of shares.A controlling interest is, therefore, not an identifiable or distinct capital asset independent of the holding of shares. *Sec. 2(14).* **[329ITR126]**

A day **Mad.** **CIT - Southern Sea Foods Ltd.**

A reading of rule 6D of the Income-tax Rules, 1962, would show that it does not specify a twenty-four hour period as comprising a day for the purpose of calculating travelling allowance. For an employee to claim allowance for a day, he need not be travelling all the twenty-four hour. *Rule-6D.***[215ITR176]**

A Hindu deity **SC** **Official Trustee of West Bengal - CIT**

A Hindu deity is an individual within the meaning of that word in section 3 of the Indian Income-tax Act, 1922, and can be treated as a unit of assessment. As a Hindu deity is a juristic person and can hold property and be in receipt of income and can also sue and be sued in a court of law, there is no reason why its income should be held to be outside the ambit of Taxation, if it can be brought within it without straining the language of the statutory provision. It would naturally be taxed through its shebaits who are in possession and management of its property. *Sec.4.* **[93ITR348]**

A mistake apparent on. SC T.S. Balaram, ITO - Volkart Brothers

A mistake apparent on the record must be an obvious and patent mistake and not something which can be established by a long drawn process of reasoning on points on which there may be conceivably two opinions. A decision on a debatable point of law is not a mistake apparent from the record. *Sec.154.* **[82ITR50]**

A mistake on the face of the record Kar. D.S. Srinivas - ITO

A mistake found on the face of the record must be an obvious and patent mistake and not something which can be established by a long drawn process of reasoning on points on which there may be conceivably two opinions. A decision on a debatable point of law is not a mistake apparent from the record. *Sec.154.***[183CTR269, 262ITR209,129Taxman 657]**

A mutual association AP. CIT-West Godavari District Rice Millers Association

A mutual association is an as association of persons who agree to contribute funds for some common purpose mutually beneficial and receive back the surplus left out in the same capacity in which they have made the contributions. Therefore, the capacity as contributors and participants remains the same. The participation envisaged in the principle of mutuality is not that the members should take the surplus to themselves. It is enough if they have a right of disposal over the surplus. *Sec.28(iii) .***[40CTR96,150ITR 394,17Taxman34]**

A proper notice Cal. Rameshwar Sirkar - ITO

The mere fact that the serving officer did not find the assessee to be served with the notice at his address is not sufficient to establish that he could not be found. It must be shown not only that the serving officer went to that place at a reasonable time when the assessee was expected to be present, but also that if he was not found, proper and reasonable attempts had been made to find him either at that address or elsewhere. A notice by affixture without reasonable attempts to find the assessee is not a proper notice.

Sec.282. **[88ITR374]**

A provision **AP. CIT - D.B.R. Mills Ltd.**

Section 36(1)(v) and the first limb of sub clause (i) of clause (b) in sub-section (7) of section 40A are identical, the only difference being that, while section 36 (1)(v) uses the expression any sum paid, section 40A(7)(b)(i) uses the expression any provision made, there by making the sweep of the latter provision far wider. A provision is an appropriation of money for a known and existing liability. It is not necessary that such liability is quantified, but it must certainly be a known and existing liability. Where a liability attaches or accrues by operation of law, the making of, or the failure to make, an entry (debit entry) in the books of account is immaterial. But where the liability does not attach, or accrue, by operation of law, an entry is necessary to show that a particular sum has been appropriated for a particular purpose. In other words, such an appropriation is necessary to show that a provision has been made by the assessee for a particular purpose which in this case is for the purpose of payment-into an approved gratuity fund.*Sec.40A(7)(b)(i).* **[72CTR148,174ITR442,40Taxman62]**

A separate business **Mad.** **Sri Ranga Vilas Ginning and Oil Mills - CIT**

If in the course of carrying on any business, speculative transactions are entered into along with non-speculative transactions, by reason of the statutory fiction in Explanation 2 to section 28 of the I.T. Act, 1961, the speculative transactions will constitute a separate business. In other words, where a trader carries on a regular business in spot dealings in a commodity and also indulges in speculative transactions in the same commodity, while in point of fact the business of the assessee may be regarded as an integrated business combining in itself not only spot transactions but also forward dealings, by virtue of Expln. 2 to section 28, the speculative transactions must be separated and regarded as constituting a separate business. *Sec.28Expl.2.* **[133ITR85]**

A settlement of the contract **SC** **CIT - Shantilal P. Ltd.**

A transaction cannot be described as a speculative transaction within the meaning of sub-section (5) of section 43 of the I.T. Act, where there is a breach of the contract and a dispute between the parties damages are awarded as compensation by an arbitration award. What is really settled by the award of such damages and their acceptance by the aggrieved party is the dispute between the parties Section 43(5), however, speaks of a settlement of the contract, and a contract is settled when it is either performed or the promise dispenses with or remits,wholly or in part, the performance of the promise made to him or accepts, instead of it, any satisfaction which he thinks fit. It is this sense of the law which must prevail in sub-section (5) of section 43 and not that of the layman. *Sec.43(5).* **[35CTR395,144ITR57,70Taxation166,14Taxman1]**

A speculative transaction / A breach of contract **AP.** **CIT - Sri Venkateswara Rice and Oil Mills**

The artificial definition of speculative transaction in section 43(5) renders every transaction, in which a contract for the purchase or sale of any commodity, including stocks and shares, is periodically or ultimately settled otherwise than by the actual delivery or transfer of the commodity or scrips,a speculative transaction. A contract is speculative if it is settled without actual delivery. If the contract is broken, i.e., for any reason one party is unable to give delivery or the other is unable to take delivery, it is a case of breach of the contract. It depends, therefore, on the facts and circumstances of the case as to whether there has actually been a breach of the contract or a settlement of the contract. A breach takes place on account of repudiation of the contract or failure to perform it or a contract may be terminated through frustration, impossibility or through any other cause which ends the contract under the Contract Act or the Sale of Goods Act. When the obligation to supply or take delivery, as the case may be, comes to an end by operation of law, it does not make the transaction speculative. However, if the parties mutually settle the contract without any intervening circumstances, then it may be a speculative transaction if all the other conditions of section 43(5) are satisfied. The material question is: Why was the contract settled? If it was settled by mutual consent to avoid delivery, then it would be speculative. If it was settled because of inability of the assessee to supply or on account of the fact that it did not have the necessary resources to give the delivery, then it would be a breach of contract. *Sec.43(5).* **[154ITR756]**

A substantial question of law **Cal.** **West Bengal State Electricity Board**

The phrase a substantial question of law means any question of law which affects the substance of the case. If the High Court is satisfied that the question is one of law, and if decided in favour of the prospective appellant will substantially affect the tax liability or some other matter of substance in the case,then the High Court is compelled to entertain the appeal. It cannot entertain appeals only on points of law which are somehow thought by it to be sufficiently serious. *Sec.260A(1).***[165CTR502,248 ITR152,128Taxman535]**

Adventure in the nature of trade **Del.** **CIT- Raunaq Singh Swaran Singh**

Profit-making would normally not be an irrelevant consideration for every honest and prudent purchaser but will not in every case make the purchase a venture in the nature of trade. To bring a transaction within this category it has to be shown that the sole intention at the time of purchase was to sell the property purchased later on at a profit. *Sec.2(13).***[85ITR220,32Taxation23]**

Abatement **Jhar.** **Abhay Kumar Shroff - CIT**

The second proviso to section 153A makes it clear that assessment or reassessment relating to any assessment year falling within the period of six assessment years pending on the date of initiation of the search under section 132 or requisition under section 132A shall abate. In other words, if on the date of initiation of search or requisition under section 132 or section 132A any assessment or reassessment proceeding is initiated relating to any assessment year falling within the period of six assessment years, it shall stand abated and the assessing authority cannot and shall not proceed with such pending assessment after initiation of search or requisition. The word abatement means the act of eliminating or nullifying or suspension or defeat of a pending action. *Sec.153 A.* **[290ITR114]**

2. The word abatement is referable to something which is pending, alive, or is subject to deduction. Abatement refers to suspension or termination of proceedings either of the main action, or proceedings ancillary or collateral to it. Proceedings which have already terminated are not liable for abatement unless the statute expressly provides for it. **CIT - Smt. Shaila Agarwal, All.** *Sec.153A.* **[246CTR266,346 ITR130,204Taxman276]**

Accommodation advance SC Amarchand Sobhachand - CIT

An accommodation advance is a neutral expression: it may be of the nature of a loan advanced in the ordinary course of business by a money-lender; it may be an advance in the money-lending or other business of the assessee but not in the nature of a loan; or it may be wholly unrelated to the business of the taxpayer. *Sec.36(1)(vii).* **[82ITR591]**

Accompanied by a fee of Rs.100/- Mad. Nagappa Chettiar - CIT

The words accompanied by a fee of Rs. 100 in section 66(1) of the Indian Income-tax Act, 1922, should not be given a too literal interpretation. Accompanied cannot mean necessarily that the sum of Rs. 100 or something representing that sum should be contained in the same envelope as the application or that both the application and the money should be delivered together at the same time. The application however is not complete without the payment of the fee. A reasonable construction of this requirement would be that the assessee should have made the payment of the fee in such time that in the ordinary course it would either be received or deemed to be received within the time allowed. *Sec.256.* **[26 ITR741]**

Accounting SC J.K. Ind. Ltd. - Union of India

Accounting is now considered to be a service activity : its primary function is to provide quantitative information, primarily of financial nature about economic entities. The primary role of accounting is to provide an effective measurement and reporting system. Accounting today includes several branches, e.g. Financial Accounting, Management Accounting and Government Accounting.

Misc. **[213CTR301,297ITR176,165 Taxman 323]**

Accrual Mad. CIT - Annapurani Veerappan

The notion of real income cannot be brought into play where income had accrued according to the accounts of the assessee and there is no indication of the assessee treating the amounts as not having accrued. After the event of accrual, on a mere ipse dixit of the assessee, no reversal of the situation can be brought about and the concept of real income cannot be so used as to make the accrued income non-income simply because, after the event of accrual, the assessee neither decides to treat it as a bad debt nor claims deduction under section 36(2) of the Income Tax Act, 1961, but still enters the same with a diminished hope of recovery in the suspense account. The extension of the concept of real income to this field, to negate accrual after the amount had become payable, is contrary to the postulates of the Income Tax Act and where interest had accrued and the assessee had debited the account of the debtor, the difficulty of recovery would not make its accrual a non-accrual.

*Sec.36(1)(vii).***[193ITR426]**

Accrue **Ker.** **T.O. Hydrose - CED**

The word accrue used in section 34(4) of the Act means, according to dictionaries,to fall (to anyone) as a natural growth or increment; to come as an accession or advantage, arise or spring as a natural growth especially interest. Accordingly to section 34(4) of the Act, interpreted in the light of this meaning, if, in the cash or property gifted either by way of an out and out gift or by way of a settlement, the donee gets absolute domination and he invests the sam in any manner he likes, then the profit or the income resulting from such investment will not form part of the estate of the deceased. In such a case, the profit or income cannot be said to have accrued to have resulted as a natural growth, accession or increment from the property gifted. On the other hand, if the property gifted yields as a natural growth or result, without the intervention of the donee, income or profits, then such income or profits can be said to have accrued from the property gifted, with the result that the income or profit will also form part of the estate left by the deceased.

[*Estate Duty Act,***81ITR745]**

2. Income must be held to accrue at the point of time when a debt becomes due. Thus, the verb accrue is inter-connected with income becoming due because it is only when something becomes due to the assessee and appears crystallised as a debt in his favour that the right to receive can be said to have emerged. Until then, there is no accrual of income, as a person cannot recover which is not due and payable. Deferment of payment or the quantification of course, does not matter. If the assessee can show that either by the covenants in the contract or by any operation of law, the right to receive did not arise to the assessee and, therefore, no debt equivalent to the amount of the income in question was created in his favour, no income can accrue. It is necessary for the income to accrue that the income must be due to be paid to the assessee. **[CIT-Eastern Investments Ltd.,Cal.** *Agri. I.T. Act, Ker.* **213ITR334].**

3. Accrue means to increase, to augment, to be added as an increase, to arise or spring as a natural growth or result. The words accrue and arise have not been defined in the Act but they appear to be synonymous and have been used for bringing in a natural result. Strictly speaking, the word accrue may not be synonymous with arise, the former would be connoting the idea of growth or accumulation and the latter, the growth or accumulation with a tangible shape so as to receivable. It is clear that income may accrue to an assessee without the actual receipt of the same.If the assessee acquires the right to receive the income,the income can be said to have accrued to him though it may be received later,on its being ascertained. **[CIT-PunjabBoneMills,P&H.,***Sec.35B.***146 CTR 63,232 ITR 795,96Taxman 55]**

Accrue and arise **Kar. CIT - A.B.V.Gowda**

Income is liable to be taxed under the Income-tax Act, 1961, on the basis of its accruing or arising to the assessee, or on the basis of its receipt by the assessee, during the relevant previous year. The words accrue, arise and is received are not synonymous. The expression is received conveys a clear and definite meaning. But the words accrue and arise depend upon the rights of the assessee to secure the income though the actual receipt of the income may not be there. Ordinarily, income is said to have accrued to a person when he has acquired an enforceable right to that income, though actual quantification and receipt may follow in due course. *Sec.5(1)(b).***[42CTR265,157ITR697]**

2. The expressions accrue and arise do not mean actual receipt of income,but indicate a stage anterior to the point of time when the income has become receivable. They indicate that the assessee has acquired the right to receive the income, though it may be received later. **[KnoWerX Education, AAR,** *Sec.5.r.w.s.9* **[217CTR50,301 ITR207,170Taxman98]**

Accrue or arise **Bom. Bhogilal Laherchand - CIT**

Though income may accrue or arise to an assessee before he actually receives it, income cannot accrue or arise to him until he acquires a right to receive the same; and unless and until there is created in favour of the assessee a debt due by somebody, it cannot be said that he has acquired a right to receive the income. *Sec.5.* **[28ITR919]**

2. Under section 5(1)(b) of the Income-tax Act, 1961 when the income accrues or arises or is deemed to accrue or arise to the assessee in India during the previous year, it is to be taxed in that year. It is important, therefore, that receipt of a particular amount in the relevant year should be an income under the provision. The relevant yardstick is the time of accrual or arisal for the purpose of Taxation, viz., in order to be chargeable, the income should accrue or arise to the assessee during the previous year. There must be a right to receive the income on a particular date, so as to bring about a creditor and debtor relationship on the relevant date. A right to receive a particular sum under agreement would not be sufficient unless the right accrued by rendering of services and not by a promise for services : where the right to receive is anterior to the rendering of services, the income would accrue on the rendering of the services. **[CIT-Dinesh Kumar Goel,Del.***Sec.5(1)(b).***[239 CTR46,331ITR10,197Taxman 375]**

Accrue, arise SC Seth Pushalal Mansinghka (P.) Ltd. - CIT

The words accrue and arise do not mean actual receipt of the profits or gains. Both these words are used in contradistinction to the word receive and indicate a right to receive. If the assessee acquires a right to receive the income, the income can be said to accrue to him, though it may be received later, on its being ascertained.*Sec.5(1)(b).* **[66ITR159]**

Accrue, arise and is received All. CIT - Govind Prasad Prabhu Nath

The words.... accrue, arise and ... received, have not been defined in the Income-tax Act. Mere receipt of income is not the sole test of chargeability. Receipt of income refers to the first occasion when the recipient gets the money under his own control. According to the Oxford English Dictionary, the meaning of the word accrue is to fall as a natural growth or increment; to come as an accession or advantage. The word arise is defined as to spring up, to come into existence. The words accrue and arise do not mean actual receipt of profits or gains. Both these words are used in contradistinction to the word receive and indicate a right to receive. Thus, it is manifest that if an assessee acquires a right to receive income, the income can be said to accrue to him though it may be received later on. Unless and until there is created in favour of an assessee a debt due by somebody, it cannot be said that he has acquired a right to receive the income or that income has accrued to him. A mere claim to income without an enforceable right thereto cannot be regarded as accrued income for the purpose of the Income-tax Act. *Sec.5(1)(b).* **[171ITR417,35Taxman513]**

Accrues **HP. H.P. Mineral & Ind. Dev.Corp. - CIT**

Taxability is attracted not only when income is actually received but also when it accrues. Income accrues when it falls due, that is to say when it becomes legally recoverable, irrespective of whether it is actually received or not and accrued income is that income which the assessee has a legal right to receive. *Sec.4*[**217CTR388,302 ITR120]**

Accrues and arises **Del. CIT-National Electric Supply and Trading Corporation (P.) Ltd.**

The expressions income is received, accrues and arises, as appearing in section 5 of the Act, have not been defined. The two words, i.e., accrue and arise do not mean actual receipt of profits or gains. Both these words are used in contradistinction to the word receive. Thus, it is manifest that if an assessee acquires a right to receive the income, the income can be said to accrue to him though it may be received later on. There is clear distinction between cases where the right to receive payment is in dispute and it is not a question of merely quantifying the amount to be received, and cases where the right to receive payment is admitted and the quantification only of the amount payable is left to be determined in accordance with settled or accepted principles. *Sec.5.*[**164CTR599, 248ITR794,114Taxman326]**

Accrues, Arises **Ker. CIT - Jai Hind Travels (P.) Ltd.**

In order that income, profits or gains may accrue to a person, it is necessary that he must have acquired a right to receive the same or a right to the income, profits or gains must have become vested in him though its valuation may be postponed or though its materialisation may depend on the contingency that the making up of the accounts would show income, profits or gains. Thus it is manifest that if an assessee acquires a right to receive income the income can be said to accrue to him though it may be received later on. A mere claim to income without an enforceable right thereto cannot be regarded as accrued income for the purpose of the Act.

Accrues, arises and is received are three distinct terms. So far as receiving of income is concerned, there can be no difficulty; it conveys a clear and definite meaning. The words accrue and arise also are not defined in the Act. The three expressions accrues, arises and is received having been used in section 5 of the Act, strictly speaking, accrues should not be taken as synonymous with arises but in the distinct sense of growing by way of addition or increase or as an accession or advantage ; while the word arises means comes into existence or notice or presents itself. The former connotes the idea of a growth or accumulation and the latter of growth or accumulation with a tangible shape so as to be receivable. Both the words are used in contradistinction to the word receive and indicate a right to receive. They represent a state anterior to the point of time when the income becomes receivable and connote a character of the income which is more or less inchoate. Income is said to be received when it reaches the assessee; when the right to receive the income becomes vested in the assessee, it is said to accrue or arise. Sec.5(1)(b).**[158CTR664,243ITR451,108Taxman 242]**

Accruing or arising **Bom.** **CIT - Chunilal B. Mehta**

Section 4 of the Indian Income-tax Act refers to the place where the profits accrued or arose and not the place where the person who is the ultimate source of profits resides or carries on business. The words accruing or arising extend the scope of the tax to income which may not be received in British India. Sec.5.**[3ITR376]**

2. The words accruing or arising in section 4(1) of the Indian Income-tax Act mean something different from received. They indicate some origin or source of growth for the income in question. **[V.G. Every in re,Cal.,**Sec.5.**5ITR216]**

Accruing, arising or received **PC** **CIT - Diwan Bahadur S.L. Mathias**

The words accruing, arising or received exhibit some variation in meaning and in a limited sense the expressions may be antithetical, but they are not completely disjunctive or mutually exclusive. Sec.5.**[7ITR48]**

Accumulated **Mad.** **CIT - M.V. Murugappan**

Accumulated normally means heaped up, stored up, or put by. Accumulation is a conscious act and the result of a decision especially when a company is concerned. Current profit is what accrues in praesenti accumulated profit relates to the past. *Sec.2 (22).*[**62ITR382]**

Accumulated profits **Ker. CIT- Damodaran**

Section 2 (6A)(e) creates a fiction by which a loan made by a company to a shareholder is treated as dividend subject to the condition that it shall be treated as dividend only to the extent to which the company possesses accumulated profits. Accumulated normally means heaped up, stored up or put by. It indicates an effort on the part of the person in that direction. It is, therefore, inappropriate to describe current profits as profit heaped up, stored up or put by. Current profit is what accrues in praesenti: accumulated profit relates to the past. Therefore, current profit cannot be included in the term accumulated profits. The amendment of section 2(6A) by the Finance Act, has not changed the meaning of the term. Accumulated profits cannot mean profits or gains of any company as determined under section 10 of the Act for the levy of income-tax. The expression accumulated profits necessarily relates to profits which could be accumulated by a company from time to time. This means that all liabilities due from the company will have to be deducted from the profits to enable the company to accumulate the same.[**CIT-Roshanlal, All.,***Sec.2(22).* **98ITR349]**

2. Accumulated profits within the meaning of clause (e) will necessarily be comprised of the amount available for being distributed as profits. Profits can accumulate even within a single year. Accumulated means earned bit by bit and accumulated. The entire amount which is available for distribution as profits on a particular date would be the accumulated profits and any amount paid as advance or loan to the shareholder to the extent of this amount of accumulated profit will be dividend within the meaning of section 2(6A)(e) of the Act. *Sec.2(22).***[85ITR590]**

3. The word accumulated profits in section 2(6A) of the Indian Income-tax Act, 1922, cannot include current profits, and, therefore, the profits earned by the company during the year in which loans are advanced to a shareholder cannot be regarded as accumulated for the purpose of section 2(6A)(e). **[CIT-Damodaran, SC,** *Sec.2 (22).***13CTR191,121ITR572,2Taxman39]**

Accumulated profit of the company before its **Bom.** **CIT-Scindia Steam Navigation Co. Ltd.**

The words attributable to the accumulated profits of the company immediately before its liquidation in section 2(6A)(c) cannot be read as equivalent to accumulated before its liquidation in the sense that the company had gathered the amount in its hand before the date of liquidation. *Sec.2(22).***[8CTR543,125ITR118]**

Acquiring **MP.** **CIT - Bright Automotives and Plastics Ltd.**

The use of the expression acquiring in section 35AB has to be given a liberal meaning rather than a strict one. In other words, in order to attract the rigour of section 35AB it may not be necessary for the assessee to actually become an absolute owner of the know-how...the expression acquire has to be used liberally and in the context of actual user of know-how be that in a capacity as condition -nal/limited owner or absolute owner or as licensee. *Sec.35AB.* **[192CTR117,273 ITR59,141Taxman582]**

Actively engaged in the conduct of the.. **Guj.** **CIT - Natwarlal Tribhovandas**

The expression actively engaged in the conduct of the business occurring in section 2(7)(iii)(b) of the Finance (No. 2) Act must be given a liberal meaning. The expression does not necessarily signify active and continuous participation in the actual transaction of the day-to-day business of the firm; it is flexible enough to take in the case of a partner who devotes time, attention and labour to some activity or assignment calculated or designed to lead to the preservation, growth or advancement of the business of the firm. *Sec.2(7)(iii)(b),Finance Act,***[87ITR703]**

Actual **Del.** **CIT - Dalmia Dadri Cement Ltd.**

When the Legislature had prefixed the word actual to the word cost in section 43(1) of the I.T. Act, 1961 it was to lay emphasis on the reality and genuineness of the cost and to exclude collusive, inflated, deflated or fictitious cost. *Sec.43(1).***[125ITR510]**

Actual cost Ker. CIT - Cochin Electric Co. Ltd.

The words actual cost mean cost accurately ascertained. It does not mean that the cost should be defrayed out of the assessee's own resources. An electric supply company is entitled to the allowance of depreciation in respect of that portion of the service line towards the construction of which the consumer has contributed amounts.

Sec.43. **[57ITR82]**

2. The words actual cost in section 10(5) mean the cost actually ascertained and do not mean that the cost should necessarily be defrayed from the assessee's own resources. **[CIT- Ambala Cantt. Electric Supply Co. Ltd.,P&H.,***Sec.43.* **82 ITR 217]**

3. The statute having specifically used the expression actual cost to the assessee, the interest payments made on the amounts borrowed for the purpose of acquiring the machinery can be taken to be an expenditure incurred by the assessee in acquiring the machinery and that will go to add to his actual cost of the machinery. **[CIT-L.G. Balakrishnan & Bros.(P.)Ltd.,Mad.,***Sec.43(6).***95ITR 284]**

4. The expression actual cost on which depreciation under section 10(2)(vi) and development rebate under section 10(2)(vib) of the Indian Income-tax Act, 1922, is based has not been defined in the Act. Such meaning will have to be assigned to it as is understood in the commercial or business world because it deals with the computation of profits and gains of a business. When a businessman sets up a factory, the cost of such factory would include the cost of the machinery and the cost of construction of the building in which the machinery is installed. The cost of the machinery may include not only the actual price of the machinery but also such incidental expenses as sales tax, octroi, railway freight, insurance, transportation charges, etc. Similarly, the cost of installation of the machinery will include not only the price of the building material, but also the wages paid to labour, the supervisory staff like technicians and engineers who are employed to set up the machinery and put it in working order. From the accountancy point of view also all preliminary expenses directly connected with acquisition of fixed business assets like buildings, plant, machinery, etc., are to be capitalised and added to the total cost. This includes interest incurred by a newly started company, which is in the process of construction and erecting its plant, on borrowings which are used to finance capital expenditure before production is commenced. Section 208 of the Companies Act, 1956, also permits the capitalisation of interest on borrowed capital.**[CIT -J.K. Cotton Spinning and Weaving Mills Ltd.,All.** *Sec.43.* **98ITR153,37Taxation182]**

5. The Explanation of the expression actual cost for the purposes of section 10(5) of the Act, in which the Legislature has stated that the expression actual cost means the actual cost of the assets to the assessee reduced by that portion of the cost thereof as has been met directly or indirectly by Government or public or local authority are applicable only to the provisions in connection with depreciation allowance under sub-section (2)(vi) and cannot be given any larger operation than what the Legislature intended it should have. The Legislature has found it necessary to make a specific provision to exclude the contributions received from Government or public or local authority in order to give a particular meaning to the expression actual cost of the assets to the assessee in connection with depreciation allowance. Therefore, ordinarily, the expression actual cost of the assets must mean the actual amount paid for the assets irrespective of the source from which the cost or part thereof has been met. **[CIT-Bombay Suburban Electric Supply Co., Bom.,***Sec.43.***106ITR752]**

6. The expression actual cost in the context of the statutory provisions, sections 33(1) and 43(A) of the Income-tax Act,1961, must be understood in the sense in which no commercial man would misunderstand. The word cost is not synonymous with price. Besides the price of machinery or plant, cost takes in many other items of expenditure such for instance as freight, warehouse or insurance charges, legal expenses incurred in connection with its acquisition and interest incurred before the commencement of production on the capital contributed or borrowed to acquire such asset. In other words, in determining the actual cost of a fixed asset, all expenditure necessary to bring such asset into existence and to put it in working condition may legitimately be taken into account. **[Arvind Mills Ltd.-CIT,Guj.,***Sec.43A,33.***112ITR64]**

7. In computing the depreciation allowance, all expenditure incurred directly or indirectly or intimately on the capital assets acquired by the assessee can be included in the term actual cost of the asset. It will not be correct to treat the words actual cost to mean only the cost paid to the vendors for the asset.The term actual cost should be liberally construed. **[CIT-Great Eastern Shipping Co.Ltd.,Bom.** *Sec.32.* **118ITR772,1Taxman393]**

8. As the expression actual cost has not been defined it should be construed in the sense in which no commercial man would misunderstand. For this purpose, it would be necessary to ascertain the connotation of the expression in accordance with the normal rules of accountancy prevailing in commerce and industry.The accepted accountancy rule for determining the cost of fixed assets is to include all expenditure necessary to bring assets into existence and to put them in working condition. **[Ballarpur Paper and StrawBoardMillsLtd.,Bom.,***Sec.43***[10CTR227,118 ITR 613]**

9. All expenditure necessary to bring assets into existence and to put those assets in working condition is part of the actual cost of the assets to the assessee. **[Shree Vallabh Glass Works Ltd.-CIT,Guj.,**Sec.43(6).r.w.s.32 .**[19 CTR 111,127ITR37]**

10. All expenditure incurred directly or indirectly or intimately on the capital assets acquired by the assessee is to be included in the term actual cost of the assets. Further, the words actual cost have to be liberally construed and should not be taken to mean the cost paid to the vendors for the asset alone. **[CIT- Nirlon Synthetic Fibres and Chemicals Ltd. Bom.** *Sec.32*.**25CTR 155, 137 ITR 1]**

11. The expressions actual cost and the actual cost of the assets to the assessee do not bear the same meaning. Section 43(1) of the Income-tax Act, 1961, specifically and without ambiguity says that actual cost means the actual cost of the assets to the assessee reduced by that portion of the cost there of, if any, as has been met directly or indirectly by any other person or authority. Hence, section 43(1) of the Income-tax Act, 1961, supersedes the rule that in computing the actual cost for the purposes of depreciation, it is immaterial whether someone else has recouped the assessee what he has spent on the asset only to the extent to which the assessee was reimbursed by the Government or any public or local authority. **[Ranchi Electric Supply Co.Ltd.–CIT,Pat.,***Sec.43(1).***150ITR95, 16Taxman231]**

12. The expression actual cost has not been defined and hence it should be construed in accordance with the normal rule of accountancy prevailing in commerce and industry. The accepted rule of accountancy for determining the cost of fixed assets is to include all expenditure necessary to bring such asset into existence and to put it in working condition.The Supreme Court has held in Challapalli Sugars Ltd. v. CIT [98ITR167(SC)], that in case money is borrowed by a newly started company which is in the process of constructing and erecting its plant, the interest incurred before the commencement of production on such borrowed money can be capitalised and added to the cost of the fixed assets created as a result of such expenditure. The principle enunciated by the Supreme Court must logically lead to the conclusion that any increased monetary liability imposed upon the purchaser of the machinery would form part of the actual cost of the purchase of the machinery. Where the assessee has to pay an additional sum on machinery purchased abroad on deferred payment basis on account of devaluation of the Indian rupee, it would be entitled to development rebate on the enhanced cost of machinery and plant due to devaluation. ....In other words, in the matter of calculation of development rebate which an assessee is entitled to under section 43 read with section 33, sub-section (2) of section 43A has no overriding effect. Sub-section (2) of section 43A applies only where the actual cost of the machinery is to be calculated under section 43A and section 43 cannot be applied. **[CIT-Coromandel Fertilisers Ltd.,AP,***Sec.43A (2),33.***[156ITR283,18Taxman411]**

13. Where a Government subsidy is intended as an incentive to encourage entrepreneurs to move to backward areas and establish industries, the specified percentage of the fixed capital cost, which is the basis for determining the subsidy, being only a measure adopted under the scheme to quantify the financial aid, is not a payment, directly or indirectly, to meet any portion of the actual cost. The expression actual cost in section 43(1) of the Income-tax Act, 1961, needs to be interpreted liberally. Such a subsidy does not partake of the incidents which attract the conditions for its deductibility from actual cost. **[CIT- Swastik Sanitary Works Ltd.,Guj,***Sec.43,32,32A.***[205CTR517, 286 ITR 544]**

14. Section 43(1) of the Income-tax Act, 1961 1961 defines actual cost as the actual cost of the assets to the assessee, reduced by that portion of the cost there of, if any, as has been met directly or indirectly by any other person or authority. Since the definition of actual cost repeats the same words again, one has to look at the normal concept of actual cost. The accepted accountancy rule for determining the cost of fixed assets is to include all expenditure necessary to bring such assets into existence and to put them in working condition.**[CIT-Widia(India)Ltd.,Kar.,***Sec.43(1).***97CTR 218,193ITR475]**

15. A plain reading of the definition of actual cost in section 43 of the Income-tax Act, 1961, makes it clear that, where the asset has been acquired by an assessee in the usual course by purchase, etc., the actual cost means the actual cost of the assets to the assessee. In the case of acquisition of an asset by way of gift or inheritance, the cost of the asset to the assessee might be nil. To take care of this situation, Explanation 2 provides that, in such cases, the written down value of such asset in the hands of the previous owner or the market value thereof on the date of acquisition, whichever is less, should be taken as the actual cost of the asset to the assessee. ..It is well-settled that actual cost is not a static figure. It will vary from time to time. At the time of first acquisition, it might be a particular figure but that does not stop there.It will go on changing. **[Ciba of India Ltd.-CIT,Bom.,**<u>*Sec.43.*</u>**114CTR105,202ITR1,120 Taxation 158,70 Taxman505]**

16. Where Government subsidy is intended as an incentive to encourage entrepreneurs to move to backward areas and establish industries, the specified percentage of the fixed capital cost, which is the basis for determining the subsidy, being only a measure adopted under the scheme to quantify the financial aid, is not a payment, directly or indirectly, to meet any portion of the actual cost. The expression actual cost in section 43(1) of the Income-tax Act, 1961, needs to be interpreted liberally. Such a subsidy does not partake of the incidents which attract the conditions for its deductibility from actual cost. **[CIT-P.J.Chemicals Ltd., SC** <u>*Sec.43(1).*</u>**121CTR201,210ITR830]**

Actual cost of a machinery or plant **Bom.** **CIT-Poona Electric Supply Co. Ltd.**

In determining the question what is the actual cost of a machinery or plant to the assessee, it is an irrelevant consideration whether the assessee has spent the whole amount or only a part of it. What the taxing authorities have to consider is what the actual cost of the article is independently of who has contributed towards that particular amount. <u>*Sec.43.*</u>**[14ITR622]**

Actual cost to the assessee **Pat.** **CIT- Ranchi Electric Supply Co. Ltd.**

The actual cost to the assessee within the meaning of section 10(5) (a) is the actual cost incurred in installing service connection irrespective of any consideration as to the amount actually contributed by the company or the amount actually recouped ultimately from the consumers. Consequently, the amount paid by the consumers on account of service connection should not be taken into account in calculating the actual cost to the company. <u>*Sec.43.*</u> **[26ITR89]**

2. The meaning of the expression actual cost to the assessee as used in section 10(5) of the Act was what the assessee had, in fact, expended or laid out for the purpose of acquiring the depreciable assets. **[Habib Hussein-CIT,Bom.**<u>*Sec.43.*</u>**48ITR859]**

3. The expression actual cost to the assessee means the amount in fact paid by the assessee and the legislature in using that expression wanted to indicate the sum of money or money's worth which it cost the assessee to acquire the asset. **[CIT-Lothian Jute Mills Co. Ltd.,Cal.,**<u>*Sec.43.*</u>**[66ITR630]**

Actual delivery Cal. Budge Investment Co. Ltd. - CIT

There must be actual delivery or transfer of the commodity itself in order to take a transaction out of the definition of speculative transaction in Explanation 2 to section 24(1). Delivery of a pucca delivery order without delivery of the goods cannot be actual delivery within the meaning of Explanation 2. <u>*Sec.73.*</u>**[73ITR772]**

2. The words actual delivery in Explanation 2 to section 24(1) mean real as opposed to notional delivery. Whether a transaction is speculative in the general sense or under the Contract Act is not relevant for the purpose of this Explanation. The definition of actual delivery in section 2(2) of the Sale of Goods Act which has been held to include both actual and constructive or symbolical delivery has no bearing on the definition of speculative transaction in the Explanation. A transaction which is otherwise speculative would not be a speculative transaction within the meaning of Explanation 2 if actual delivery of the commodity or the scrips has taken place; on the other hand, a transaction which is not otherwise speculative in nature may yet be speculative according to Explanation 2 if there is no actual delivery of the commodity or the scrips. The Explanation does not invalidate speculative transactions which are otherwise legal but gives a special meaning to that expression for purposes of Income-tax only. **[Davenport and Co. P. Ltd.-CIT,SC,** <u>*Sec.73.*</u>**100ITR715]**

Actually SC Madeva Upendra Sinai -Union of India

The key word in clause (b) of section 43(6) of the Income-tax Act, 1961, is actually. It is the antithesis of that which is merely speculative, theoretical or imaginary. Actually contra-indicates a deeming construction of the word allowed which it qualifies. The connotation of the phrase actually allowed is thus limited to depreciation actuary taken into account or granted and given effect to, i.e., debited by the Income-tax Officer against the incomings of the business in computing the taxable income of the assessee; it cannot be stretched to mean notionally allowed or merely allowable on a notional basis. <u>*Sec.43(6).*</u>**[98ITR209]**

Actually allowed **Cal.** **CIT -Suman Tea & Plywood Industries (P.) Ltd.**

The provisions of section 43(6)(b) of the Income-tax Act, 1961, are in pari materia with section 10(5)(b) of the Indian Income-tax Act, 1922. Section 43(6) defines the expression written down value as meaning the actual cost to the assessee less all depreciation actually allowed. The key word is actually. It is the antithesis of that which is merely speculative, theoretical or imaginary. Actually contra-indicates a deeming construction of the word allowed which it qualifies. The connotation of the word actually allowed is thus limited to depreciation actually taken into account or granted and given effect to, i.e., debited by the Income-tax Officer against the incomings of the business in computing the taxable income of the assessee; it cannot be stretched to mean notionally allowed or merely allowable on a notional basis. *Sec.43(6).***[115CTR156, 204ITR719,71Taxman622]**

2. The words actually allowed with reference to depreciation do not mean notionally allowed and depreciation cannot be granted even when it is not claimed. **[Guindy Machine Tools P.Ltd.–CIT, Mad.,** *Sec.32(1)(iii).***176 CTR 56,254ITR780,170 Taxman 382]**

3. The words actually allowed in section 43(6)(b) of the Income-tax Act, 1961, mean limited to depreciation actually taken into account or granted and given effect to, i.e., debited by the Assessing Officer against the incomings of the business in computing the taxable income of the assessee. **[CIT- Doom Dooma India Ltd.,SC,** *Sec.43(6) (b).***222CTR,105,310ITR392,178Taxman261]**

Actually paid **Mad.** **Pereira and Roche - CIT**

Actually paid in sub-section (5) is not to be understood in a physical sense, but in the sense that the expenditure is actually incurred. *Sec.37.***[61ITR371]**

2. Reading section 36(1)(ii) of the Income-tax Act, 1961, in the light of the definition of paid in section 43(2) of the Act, an assessee need not have actually paid the bonus for claiming deduction and it is sufficient if the liability for payment of bonus was incurred according to the method of accounting on the basis of which profits and gains were completed by the assessee. However, simply because an amount is shown as provision for payment of bonus, it cannot come within the definition of paid in section 43(2) of the Act and it cannot be taken to be a sum paid to employees as bonus within the meaning of section 36(l)(ii) of the Act until it is shown that the amount was actually paid or that a legal liability for payment had been incurred during the accounting year. **[Addl.CIT-Kale Khan Mohammad Hanif, MP.,**<u>*Sec.36 (1)(ii).*</u> **114ITR812]**

3. The expression employed in section 43B of the Income-tax Act, 1961, is actually paid. In view of the non obstante clause contained in the section, it is not permissible to refer to the expression paid s defined under section 43(2) or under any other provisions of the Act. Hence, the plain meaning of the words actually paid is required to be taken into consideration. The word actual has been defined as something real in opposition to constructive or speculative. Actually means really, truly in fact. Therefore, the plain meaning of the expression actually paid means that the sum should have been actually paid to the coffers of the Revenue, i.e., really paid, and not constructively. Importing any other meaning would amount to doing violence to the plain meaning of the statute. When the words of the statute are clear and unambiguous there is no need to adopt any other meaning or provision of the Act. **[Mugat Dyeing and Printing Mills-ACIT,Guj.,**<u>*Sec.43B.*</u>**207CTR606,290ITR282]**

Adequate consideration AP. Potti Veerayya Sresty - CIT

Consideration to support a contract under the provisions of the Indian Contract Act is different from adequate consideration under section 64(iii) of the Income-tax Act. Since the Income-tax Act insists that the consideration for transfer must be adequate, there must be some means to measure the adequacy of the consideration. That is to say, the consideration that supports the transfer should be one, the value of which can be measured in terms of money or money's worth. Religious or spiritual benefit was not consideration which could be measured in terms of money or money's worth. <u>*Sec.64(iii).*</u>**[85ITR194]**

2. The expression adequate consideration cannot be equated to sufficient consideration, good consideration or valid consideration. Adequate consideration, within the meaning of section 4(1)(a)(i), must be construed as valuable consideration capable of being compared and measured with money or money's worth. **[CWT-Khan Saheb Dost Mohd. Alladin, AP.,**<u>*Sec.4(1)(a)WT.Act.*</u> **91ITR179]**

3. The expression adequate consideration has to be construed in a broad sense. In order that the court may hold that a particular transfer is not for adequate consideration, the difference between the true value of the property transferred, and the consideration that passed for the same, must be appreciable in the context of the facts and figures of the particular case. **[CGT-Cawasji Jehangir Co.(P.)Ltd.,Bom.,*Sec.4 G.T.Act.*106ITR390]**

4. Adequate consideration referred to in the Gift-tax Act,1958, is not the same as the market value and adequacy is to be determined with reference to all the circumstances and keeping in view the nature of the transaction whether it is bona fide and whether the consideration stated was the whole amount of consideration. **[CGT-Nelson and Co.,Mad.,*Sec.4(1)(a)G.T.Act.*245ITR347116 Taxman 206]**

Admission, admit/ admitted **MP.** **Shree Amarlal Kirana Stores - CIT**

The word admission, admit or admitted though not defined under any Act/rule has acquired a definite meaning in the judicial system. No sooner than the appeal/ petition/ reference as the case may be is filed, is it placed for hearing before the appellate authority/court for deciding whether it involves any arguable case or not. If the appellate authority/court is of the opinion after hearing the appellant/ writ petitioner that the case involves an arguable point, then in such event, the appellate authority/court admits the appeal/writ petition and directs issuance of notice to the respondent for its bi parte hearings. Such order of admission is a judicial order passed by the authority/court after hearing the appellant/writ petitioner on the question involved and after perusing the entire record of the case. There may be cases where the appellate authority/court finds no merit in the appeal/ petition/ reference after hearing the appellant/writ petitioner on the question involved and hence, may proceed to dismiss the case at its threshold and without calling upon the respondent to support the order impugned in the appeal/ petition/reference. In such eventuality, the question of admission of the case does not arise and the lis results in its final termination at its threshold. However, if the appellate authority/court is satisfied that the appeal/writ/reference involves an arguable point then the case is admitted for final hearing and notice is issued to the respondent. *Sec.249.***[180CTR355,259ITR572,126Taxman512]**

Advance tax **AP.** **CIT - Sreerama and Co.**

.......Section 218 says that if an assessee does not pay advance tax on the specified date, he will, be deemed to be in default. Section 219 provides for giving credit for advance tax so paid. Thus, though the word advance tax is not specifically referred to, still the scheme of the Act implies that the word tax includes advance tax. Sec.221. **[101ITR531]**

2. The expression advance tax, whether understood as defined in section 207(2) of the Income-tax Act, 1961, or as could be understood by the common man, will mean only the tax paid before the assessment as required by the Act. The Act itself contains an elaborate machinery for payment of advance tax and,hence, wherever the expression advance tax occurs in the Act, it can mean only that advance tax which was paid as required to be paid pursuant to the provisions contained in the Act and, accordingly, the tax paid by an assessee pursuant to a provisional order of assessment made under section 141 of the Act is not comprehended by the expression advance tax, if any, paid occurring in section 139(1). **[Rasiklal Kamdar -CIT,Mad.,***Sec.207.***109ITR56]**

3. It would not be reasonable to say that any installment of advance tax not paid on or before the prescribed date but paid after the due date but during the relevant financial year shall not be treated as advance tax and shall not be deducted as contemplated by the aforesaid Explanation. Therefore, the advance tax paid beyond the due date but during the relevant financial year is also to be deducted from the tax payable, namely, assessed tax, for the purpose of calculating the penalty under section 271(1)(a). **[CIT-P. RamagoudaSatyamReddyandCo.,AP.***Sec.208,219.***70CTR38, 172ITR491,36Taxman314]**

4. Advance tax is a pre-assessment collection of tax either by deduction at source or by payment of advance tax which has to be adjusted towards Income-tax levied on the total income. **[Brij Lal-CIT,SC,***Sec.2(40).***328ITR477]**

Advancement of an object of general public utility **Lah.** **Trustees of the Tribune**

A trust of a public character in which no immediate or ultimate personal benefit of any kind is reserved to any individual, created with the object of founding and maintaining a paper which conveys news and opinions on matters of general interest to the public, discusses matters of public interest and educates the minds of its readers on lines which the founder considered to be beneficial to the public at large is a trust for the advancement of an object of general public utility as that expression is understood in law, and the income in question was exempted from assessment. *Sec.2(15).* **[3ITR246]**

Advancement of any other object of .. **Cal.** **CIT-Indian Sugar Mills Association**

The expression the advancement of any other object of general public utility not involving the carrying on of any activity for profitin the definition of charitable purpose in section 2(15) of the Income-tax Act, 1961, plainly indicates that it is not the object of general public utility which would involve the carrying on of any activity for profit but the advancement of that object. In other words,the advancement of any other object of general public utility would be a charitable purpose provided that its advancement does not involve the carrying on of any activity for profit. *Sec.2(15).* **[111ITR429]**

Advancement of general public utility **Cal.** **Bengal National Chamber of Com. & Ind. - CIT**

In order to find out whether there is an advancement of general public utility not involving the carrying on of any activity for profit, the true test is to find out whether (a) the object of the assessee is one of general public utility; (b) the advancement of the object involves activities bringing in moneys; and (c) if so, whether such activities are undertaken (i) for profit or (ii) irrespective of profit. Even if (a) and (b) are answered affirmatively if c(i) is answered affirmatively exemption cannot be claimed. Whether an activity is one for or irrespective of profit depends on the facts and circumstances of the case and the real nature of the activity-whether it is one ordinarily carried on by ordinary people for gain; whether there is a built- in prescription in the constitution against making a profit;whether there has been, in practice, profit from this venture, although this last is a weak test. *Sec.2(15).***[111ITR514]**

Adventure in the nature of trade **Guj.** **CIT- Gordhandas Trikambhai Patel**

The words adventure in the nature of trade postulate existence of certain elements in the adventure which in law would invest it with the character of trade or business; and that would make the question whether a transaction is in the nature of trade and its decision one of mixed law and fact. When the I.T. Act refers to an adventure in the nature of trade, it clearly suggests that the transaction in question cannot properly be regarded as trade or business. It is allied to transactions that constitute trade or business but may not be trade or business itself. Sec.28.**[118ITR81, 52Taxation208]**

2. When section 2(4) of the Indian Income-tax Act, 1922, refers to an adventure in the nature of trade, it clearly suggests that the transaction cannot properly be regarded as trade or business. It is allied to transactions that constitute trade or business but may not be trade or business itself. It is characterised by some of the essential features that make up trade or business but not by all of them; and so even an isolated transaction can satisfy the description of an adventure in the nature of trade.**[Estate Investment Co. Ltd.-CIT,Bom.,**Sec.28.**9CTR272,121ITR 580]**

3. In deciding the character of a solitary transaction of purchase and resale, several factors are relevant, such as, whether the purchaser was a trader and the purchase of the commodity and its resale were allied to his usual trade or business or incidental to it; the nature and quality of the commodity purchased and resold; any act subsequent to the purchase to improve the quality of the commodity purchased and thereby make it more readily resaleable; any act prior to the purchase showing a design or purpose; the incidents associated with the purchase and resale, the similarity of the transaction to operations usually associated with trade or business; the repetition of the transaction; the element of pride of possession. Where a transaction is not in the line of business of the assessee but is an isolated or single instance of a transaction, the onus is on the Revenue to prove that the transaction was an adventure in the nature of trade. **[Ashok Kumar Jalan - CIT,Bom.,**Sec.28.**187ITR316]**

4. In order to treat a transaction as being within the purview of adventure in the nature of trade, it is to be seen whether the property had been purchased or acquired by the assessee with the intention to sell the property to earn profit by involving in several transactions of sale or to earn profit on its investment. **[CIT-Mohakampur Ice and Cold Storage, All.,**Sec.28.**281ITR 354, 149 Taxman 593]**

Advertisement Cal. Sarda Plywood Industries Ltd. - CIT

The word advertisement has been stated to mean a notice given in a manner designed to attract public attention. Any expenses incurred by way of advertisement must be considered from the point of view of the assessee and not from any other angle. Sec.37 .**[238ITR354]**

2. ......Advertisement means to make something known to the public or a segment of the public, to announce publicly by a printed notice or broadcast to call public attention to, especially, by emphasizing, desirable qualities so as to arouse a desire to buy and patronize. To this meaning of the term the definition adds notices, circular, label, wrapper, document, hoarding or any other audio or visual representation made by means of light, sound, smoke or gas. An advertisement is generally of goods and services and is an information intended for potential customers. **[Google Online IndiaP.Ltd.,AAR**_Sec.37._**200CTR 245,280ITR211]**

Advertisement, publicity, sales promotion **Kar.** **Smith Kline and French (India) Ltd.- CIT**

Each of the three words advertisement, publicity and sales promotion cannot always be confined to distinct and different concepts. Some aspects of one word would naturally overlap with the meaning attributed to the other words. No doubt, in a commercial sense, the purpose of these activities is to gain goodwill and a market but the mode of achieving this object cannot be confined to the act of media propaganda and a direct approach to the consumers by publicising the product through newspaper advertisements, posters or some other similar methods. The nature of the advertisement or publicity depends upon the nature and quality of the article in question. An inducement to the public to buy a particular commodity may be formulated in a mode most suitable to the article in question. _Sec.37._**[97CTR47,193ITR582, 59Taxman387]**

Affairs **SC** **Vr. N.M.Subbayya Chettiar - CIT**

The word affairs must mean affairs which are relevant for the purpose of the Income-tax Act and which have some relation to income. _Sec.6._**[19ITR168]**

2. The word affairs in section 4A(b) means the affairs of a Hindu undivided family which are capable of being controlled and managed by the family as such. Where a coparcener enters into partnership with strangers, the Hindu undivided family exercises no controlling power of management over the partnership. **[CIT - Nandlal Gandalal,SC,** _Sec.6(3)._**40ITR1]**

3. The word affairs means affairs which are relevant for the purposes of the I.T.Act, and which have some relation to the income sought to be assessed. **[CIT-Bank of China,Cal.,**_Sec.6(3)._**154ITR 617]**

4. The word affairs meant affairs which were relevant for the purpose of the Act which have some relation to income. **[Ms. Meenu Sahi Mamik, AAR,**<u>*Sec.6(3).*</u>**287 ITR514]**

Affect **Mad. CIT - Fab Exports (P.) Ltd.**

The term affect here does not seek to confer a benefit on the assessee nor does it seek to deprive the assessee of any benefit available to the assessee under the Act. It seeks to build as it were, a wall, between what is to be determined in relation to the sections mentioned in sub-section(2) and what has been provided for in sub-section (1). <u>*Sec.115 J*</u> **[176CTR129,258 ITR56,131Taxman790]**

Agency **Ker. CIT - Kerala Nut Food Co.**

....the word agency has not been defined in the Act, but accoridng to the Concise Oxford Dictionary it meant function of an agent or representative; business establishment of an agent. <u>*Sec.35B(1)(b)(iv).*</u>**[151CTR202,239ITR127]**

Agent **P&H. Hazoora Singh - CIT**

The agent of a non-resident is a representative assessee in respect of the income of the non-resident irrespective of the fact whether he has been treated as agent under section 163 of the Income-tax Act, 1961, or not. Section 163 does not define the term agent but only states that the persons mentioned in it will be included in the term agent in relation to a non-resident.Section 182 of the Indian Contract Act,1872, defines agent as a person employed to do any act for another or to represent another in dealing with third persons. For the appointment of an agent, it is not necessary that there must be written authority. <u>*Sec.163.*</u>**[52CTR 531,160ITR746,25Taxman 211]**

2. The word agent is defined in the Concise Oxford English Dictionary, inter alia, as under: 1. a person that provides a particular service, typically one working transactions between two other parties. A person who manages financial or contractual matters for an actor, performer, or writer, 2. a person who works in secret to obtain information for a Government. 3. a person or thing that takes an active role or produces a specified effect. Grammar-the doer of an action, 4. Computing-an independently operating internet programme, typically one set up to locate information on a specified subject and deliver it on a regular basis....The meaning of the term agent is given in Black's Law Dictionary, inter alia, as follows: The etymology of the word agent or agency tells us much. The words are derived from the Latin verb, ago, agree; the noun agens, agentis. The agent denotes one who acts a doer, force or power that accomplishes things...... section 182 of the Indian Contract Act defines agentas a person employed to do any act for another or to represent another in dealing with third parties. **[Dun and Bradstreet Espana S.A.,AAR***Sec.195,44D.***193CTR9,272ITR99,142Taxman 284]**

3. Section 140 of the Income-tax Act, 1961, specifies the persons who can sign the return. In the case of an association it has been provided in clause (e) that the return can be signed by any member of the association or the principal officer thereof. The principal officer has been defined under section 2(35). From a reading of the provision it is clear that the return can be signed by the agent also. Any person can appoint an agent orally or in writing or agency can be inferred by implication **[CIT-Rudra Bilas Kisan Sahkari Chini Mills, All.,***Sec.2(35),140***197CTR397,280ITR249,145 Taxman 497]**

Aggregate of the amount... on which tax is not payable.. Ker. CIT - Dat Pathe

The portion in rule 2(3) of the Rules reading as reduced by the aggregate of the amount or any portion thereof on which Income-tax or super-tax is not payable and the amount in respect of which a deduction of Income-tax or super-tax has been granted under any provision of the Act must refer to the income which is included in the total income as envisaged by section 66 of the Income-tax Act, 1961, but on which no Income-tax or super-tax is payable and in relation to which deductions were permitted under section 110 of the same Act, as well as under any other provisions of the Income-tax Act, 1961. It is clear from these provisions of the Income-tax Act, and particularly from the expression total income used in rule 2(3) of the Rules that the total income must be the total income as computed under the provisions of the Income-tax Act. *Sec.110. r.w.rule2(3).***[83ITR823]**

Agrees Bom Titanor Components Ltd. - CIT

The word agrees used in section 158A(1) does not mean that the Assessing Officer or the appellate authority has been given any unbridled authority not to agree. The subsequent provisions clearly require the authority to consider the facts and thereafter to either admit the declaration filed by the assessee or reject it. The appellate authority is, therefore, required to judicially evaluate the reasons for not agreeing given by the Assessing Officer and pass a reasoned order keeping in mind the design in adding section 158A to the Act. *Sec.158A(1).***[224CTR566,323ITR266,184Taxman10]**

Agricultural implement Bom. CIT - Kirloskar Brothers Ltd.

An agricultural operation extends to all activities which makes the agricultural produce marketable. Consequently, an implement used to make the agricultural produce marketable must be held to be an agricultural implement. *Sec.80E.r.w.Sch.***[181ITR523,48 Taxman146]**

Agricultural income Oudh Maharaja Sir Pateshwari Prasad Singh - CIT

Income from sale of forest trees, khader, mohua and narkul (wild grass and reeds) all of which are of spontaneous growth, is not agricultural income even if the land on which they have grown is assessed to land revenue...... *Sec.2(1A).***[15ITR181]**

2. Income from the sale of forest produce such as timber, tendu leaves, mohua flowers, harra nuts, etc., derived from a forest which is not a cultivated one but is of spontaneous growth is not agricultural income and is not exempt from Taxation under section 4(3)(viii) of the Indian Income-tax Act. Agriculture is the art or science of cultivating the ground. The essence of agriculture, even when it is extended to include forestry, is the application of human skill and labour; without that it can be neither an art nor a science. It is essential that the income should be derived from some activity which necessitates the employment of human skill and labour and which is not merely a product of man's neglect or inaction except for the gathering- in of the spoils. Not only must the assessee labour to reap the harvest, but he must also labour to produce it. Clause (a) of section 2(1) refers to income derived from land which is used directly for agricultural purposes and clause (b) to by-products, such as the selling of milk, the pasturing of cattle, etc., provided the endeavour is agricultural and provided it is reasonably connected with land used for agricultural purposes. **[Beohar Singh Raghubir Singh-CIT,Nag.,***Sec.2(1A).***16 ITR 433]**

3. Agricultural income cannot be said to accrue to every person into whose hands the produce of the land passes. It is only the owner, landlord or ryot, or persons having a derivative interest in the land from these persons that can be said to derive income from the land by the performance of agricultural operations in it. **[CIT -Maddi Venkatasubbayya,Mad.,**<u>*Sec.2(1A).*</u>**20ITR151]**

4. The term agricultural income was defined in section 2(1) of the Indian Income-tax Act, 1922. The same definition of agricultural income has been adopted in section 2(1A) of the Income-tax Act, 1961. From a perusal of both clauses (a) and (b) of the definition of agricultural income, it is clear that the income must be derived from and which is used for agricultural purposes.**[CIT-Green Gold TreeFarmersP.Ltd.,Uttr.,**<u>*Sec.2(1A).*</u>**299ITR262,167Taxman151]**

Agricultural land Guj. Rasiklal Chimanlal Nagri - CWT

The expression agricultural land has nowhere been defined in the Constitution and has therefore, to be understood in the plain, ordinary meaning of the expression according to English language bearing in mind the fundamental principle of construction that the expression occurs in a head of legislative power under which the Wealth-tax Act has been enacted and it should, therefore, receive the widest and most liberal meaning. Whether a land is agricultural land or not cannot depend on the fluctuating or ambulatory intention of the owner of the land. The criterion must be something more definite and more objective, something related to the nature or character of the land and not varying with the intention of the owner as to the use to which he wants to put the land at a particular point of time. The intention as to user is, however, not altogether an irrelevant consideration and is a factor which would bear on the nature of or character of the land, but it does not afford a sole or exclusive criterion for determining whether a land is agricultural land or not. Where the land is actually paid to use, there is usually not much difficulty in ascertaining the nature or character of the land. If the land is used for agricultural purposes, ordinarily it would be correct to say that the land is agricultural land and vice versa. But even this test may not always furnish a correct answer, for there may be cases where land admittedly non-agricultural (such as a building site) may be used temporarily for agricultural purposes. In a case where the land is not being actually put to any use, the trust test to be applied for the purpose of determining whether a particular land is agricultural land or not, is not whether the land is capable for being used for agricultural purpose, but whether, having regard to the various relevant factors, the general nature or character of the land is such that it can be regarded as agricultural land. <u>*Sec.2(e)(i)WT.Act.*</u>**[56ITR608]**

2. In considering the meaning of agricultural landas used in the Wealth-tax Act, we should restrict ourselves to the ordinary meaning of the term agricultural landand not get lost in the woods to find out the tree. As stated earlier, agricultural land means land pertaining to or connected with agriculture. In order to understand which lands pertain to or are connected with agriculture, we should find out whether a particular land is used or earmarked for agricultural operations. **[Tea Estates India Pvt. Ltd. - CWT,Cal.** *Sec.2(e)(i)WT.Act.***59 ITR 428]**

3. In order to come within the category of agricultural land, the land must not only be capable of being used for agricultural purposes but should have been actually used as such at some point of time. A temporary non-user for agricultural purposes will not affect the character of the land but a permanent abandonment of user for agricultural purposes will affect the character of the land as agricultural land. The actual conversion of the land for non-agricultural purposes will also affect the character of the land as agricultural land. Whether such a conversion has taken place will depend on the facts of each case. **[Shiv Shankar Lal-CIT, Del.,** *Sec.2(14)(iii),45.***94 ITR433]**

4. Simply because agricultural land has not been defined in the Wealth tax Act, it is not correct to give the expression as wide a meaning as possible. The correct rule is to find out the exact sense in which the words have been used in the particular context and give an interpretation in consonance with the purpose of the statute. The object of the Act is to tax surplus wealth and it is clear that all land was not excluded from the definition of assets. Therefore, it is imperative to give reasonable limits to the scope of the expression agricultural land and give it a restricted meaning. The determination of the character of the land, according to the purpose for which it is meant or set apart and can be used, is a matter which ought to be determined on the facts of each particular case. What is really required to be shown is the connection with an agricultural purpose and user and not the mere possibility of user of land, by some possible future owner or possessor, for an agricultural purpose. It is not the mere potentiality, which will only affect its valuation as part of assets, but its actual condition and intended user which has to be seen for purposes of exemption from wealth-tax. One of the objects of the exemption is to encourage cultivation or actual utilisation of land for agricultural purposes. If there is neither anything in its condition, nor anything in the evidence to indicate the intention of its owners or possessors so as to connect it with an agricultural purpose, the land could not be agricultural land for the purposes of earning an exemption under the Act. Entries in revenue records are,however, good prima facie evidence. **[CWT-Officer-in-charge(Court of Wards), SC,**<u>*Sec.2(e)(i)WT.Act.*</u> **105 ITR133]**

5. The expression agricultural land has not been defined in the Act. There are various criteria which may be applied to find out if the land in a particular case is agricultural or not. Some of them are: (i) Classification and assessment of the land to land revenue.(ii) Whether agricultural operations are carried on. (iii) Intention of the owner. A temporary user of the land either for agricultural or non-agricultural purposes is not important. The real intention of the owner has to be ascertained. (iv) Character of adjoining lands. **[Addl.CIT-Tarachand Jain,Pat.,** <u>*Sec.2(14).*</u>**[123ITR567]**

6. Though the expression agricultural land has to be given a wide connotation it must be confined to denote land which is either actually under cultivation or which is appropriated or set apart for a purpose which could be regarded as agricultural and for which it could be reasonably used without any alteration of its existing character. If there is no evidence to show that forest land had been cleared and prepared or earmarked for agricultural purpose, it would be too unreal or too soon to hold that it had become agricultural land and the question in every case has to be decided on evidence of actual or intended user for which the land may have been prepared or set apart. The burden is on the assessee to show that forest land covered with wild and natural growth had been converted into land which was actually used for agricultural purpose or had keen prepared or set up for such a purpose. **[CWT-T.N.K.GovindarajuChettiar,Mad.,**<u>*Sec.2(e)(i)WT.Act*</u>**149 ITR588, 18Taxman217]**

7. The expression agricultural land is not defined in the Income-tax Act, 1961. ... Actual user of the land for agricultural purpose or absence thereof at the relevant time is undoubtedly one of the crucial tests for the determination of the issue. It is well-settled that the nature and character of land may undergo a change depending upon its situation, growth of the locality or zone in which it is situate and its potentiality. The fact that the land is sold or transferred to a non-agriculturist for a non-agricultural purpose or that it is likely to be used for non-agricultural purposes soon after its transfer is also a relevant factor germane to the determination of the issue. Merely because the land was used for agricultural purposes in the remote past or it continues to be assessed to land revenue as agricultural land is not decisive. In order to ascertain the true character and nature of the land it must be seen whether the land had been put to use for agricultural purposes for a reasonable span of time prior to the relevant date and further as to whether on the date of the transfer the land in question was intended to be put to use by the purchaser for agricultural purposes for a reasonable span of time in future.**[Gopal C. Sharma-CIT,Bom.**<u>*Sec.2(47).*</u>**116CTR377,209ITR946,125Taxation551,72Taxman353]**

Agricultural primary commodities **Guj.** **CIT-Gujarat State Export Corporation**

The expression agricultural primary commodities has not been defined in the Act. However, the phrase envisages that there should be an agricultural commodity, and further, it should be a primary commodity. The provision does not require that the commodity, namely, the goods which are exported cease to be agricultural commodities. The only prohibition or exception carved out qua the goods or merchandise which are exported is that such goods should not be agricultural primary commodities. A primary product is one as it actually grows. <u>*Sec.80HHC.*</u>**[199CTR217,280ITR62]**

Agricultural product **SC** **CIT - Cynamid India Ltd.**

Section 35C of the Income-tax Act, 1961, was designed to encourage development of agriculture. The term agricultural product or product of agriculture is required to be construed liberally so as to include not merely the primary product as it actually grows, but also a product which undergoes a simple operation so as to make it more saleable or more usable. <u>*Sec.35C.*</u>**[153CTR201,237ITR585, 104Taxman94]**

Agricultural purposes **Guj.** **Chhotalal Prabhudas - CIT**

If land is actually used for agricultural purposes at least prima facie, it can be said to be land which is either actually used or ordinarily used or meant to be used for agricultural purposes.If it is actually used at the relevant date for agricultural purposes and there are no special features, as for example, a building site being actually used as a stop-gap for agricultural purposes, it would be agricultural land. Potential use of the land as agricultural land is totally immaterial. Entries in the record of rights are good prima facie evidence regarding land being agricultural and if the presumption raised either from actual user of the land or from agricultural use of the land is to be rebutted, there must be material on the record to rebut the presumption. The approach of the fact-finding authorities, namely, the Income-tax authorities and the Tribunal, should be to consider the question from the point of view of presumption arising from entries in the record of rights or actual user of the land and then consider whether that presumption is dislodged by the presence of other factors in the case. *Sec.2(2)(i)WT.Act.***[116ITR631]**

2. Prima facie land could be said to be agricultural which is either actually used, ordinarily used, or is meant to be used, for agricultural purposes. If it is actually used at the relevant date for agricultural purposes and there are no special features, as for example, a building plot being used for agricultural purposes as a stop-gap arrangement, it would be agricultural land. **[Chandravati Atmaram Patel-CIT,Guj.,***Sec.2(14)(iii).***7CTR211,114ITR302]**

Agriculture Mad. CIT - K. E. Sundara Mudaliar

Irrespective of the nature of the produce or product of the land, whatever is grown on land aided by human labour and effort, whatever does not grow wild or spontaneously on the soil without human labour or effort, would be an agricultural product, and the process of producing it would be agriculture within the meaning of that expression in section 2 of the Indian Income-tax Act. In short, the word is used to denote the raising of valuable or useful products deriving nutriment or sustenance from the soil with the aid of human labour or skill. *Sec.2(1A).***[18ITR259]**

2. The term agriculture is not confined merely to tilling the earth of ploughing of the soil but is used in a wide sense to denote the raising of useful or valuable products which derive nutriment from the soil with the aid of human skill and labour; it would include horticulture, arboriculture and sylviculture where the growth of the trees is effected by the expenditure of human effort, skill and attention in such operations. **[Jyotikana Chowdhurani-CIT, Ass.,***Sec.2(1A).***26ITR424]**

3. Agriculture in its primary sense denotes the cultivation of the field and is restricted to cultivation of the land in the strict sense of the term, meaning thereby tilling of the land, sowing of the seeds, planting and similar operations on the land. These are basic operations and require the expenditure of human skill and labour upon the land itself. Those operations which the agriculturist had to resort to and which are absolutely necessary for the purpose of effectively raising produce from the land, operations which are to be performed after the produce sprouts from the land, e.g., weeding, digging the soil around the growth, removal of undesirable under growth, and all operations which foster the growth and preservation of the same not only from insects and pests but also from depradation from outside, tending, pruning, cutting, harvesting and rendering the produce fit for the market, would all be agricultural operations when taken in conjunction with the basic operations. The human labour and skill spent in the performance of these subsequent operations cannot be said to have been spent on the land itself. The mere performance of these subsequent operations on the products of the land, where such products have not been raised on the land by the performance of the basic operations, would not be enough to characterise them as agricultural operations; in order to invest them with the character of agricultural operations these subsequent operations must necessarily be in conjunction with and in continuation of the basic operations which are the effective cause of the products being raised from the land. The subsequent operations divorced from the basic operations cannot constitute by themselves agricultural operations. Only if this integrated activity which constitutes agriculture is undertaken and performed in regard to any land can that land be said to have been used for agricultural purposes and the income derived there from be said to be agricultural income derived from the land by agriculture, under section 2(1) of the Indian Income-tax Act,1922. **[CIT-Raja Benoy Kumar Sahas Roy, SC,** *Sec.2(1A)*.**32ITR 466]**

4. The term agriculture should not be given such a wide meaning as to include activities in relation to land like rearing of livestock,dairy farming, butter and cheese making, etc. **[State of Orissa-Ram Chandra Choudhury,Ori,** *Agricultural Income-tax Act* **[46ITR246]**

5. Agriculture means the cultivation of the land meaning thereby tilling of the hand, sowing, planting and similar operations on the land. Only when the integrated activity of agriculture is undertaken and performed on any land, can that land be called agricultural land. The mere presence of trees on the land will not make is agricultural, especially when mere situated in the heart of a town and is surrounded by residential buildings. The question whether a land is agricultural land or not, does not also depend upon the intention of the owner to use the land or on the fluctuating or ambulatory intention of the owner. The criterion must be something more definite and more objective, something related to the nature or character of the land. **[Syed Rafiqur Rahman -CWT,Pat.,** *Sec.2(e)WT.Act.***75 ITR 318]**

Agriculture and cultivation **Cal.** **CIT(Agr.) - Raja Jagadish Chandra Deo Dhabal Deb**

The ordinary dictionary meanings of the two words agriculture and cultivation are different.Agriculture is of much wider import than cultivation. *Sec.2.*[**17ITR426]**

Aircraft **AAR** **Lloyd Helicopters International Pvt.Ltd.**

Etymologically, the expression aircraft comprehends within it any structure or machine designed to travel through air with the possible exception of hovercrafts. *Sec.44BB.*[**166CTR226,249ITR 162,115Taxman334]**

Aircraft, aero engines **Bom.** **CIT - Kirloskar Oil Engines**

A conjoint reading of items D(1) and E(1) of the depreciation table appended to the Income-tax Rules, (Part I of Appendix I), makes it clear that the subject-matter of the two items is quite different and distinct. Though both items are given under the heading Aeroplanes, item D(1) prescribes the rates of depreciation on Aircraft and aerial photographic apparatus whereas item E(1) prescribes the rate of depreciation on Aero-engines. Aircraft and aero engines are two quite different and distinct machines. An aircraft is a machine capable of flight, whereas an aero-engine is only the power unit of an aircraft. An aero-engine, therefore, cannot be termed as aircraft or, to put it differently, an aircraft can never be described as an aero-engine even if it is heavier-than-air. An aircraft which is heavier-than-air is also aircraft. In Chamber's Science and Technology Dictionary,the word aircraft has been described only as a mechanically driven heavier-than-air flying machine with wings of fixed or variable sweep angle. It has not been described to mean gliders, balloons and other flying machines. Similarly, in the definition of aircraft in the Aircrafts Act, 1934, balloons, airships, kites, gliders and flying machines have been added by specific inclusion. It is, therefore, not correct to say that aircraft, which are heavier-than-air are not aircrafts but are aero-engines. All aircrafts, whether lighter-than-air or heavier- than-air, are aircrafts. No aircraft can ever be termed as an aero-engine because an aero-engine is not an aircraft or aeroplane at all. It is only the power unit of an aircraft. It is thus clear from the above discussion that aircraft does not mean only crafts like balloons, airships, helicopters but also means aircrafts heavier-than-air. *Sec.32.*[151CTR544,230ITR88]

All the provisions of this Act shall apply.. **All.** **Addl.CIT-Hasmat Rai Raj Pal**

The words all the provisions of this Act shall apply accordingly used in section 297(2)(d)(ii) mean only the procedural provisions. It follows that the substantive provisions of the Act of 1961 will not apply to the completion of an assessment relating to an assessment year prior to the coming into force of the Act of 1961 provided other conditions contained in section 297(2)(d)(ii) are satisfied. *Sec.297 (2)(d)(ii).*[61CTR256,167ITR794,32Taxman72]

Allotment of shares **SC** **Khoday Distilleries Ltd. -CIT**

The words allotment of shares were used to indicate the creation of shares by appropriation out of unappropriated share capital to a particular person. There was a vital difference between creationand transferof shares. Allotment is not transfer. There is no element of existing right in the case of allotment as required by section 2(xii). *Sec.2(xii), xiv)G.T.Act.*[220CTR228,307ITR31]

Allowance **All.** **Cawnpore Textiles Ltd. - CIT**

A reading of the provisions of section 40A(5) of the Income-tax Act,1961, makes it clear that all expenditure incurred by the assessee which results directly or indirectly in the provision of any perquisite to an employee or use of any asset by an employee is to be taken into account for computing the value of disallowance and even rent-free accommodation provided to the employee is treated to be a perquisite. The expression allowance in section 40(a)(v) and section 40A(5)(a)(ii) of the Act takes in depreciation allowance and the ceiling on expenditure provided under these provisions applies to depreciation allowance on all assets belonging to the employer-assessee used by an employee.*Sec.40.*[276ITR612,144Taxman590]

Already pending **AAR** **Monte Harris - CIT**

The words already pending should, therefore, be interpreted to mean: already pending as on the date of the application and not with reference to any future date. No applicant can afford to ignore the mandatory provision requiring a return of income to be filed within the prescribed time merely because he has filed an application before the Authority. Sec.245.**[128CTR59,218ITR413, 82Taxman365]**

Amendment **Del. J.N. Sahni - ITAT**

The expression amendment must be assigned its true meaning. While an order of amendment is passed, the order remains but when an order is recalled it stands obliterated. What cannot be done directly, cannot be done indirectly. Sec.254.**[174CTR367,257 ITR16,123Taxman569]**

Amount of scrap value, if any **Mad. CIT-Ashoka Betelnut Co. P. Ltd.**

....when the asset was discarded, demolished or destroyed, the asset which was no longer put to use or which had suffered damage or destruction, may even after being so discarded, damaged or destroyed still have value as scrap. It was immaterial whether the assessee had chosen to convert that scrap value into cash. What was required to be seen was the value of scrap and not the amount realised by the assessee. The fact that the assessee by choosing to gift away the salvaged material after the building had been demolished, chose not to realise the price from the person to whom the articles were given did not deprive the article of its value. The scrap value had to be taken into account while determining the extent to which the assessee could claim the benefit under section 32(1)(iii) of the Income-tax Act,1961. Sec.32(1)(iii).**[178CTR349,259 ITR733,125Taxman321]**

Amounts **SC** **Sandvik Asia Ltd. - CIT**

The expression amount in the earlier part of section 244(1A) refers not only to the tax but also to the interest; it is a neutral expression and it cannot be limited to the tax paid in pursuance of the order of assessment. Sec.244(1A).**[280ITR643]**

An annual charge **Gau.** **CIT - Satyanarayan Sikaria**

Under the arrangement as detailed in Chapter IV of the Act,an annual charge is to be understood as an annual payment payable by the owner to a third person, that is secured by creation of a charge on the property under assessment. An annual charge in the context signifies something more than yearly payment. It connotes a liability to pay and a charge on the house property with the view to off-load that liability. The term liability here is to be understood as debts and obligations, either absolute, contingent, express or implied, which one is bound in law and justice to perform. <u>*Sec 24(1)(iv).*</u>**[160CTR369,238ITR855,152Taxation472,108 Taxman41]**

An application made Mys. Shanta Bai Devarao - CIT

An application under section 66(1) of the Indian Income-tax Act, 1922, can be deemed to have been made within the meaning of the section only when that application is received in the office of the Appellate Tribunal. An application which is posted before the expiry of the period of limitation prescribed in section 66(1) but which is received by the Tribunal after the expiry of the period cannot be held to have been made within the prescribed period and is liable to be rejected as time- barred. <u>*Sec.256.*</u>**[46ITR272]**

An approved gratuity fund provided the same is under an trust Del. Sony India P. Ltd. - CIT

Section 36 of the Income-tax Act,1961, provides for deductions that are admissible while computing the income referred to in section 28. One of the deductions which is made admissible under clause (iv) of section 36 is any sum paid by the assessee by way of contribution towards a recognised provident fund or an approved superannuation fund. Clause (v) of section 36(1) similarly provides for deduction of any sum paid by the assessee by way of contribution towards an approved gratuity fund provided the same is under an irrevocable trust. A plain reading of section 36(1)(iv) and (v) makes it manifest that deductions there under are admissible only if the employer pays the contributions towards a recognised provident fund, an approved superannuation fund or an approved gratuity fund. <u>*Sec.36(iv)(v).*</u>**[285ITR213]**

An article ... of personal use Raj. Hanuman Mal Sekhani - CWT

Articles in respect of which exemption can be allowed under section 5(1)(viii) of the Wealth-tax Act, 1957, must be of such a nature as could ordinarily be put to use by the assessee personally or for his household purposes. An article can be of personal use if it is used or could be used by the person of its possessor and must have more or less intimate relation with the assessee. Sec.5(1)(viii)W.T.Act. **[57CTR185,168ITR364,34Taxman21]**

An asset in the hands .....on the relevant valuation date **Guj.** **CWT - Arvindbhai Chinubhai**

Even assuming that the mere likelihood of obtaining a refund, in future, of tax paid for the relevant assessment year could be characterised as an asset in the hands of the assessee on the relevant valuation date especially when the assessment proceedings for the relevant assessment years are pending on the valuation date, it would be impossible to comprehend and predicate with any degree of certainty as to what would be the actual amount of tax refund, if any, which would be available to the assessee in future and which could be treated as his asset on the valuation date. The right of the assessee to receive a particular amount of refund vis-a-vis tax paid for a given assessment year is a mere possibility. Equally so is the corresponding liability of the department to refund to the assessee a particular amount. So long as the assessment proceedings are not processed and finalised by the department, the alleged right and the corresponding liability remain in a fluid state. Therefore, even assuming that the likelihood of obtaining refund of the amount may be an asset, it is not capable of evaluation as on the valuation date and such an asset is not capable of being ascertained and cannot be treated as an asset for the purpose of arriving at the net wealth of the assessee during the relevant assessment year. *Sec.2(e)WT.Act.r.w.s.Rule1D* **[24CTR228, 133ITR800]**

An association of persons **Pat.** **Mahendra Kumar Agrawalla - ITO**

An association of persons must be one in which two or more persons join in a common purpose or common action, the object of which is to produce income, profits or gains. The question whether there has been a combination on behalf of the petitioner and his brother to produce income, profits or gains or not is a pure question of fact. It could not be urged that an association of persons to produce income cannot be formed on behalf of a minor. The mere fact of common management of the colliery would be no justification for the assessment of the owners as an association of persons. *Sec.4.* **[103ITR688]**

2. An association of persons is a unit of assessment under the Indian Income-tax Act. The precise meaning of the expression has come up for consideration on several occasions before the courts.The word association indicates plainly the voluntary combination for a common endeavour and not a mere legal status resulting from operation of law.Co-owners, co-heirs or co-legatees do not constitute such an association in respect of the income of the joint or common asset by reason only of their jural relationship. But, if they unite themselves with the objective of earning income, they constitute an association of persons for assessment purposes and they cannot take advantage of their legal position to resist assessment on that basis. The essential criterion that attracts the label of association of persons in the Income-tax Department is the unity of the income-making purpose rather than the unity of title in the income-yieldingasset. **[CIT-Chandmal Rajgarhia,Pat.**<u>*Sec.2(31)r.w.s.4.*</u> **126CTR 321,213ITR789]**

3. An association of persons is one in which two or more persons join in a common purpose or common action and as the words occur in a section which imposes a tax on income, the association of those persons must be one, the object whereof is to produce income, profits or gains and share the same. The question whether an association of persons has engaged itself in an activity with a common object to produce income, profits and gains and share the same would depend upon the facts and circumstances of each case. The jointness of the enterprise aimed at production and the sharing of income, profits or gains are the essential ingredients to call a body of persons an association of individuals. **[State of T.N.-Thiruvalargal SingaraEstate.,Mad.**<u>*Sec.2(31)r.w.s.4.*</u>**217ITR199]**

4. An association of persons is not something distinct and separate from a body of individuals. The latter expression has been added to obviate any controversy as to whether only combinations of human beings are to be treated as a unit of assessment. When several individuals are found to have joined together for the purpose of making profit, the group of individuals may be conveniently described as a body of individuals. An association of persons or a body of individuals, whether incorporated or not, has been brought within the net of Taxation with the intention clearly to hit combinations of individuals or other persons who were engaged together in some joint enterprise. The combinations may or may not be incorporated. A profit-yielding joint venture has to be taxed as a single unit. **[Meera and Company-CIT,SC,**<u>*Sec.2(31)r.w.s.4.*</u>**139 CTR492,224ITR635,91Taxman219]**

An attempt to commit a particular offence **Mad.** **ITO - Dharamchand Surana**

An attempt to commit a particular offence is made when the person : (i) intends to commit that particular offence; and (ii) having made preparations and with the intention to commit the offence, he does an act towards its commission; such an act need not be the penultimate act towards the commission of that offence but must be an act during the course of committing that offence. Section 276C of the Income-tax Act,1961, deals with the attempt to evade tax which is chargeable or imposable indicating that it need not be relating to the tax that was paid already. *Sec.276C.***[216ITR678,96Taxman 317]**

An authorised agent of the assessee **All.** **Addl.CIT–Premkumar Rastogi**

A person who is not an authorised agent of the assessee or an agent or manager personally carrying on the assessee's business or an adult member of his family but who has merely accepted notices in the past on his behalf cannot be deemed or treated to be an authorised agent of the assessee. *Sec.282.***[24ITR381]**

An error apparent on the face of the record **Mad.** **Express Newspapers Limited - DCIT**

A patent, manifest and self-evident error which does not require elaborate discussion of evidence or argument to establish it, can be said to be an error apparent on the face of the record. *Sec.254(2).* **[228CTR405,320ITR12,186Taxman111]**

An incorrect assumption of facts **SC** **Malabar Industrial Co. Ltd. - CIT**

An incorrect assumption of facts or an incorrect application of law will satisfy the requirement of the order being erroneous. In the same category fall orders passed without applying the principles of natural justice or without application of mind. *Sec.263.***[59CTR1, 243ITR83,109Taxman66]**

An order of any court **Kar.** **T. M. Kousali - ITO**

An order of any court means an order of any and every court in the country. The hierarchy and status of the court in the country is not decisive. All that this provision provides is that it must be a court and there must be an order of a court. The nature of the court and the nature of the order made by the court have no relevance. If there is an order of a court,whatever be its status, then the bar of limitation is automatically lifted. *Sec153(3)(ii).***[155ITR739]**

And-or **Mad.** **CIT - Puthuthotam Estates (P) Ltd.**

The circumstances under which the word and may be construed as or and vice versa should be somewhat rare. Otherwise if the two are taken to be interchangeable terms, then it would result in Parliament throwing into the statute the two expressions indiscriminately and leave them to the courts to sort out the meaning. In ordinary usage and is conjunctive and or is disjunctive. *Sec.33A.* **[18CTR3,127ITR481]**

And also **Bom** **CIT - Jet Airways (I) Ltd.**

Parliament having used the words assess or reassess such income and also any other income chargeable to tax which has escaped assessment, the words and also cannot be read as being in the alternative. On the contrary, the correct interpretation would be to regard those words as being conjunctive and cumulative. It is of some significance that Parliament has not used the word or. The Legislature did not rest content by merely using the word and. The words and as well as also have been used together and in conjunction. Evidently, what Parliament intends by use of the words and alsois that the Assessing Officer, upon the formation of a reason to believe under section 147 and the issuance of a notice under section 148(2) must assess or reassess :(i) such income;and also (ii) any other income chargeable to tax which has escaped assessment and which comes to his notice subsequently in the course of the proceedings under the section. *Sec.147r.w.s.148.***[239 CTR183,331ITR 236,195Taxman117]**

Annual payments **Cal.** **CIT - State Bank of India**

Annual payments as used in taxing statutes must be taken to mean payments, in whatever kind of installments paid, made every year in discharge of a liability incident to that year, if it has to be made during more than one year, whether consecutively or otherwise. A payment is annual if it has the quality of recurrence in different years, although it might not be in every one of the succession of years. It is also not necessary. *Sec.24.*[**31ITR545**]

Annual value **Bom.** **Panalal Silk Mills Pvt. Ltd. - CIT**

Section 24(1)(ix) does not speak of either gross or net annual value. All that this clause provides is that where the property is let and was vacant during a part of the year, that part of the annual value which is proportionate to the period during which the property is unoccupied is deductible from income chargeable under the head Income from house property. It is section 23 which gives the meaning of the expression annual value. The scheme of section 23 is that if the property has not been let, then the annual value shall be deemed to be the same for which the property might reasonably be expected to be let from year to year. This, of course, is a hypothetical figure which has to be estimated by the Assessing Officer where the property is not actually let. Where the property is let and the annual rent received or receivable by the owner is in excess of the amount which could be calculated on hypothetical basis as above, then the annual value of the property shall be deemed to be the amount so received or receivable. The proviso to section 23(1) provides that where the property is in the occupation of a tenant, the taxes levied by any legal authority in respect of the property shall, to the extent the taxes are borne by the owner, be deducted in determining the annual value of the property. This section does not speak of gross or net annual value. It merely indicates two different methods of computing the annual value which are applicable to two different situations. The amount determined in accordance with section 23 would be the annual value of the property and it is only a proportionate amount out of this annual value, proportionate to the period of non-occupancy, which is permissible as a deduction under section24(1)(ix). *Sec.24 (1)(ix).*[**194ITR270,59Taxman221**]

Annual value of property **Cal.** **Liquidator, Mahmudabad Properties Ltd. - CIT**

Under sections 22 and 23 what is charged is the annual value of the property. That annual value is deemed to be the sum for which the property might reasonably be expected to let from year to year. These are deeming provisions and are based on the idea of hypothetical tenancy. These provisions also show that the property has to be considered as it is at the time of the determination of its annual value.*Sec.22.* **[83ITR470]**

Annually recurring payment **CIT Bom.** **Dalip Singh Bakshi**

If a payment recurs over a long period of years, it can be called an annually recurring payment and notwithstanding the fact that the mode of payment is monthly or the intervening period between two payments is less than a year. Thus, if any amount is payable monthly and such payment is to recur year after year the liability must be treated as one for making an annual payment.... *Sec.24(1).***[12CTR10,122ITR96,53Taxation12]**

Annuity **All.** **P.K.Banerji - CWT**

An annuity, in the sense of the law, means a fixed sum of money payable every year and it is not dependent upon variations in the net income of the property. But the amount of an annuity may not be absolutely fixed and may be variable from year to year, but so long as the variation in the amount to be received is in no way dependent upon or related to the general income of the estate, it will be still described as an annuity. An annuity confers no interest in any particular part of the property charged, but simply a security extending over the whole. *Sec.2(e) W.T.Act.***[83ITR117]**

2. The word annuity in clause (iv) of section 2(e) of the Wealth-tax Act, 1957, must be given the signification which it has assumed as a legal term, owing to judicial interpretation and not its popular and dictionary meaning. The hallmark of an annuity is (i) it is a money payment, (ii) made annually, (iii) it is of fixed sum, and (iv) usually it is a charge personally on the grantor. An annuity involves the conversion of capital into income. Section 2(e)(iv) excludes annuity from assets where the terms and conditions of the grant precluded the commutation of any portion thereof. It connotes that there should be, in the contract of annuity, a specific term against commutation, if the annuity is to be excluded in the computation of net wealth. In the alternative, a prohibition against commutation should clearly emerge from the scheme of the settlement and from the terms and conditions relating to the annuity itself. Annuity shall be excluded from the assets only if the terms and conditions relating to the grant make it impossible to interchange it for a lump sum amount. It is generally a question of fact depending on the terms and conditions of the grant giving the annuity. The question of commutability may be inferred not only from the specific conditions but also from all relevant terms creating the annuity and other related conditions. Thus, non-commutability may arise either on account of specific prohibition in the terms and conditions contained in the grant or it may otherwise be so inferred from the circumstances or the law applicable. **[Parmanandbhai Patel - CWT,MP.,**_Sec.2(e)(iv)WT.Act._**74CTR90,177ITR339]**

3. A right to an annuity is not includible in the asset of the assessee for the purpose of wealth-tax where the terms and conditions relating thereto preclude the commutation of any portion thereof into a lumpsum grant. The expression annuity has not been defined in the Wealth-tax Act. The principles relating to annuity are: (1) An annuity is a right to receive de anno in annum a certain sum payable by the grantor personally or out of corpus or income of the estate or a fund specifically set apart. The amount receivable by the annuitant must be a predetermined fixed amount and the variation in the quantum of the said amount, if any, should not be dependent upon the increase or fall in the general income of the estate or the trust fund. (2) Such predetermined or fixed amount may be payable by the grantor personally, i.e., secured merely by a personal covenant of the grantor. (3) An annuity is a right to receive a specified sum and not an aliquot share in the income arising from any fund. If the beneficiary is entitled to a share in the estate or a share in its income in entirety or in a particular proportion as contrasted with a mere right to receive a predetermined fixed sum per annum, such an interest does not partake of the character of an annuity. (4) If the amount of annuity payable fluctuates or varies depending upon the general income of the estate and its quantum tends to increase or decrease depending upon the quantum of general income of the estate as such, the provision made for payment of such an amount would not partake of the character of an annuity. This principle, however, has no application to a situation where the annuity may have to abate proportionately by operation of law. (5) Merely because the predetermined and fixed amount of annuity is made payable out of the income of the trust which is one of the permissible modes of creating an annuity, it does not follow that the right to receive such sum is dependent upon or related to the general income of the estate and the right to receive such predetermined amount is not an annuity. (6) The terms and conditions precluding commutation of an annuity into a lumpsum may be express or implied. If the terms and conditions of the trust deed or any other relevant deed provide for payment of a continuing annuity to the beneficiary throughout his life and provide for handing over of the entire corpus to another beneficiary on the death of the surviving annuitant, it necessarily follows that the right to receive the periodical amount cannot be commuted into a lumpsumgrant.**[Gopikumari Birla-CWT,Bom.,**<u>*Sec.2(e)(iv)WT.Act.*</u> **192ITR318,59Taxman366]**

4. In order to constitute an annuity the payment to be made periodically should be a fixed or predetermined one and it should not be liable to any variation depending upon or on any grant relating to the general income of the fund or estate which is charged for such payment. Payment of annuity thus should not be dependent upon the income of the corpus. From the definitions in section 2(e) of the Wealth-tax Act,1957, it is clear that the right to annuity is an asset within the meaning of section 2(e) right from the commencement of the Act. Such right to annuity was excluded from the definition of asset by virtue of item (iv) of sub-clause (1) of clause (e) of section 2 in cases where the terms and conditions relating thereto precluded the commutation of any portion thereof into a lump sum grant. If this condition was fulfilled, the right to annuity was excluded from the definition of asset. This exclusion was subjected to a further condition with effect from April 1,1975, by the Finance Act, 1974, that it is not an annuity purchased by the assessee or purchased by any other person in pursuance of a contract with the assessee. **[CWT-Ajit alias Hamid Alikhan, Bom.,** <u>*Sec.2(e)(iv)WT.Act.*</u>**128CTR230,215ITR454]**

5. ....the amount standing to the credit of a subscriber under the Compulsory Deposit Scheme (Income-tax Payers) Act,1974, and which is to be received back by the subscriber in five equal installments every year together with interest constitutes annuity within the meaning of section 2(e)(2)(ii) of the Wealth-tax Act, 1957 and hence is not an asset within the meaning of section 2(e). **[Udai Chand Jain- CIT, All.**<u>*Sec.2(e)(2)(ii)WT.Act.*</u>**142CTR155,228 ITR190,98Taxman209]**

Annuity and property SC Ahmed G.H. Ariff - CWT

The right of a beneficiary to receive an aliquot share of the net income of properties comprised in a wakf-alal-aulad created by a Muslim governed by the Hanafi school of Mahomedan law is property and is covered by the definition of assets in section 2(e) of the Wealth-tax Act,1957, and the capitalised value of the right is assessable to wealth-tax. This would be so even if the aliquot share of the income was intended merely for the maintenance and support of the beneficiary. Such a right is not a mere annuity. The word annuity in clause (iv) of section 2(e) must be given the signification which it has assumed as a legal term owing to judicial interpretation and not its popular and dictionary meaning. <u>*Sec.2(e)WT.Act.*</u> **[76ITR471]**

Any **Cal.** **CIT - Arvind Investments Ltd.**

The opening words of the Explanation to section 73 are where any part of the business of a company. Any is a word which excludes limitation or qualification. A restricted meaning should not be given to the phrase any part of the business. <u>*Sec.73.*</u>**[94CTR263,192ITR 365,58Taxman216]**

2. A perusal of section 40(b) of the Income-tax Act, 1961, shows that any payment of the nature described, made by the firm to any of the partners of the firm would not be deductible. The word any is a word of wide import and should be given its full meaning in the context of the provision. It will be seen that there is no indication whatsoever to differentiate between the nature of remuneration or between the purpose for which remuneration was given to a partner. As the provision reads, it imposes an absolute embargo against deductions in respect of any payment made by the firm, of the nature enumerated, to any partner of the firm. Explanation 2 which has been added clarifies that interest paid by the firm to an individual who is a partner in a firm in a representative capacity shall not be taken into account for the purpose of the said clause. The wording of the provisions of section 40(b) clearly indicate that they are intended to cover all the payments of the nature described to a partner of the firm, by the firm and do not indicate that only such payments as are made to a partner in his capacity as a partner and not other payments made to such partner are covered by the said provision. There is nothing in the said provision to indicate that any category of salary, remuneration, etc., though paid by a firm to a person who is a partner, are to fall outside the scope of the above provision. **[CIT-Yoganand Textiles,Guj.,**<u>*Sec.40(b).*</u>**202 ITR 869]**

3. The power under section 127 of the Income-tax Act, 1961, to transfer cases from one Assessing Officer to another can also be exercised in respect of block assessments. The word any in that section must be read as all. **[K.P. Mohammed Salim-CIT, SC,** <u>*Sec.127.*</u> **216CTR97,300ITR302,169Taxman465]**

4. Grammatically, the word any can relate either to only one person particularized by the circumstances or situation or to several or limitless indistinguishable or indeterminate persons. Any person, or a person or the person can be used synonymously in the singular.

This is the user which commends itself to protect against the peril of the provision being perceived as ultra vires Chapter III of the Constitution. **[S.R.Batliboi&Co., Del.,**<u>*Sec.132.*</u>**224CTR369,315 ITR 137,181Taxman9]**

Any assessment made under this act **SC** **Illuri Subbayya Chetty and Sons - State of A.P.**

The expression any assessment made under this Act is wide enough to cover all assessments made by the appropriate authorities under the Act whether the said assessments are correct or not. It is the activity of the Assessing Officer acting as such officer which is intended to be protected and as soon as it is shown that, exercising his jurisdiction and authority under the Act, an Assessing Officer had made an order of assessment,that clearly falls within the scope of such prohibition. <u>*Sec.293.*</u>**[50ITR93]**

Any case **Ker.** **K.Kader Haji - CIT**

Section 127 used the expression any case. The Explanation contains the word case in relation to any person whose name is specified in any order or direction issued there under and means all proceedings under the Income-tax Act in respect of any year. Therefore, the expression any case means all proceedings under the Act in respect of any year. The expressions any year and any case exclude limitation or qualification, which points in a distributive construction. The words any case and any year connote wider generality. Considering the object and purpose of section 127 and to facilitate effective and co- ordinated investigation, the Explanation to section 127 has to be liberally construed and therefore block assessment would fall within the expression case enabling the Commissioner to transfer those cases relating to block assessment to the subordinate officer. <u>*Sec.127.*</u>**[189CTR313,268ITR465,140 Taxman527]**

Any deposit **All.** **Chaubey Overseas Corporation - CIT**

The words any deposit have been used to cover all kinds of deposits including trade deposits and genuine deposits. <u>*Sec.269T.*</u>**[303ITR9, 170Taxman9]**

Any expenditure **Mad.** **CIT - Bharat Overseas Bank Ltd.**

Section 35B of the Income-tax Act, 1961, refers to any expenditure. Expenditure refers to an outgoing. If the monies had been expended, all such monies would qualify for the weighted deduction, provided the expenditure was incurred for any one of the purposes referred to in section 35B. It is significant that the section refers to any expenditure. The use of the word any preceding the word expenditure would also indicate that the focus is on the expenditure actually incurred by way of outgoing, and once it is established that such expenditure had been incurred, there is no occasion for making any deductions there from. *Sec.35B.***[242ITR314,117 Taxman759]**

2. The expression any expenditure has been used in section 37 of the Income-tax Act,1961, to cover both expenses incurred as well as an amount which is really a loss even though such amount has not gone out from the pocket of the assessee. **[CIT- Woodward Governor India P. Ltd.,SC,** *Sec.37.***223CTR1, 312 ITR 254,179 Taxman 326]**

Any income **Mad.** **DIT(Exemp.)-A.M.M.Hospitals and Medical Benefit Society**

Section 10(22A) of the Income-tax Act,1961, exempts from levy of tax any income of a hospital or other institution for the reception and treatment of persons suffering from illness or mental defectiveness or for the reception and treatment of persons during convalescence or of persons requiring medical attention or rehabilitation, existing solely for philanthropic purposes and not for purposes of profit. The ratio of the decision of the court in CIT v. A. M. M. Arunachalam Educational Society (243 ITR 229) (Mad) is equally applicable to the income referred to in section 10(22A). Any income in that provision in effect means all income. The material requirement of section 10(22A) is that the hospital or institution referred to therein of the nature set out in that provision should exist solely for philanthropic purposes and not for the purposes of profit. *Sec.10(22A).***[187CTR567,262ITR241,140Taxman81]**

Any income by way of interest **Mad.** **CIT-S.K.Sundararamier & Sons**

The expression any income by way of interest in section 194A of the Act would also seem to indicate that section 194A of the Act takes in all interest income except those which are specifically excluded from the scope of section 194A of the Act. *Sec.194A.***[240ITR740]**

Any individual **SC** **CIT - Shri Om Prakash**

Under section 4 of the Income-tax Act,1961 the charging section, the total income of the previous year or years of every person is charged for any assessment year at the rate or rates prescribed by the Finance Act. A plain reading of the definition of person in section 2(31) shows that both an individual and a Hindu undivided family are, inter alia, constituents of the meaning of the term person. The expression any individual is narrower than the terms person and assessee defined in section 2(7); an individual is a person but every person need not be an individual. So also an individual may be an assessee but every assessee need not be an individual. Had Parliament intended to give a wider meaning to the word individual in section 64(1)(i) and (ii) so as to include the karta of a Hindu undivided family it would have drafted the provision differently. It is thus clear that individual in section 64(1) does not take in karta of the Hindu undivided family within its import. *Sec.2(7),2(31).***[155CTR206,238ITR1044,105Taxman619]**

Any order **Mad.** **CIT - Shriram Engg. Construction Co.Ltd.**

Section 263 of the Income-tax Act,1961, refers to any order. The words any order would only mean any order under the Act, and if the Commissioner considers that any order passed by the Assessing Officer is erroneous in so far as it is prejudicial to the interests of the Revenue the Commissioner has the power to revise it. *Sec.263.* **[330ITR568]**

Any other case **P&H.** **CWT - Anil Tayal (HUF) (No. 1)**

The expression any other case in clause (b) of section 16A(1) is wide enough to include a case where no return has been filed by the assessee. If a narrow view is taken that a reference under section 16A(1) can be made only where the return has been filed, then the expression any other casein clause (b) of section 16A(1) and the expression where no such return has been made in sub-section (4) of section 16A would become redundant. *Sec.16A(1).W.T.Act.***[195CTR 420,238ITR1044,146Taxman239]**

Any other object of general public utility **SC** **CIT-Gujarat Maritime Board**

The expression any other object of general public utility in section 2(15) of the Income-tax Act,1961, is of the widest connotation. That expression would prima facie include all objects which promote the welfare of the general public. <u>*Sec.2(15).*</u>**[295ITR561]**

Any other person **Mad.** **CIT-K.P.Shaik Mohamed Rowther and Co. (P) Ltd.**

Sub-rule (2) of rule 6D of the Income-tax Rules,1962, places a ceiling on the travelling expenses incurred by an employee or any other person to be allowed as expenditure. In view of the expression any other person used in the sub- rule, even if a director or a managing director is not an employee of the company, the ceiling prescribed by sub-rule (2) of rule 6D applies. A director or a managing director of a company will come within the expression any other person mentioned in rule 6D(2) and deduction can be allowed for the travelling expenses incurred by them within the ceiling prescribed there under<u>*Rule.6D(2).*</u>**[161CTR 453,242ITR245]**

2. The expression any other person in rule 6D(2) is broad enough to include professionals within its ambit. **Industrial Cables (India) Ltd.-CIT, PH,** <u>*Rule.6D(2).*</u>**[272ITR314]**

Any other receipt of a similar nature **Kar.** **CIT-Motor Industries Co. Ltd.**

The expression any other receipt of a similar nature, has to be understood in the context of the words preceding the expressions brokerage, commission, interest, rentor charges. ... Though the object of enacting section 80HHC of the Act is to provide incentive to export business and thus earning foreign exchange, the benefit should go to only exporters and should not be misused in getting the benefit when there is no element of export involved in the income. Keeping in mind the legislative intent, it is clear that such incomes which have no direct nexus with the export turnover are liable to be deducted in arriving at the profits of the business. It is only when the assessee has an independent income which has no nexus with the income derived from export, which is in the nature of brokerage, commission, interest, rent or charges, and by inclusion of that income the profits of the business result in distortion, that, such income should be excluded. <u>*Sec.80 HHC.*</u>**[239CTR541,331ITR79]**

Any other sum chargeable under the provisions of the Act **AAR** **Imt Labs (India) P. Ltd.**

That the expression any other sum chargeable under the provisions of the Act, in section 195(1) would mean a sum on which Income-tax is leviable. The only consideration would be whether payment of the sum to the non-resident was chargeable to tax under the provisions of the Act. The sum might or might not be income; income might be hidden therein or otherwise embedded therein. The scheme of tax deduction at source applies not only to the amount paid, which wholly bears income character but also to gross sums, the whole of which might not be income or profit of the recipient. *Sec.195(1).* **[206CTR396,287ITR450,157Taxman189]**

Any person **All.** **Dhiraj Mal - CIT**

The expression any person, in its widest connotation, may include any person, whether connected or not with the assessee; but this construction cannot be accepted for purposes of section 34(3), for the said expression is necessarily circumscribed by the scope of the subject-matter of the appeal or revision, as the case may be; that person must, therefore, be one who would be liable to be assessed for the whole or a part of the income that went into the assessment of the year under appeal or revision. *Sec.147.* **[79ITR242]**

2. Under the provisions of section 34(3) of the Indian Income-tax Act, 1922, the bar of limitation will not apply where reassessment is made on any person consequent on a direction or finding given by an appellate or revisional authority. The expression any person in the setting in which it appears must be confined to a person intimately connected with the assessments of the year under appeal. The intimate connection must be such that the assessment of one must depend on the assessment of the other as in the case of a partner and the partnership firm, and of an individual and members of the joint Hindu family. Merely because in the course of assessment a finding is recorded that the income belonged to somebody else, an intimate connection is not automatically established by the fact that the same income is sought to be assessed in the hands of another. It should be possible to say that the interests of the person whose income is sought to be reassessed on the basis of a finding recorded in the assessment proceeding of another are so closely intertwined that no separate issues will really arise at a later stage when the assessment is reopened acting on the finding or a direction. If impliedly a person is represented in the earlier proceedings, only then the question of intimate connection can be said to be established. Merely because the same income becomes the subject-matter of the two assessments an intimate connection cannot be established. **[CIT- Homi Mehta and Sons P. Ltd.,Bom.,** *Sec.147.***27CTR238,137ITR213]**

3. ....an application for a tax clearance certificate could be made and issued only to the transferor of the property and not to the transferee because the words any person occurring in section 230A could have reference only to the transferor and could not include the transferee as well. **[Smt.Fouzia Shahi Nazeer - B.K. Lingappa, Kar.,** *Sec.230A.***182ITR342]**

4. Section 10(22) of the Income-tax Act,1961, provides that in computing the income of a previous year of any person, any income of a University or educational institution is exempted provided it exists solely for educational purposes and not for purposes of profit. For the purpose of exemption under this provision, it is not necessary that the assessee should be a University or the educational institution. The character of the assessee is not material. He may be any person including a person engaged in business or profession and even then he would be entitled to exemption under section 10(22) of the Act in respect of any income falling within the scope of clause (22) of section 10 of the Act. The availability of exemption should be evaluated every year to find out whether the institution existed during the relevant year solely for educational purposes and not for purposes of profit. **[CIT-Vidya VikasVihar,Bom.,** *Sec.10(22).***187CTR446,265ITR489]**

Any person who has not hitherto being assessed **Mys.** **K.Y.Pilliah - CIT**

The words any person who has not hitherto been assessed in section 18A(3) of the Indian Income-tax Act,1922, means any person who has not been actually assessed up to that date, and includes persons against whom assessment proceedings are pending but who have not been assessed up to the date. *Sec.210.***[53ITR705]**

Any person who is responsible for paying **Mad.** **Shital N. Shah - ITO**

The words any person who is responsible for paying found in section 194A of the Income-tax Act, 1961, have to be read in conjunction with section 204 of the Act which furnishes the meaning of person responsible for paying. This provision makes it abundantly clear that, if the payer is a company, the company itself including the principal officer thereof will be the person responsible for paying. If that be so, it is fairly apparent that the company itself including the principal officer thereof were liable for prosecution for the alleged contravention. Section 2(35) would then step in to find out as to who the principal officer would be. Section 2(35) makes it clear that the partners of the firm do not fall within that fold unless the Income-tax Officer had served a notice on any of them of his intention of treating them as the principal officer of the firm, connected with the management or administration thereof. *Sec.194A.*[88 CTR103, 188ITR376]**

Any policy of insurance SC CWT - Yuvaraj Amrinder Singh

While granting the exemption under section 5(l)(vi), the Legislature has used the expression any policy of insurance, which is one of very wide import. The exemption is not confined to rights or interests in life insurance policies alone, much less any particular species of life insurance policies, but it extends to rights or interests of an assessee in other types of insurance policies also. The exemption contemplated by section 5(l)(vi) covers interests of an assessee in all types of insurance policies and the expression any policy of insurance in the section would a fortiori attract within its ambit or scope a deferred annuity policy based on, human life, it being a species of life insurance policies and, therefore, unless there is some warrant to cut down the ambit or scope of that expression, the right or interest of an assessee in such a policy would be exempt from the charge of wealth-tax unless of course any moneys there under have become due and payable to the assessee on the valuation date. *Sec.5(l)(vi) WT.Act.***[49CTR211,156ITR525,23 Taxman25]**

Any remuneration, benefit or amenity Guj. Gujarat Steel Tubes Ltd. - CIT

The phrase any remuneration, benefit or amenity is of wide amplitude and it covers benefit or amenity in cash or in kind. *Sec.40(c)(i).***[210ITR358]**

Any residential accommodation in the nature of a... Guj. CIT- Gaekwar Mills Ltd.

Under sub-section(3) of section 37 of the Income-tax Act,1961, any expenditure incurred by an assessee after March 31,1964 ,on maintenance of any residential accommodation including any accommodation in the nature of a guest house was allowable only to the extent and subject to such conditions, if any, as may be prescribed. Sub-section (4)(i) of section 37 lays down that no allowance shall be made in respect of any expenditure incurred by the assessee after the 28th of February, 1970, on the maintenance of any residential accommodation in the nature of a guest house. What clause (i) of section 37(4) refers to is not a guest house, but any residential accommodation in the nature of a guest house. The expression any residential accommodation in the nature of a guest house will have a much wider meaning than guest house. *Sec.37.* **[99CTR19,193ITR734,62Taxman333]**

Any similar payment SC Continental Construction Ltd. - CIT

The true clue to the interpretation of the expression any similar payment in section 80-O lies not in the preceding three words but really in the second part of the section. The essence of the exemption lies, not in consigning the receipt to one of these pigeon-holes but in examining whether the receipt is a payment in consideration of one of the two situations envisaged in the section. To illustrate:where the assessee is the owner of a patent or invention, he may generally permit another to make use of the patent or the invention in consideration of a royalty payment. Or, again, where the assessee is in possession of technical know-how, he may be prepared to allow another to make use thereof in consideration of a feeto the assessee.He may also stipulate a consideration in the form of a commission based on the sales of the products the other party is able to manufacture with the aid of such invention or know-how. Again, an assessee may have achieved some speciality and he may agree to lend his services to some other person and stipulate a consideration therefore which may be variously described. The nature of the asset, right, information or services which can be brought under this provision may be varied and the consideration stipulated for allowing another to avail of the assessee's asset,knowledge or services can likewise assume multifarious forms. The word similar connotes that the payment made to the assessee need not be in the nature of royalty, commission or fees only; it could be any payment of like nature, i.e., made in consideration of the use or supply of such an asset, knowledge or services in the same manner as royalty, fees or consideration could be. *Sec.80***[AIR1992SC803,101CTR 386,195ITR81,107Taxation132,60Taxman429]**

Any speculation business Bom. CIT - Lokmat Newspapers P. Ltd.

The expression any speculation business means a speculation business of the assessee in respect of which profits and gains for the assessment year in question have arisen and there is no justification to restrict the content of that speculation business where profits have arisen by excluding a business involving actual delivery of shares. No such restriction is found in the Explanation. In other words, once the assessee is carrying on a speculation business and the profits and gains have arisen from that business during the course of the assessment year, the assessee is entitled to set off the losses carried forward from a speculation business arising out of a previous assessment year. *Sec.43(5).***[230CTR 521,322ITR43]**

Any such debt or part thereof Raj. CIT - Bank of Rajasthan Ltd.

The use of the words any such debt or part thereof clearly indicates that the exclusion of the provision in the proviso to section 36(1)(vii) will apply only in cases, where a provision for bad and doubtful debts has been made in the relevant accounting year on the bad and doubtful debts, which were outstanding at the commencement of the relevant accounting year and/or were also outstanding at the end of the relevant year. The aggregate average advances with reference to which the deduction in respect of the provision for bad and doubtful debt can be allowed necessarily implies that such a provision has to be made in respect of the loans and advances made at the end of the year. Thus, it is evident that the clauses, i.e., sections 36(1)(vii) and 36(1)(viia) are separate and they are distinct and independent. *Sec.36(1).***[174CTR400,255ITR599,124 Taxman781]**

Any sum Cal. CIT - Samnugger Jute Factory Co. Ltd.

The words any sums in section 15B(1) of the Indian Income-tax Act,1922, must be sums assessable in their nature, being parts of the assessable income of the relative accounting year and sums brought into the assessment and about to be brought to charge. *Sec.80G.***[24ITR 265]**

Any sum payable **Mad.** **Kalpana Lamps and Components Ltd. - DCIT**

The expression any sum payable under clause (a) means a sum for which the assessee has incurred liability in the previous year even though such sum might not have been payable within that year under the relevant law. The year in which the liability was incurred is taken as the year in which the sum was paid. That provision is meant for assessees who have received the deferral benefit under the schemes framed by the State Governments with regard to sales tax. Sec.43.**[175CTR549,255ITR491,125Taxman1045]**

Any unit of residential accommodation **Ker.** **CIT- Abad Hotels India (P.) Ltd.**

The word any occurring in section 3 of the Expenditure-tax Act, 1987, cannot be restricted to one unit of residential accommodation. The words any unit of residential accommodation means all the units of accommodation. *Exp.Tax Act.***[193CTR408, 272ITR331,142Taxman29]**

Any work **Ker.** **CIT - Indian Textile Paper Tube Co. Ltd.**

There is nothing in sub-section (1) of section 194C of the Income-tax Act, 1961, to hold that a contract to carry out a work or a contract to supply labour to carry out a work should be confined to works contracts and that the words any work occurring in the sub-section mean any work and not a works contract. Therefore, a transport contract simplicter (i.e., a transport contract for mere carriage of goods without loading and unloading facility) would amount to carrying out any work within the meaning of section 194C(1) of the Act and, therefore, deduction of tax at source at the rate of two per cent. from the amounts credited to the account of the contractor has to be made by a person responsible for paying any sum for the transport contract. *Sec.194C.***[150CTR528,236ITR993, 103Taxman29]**

Apparatus and appliances **Bom.** **CIT-I.B.M.World Trade Corporation**

The word apparatus is a word of much wider import than the word appliances and the definition of plant in section 43(3) of the Income-tax Act, 1961, shows that scientific apparatus is included in the word plant. But while a given appliance may be in the form of an apparatus, the converse will not always be true and every apparatus will not necessarily be an appliance. The word appliances is qualified by the word office in section 33 and the words have to be construed in the context of appliances which are generally used in an office as an aid or a facility for the proper functioning of the office. Sec.43.**[130ITR739]**

Apparent Del. Hotz Hotels Pvt. Ltd. - CIT

In order to bring in application under section 154 of the Income-tax Act, 1961, the mistake must be apparent from the record. Section 154 does not enable an order to be reversed by revision or by review, but permits only some error which is apparent on the face of the record to be corrected. Where an error is far from self-evident, it ceases to be an apparent error.It does not cover any mistake which may be discovered by a complicated process of investigation, argument or proof. The word mistake, in Taxation laws, has a special significance and it is inherently indefinite in scope. The word apparent means that it must be something which appears to be so ex facie that it is incapable of argument or debate. Sec.154. **[264CTR319,248ITR647,118Taxman94]**

2. In order to attract the power to rectify under section 254(2) of the Income-tax Act, 1961, it is not sufficient if there is merely a mistake in the order sought to be rectified. The mistake to be rectified must be one apparent from the record. A decision on a debatable point of law or disputed question of fact is not a mistake apparent from the record. The plain meaning of the word apparent is that it must be something which appears to be so ex facie and is incapable of argument or debate. **[Smt. Baljeet Jolly-CIT,Del.**Sec.254.**164 164CTR37,250ITR113,113Taxman 38]**

Apparent consideration P & H. CWT - Smt. Patwant Kaur

Strictly speaking apparent consideration would mean the whole or part of the consideration for such transfer, which is payable on any date or dates falling after the date of such agreement for transfer, and the value of such consideration payable after such date shall be deemed to be the discounted value of such consideration as on the date of such agreement or transfer, determined by adopting such rate of interest as may be prescribed in this behalf. <u>*Sec.7(4) WT.Act.*</u>**[192CTR479,275ITR481,142Taxman406]**

Apparent from the record Bom. Arvind N. Mafatlal - ITO

The expression apparent from the record should not be equated with the expression apparent on the face of the record. The mistake to be rectified should, however, be a mistake patent on the record and not a mistake which may be discovered by a process of elucidation,argument or debate. <u>*Sec.154.*</u>**[32ITR350]**

Applied AP. CIT - Nizams Charitable Trust

In order to secure exemption under section 11(1)(a) of the Income-tax Act,1961, it is sufficient if the assessee provides or sets apart the funds for a charitable purpose and it is not necessary that the assessee must have spent the amount specified for a charitable purpose during the relevant assessment year. That this is the intention of the Legislature is evident from the word used, namely, applied in section 11(1)(a). If the intention of the Legislature was otherwise, nothing prevented the Legislature from using the word spent instead of applied in section 11(1)(a). <u>*Sec.11(1).*</u>**[131ITR497]**

2. The word applied is wider in import than the word expenditure. According to Webster's Third New International Dictionary, the word applied means to put to practical use; engaged in for a utilitarian or contributory purpose; employed in the decoration, design or execution of useful objects. The word expenditure means disbursement. Expend means to pay out or distribute; to spend. The Supreme Court in the decision in Indian Molasses Co.(P.) Ltd v.CIT 37 ITR66 held that the word expenditure means put out or away; spending something which is gone out irretrievably. Considering these two words, the word applied is of a wider import.The money or amount will not go irretrievably when it is applied to a purpose. **[CIT-St. George Forana Church, Ker.,** <u>*Sec.11.*</u> **170 ITR 62]**

Apportionment SC Continental Construction Ltd. - CIT

For purposes of income- tax, the principle of apportionment has always been applied in different contexts. Consolidated receipts and expenses have always been considered apportionable in the contexts: (a) of the capital and revenue constituents comprised in them; (b) portions of expenditure attributable to business and non-business purposes; (c) of places of accrual or arisal; and (d) of agricultural and non- agricultural elements in such receipts or payments. Sec.80-O.**AIR1992SC803,101CTR386,195ITR81,107 Taxation132,60 Taxman 429]**

Appropriate authority **Mad.** **Venkata Seshavatharam - C.Venkata Rangayya**

The expression appropriate authority in clause (m) of section 54(3) of the Indian Income-tax Act, 1922, includes a Civil Court where it is called upon to determine judicially whether a person has or has not been assessed to Income-tax in a given year. Sec.137. **170ITR62,36 Taxman42]**

Appurtenance **Mad.** **M.Ramalakshmi Reddi - CIT**

What is necessary for the enjoyment of the building is alone covered by the expression appurtenance. The touchstone of appurtenance is dependence of the building on what appertains to it for its use as a building. What is integral is not necessarily appurtenant.A position of subordination, something incidental or ancillary or dependent is implied in appurtenance. Sec.22.**146CTR154,232ITR281, 100Taxman509]**

Appurtenant **P & H.** **Gowardhan Das and Sons - CIT**

Section 22 of the Income-tax Act, 1961, provides that it is the income from property consisting of any building or land appurtenant thereto which is assessed under section 22 and not the income from renting out of open land or some kutcha plinth only. What is covered by the expression appurtenant is the land which is necessary for enjoyment of the building and not the land only. Sec.22. **[207CTR47,288ITR481,158TaxmanTaxman465]**

Arising out of the order passed by it **All.** **Ram Nath RamPrasad - CIT**

A question cannot be said to arise out of an order passed by the Tribunal unless it was raised before it; in other words, every question of law applicable to the relevant facts proved or admitted before the Tribunal is not a question of law arising out of the order passed by it. It is not enough that the question of law may be applied to the relevant facts found or proved before the Tribunal; it must have been raised before it. The view that every question of law applicable to the relevant facts found or proved is a question arising from the Tribunal's order is wrong. *Sec.256.*__54ITR777]__

Arrears Mad. T.M.K. Abdul Kassim - First Addl.ITO

Before any arrears of tax can emerge there must be a legal obligation to pay an amount which has been ascertained or is easily ascertainable and a default in the performance of that obligation.

*Misc.***[33ITR466]**

Articles Bom. CIT - Hindustan Antibiotics Ltd.

The word articles used in the expression has begun or begins to manufacture or produce articles in section 15C(2)(ii) must be interpreted regard being had to the object for which the section was enacted. The provision was enacted with a view to encouraging the establishment of new industrial undertakings and the object was sought to be achieved by granting exemption from tax on profits derived from such undertakings during the first five years. The object of the section pre-supposes that profits are capable of being earned. Hence, until an assessee reaches a stage where it is in a position to decide that a final product which can be ultimately sold in the market can be manufactured it cannot be said to have started manufacture of the articles. If it becomes necessary for an assessee to produce a trial product at an earlier stage to verify whether it can be used ultimately in the manufacture of the final article, the commencement of operation for the manufacture of the trial product would not constitute commencement of manufacture of articles for the purposes of section 15C. *Sec.80J.***[93ITR548]**

2.There could be some expansiveness in the expression articles in contrast to the word goods. But in the context of manufacturing or producing articles and deriving profits of business, articles should be synonymous with the word goods. Goods cannot include an immovable property and, therefore, the word articles cannot likewise include an immovable property including a house property.

[CIT- Madgul Udyog,Cal.,*Sec.80J.***208ITR541]**

Articles or things P&H. Ramesh Chaner - CIT

The articles or things referred to in sub-section (3) of section 132 are those which the authorised officer was empowered to search for and seize and no other. They must be articles or things which it may be necessary to search for before they can be seized. The power conferred under section 132(1) is contemplated in relation to those cases where the precise location of the article or thing is not known to the Income-tax department and, therefore, a search has to be made for it, and where it will not ordinarily be yielded over by the person having possession of it and, therefore, it is necessary to seize it. It is such article or thing alone which can be the subject of an order under section 132(3). Therefore, it would not include a case where it is already known that the article or thing is kept in a certain building or place and will ordinarily be yielded up by the person holding custody of such article or thing. <u>*Sec.132.*</u>**[93ITR244]**

Assessable income All. Juggilal Kamlapat Cotton Spinning and Weaving Co. Ltd. - CIT

Assessable income means income assessable to Income-tax under the Indian Income-tax Act and has the same meaning wherever it is used in, sub-section (1) of section 23A. There is no warrant for the suggestion that the words assessable income in the first part of that section mean income which the company considers to be its assessable income and treats it as such in its balance-sheet and accounts. <u>*Sec.104,109.*</u>**[28ITR78]**

2. The expression assessable income which was used in section 23A(1) of the Indian Income-tax Act,1922, before it was amended in, meant income liable to be assessed to income-tax. **[CIT - Homi Mehta and Sons Ltd.,Bom.,**<u>*Sec.104.*</u> **46ITR1135]**

Assessed All. CIT - Roop Narain Ram Chandra (P.) Ltd.

The word assessed in section 18A(3) should be read in its ordinary sense as comprising every kind of assessment including a provisional assessment under section 23B of the Act. A person on whom a provisional assessment has been made is not a person who is not hitherto assessed within the meaning of section 18A(3) and is under no obligation to file an estimate of his income under that provision. <u>*Sec.210.*</u>**[87ITR13]**

2. When section 30(1) of the Indian Income-tax Act, 1922, uses the words any assessee . . . . denying his liability to be assessed under this Act, the word assessed is used in a comprehensive sense to mean subjected to the whole procedure for ascertaining and imposing liability on the taxpayer. There is nothing in section 30(1) to indicate that a narrow meaning should be given to the word assessed. On the contrary, the words under this Act, clearly show that the reference here is to the whole procedure laid down in the

Act for imposing liability on the taxpayer. The denial of liability to be assessed may be in respect of the whole income or any part of the income. It may be based on any ground, whether of fact or law, and it may be total denial of liability or denial of liability under particular circumstances. But, the denial must be of the liability to be assessed under the Act and not merely under any particular provision of the Act. When an assessee claims that he is not liable to be proceeded against under section 35(1), he is not denying his liability to be assessed under the Act. His objection is only against a proceeding for assessment under the particular provision of the Act. **[Mandal Ginning and Pressing Co. Ltd. & others - CIT, Guj.,** *Sec.256.***[87ITR13]**

3. The word assessed in section 212(3) refers only to the computation of the income or to the determination of the amount of tax payable.Merely because the computation of income or the determination of the amount of tax has not resulted in any tax liability it does not mean that the process of computation of income or the determination of the amount of tax is not an assessment. **[CIT - N.D.Georgopoules,Mad.,** *Sec.212.***[125TR630]**

4. Considering the object to be achieved by the insertion of sub-section (9), the word assessed should not be given a literal meaning to mean completed assessment. From a joint reading of sub-sections (1) and (9) of section 171 the intention of the Legislature was that once a Hindu undivided family has been assessed as a Hindu undivided family it shall continue to be assessed as such unless an order recognizing partition is passed by the authority concerned. The assessment order passed is relatable to the assessment year involved therein and it will be treated to have been passed on the last date of the assessment year, for the purpose of section171 of the Act. **[CIT- Gyan Chand and Sons,All.,***Sec.171.***202CTR605,303 ITR267,146Taxman429]**

Assessed tax P&H. CIT - P.B.Nanda

A reading of section 273 of the Income-tax Act, 1961, along with section 215(5) shows that for purposes of calculation of penalty on default committed under section 273(a), assessed tax means the tax on the basis of regular assessment reduced by the deductions permissible in accordance with the provisions of sections 192 to 194, section 194A, section 194C, section 194D and section 195 and further subject to advance tax. If the Legislature wanted to deduct also the amount paid on self-assessment under section 140A, then it would have been so stated in section 215(5). Therefore, penalty under section 273(a) is to be levied with reference to the net tax payable on completion of the assessment after giving credit only for the advance tax paid and the tax deducted at source but not for the tax paid on self- assessment under section 140A. *Sec.215.* **18CTR105,125ITR429]**

2. The words assessed tax, which has been defined in the Explanation to the section, provides that assessed tax, means tax as reduced by the sum, if any, deducted at source under Chapter XVII-B or paid in advance under Chapter XVII-C. **[CIT-Hari Chand Hans Raj,P&H.,***Sec.271, 272.***21CTR219,128ITR467]**

3. The expression assessed tax used in section 215(1) of the Income-tax Act, 1961, has to be understood as the tax finally assessed as reduced by the amount of tax deductible in accordance with the provisions of section 194A of the Act. So long as section 215 of the Act permits levy of interest only on the difference between the assessed tax and advance tax actually paid, the amount of tax deductible at source under section 194A will have to be taken note of and this has been specifically provided for in sub-section (5) of section 215. Consequently, in construing sub-section (5) of section 215, it is not possible to understand the expression deductible occurring therein as deducted. Where the statute provides for deduction of tax at source in respect of a particular income, the assessee concerned need not pay advance tax in relation to the said income. **[CIT-Madras Fertilisers Limited,Mad.,** *Sec.215.***149 ITR703]**

4. While computing the assessed tax as defined under section 215(5) of the Income-tax Act,1961, for the purpose of determining the quantum of minimum penalty imposable for the default falling under section 273(1)(b) tax deductible at source and deducted within the financial year has to be taken into consideration. The assessee is not liable to pay any advance tax on the income which is subject to deduction of tax at source. The advance tax is, paid in a financial year; the credit for the tax so paid in advance is given to the assessee at the time of regular assessment in the assessment year immediately following the year in which the advance tax is paid. It is, therefore, clear that tax deductible in the context of section 215(5) means tax deducted at source. The amount of tax so deducted at source under the provisions of sections 192 to 195 is, so far as the affected person is concerned, to be treated as income received by him. For the purpose of computation of his total income, gross salary, gross dividend or gross interest, etc., i.e., the amount actually received plus the amount of tax deducted at source, will have to be considered. Unless tax is deducted at source during the relevant previous year by the payer, the payee cannot claim the benefit of such deduction while filing the estimate of advance tax. **[CIT - Borhat Tea Co. Ltd.,Cal.**<u>*Sec.215.*</u>**193ITR134]**

5. .... assessed tax means tax as reduced by the sum, if any, deducted at source under Chapter XVII- B or paid in advance under Chapter XVII-C,... **[CIT-Adhyapak Prakeshan Mandir, All.,** <u>*Sec.271.*</u>**[197ITR714]**

6. The words assessed tax occurring in sub-section (1) of section 215 of the Income-tax Act,1961, dealing with interest payable by the assessee are to be read in the light of the special meaning given to them under sub-section (5) of section 215 and accordingly, assessed tax would mean not the full amount of the assessed tax determined on the basis of the regular assessment, but the amount reduced there from to the extent of tax deductible in accordance with the provisions of sections 192 to 194, 194A, 194C, 194D and 195 so far as it related to income subject to advance tax. **[CIT-Ranoli Investment P.Ltd.,Guj.,**<u>*Sec. 215.*</u>**146CTR745,235ITR433]**

7. The expression assessed tax is defined to mean the tax assessed on regular assessment which means the tax determined on the application of section 115J /115JA in the regular assessment. Interest under section 234B is payable on failure to pay advance tax in respect of tax payable under section 115JA. **[JCIT -Rolta India Ltd., SC,**<u>*Sec.115J.*</u>**237CTR329,330ITR470,196Taxman594]**

Assessed to income- tax **Mad.** **Venkadadri Somappa - Narasepally Venkataswami Chetty**

.... to say that a person has been assessed to Income-tax may properly be paraphrased by saying that he has paid income- tax, or that his income has been subject to income-tax, or has been reduced by the amount of the income-tax. <u>*Sec.4.*</u>**[9ITR284]**

Assessee **Lah.** **Probynabad Stud Farm**

Uder the Indian Income-tax Act,1922 assessee is the person who is in actual receipt and control of the income. <u>*Sec.2.*</u>**[4ITR114]**

2. Where a person carrying on a business, profession or vocation has been succeeded in such capacity by another person and an assessment is made on the successor under section 26(2) of the Indian Income-tax Act, 1922 on the profits of the predecessor, the word assessee in section 10(2)(vi) of the Act must be construed as referring to such predecessor, and depreciation allowance should be calculated on the original cost to the predecessor and not to the successor.**[CIT - Mazagaon Dock Ltd.,Bom.**<u>*Sec.32.*</u>**6ITR124]**

3. Though the words assessee in section 10(2) must, when there is a successor to the business charged to tax, be read in certain of the paragraphs as including both predecessor and successor, it does follow as a consequence that the unabsorbed depreciation of the predecessor must be added to that of the successor or that even in a case when the only business concerned is that which is transferred, the business when transferred carries to the purchaser its unabsorbed depreciation. **[Indian Iron and Steel Co.Ltd.-CIT, PC,**<u>*Sec.32.*</u>**11ITR328]**

4. ..for the purpose of fixing the liability for tax, the provisions of the Partnership Act would be available or the liability might pass to the partners under section 44, but the partners would still not be covered by the word assessee as used in section 29. **[Moti Lal Purshottam Das - ITO,All.,**<u>*Sec.156.*</u>**27ITR202]**

5. The generality of the definition of assessee in section 2(2) of the Indian Income-tax Act, 1922, is sufficient to include even a legal representative who is to pay the tax, though out of the assets of the deceased person. **[Addl.ITO-E-Alfred,SC,** <u>*Sec.2(7).*</u>**39ITR497]**

6. On the death of an assessee, his legal representative becomes only an assessee; he would not become an assessee in default until the tax is again demanded from him under section 23, and is not paid within the time allowed by section 23. **[Raja Pid Naik-Agricultural ITO,Mys.,**<u>*Sec.2(7).*</u>**44ITR442]**

7. Under section 33B read with the definition of assessee in section 2(2) of the Act proceedings under Section 33B can be taken against the legal heirs of a deceased assessee provided notice of the same is issued to the person liable to pay the tax assessed on the deceased. **[CIT- Sunderlal,SC,**<u>*Sec.2(7).*</u>**96ITR310]**

8. Section4(1)(b) of the Wealth-tax Act,1957, has to be interpreted by taking into account the definition of the expression assessee contained in section 2(c) of the Act which provides that assessee means a person by whom wealth-tax or any other sum of money is payable under this Act. There is nothing in section 4(1)(b) which would suggest that the expression assessee is employed therein in a different or restricted sense so as to apply to assessees who are individuals only. The karta of Hindu undivided family who is a partner in a firm would fall within the expression assessee in section 4(1)(b).**[Sudhakar Manibhai and Kulinsingh Manibhai-CWT,Guj.,** <u>*Sec.4(1)WT.Act.*</u>**[6CTR473,111ITR384]**

9. There is nothing in section 36(2) of the Act to indicate that assessee refers to the original creditor and does not include a transferee or assignee of the debt. The condition which has been expressly incorporated in section 36 and which did not find a place in section 10(2)(xi) of the Act of 1922 is that the amount of the debt or part thereof should have been taken into account in computing the income of the assessee in a previous year.The emphasis is not on the assessee being the original creditor but the taking into account of the debt in computing the income of the same business. **[T.N.Shah (P.) Ltd.-Addl.CIT,All.,***Sec.36* **8CTR 207,120ITR354]**

10. The reference to the expression assessee in section 36(2)(i)(a) of the I.T. Act, 1961, does not require as a statutory condition the identity of the same assessee both at the time when the debt comes into being and at the time when the debt became bad and claimed as bad and, consequently, the claim for deduction of the amount as a bad debt can and must be considered, other things being equal, in the hands of the person who happens to have the debt and in the particular year in which the allowance is claimed as a bad debt. On a proper construction and understanding of the provisions of sections 36(1)(vii) and 36(2)(i)(a), the bad debt may have to be allowed in cases where there is continuity of the business in which the debt had originally occurred irrespective of whether there had since been any change in the person carrying on the business. **[E.A. V.KrishnamurthyandSon-CIT,Mad.,***Sec.36.* **152ITR640]**

11. ......the word assessee in section 2(7) of the Income-tax Act, 1961, is defined to mean a person by whom any tax or a sum is payable under the Act. The said definition includes other persons also. The word person in section 2(31) is defined by the Act to include a Hindu undivided family, company, firm, etc. Thus the word assessee would include not only an individual but also a Hindu undivided family and others who are included within the meaning of the word person. **[CIT-Gunvantlal Ratanchand, Guj.,** *Sec.2(7).* **[208ITR1028]**

12. There is nothing in section 36(2) to indicate that assessee refers to the original creditor and does not include a transferee or assignee of the debt. The condition which has been expressly incorporated in section 36 and which did not find a place in section 10(2)(xi) of the Indian Income- tax Act,1922, is that the amount of the debt or part thereof should have been taken into account in computing the income of the assessee in a previous year. The emphasis is not on the assessee being the original creditor but the taking into account of the debt in computing the income of the same business. If, in a given case, the income of a business is computed by taking into account a certain debt, it does not appear reasonable that, in the absence of any statutory prohibition, allowance on account of the debt having become bad should be denied only because the assessee's identity has changed though the identity of the business continues. **[CIT-Hotel Ram Bagh Palace,Raj.,**<u>*Sec.36.*</u>**209ITR638]**

13. The assessee has been defined under section 2(7) of the Act. Assessee would mean a person by whom any tax or any sum of money is payable under this Act. Therefore, what is clear to us is that in terms of the statute, only an assessee who is liable to pay tax in terms of the order alone is provided with a right of appeal in terms of the statute though to a certain extent the association may be an aggrieved party but the association is not an assessee in terms of the definition. **[Mico Employees Association - ACIT(TDS),Kar.,**<u>*Sec.2(7).*</u>**292ITR567]**

Assessee in default **Kar.** **S.N.Santhalingam - ITO**

The scheme of the Income-tax Act is that there should first be an assessment. A notice of demand should be served on the assessee in regard to whom assessment is made and if default is made in payment of the amount demanded then such assessee would become an assessee in default. <u>*Sec.222.*</u>**[121ITR868]**

2. The definition of assessee given in section 2(7) of the Income-tax Act,1961, includes a person deemed to be an assessee in default under any of the provisions of the Act. Clause (x) of section 226(3) makes the garnishee in default of compliance with the notice under section 226(3) a person deemed to be an assessee in default. Though the definition of assessee defined in section 2(7) includes an assessee in default, this has to be read in the context of section 226, which bears a different connotation in view of the provisions contained in the Second Schedule to the Income-tax Act. The provisions refer to the expression defaulter at every stage. The word defaulter has been expressly defined in rule 1(b) of the Second Schedule to mean the assessee mentioned in the certificate. **[Shaw Wallace and Co. Ltd.-Union of India,Cal.,**<u>*Sec.2.r.w.s.226.*</u> **189CTR1,267 ITR 248]**

Assessee is in default **Mys.** **Raja Pid Naik -Agricultural ITO**

On the death of an assessee, his legal representative becomes only an assessee; he would not become an assessee in default until the tax is again demanded from him under section 23, and is not paid within the time allowed by section 23. <u>*Sec.143.*</u>**[69 ITR401]**

Assessing Officer **Cal.** **Reckitt Colman of India Ltd. - ACIT(TDS)**

The expression Assessing Officer in section 2(7A) is not confined to the Assessing Officer making the regular assessment but includes others also who may come within the purview of section 2(7A). Rule 36A of the Income-tax Rules,1962, has been inserted by way of amendment effective from July 12, 1988 .On a plain reading of rules 36A, 37 and 37A, it clearly appears that as per the powers delegated by or under section 206 of the Act which is under Chapter XVII concerning tax deduction at source, the officer authorised has the jurisdiction to receive returns relating to tax deducted at source. The Assessing Officer has the jurisdiction under section 131 to call for the records or evidence or document or discovery as the civil court has got under the Code of Civil Procedure when trying a suit. Now under the same section 131, this Assessing Officer has been given power to impound books of account or other documents. This Assessing Officer has been given power to charge interest and to levy penalty. The authority having power to charge interest and levy penalty on a particular return cannot be said to be an officer who is meant for the purpose of accepting only TDS returns by affixing a rubber stamp. *Sec.2(7A).***[172CTR499,252ITR550, 124Taxman496]**

2. From the definition of Assessing Officer, the court cannot come to the conclusion that if the income computed by the Assessing Officer is enhanced by the Commissioner (Appeals), the said order has to be treated as an order passed by an Assessing Officer. It has to be treated as an order passed by the appellate authority and not by the original authority. When the words, Assessing Officer are used in section 255,the court cannot hold the order of the appellate authority as an order of the Assessing Officer. **[CIT-Mahakuteshwar Oil Ind., Kar.,***Sec.255.***298ITR390]**

Assessment **PC** **Maharajadhiraj of Darbhanga - CIT**

The word assessment in section 26(2) is not confined to the definite act of making an order of assessment and where a notice has been given under section 22(2) to a person to furnish within the time specified a return in the prescribed form, the process of assessment has begun and continues until some order of assessment is made.

*Sec.2(8).***[2ITR345]**

2. The word, assessment is used in Income-tax Acts as meaning sometimes the computation of income, sometimes the determination of the amount of tax payable, and sometimes the procedure laid down in the Act for imposing liability upon the taxpayer. **[CIT-Khemchand Ramdas,PC,** *Sec.2(8)*.**6ITR414]**

3. The word assessment is used in the Act as meaning sometimes the computation of income, sometimes the determination of the amount of tax payable, and sometimes the whole procedure laid down in the Act for imposing liability on the taxpayer. **[Kunwar Bishwanath Singh -CIT,All.,***Sec.2(8).***10ITR322]**

4. The word assessment can bear a very comprehensive meaning; it can comprehend the whole procedure for ascertaining and imposing liability upon the taxpayer and there is nothing in the context of section 297 of the 1961 Act, which compels the court to give the expression procedure for the assessment a narrower meaning. The expression proceedings for the assessment of that person in clause (a) of section 297(2)(a) of the Income-tax Act, 1961, has a very comprehensive meaning. **[Kalawati Devi Harlalka - CIT,SC,***Sec.297.***66 ITR 680]**

5. The word assessment in the first proviso of section 34(3) meant not merely the computation of the income of the assessee but also the determination of the tax payable by him. **[CIT- Balkrishna Malhotra,SC,***Sec.147*.**80ITR759]**

6. There is nothing in the context of the provisions of clauses (f) and (g) of section 297(2) of the Income-tax Act, 1961, to warrant that the expression assessment therein can only refer to the original assessment and not to reassessment under section 34 of the 1922 Act or section 147 of the 1961 Act, as the case may be. The expression assessment in these two provisions should be given the meaning to be found in section 2(8) of the 1961 Act and would include reassessment. **[CIT-Bashirkhan Ismailkhan, Bom.,** *Sec.297.***109 ITR 629]**

7. ...assessment has been defined in the Act to mean reassessment, and regular assessment has not been defined to mean an initial assessment. **[Kashiram Tea Industries Ltd.-ITO,Cal.,** *Sec.2(8).***132 ITR783]**

8. The word assessment in section 153 of the Income-tax Act, 1961, prior to its amendment with effect from April 1,1973, means not merely the computation of the income of the assessee but also the determination of the tax payable by him. **[Mohendra J. Thacker and Co.-CIT,Cal.,***Sec.2(8).***29CTR1,139ITR793]**

9. The word assessment has been defined in section 2(8) of the Act of 1961 as including reassessment and the definition provision would apply if there is nothing repugnant in the subject or the context and there is nothing to indicate that the definition under section 2(8) of the Act cannot be read into section 297(2)(g) of the Act. Thus, if a reassessment for the year ending on March 31,1962, or any earlier year, was completed on or after April 1,1962, proceedings for penalty could be initiated and penalty also imposed or levied under the provisions of the Act of 1961. **[CIT - C.J. Sheth,Mad.,**<u>*Sec.297.*</u>**76CTR132,179ITR30,44Taxman384]**

10. ...the expression assessment is not used in the Income-tax Act merely in the sense of computation of income. The determination of the tax payable by the assessee is as much mandatory as the determination of his income. Section 143(1)(a) of the Income-tax Act, 1961, enjoins upon the Income-tax Officer not only a duty to assess the income of an assessee, but also a duty to determine the sum payable or refundable to the assessee on the basis of such assessment. It is, therefore, obvious, that an assessment under section 143(1) comprises computation of income and computation of tax on such income. A bare reading of the provisions of sections 143, 154 and 156 would show that where the Income-tax authority amends any assessment order which has the effect of enhancing the assessed income, he is bound to issue and serve on the assessee a notice of demand under section 156 of the Act. Such notice would indicate the tax demanded. It is, therefore, apparent that the rectification of an assessment order would include rectification of computation of income as also rectification of computation of tax earlier made during the regular assessment and such rectification of mistake would take effect from the date the demand notice under section 156 is served on the assessee. **[CIT-Miss. Swarn Taneja.,MP.,**<u>*Sec.143.*</u>**87CTR66,186ITR348]**

11. In the Income-tax Act,1961 the word assessment may carry different meanings depending on the context in which it is used. The words assessment and reassessment are wider in scope than the word regular assessment. Regular assessment is narrower in its scope and is limited only to the assessment under section 143 or section 144 of the Income-tax Act, 1961 . A fresh assessment made under section 263 under the direction of the CIT is not a regular assessment. It does not fall within the purview of sections 143 and 144 read with clause (40) of section 2 of the Income-tax Act. Assessment made under section 263 has been termed fresh assessment. **[CIT-Multi metals Ltd.,Raj.,***Sec.2(40).***88CTR1,187 ITR98]**

12. The assessment made for the first time though not on the basis of return voluntarily filed in the manner required by the statutory provision and within the time prescribed therein is a regular assessment. The word assessment as defined in the Act includes reassessment. **[CIT-S.Antony,Mad.,**<u>Sec.2(8).</u>**125CTR421,242 ITR363]**

13. The term assessment in the provision prescribing the period of limitation in the Wealth-tax Act, 1957, has been used in comprehensive sense which includes the integrated process of computation of net wealth as well as computation of tax liability thereon. The two actions need not be simultaneous; they may be taken separately and at different times but the assessment is complete only when both the processes are over, namely, determination of net wealth and determination of tax payable on such net wealth.It is only when both the processes are complete that the assessment can be said to be completed. As a result, both the processes must take place prior to the expiry of the period of limitation. **[Prabhavati B. Solanki-CWT,Guj.**<u>*Sec.16.r.w.s.17A.WT.Act*</u>. **159 CTR 406,243ITR827,105 Taxman563]**

14. Assessment is used as meaning sometimes the computation of income, sometimes the determination of the amount of tax payable and sometimes the whole procedure laid down in the Act for imposing liability upon the taxpayer. In the scheme of things, the intimation under section 143(1)(a) of the Income-tax Act, 1961, cannot be treated to be an order of assessment. **[CIT - Punjab National Bank,Del.,**<u>*Sec.143.*</u> **166CTR340, 249ITR 763,116 Taxman310]**

15. The word assessment is used in the Income-tax Act, 1961, in a number of provisions in a comprehensive sense and it includes all proceedings starting with the filing of the return or issue of notice and ending with the determination of the tax payable by the assessee. Proceedings for the assessment of a firm consist of computation of the income of the firm, determination of tax payable by the firm, apportionment of the income of the firm between its partners in the case of a registered firm, and in appropriate cases the imposition of tax on the firm after including the share of the income of certain partners in the income of the firm, even though the firm is registered. The proceedings under section 182 do not come to an end merely on computation of the income of the firm and the determination of tax payable by the firm on that income. When proceedings are taken under section 155 of the Act to give effect to the order of the appellate authority and in pursuance thereof treat a firm as registered under the Act,which was earlier treated as unregistered by the Income-tax Officer, it is a proceeding for assessment. In the proceedings under section 155, the Income-tax Officer amends the order of completed assessment of a partner in a firm on the assessment or reassessment of the firm, on the ground that the share of the partner in the income of the firm has not been correctly included in the income of the partner.Such proceedings are amendment proceedings and, therefore, clearly form part of the proceedings for assessment. **[CIT - Sanjai Kumar Gupta,All.,** *Sec.155.***276ITR73,146Taxman462]**

16. Under the first proviso to the newly substituted section 143(1), with effect from June 1,1999,except as provided in the provision itself, the acknowledgment of the return shall be deemed to be an intimation under section 143(1) where (a) either no sum is payable by the assessee, or (b) no refund is due to him. It is significant that the acknowledgment is not done by any Assessing Officer, but mostly by ministerial staff. It cannot therefore be said that an assessmentis done by them. **[ACIT-Rajesh Jhaveri Stock Brokers P. Ltd.,SC,** *Sec.143.***210CTR30,291ITR500]**

17. The meaning of the expression assessment in a given provision must be determined on an examination of the relevant provisions in question and the fact that it is used in the narrower sense elsewhere does not mean that it is so used in the provision under examination. **[Brij Lal-CIT,SC,** *Sec.2(40).***328ITR477]**

Assessment made under this .. **PC** **Raleigh Investment Co. Ltd.- Governor- Gen. in Council**

The meaning of the phrase assessment made under this Act is an assessment finding its origin in an activity of the assessing officer acting as such; the circumstance that he has taken into account an ultra vires provision of the Act is in that view immaterial in determining whether the assessment is made under this Act. The phrase describes the provenance of the assessment, it does not relate to its accuracy in point of law. *Sec.5.r.w.s.293.***[15ITR332]**

Assessment or reassessment **Del.** **CIT - Srikishan Dass**

The words assessment or reassessment in the first limb of section 35(5) of the Indian Income-tax Act, 1922, are wide enough to include rectification and the Income-tax Officer is competent to rectify the assessment of a partner made under the 1922 Act, by invoking the provisions of section 35(5) to give effect to the consequences of an order or rectification passed under section 154 of the Income-tax Act,1961, in the case of the firm. <u>*Sec.154.*</u>**[18CTR297,125ITR730]**

Assessment to be made under section 143(3) **Mad.** **CIT - Sundaram Spinning Mills**

The expression assessment to be made under section 143(3), used in section 144B of the Income-tax Act, 1961, does not cover a reassessment and this expression is confined only to the assessments to be made originally and, therefore, the procedure prescribed under section 144B is not available to reassessments to be made under section 147. Consequently, the extended time limit provided for in Explanation (1)(iv) to section 153 is not available to the Department. <u>*Sec.144B.*</u>**[141CTR109,225ITR214,96Taxman287]**

Assessment year **Mad.** **Rockweld Electrodes (India) Ltd. - CIT**

The concept of assessment year must be understood in the manner prescribed under the provisions of the Income-tax Act and not with reference to a particular assessee. <u>*Sec.2(9).*</u>**[83CTR45,185ITR62]**

2. Section 2(9) of the Income-tax Act,1961,defines the assessment year to be the period of 12 months commencing on the first day of April every year. It is a standard period of 12 months commencing on April 1 of every year. It does not depend upon one or other assessee and whether or not he had a previous year relevant to a particular assessment year.It is as invariable as the calendar year. The assessment years mentioned in sections 33 and 80J must be read in this light. **[Premier Cable Co. Ltd.-CIT,SC,** <u>*Sec.(9).*</u>**153CTR172,237ITR202,103Taxman640]**

3. The concept of assessment year must be understood in the manner prescribed under the provisions of the Income-tax Act and not with reference to a particular assessee. Whenever the assessee is permitted to get relief for a specified number of consecutive assessment years, the assessment years should be taken in the natural sequence. Irrespective of the change of the assessment year of the assessee, the period of four years will have to be reckoned on the basis of the definition of assessment year as contained in section 2(9) of the Income-tax Act, 1961. **[CIT-Nuforam Industries,Guj.,** <u>*Sec.2(9)r.w.s.80J.*</u>**171CTR171,252ITR697,128Taxman571]**

Asset **All.** **Lucknow Producers'Co-operative Milk Union Ltd.- CIT**

The term asset has not been defined in the I.T. Act, 1961. The term asset used in Explanation 2 to section 43(1) is to be understood as meaning an asset, the written down value of which may be determined at any particular point of time in accordance with the provisions of the Act. It is obvious that there can be no question of written down value of cash either in the hands of the donor or of the donee, because it is not subject to depreciation in the sense envisaged in the Act. Explanation 2 cannot, therefore, be taken to cover the gift or grant of money. It should be confined to gift of an asset only as is clear from the language. *Sec.43(1).***[34CTR81, 143ITR60]**

2. Life interest in property is an asset within the meaning of section 2(e) of the Wealth-tax Act,1957. **[CWT-N.Lakshmikutty Amma, Ker.,***Sec.2(e).WT.Act.***180ITR 289,46Taxman184]**

3. The expression asset is defined in clause (e) of section 2, the relevant portion whereof reads thus: (e) assets includes property of every description, movable or immovable. **[CWT-Prince Muffakkam Jah Bahadur, AP.,***Sec.2(e).WT.Act.***76CTR56,186 ITR 421,44Taxman34]**

4. The definition of asset in section 2(e) of the Wealth-tax Act, 1957, is an inclusive definition. All properties of every description, movable or immovable, are included therein except those specifically excluded. **[CWT-Vidur V. Patel, Bom.,***Sec.2(e)WT Act.***[124CTR343,215ITR30,79Taxman289]**

Asset held by the assessee in such business **Mad.** **CWT- K.M. Desikar**

The word assessee in the phrase asset held by the assessee in such business in section 7(2)(a) of the Act has to be understood as including an individual, a firm or an association of persons and not merely a partnership. *Sec.7.WT.Act.***[92ITR101]**

Asset or advantage of an enduring nature **Mad.** **CIT-Carburettors Ltd.**

The expression asset or advantage of an enduring nature was evolved to emphasise the element of a sufficient degree of durability appropriate to the context. What is relevant is the purpose of the outlay and its intended object and effect, considered in a common sense way having regard to the business realities. *Sec.37.* **[141CTR595,221ITR680,139Taxation331]**

Assets **All. Ram Lakhan - ITO**

The word assets in section 24B(1) of the Indian Income-tax Act should be interpreted to include not merely the corpus but also the usufruct of the assets left by the deceased and it should make no difference that the usufruct accrued only after the date of the death of the deceased. Rents falling due after the death of a Hindu father, of property owned by the father, are assets of the father in the hands of his sons who have inherited such property after his death, and can, therefore, be attached for recovering Income-tax payable by the father. *Sec.226.* **47ITR311]**

2. The word assets in entry 86 of List I of the Seventh Schedule to the Constitution of India is not qualified by any limiting expression and, on the other hand, by excluding agricultural lands alone the framers of the Constitution made it absolutely clear that all other assets would come within the scope of that entry. Therefore buildings and lands cannot be held to be excluded from the scope of the entry. Parliament was competent to levy wealth-tax on the capital value of lands and buildings. **[Vysyaraju Badri Narayanamurthy-CWT.Ori.,** *Sec.3.* **56ITR298**

3. Section 2(e) defines assets to include property of every description; it, however, excludes certain items from the purview of the charge by excluding them from the definition of assets .A harmonious reading of section 2(e)(iv) with section 5(1)(vi) and section 5(1)(vii) would be that while non-commutable annuities are wholly outside the purview of wealth-tax, commutable annuities are exempt under sections 5(1)(vi) and 5(1)(vii) to the limited extent mentioned in each. It is well settled that when such a harmonious construction is possible and which furthers the object of the Act, namely, to promote thrift and channelise private savings for national use, the same must be preferred to the construction which leads to a conflict between sections 2(e)(iv) and 5(1)(vi). **[CWT - Yuvaraj Amrinder Singh, SC,** *Sec.2.r.w.s.5.WT.Act.* **49 CTR211, 156TR525,23Taxman25]**

4. The definition of the term assets in section 2(e) of the Wealth-tax Act,1957, is an inclusive definition and anything that amounts to property would fall within its ambit unless specifically excluded by the Act. Property is a term of the widest import and subject to any limitations or qualifications which the context may require, it signifies every possible interest which a person can acquire, hold or enjoy. **[CWT-Sudha P.Patel,Bom.,** *Sec.2(e).WT.Act.* **196ITR8]**

5. From the inclusive definition given under section 2(e) it is clear that the word assets includes property of every description, movable or immovable, excluding the assets enumerated in the clause or sub-clause of that section. Section 2(e) takes in property of every description and that expression is qualified by the words movable or immovable. Properties which do not ordinarily answer the test of movability or immovability such as intangible rights or incorporeal rights, cannot be regarded to be assets within the meaning of section 2(e).A debit balance in the capital account of a firm cannot be said to be a tangible right and cannot be regarded as an asset of the industrial undertaking belonging to the firm. **[CIT-Vidya Charan, All.,** *Sec.2(e)WT.Act.***133CTR375,220ITR78,86Taxman 133]**

6. The definition of assets in section 2(e) states that it includes property of every description, movable or immovable, but does not include the assets specified. Stock-in-trade was not an asset excluded from the definition in the assessment year 1981- 82. Even if the assessee is a dealer in shares, such shares constitute assets. **[CWT-C.D.R.Laxmidevi,Guj.,***Sec.2(e).WT.Act.***193CTR222,274 ITR257, 144Taxman600]**

7. Section 2(e)(v) of the Wealth-tax Act,1957, as it stood after amendment by the Finance Act,1969,excluded any interest in property where the interest is available to an assessee for a period not exceeding six years from the date the interest vests in the assessee from the term assets under the Act. **[CWT -Smt. Kiran Devi,All.,***Sec.2(e).WT.Act.***[277ITR344,147Taxman40]**

Association Mad. Estate of Khan Sahib Mohd.Oomer Sahib - CIT

Association does necessitate the exercise of volition of those who form the association. The exercise of that violation can be by or on behalf of those who form the association. An association of persons can include minors. To render an association of persons a taxable unit, for which section 3 of the Income-tax Act provides, the objects of the association must be to produce income, profits or gains.To associate is to join in a common purpose or action. *Sec.2(31).* **[33ITR767]**

Association of individual Mad. State of Madras - Karuppan Chettiar

A common object to earn income, a joint endeavour towards that end and a resultant income being the fruit of the joint enterprise are the essentials which should be present before individuals can be clubbed together as an association to be taxed. When tenants-in-common of a property divide the income in the ratio of their definite shares without a division of the corpus by metes and bounds, they cannot be said to have earned their respective income by a joint endeavour. A joint management of the undivided property held in shares either by one of the sharers or by a duly appointed attorney and would not constitute them as association of individuals.

Agri.I.T.Act.[61ITR488]

Association of individuals All. Mohammad Aslam - CIT

The expression association of individuals in section 3 of the Indian Income-tax Act is -ejusdem generic- with the word immediately preceding, viz., firm and before there can be an association of individuals within the meaning of the section it must first be shown that the association has at least some of the attributes of a firm or partnership though not in the strictly legal sense of the term. Persons having specified but undivided shares in property which produces income do not come within the expression association of individuals in section 3 of the Act and are not liable to be assessed to income-tax as such. Even the appointment by a body of co-owners of a common collecting agent will not convert such body of co-owners into an association of individuals within the meaning of section 3. *Sec.2(31).*[4ITR412]

2. The words association of individuals in section 3 of the Indian Income-tax Act,1961 must be construed in their plain and ordinary meaning. They are not ejusdem generis with the word immediately preceding, namely, firm. The only limit imposed on the words is such as necessarily follows from the fact that the words appear in an Act imposing a tax on income, profits and gains, so that the association must be one which produces income, profits or gains.

[CIT - Laxmidas Devidas,Bom.,*Sec.2(31).*[5ITR584]

Association of persons SC G.Murugesan & Bros.-CIT

For forming an association of persons, the members of the association must joint together for the purpose of producing an income. An association of persons can be formed only when two or more individuals voluntarily combine together for a certain purpose. Hence volition on the part of the members of the association is an essential ingredient. Even a minor can join an association of persons if his lawful guardian gives his consent. In the case of receiving dividends from shares, where there is no question of any management, it is difficult to draw an inference that two or more shareholders functioned as an association of persons from the mere fact that they jointly own one or more shares, and jointly receive the dividends declared. These circumstances do not by themselves go to show that they acted as an association of persons. In the case of an association of persons it is always open to its members to withdraw from the association. No one can be compelled to continue as a member of the association.No particular form need be observed for withdrawing from an association. *Sec.2(31).***[88ITR432]**

2 .The word associate means to join in common purpose, or to join in an action. Therefore, association of persons, as used in section 3 of the Indian Income- tax Act, means an association in which two or more persons join in a common purpose or common action, and as the words occur in a section which imposes a tax on income, the association must be one the object of which is to produce income, profits or gains. **[CIT-Indira Balkrishna,SC,***Sec.2(31).***39ITR546]**

3. The word associate means, according to the Oxford Dictionary, to join in common purpose or to join in an action. Therefore, an association of persons must be one in which two or more persons join in a common purpose or common action, and when the words occur in an enactment which imposes a tax on income, the association must be one the object of which is to produce income, profits or gains. **[NarayanNepak-CIT,Ori.,***Sec.2(31).***24CTR76,136 ITR133]**

4. The expression association of persons as used in section 4 of the Income-tax Act,1961,means an association in which two or more persons join in a common purpose or common action, and, as the words occur in a section which imposes a tax on income, the association must be one the object of which is to produce income, profitsorgains. **[CIT-S.B.SugarMills,All.,***Sec.2(31).***44CTR129,156 ITR273,22 Taxman 277]**

5. An association of persons must be one in which two or more persons join in a common purpose or a common action, and as the words occur in a section which imposes a tax on income, the association must be one the object of which is to produce income, profits or gains. **[CIT-Friends Enterprises, AP.,***Sec.2(31).***66CTR 143,171ITR269,34Taxman411]**

6. It is now well-settled that the word individual does not necessarily and invariably always refer to a single natural person. A group of individuals may as well come in for treatment as an individual under the tax laws if the context so requires. The word association means to join in any purpose or to join in action. Therefore, association of persons as used in section 2(31)(v) of the Income-tax Act,1961, means an association in which two or more persons join in a common purpose or common action. The association must be one the object of which is to produce income, profits or gains. In the case of a discretionary trust, neither the trustees nor the beneficiaries can be considered as having come together with the common purpose of earning income. The beneficiaries have not set up the trust. The trustees derive their authority under the terms of the trust deed. Neither the trustees nor the beneficiaries come together for a common purpose. They are merely in receipt of income. The mere fact that the beneficiaries or the trustees, being representative assessees, are more than one, cannot lead to the conclusion that they constitute an association of persons. **[CIT-Shri Krishna Bandar Trust,Cal.,*Sec.2(31).*201ITR 989]**

7. The word associate means to join in common purpose, or to join in an action. Therefore, association of persons as used in section 3 of the Income-tax Act, 1961, means an association in which two or more persons join in a common purpose or common action, and the association must be one the object of which is to produce income, profits or gains.**[CIT-N.K.Patni,All.,*Sec.2(31).*148CTR411,234ITR 12]**

8. An association of persons can be formed only when two or more individuals voluntarily combine together for a certain purpose. Hence, volition on the part of the members of the association is an essential ingredient. **[CIT-Sehgal Oil and General Mills,P&H. *Sec.2(31).*203CTR,215,303ITR102]**

At any time Cal. CIT - Srimati Minabati Agarwalla

The phrase at any time,with regard to the voluntary submission of returns under section 22(3), must mean at any time within the period prescribed by other provisions of the statutes; once the period mentioned in section 34(3) is over, there is an absolute bar to any order of assessment or reassessment unless it comes within the exceptions mentioned in the proviso to sub-section (3) of section 34. The view that since the liability of an assessee exists even when the machinery for assessment and realisation of the tax becomes non-available due to the expiry of the period mentioned in the machinery sections, which include section 34 of the Act, it is open to an assessee to waive the machinery and to invite the Department to realise from him the tax which has already accrued, is not correct.

Sec.139.[79ITR278]

At the time of making an assessment **E.P.** **Rajmal Paharchand- CIT**

The expression at the time of making an assessment in section 25A means in the course of the process of assessment. It is not restricted to the time of making the final order determining the assessment. The power of the Income-tax Officer to pass an order under sub-section (1) of section 25A, therefore, arises when, at the time of making an assessment under section 23, a claim is made by a member that a partition has taken place but not necessarily during the accounting year. Sec.143.[18ITR1]

At the time of credit of such income to the account of **Guj.** **CIT-Ranoli Investment P. Ltd.**

The words at the time of credit of such income to the account of the payee in section 194A would take within their sweep, the interest debited to interest account or any other nominal account when the debit is for a specific amount calculated with reference to the deductor's liability to a particular creditor in accordance with the terms and conditions of the loan. The time of deduction would be when the interest is credited. The liability of the deductor would arise for failure to make deduction at the time of credit notwithstanding that it came to be made later on at the time of actual payment. Deduction made at such belated stage of payment, would not be tax deducted at source properly so-called and such subsequent deduction even when deposited with the Government, cannot be treated as tax deducted at source. Sec.194A.[146CTR 745,235ITR433]

At the time of making **PC** **Maharajadhiraj of Darbhanga - CIT**

The words, at the time of making an assessment in section 26(2) mean in the course of the process of assessment. Sec.187.r.w.s.143. [2ITR345]

Attributable **SC** **Hero Exports - CIT**

The word attributable is wider than the word derived. The word attributable in section 80HHC(3)(b) in the main section itself indicates that apportionment (principle of attribution) is not omitted from that provision.The question of allocation of costs also arises in cases falling under section 80HHC (3) (b). Firstly, clause (e) of the Explanation which refers to allocation of costs applies to section 80HHC(3)(a), section 80HHC(3)(b)and section 80HHC(3) (c). Secondly, section 80HHC(3)(b) equates export profits to export turnover less direct and indirect costs attributable to the exports of trading goods. Therefore, the principle of attribution is retained. Thirdly, keeping in mind the provisions of section 80HHC(3)(b) read with clauses (d) and (e) of the Explanation, it is clear that the Legislature intended allocation of costs between export turnover and total turnover. It may be that, in most cases apportionment may not apply; but in certain cases falling under section 80HHC(3)(b) the ratio still applies. For example, where the assessee exports all boughtout items but brings back only a part of the export proceeds into India, the ratio will apply and, therefore, if one is to read clause (e), it retains the words indirect costs to be allocated in the ratio of export turnover to total turnover. *Sec.80HHC.***[295ITR454]**

Attributable to **SC** **Cambay Electric Supply Industrial Co.Ltd. - CIT**

In computing the total income of the assessee carrying on the business of an industry specified under section 80E of the Income-tax Act, 1961 1961 [as it stood prior to its amendment by the Finance (No. 2) Act], for the purpose of the special deduction permissible thereunder, the balancing charge arising as a result of the sale of old machinery and buildings and worked out in accordance with section 41(2), irrespective of its real character, has to be taken into account and included as income of the business. In other words, the balancing charge will have to be taken into account before computing the deduction of 8 per cent. under section 80E.The legal fiction under section 41(2) and the grant of special deduction under section 80E in the case of specified industries are so closely connected with each other than taking into account the balancing charge (i.e., the deemed profits) before computing the 8 per cent. deduction under section 80E(1) will amount to extending the legal fiction within the limits of the purpose for which the fiction has been created. The legislature has deliberately used the expression attributable to, having a wider import than the expression derived from, thereby intending to cover receipts from sources other than the actual conduct of the business of the specified industry.

Sec.80E. **[7CTR50,113ITR84]**

2. Section 80P(2)(c) of the Income-tax Act, 1961, exempts income of co-operative societies to the extent mentioned in the section if the profits or gains are attributable to the activity in which the co-operative society is engaged. The expression attributable to is much wider than the expression derived from and it covers receipts from sources other than the actual conduct of the business of the assessee.**[CIT-Co-op.CaneDev.UnionLtd.,All.,***Sec.80P(2).***10CTR 282,118 ITR770]**

3. In section 80-I the Legislature has deliberately used the expression attributable to which has a wider import than the expression derived from and has thereby shown its intention to grant the relief in respect of receipts from sources other than the actual conduct of the business of the specified priority industry.

[CIT-AshokLeylandLtd.,Mad.,*Sec.80I.***130ITR900]**

4. The expression attributable to occurring in section 80P(2)(a)(i) has to be understood in a broad sense and it cannot be equated to the expression derived from.**[CIT-Madurai District Central Co-op.Bank Ltd.,Mad.***Sec.80P(2).***148 ITR196, 17Taxman247]**

5. The expression attributable to occurring in section 280ZB of the Income-tax Act, 1961, is of wide import and any profits or gains having reasonable nexus to the activity of manufacture or production of specified articles must be considered attributable to such manufacture or production. The Legislature has deliberately used the expression attributable to instead of derived from and has thereby shown clear intention to grant relief even in respect of receipts from sources other than actual conduct of the business of manufacture or production of the specified item.

[Khandelwal Ferro Alloys Ltd.-ITO,*Sec.280ZB.***39CTR130,152 ITR20,16 Taxman 135]**

6. As the term attributable to is wider than derived from, the relief is not confined only to the profits of the priority industry, strictly so called. It would have a wider ambit. In order to be eligible for deduction under section 80-I income could be from another source as long as that other source is in one form or the other connected with the business of the priority industry. **[CIT-Seshasayee Paper andBoardLtd.,Mad.,***Sec.80-I.* **[207ITR80]**

7. The expression attributable to is much wider than the expression derived from. The expression attributable to suggests that the

Legislature intended to cover the receipts from sources other than the actual conduct of the business of the assessee.**[CCIT(Admn.) - Kisan Sahkari Chini Mills Ltd.,All.,**<u>*Sec.80P(2).*</u>**273ITR42]**

Audit AP. A.S.Sarma - Union of India

Audit is a systematic and meticulous scrutiny of accounts with reference to the receipts and disbursements and vouchers and allied matters and submission of report surveying the financial outlay and highlighting the lapses or infirmities in the maintenance of accounts or utilisation of funds. The report submitted by the auditor on an in-depth probe into the transactions with reference to accounts and other matters is expected to mirror the correct state of affairs. <u>*Sec.44AB.r.w.s.142(2A)*</u>**[67CTR131,175ITR254,35Taxman411]**

Authorised representative PH. Taxation Bar Association - CIT

Clause (vi) of section 288(2) of the Income-Tax Act, 1961, defines authorised representative to mean any person who has acquired such educational qualifications as the Board may prescribe for the purpose. A perusal of r. 51 of the I.T. Rules, 1962, which specifies the qualifications required by clause (vi) of section 288(2), shows that a person who possesses a degree in commerce or law, conferred by any of the Universities mentioned therein, would satisfy the requirement of clause (vi). R. 51 of the Rules does not draw any distinction between a degree in law (academic) and a degree in law (professional). The only requirement of the rule is that a person should possess a degree in commerce or law. <u>*Sec.288(2).*</u>**[141ITR82]**

Authority All. U.P. State Warehousing Corporation - ITO

Ex hypothesi,the term authority, as occurring in section 10(29), cannot possibly be construed as referring to an authority having quasi-governmental powers. Any legal entity or juristic personality constituted by law for the purpose of marketing commodities would be an authority within the meaning of section 10(29). The petitioner which was established under the Warehousing Corporations Act, was an authority constituted under law within the meaning of section 10(29). <u>*Sec.10(29).*</u>**[94ITR129]**

2. The mere fact that a juristic personality is created under any particular statute or law having as its direct or indirect object marketing of commodities will not make the entity an authority within the meaning of clause (29) of section 10 of the I.T. Act, 1961. Such juristic entity or personality must have some other attributes which would entitle it to claim to be an authority. To be an authority, the juristic entity or personality has to possess one or more of the following characteristics: (a) exercise of governmental or quasi-governmental power; (b) performance of governmental or quasi-governmental functions; (c) power to have its directions enforced with punishment, i.e., having a right to command and to be obeyed; (d) exercise of general or particular statutory power; (e) power to make regulations, rules or bye- laws having the force of law. A joint stock company, incorporated primarily for carrying on ordinary business or commercial activities cannot come within the meaning of authority under section 10(29) of the Act. **[Singhal Brothers P. Ltd.-CIT, Cal.,** Sec. 10(29). **124 ITR147]**

3. The term authority has not been defined in the Act. The term authority has been used in section 43(1) in its wider sense and would take within its ambit all constitutional, governmental or statutory authorities, which exercise power, whether original, delegated or derivative. Merely because the term Government has not been used in section 43(1), while this term was used in the Explanation to section 10(5) of the Indian I.T.Act of 1922, it would not be reasonable to conclude that Parliament intended to exclude Government from the purview of section 43(1).In the Explanation to section 10(5) of the old Act, the words used were by Government or by any public or local authority. The word authority was qualified by the words public or local which were restrictive in nature. In section 43(1) of the 1961 Act, the restrictive words have been omitted and only the word authority has been used. Parliament has omitted the word Government from section 43(1) because it felt that the wide connotation of the word authority would take within its purview Government as well, besides other kinds of authorities. A State government would, therefore, be an authority for the purposes of section 43(1) where an assessee purchased capital assets with the help of a grant from the State Government. **[Lucknow Producers' Co-operativeMilk Union Ltd.-CIT,All.,** Sec.43(1).**34CTR81, 143ITR60]**

Automotive / Locomotive **Cal.** **CIT - Birla Jute and Industries Ltd.**

Vehicles used on roads are described as automotive while those used on rails are described as locomotive. These words and expressions have separate distinct meanings, which have been in use for a long period of time. In common parlance road and rail are two different modes of transport. Even in law, rail and road have been treated differently. Separate statutes have been legislated for governing the field of road transport and that of rail transport. The definition of vehicle contained in the Road Transport Corporations Act has a limited application. Such a definition is confined to the scope and purpose of the said Act, which is related to the incorporation of a State owned Road Transport Corporation. The same is a special statute and has no general application. Therefore, the definition given therein cannot be borrowed for the purpose of the Income- tax Act. That apart road transport vehicle is defined in the Road Transport Corporations Act,1964, to mean a service to carry passengers or goods or both by road in vehicles for hire or reward. It clearly means a service by road, whereas vehicle is defined in the Act to mean a vehicle capable of being used for road transport and includes tram car, trolley vehicle or trailer. It does not include rail. The stress is on road and includes tram car but not rail or train. If a statute in an inclusive definition refers to one item distinct from the other and omits that other item, then the intention of the Legislature appears to be clear. While including one by special reference without referring to the other the omission is deliberate. If rail transport were also intended to be excluded through the exception contained in clause (b) of the proviso then instead of using the expression road transport vehicle it could have used the expression transport vehicle which would have included all kinds of transport vehicles including rail or road or it could have used rail and road transport vehicles. Instead, it has used road transport vehicle which specifically states that it is only the vehicle, which is used as transport on road. Definitely a transport used on roads cannot be used on rails. Similarly, a transport used on rail cannot be used on road. The expression road transport cannot include rail transport. _Sec.32._ **[180CTR339,260ITR55,133Taxman 337]**

Auxiliary **AAR** **UAE Exchange Centre LLC**

The word auxiliary in common English usage means helping, assisting or supporting the main activity. _Misc._ **[89CTR467,268 ITR9,139Taxman82]**

Avoidance of tax, **Guj.** **CIT - Sakarlal Balabhai**

The words avoidance of tax are not colourless words but are strong and compelling words connoting a positive volition or a deliberate intention on the part of the assessee to avoid tax. The section being punitive in character must be construed strictly. _Misc._ **[69ITR186]**

B

Bad and doubtful debts **Cal.** **Hongkong and Shanghai Banking Corporation - CIT**

The expression bad and doubtful debts in section 10(2)(xi) does not contemplate two kinds of debts,but refers to the same class of debt namely,a debt which is bad and doubtful, i.e., a debt, of which the chance of recovery is nil or slender. A doubtful debt does not mean a debt which cannot be held to be irrecoverable. Such a debt may also be held to be irrecoverable wholly or in part. *Sec.36.*[**28ITR199**]

Bad debt **Ker.** **Travancore Tea Estates Co. Ltd. - CIT**

The debt may become bad when it is proved to be irrecoverable on account of the fact that the debtor is in a bad financial position or that it had become irrecoverable. So long as there is a ray of hope to recover the debt, however dim it may be, and so long as the debt is in the process of realisation,it cannot be said that it has become irrecoverable. *Sec.36.*[**197ITR528,63Taxman558**]

Bad debts **Guj.** **Vithaldas H. Dhanjibhai Bardanwala - CIT**

Under section 36 of the Income-tax Act, 1961, it is clear that before any claim for allowance for a bad debt is held established by the Income-tax Officer it must appear that the concerned bad debt was written off as irrecoverable in the account books of the assessee for the relevant previous years.This requirement has become a condition for the grant of claim for bad debt allowance. To that extent, there is a clear departure from the scheme in the earlier Act. Still, so far as the exact requirement of the writing-off of the concerned debt as irrecoverable in the account books of the assessee is concerned, the language used in both the Acts, viz., the Act of 1922 and the Act of 1961, is almost identical. The only requirement of section 36(2)(i)(b) is that the concerned bad debt must have been written off as irrecoverable in the accounts of the assessee. If the debit entries posted by the assessee indicate the said fact the requisite statutory condition has got to be treated as fully complied with. Once the assessee has posted entries in the profit and loss account and corresponding entries are posted in the bad debt reserve account that would be sufficient compliance with the provisions of the statutory requirement for writing off as irrecoverable the concerned debt in the books of the assessee. No further requirement can be spelt out from the express language used by the Legislature. It is not necessary that the assessee must also post corresponding entries in the ledger account of the concerned parties and should close those accounts. *Sec.36.*[**21CTR 190,130ITR95,6Taxman105]**

2. The expression bad debts relates to debts of which the chance of recovery is remote.Unless and until a debt has been written off as irrecoverable amounts to the assessee as provided under section 36(2)(i)(b) of the Income-tax Act, 1961, the assessee cannot claim any deduction for bad debts. Irrecoverable in section 36 (2)(i)(b) means the amount is not recoverable.The expression provision for bad and doubtful debts cannot be construed as writing off bad debts. [**Kerala State Industrial Development Corporation Ltd.-CIT, Ker.,** *Sec.36.* **201 CTR 144,281ITR413,151Taxman127**]

Banking **Bom.** **CIT - Ahmednagar District Central Co- Op. Bank Ltd.**

Banking is not restricted to receiving deposits for the purposes of lending. Banks offer various facilities to its customers.Section 6(1)(a) of the Banking Regulation Act, 1949, is an enabling provision. It provides for various forms of business akin to banking. Section 6(1) states, inter alia, that in addition to the banking business,a banking company may engage in specified forms of business enumerated in section 6(1)(a) to (o). Section 6(1)(b) states that in addition to the business of banking, a banking company may engage itself as agent for Government or local authority or any other person for giving receipts and discharges. *Banking Regulation Act.*[**185CTR336,264ITR38,132Taxman226]**

Banking activity **Guj.** **Gujarat State Co-operative Bank Ltd. - CIT**

Parking of such surplus funds, in excess of banking reserves,cannot be considered to be a banking activity or attributed to banking activity, though it may be part of banking business in the wider sense. *Sec.80(P).*[**167CTR34,250ITR229,119Taxman160]**

Became due Del. P.C.Gulati Liquidator - CIT

Under section 41, the point of time at which the excess realised over the written down value of assets sold has to be taxed, is not the previous year in which the sale took place or the previous year in which the money was received,but the previous year in which the money became due.No amount can be said to be due, within the meaning of that section, till it has become ascertained. *Sec.32(1), 41(2).* **[86ITR501]**

Before Bom. CIT - Vijaya Hirasa Kalamkar (HUF)

.......the word before will have to be construed as up to or as not after. There are various provisions in the Income-tax Act wherein the expression before has been used (section 139(1)(b); section 184; section 212). The expression has always been taken to mean up to.

VDIS. **[229ITR772]**

Begins to manufacture All. Kanodia and Sons - CIT

The word manufacture can be used both as a verb and as a noun. When it is used as a verb it refers to the processor a centering action having a starting point and an end point. The expression manufacture when used as a noun, has a fixed point, namely,when the process culminates into the end-point, i.e., the process ends up in the final result. The expression to produce article is used in the very same section. This clearly shows that the Legislature has in mind the production of articles. There is no doubt, that the use of the word or between two expressions, viz., begins to manufacture and begins to produce articles shows that they are interchangeable and denote the same meaning. The Legislature has not used the expression to include process of manufacturing, as is evident on comparing the expression used in section 80P(2)(f). The intention is clear, namely, the expression begins to manufacture is used as a noun and that is when the undertaking achieves the end product out of the process of manufacture.*Sec.80J.* **[201CTR442,281ITR255]**

Being property of the assessee Del. CIT- Hindustan Cold Storage & Refrigeration P. Ltd.

The words being the property of the assessee appearing in section 10(2)(vi) of the 1922 Act had the same meaning as the words owned by the assessee appearing in section 32(1) of the 1961 Act, and these words merely clarified the position that already existed under section 10(2)(vi) of the 1922 Act. Sec.32.[103ITR455]

Belonging SC CWT - Bishwanath Chatterjee

Under section 3 read with section 2(m) of the Wealth-tax Act, 1957, liability to wealth-tax arises out of ownership of the asset and not otherwise. Mere possession or joint possession, unaccompanied by the right to, or ownership of, property would therefore, not bring the property within the definition of net wealth for it would not then be an asset belonging to the assessee. Sec.3,2(m).W.T.Act. **[103ITR536]**

2. It is not necessary that a person should be the exclusive owner of the property so as to claim the benefit of clause(iv) of section 5(1) of the Wealth-tax Act, 1957. A part ownership or co-ownership may be sufficient for this purpose.What is probably required is that he should be having some rights of ownership and not merely a right of possession.The word belonging is capable of signifying possession of an interest less than that of full ownership. **[CWT-Jugal Kishore Bhagat,Cal.,**Sec.5(1)(iv).WT.Act.**197ITR250]**

Belonging to P&H. CIT -Smt. Badhurani Deepinde Kaur

In Nawab Sir Mir Osman Ali Khan (Late) v. CWT (162 ITR 888), the Supreme Court stated that the liability to wealth-tax arises because of the belonging of the assessee and not otherwise. Mere possession or joint possession unaccompanied by the right to be in possession or ownership of property would therefore not bring the property within the definition of net wealth for it would not have been an asset belonging to the assessee. The Supreme Court observed that it had to be borne in mind that unlike the provisions of the Income-tax Act section 2(m) of the Wealth-tax Act, 1957, used the expression belonging to and as such indicates something over which a person has dominion and lawful dominion. Sec.2(m).WT.Act. **[184CTR359, 262ITR403,133Taxman591]**

2. The words belonging to mentioned in section 281B of the Income-tax Act, 1961, have to be read and understood in the context of the definition of transfer provided elsewhere in the said Act itself.. **[Electro Zavod (India)Pvt.LTd.-CIT,Cal.,**Sec.281B.**278ITR187]**

Benami SC Sree Meenakshi Mills Ltd. - CIT

The word benami is used to denote two classes of transactions which differ from each other in their legal character and incidents. In one sense, it signifies a transaction which is real, as for example, when A sells properties to B but the sale deed mentions X as the purchaser. Here the sale itself is genuine;but the real purchaser is B, X being his benamidar. This is the class of transactions which is usually termed as benami. But the word benami is also occasionally used, perhaps not quite accurately ,to refer to a sham transaction, as for example, when A purports to sell his property to B without intending that his title should cease or pass to B. The fundamental difference between these two classes of transactions is that, whereas in the former there is an operative transfer resulting in the vesting of title in the transferee, in the latter there is none such,the transferor continuing to retain the title notwithstanding the execution of the transfer deed. It is only in the former class of cases that it would be necessary, when a dispute arises as to whether the person named in the deed is the real transferee or B, to enquire into the question as to who paid the consideration for the transfer. But in the latter class of cases,when the question is whether the transfer is genuine or sham,the point for decision could be, not who paid the consideration but whether any consideration was paid. *Misc.* **[31ITR28]**

Benami transactions **Ker.** **Bhargavy P. Sumathykutty - Janaki Sathyabhama**

The Benami Act contains a definition of benami transactions in section 2 thereof. The definition takes in only the first category or tripartite benami transactions. Section 3 of the Benami Act contains a prohibition against any person entering into any benami transaction. It also contains a penal provision. By expressing that the definition is intended for the Act and by employing the restrictive term means in preference to the word includes, Parliament has conveyed its intention that the word benami transaction is not to be confined to one section alone and that the definition would contain only one category-the tripartite-of benami transactions. The employment of the words unless the context otherwise requires has no special impact on the statute so far as benami transactions are concerned. *Benami Act.***[217ITR 129]**

Beneficially held **Bom.** **Raghuvanshi Mills Ltd. - CIT**

The expression beneficially held does not, in the context in which it is used, imply only the relationship of trustee and cestui que trust.

Although the expression may include that relationship, it is not used as an equivalent thereof. Misc.**[56ITR470]**

Beneficiary Pat. Province of Bihar - F.R.Hayes

The definition of beneficiary as given in the Explanation to section 11 must be held to be controlled by its accepted legal meaning and that the section could not have been intended to apply to income received by full owners of property through their servant, agent or manager. *Bihar Agricultural Income-tax Act.***[14ITR326]**

Benefit or perquisite obtained Mad. CIT - Adaikappa Chettiar

In order to attract section 2(6C)(iii) of the Indian Income-tax Act, 1922, the benefit or perquisite obtained should be by some of arrangement with company. The words benefit or perquisite obtained from a company would take in only such benefit or perquisite which the company had agreed to provide and which the person concerned could claim as of right based on such agreement. A mere advantage derived from the company without its authority or knowledge will not amount to a benefit or perquisite obtained. The word obtained occurring in the said section must be agreement oriented and cannot merely mean taken. *Sec.17.***[91ITR90]**

Benefits of partnership All. Smt. Nisha Rani Agrawal - CIT

Benefits of partnership means a share in the profits, if earned. On the other hand, interest on invested capital or accumulated capital, whether that accumulation is of profits or otherwise, is an obligation upon the firm, which is independent of whether the firm earns profit in the subsequent years or not. Thus, the interest accrued or paid on invested or accumulated capital, being independent of the earning of profits by the firm, cannot be said to be a benefit of partnership. Section 64(1)(iii) of the Income-tax Act, 1961, permits adding or clubbing to the income of the assessee, only that income of a minor child of the assessee which arises from the benefits of partnership. Interest income, not being benefits of partnership, cannot be so clubbed under section 64(1)(iii). *Sec.64(1).***[294ITR46]**

Block of assets P&H Max India Ltd

The term block of assets refers only to depreciable assets. *Sec.50.***[319ITR68,178Taxman196]**

Block period **Mad.** **Lakshmi Jewellary - DCIT**

The expression block period has been defined in section 158B of the Income-tax Act, 1961. The opening part of this definition specifies that block period means the previous years relevant to ten assessment years, which precede the previous year in which the search was conducted under section 132 or any requisition was made under section 132A. Having so specified,a further addition is made to the block period by the latter part of the definition,which states that it would also include in the previous year in which the search was conducted or requisition made,the period up to the date of the commencement of such search or the date of requisition. The definition is in two parts. First, it says that it means the previous years relevant to ten assessment years preceding the previous year in which the search was conducted or requisition made. Thereafter, an addition is made to that period by stating that the block period also includes in the previous year in which the search was made, the period up to the date of search. The contention that the word includes should be related to the specification to the previous years relevant to ten assessment years is difficult to accept. Reference to the previous years relevant to the ten assessment years is followed by the words preceding the previous year. It is, therefore, clear that no part of the previous year in which the search was conducted can be regarded as forming part of the previous years relevant to the ten assessment years referred to in the opening part of the definition. The words and includes which occur in the later part of the definition are words which can only be regarded as relevant to the term block period as the inclusion of the additional period to which reference is made, is after the words. The words and includes can only be for the purpose of regarding that period also as forming part of the block period. <u>*Sec.158B.*</u>**[172CTR719,252ITR712,169 Taxation141,129Taxman890]**

2. The basic ingredient of the term block period under section 158B of the Income -tax Act, 1961, is that it relates to a certain number of years relating to and relevant to the search conducted under section 132 or requisition made under section 132A. **[CIT-Smt.Chitra Devi Soni,Raj.,** <u>*Sec.158BB*</u> **214CTR118,313ITR 174,170 Taxman 164]**

Body of individual /An association of ... **AP.** **Deccan Wine and General Stores - CIT**

The expression body of individuals must have a distinct meaning of its own, otherwise Parliament would not have introduced it.It may have some characteristics common with an association of persons but it cannot be the same thing as, or a mere species of, an association of persons. Both expressions are placed alongside each other in the definition because of the common characteristics possessed by them, both deal with combinations, whether of persons or individuals, who have a common interest and who are engaged in income-producing activity. An association of persons does not mean any and every combination of persons. It is only when they associate themselves in an income-producing activity that they become an association of persons. The expression body of individuals should receive a wide interpretation, perhaps not wide enough to include combination of individuals who merely receive income jointly without anything further as in the case of co-heirs inheriting shares or securities, but certainly wide enough to include a combination of individuals who have a unity of interest but who are not actuated by a common design and one or more of whose members produce or help to produce income for the benefit of all. <u>*Sec.2(31).*</u>**[106 ITR111]**

Body of individual whether incorporated or not **All.** **CGT -S.B. Sugar Mills**

A firm is not a separate legal entity under the general law, but is a collective name of a body of persons who have entered into a partnership.Therefore, a firm would fall within the category of body of individuals or persons,whether incorporated or not,contained in the later part of the definition. <u>*Sec.3.r.w.s.21.G.T.Act.*</u>**[120ITR126]**

Body of individuals **P&H.** **Meera and Co. - CIT**

The expression body of individuals must have a distinct meaning of its own irrespective of the fact that it may have some characteristics common with an association of persons. It could not be the same thing as, or a mere species of, an association of persons. An association of persons does not mean any and every combination of persons. It is only when they associate themselves in an income producing activity that they become an association of persons. The expression body of individuals should receive a wider interpretation, perhaps not wide enough to include a combination of individuals who merely receive income jointly without anything further as in the case of co-heirs inheriting shares or securities, but certainly wide enough to include a combination of individuals who have a unity of interest but who are not actuated by a common design and one or more of whose members produce or help to produce income for the benefit of all. A body of individuals need not necessarily be the result of an agreement,arrangement or design. It might arise out of a certain situation. *Sec.2(31).***[8CTR28,120ITR 564]**

2. Joining together for the purpose of producing income is the requisite for the formation of an association of persons. Such coming together or combining is a consensual act and depends upon the volition of the parties. Merely because certain persons are constituted joint owners such as by inheriting the property of a person on his death, they do not become an association of persons, for,in that event,the joining is the result of an operation of law and not of volition of parties. Some element of volition is necessary even for body of individuals. A body of individuals is different from individuals and if they are treated as a person because they are a body there must be joining by volition. **[CIT-T.V.Suresh Chandran, Ker.,** *Sec.2(31).***13CTR366,121ITR985]**

3. The expression body of individuals (BOI) derives colour from the context in which it occurs. There is no simple formula to determine what constitutes a body of individuals. A BOI need not necessarily be the result of an agreement,arrangement or design; it might arise out of a certain situation. Though a BOI is not identical with an association of persons (AOP), they have some similarities. An AOP may consist of non-individuals but a BOI has to consist only of individuals or human beings. A mere collection of individuals without a common tie or common aim, cannot be taken to be a body of individuals. **[CIT-Pabbati Shankaraiah,AP.,***Sec.2(31).***38 CTR37,145ITR702,15Taxman415]**

4. ...the words, body of individuals occurring in the definition section should be understood in the context and collection of the words and not in isolation. In construing the words body of individuals occurring in section 2(31) of the I.T. Act, 1961, alongside the words association of persons, the aforesaid well-settled principle of interpretation laid down by courts should be borne in mind. It is now well-settled by decisions of the Supreme Court that in order to constitute an association of persons, persons must join in a common purpose or action and the object of the association must be to produce income.It is not enough that the persons receive the income jointly. The word body of individuals occurring in section 2(31) of the Act alongside association of persons should be understood in the said background and context. **[CIT - A. P. Parukutty Mooppilamma, Ker.,***Sec.2(31).***38CTR 354,149ITR131,18Taxman275]**

5. The expression body of individuals must be given a definite meaning of its own. Association of persons and body of individuals cover two different combinations of persons and it will not be proper to try to apply the principle of ejusdem generis to give the same restricted meaning to the newly introduced expression body of individuals as had been given by the Supreme Court to the association of persons on interpretation of the word association. The fact that both the expressions are placed under a common item does not justify such restrictive interpretation. The expression body of individuals must receive a wider interpretation than association of persons, perhaps not wide enough to include combination of individuals who merely receive income jointly without anything further as in the case of co-heirs inheriting shares or securities, but certainly wide enough to include a combination of individuals who have a unity of interest but who are not actuated by a common design and one or more of whose members produce or help to produce income for the benefit of all.....The fact that a particular combination of persons has been held to be a body of individuals or association of persons in respect of a particular activity, does not mean that it will be so in respect of all activities or incomes there from. The status may vary from activity to activity. **[CIT- Modu Timblo, Bom.,***Sec.2(31).***116CTR442,206ITR647,118 Taxation, 170Taxman585]**

6. Body of individuals is a new classification in the fiscal statute. Right from 1939, if not from 1922 onwards, the I.T. Act had known and recognised the distinct category of taxpayers going by the name of association of individuals or association of persons. Although the expressions were left undefined, courts used to associate this class of persons with certain attributes, chief amongst which was their being associated in a common endevavour for producing taxable income. This at once excluded from the category those who found themselves thrown together by the accident of their birth, by the accident of another's death, by the accident of testamentary dispositions, and so on. To hold, therefore, that 'a body of individuals' must be equated to an 'association of persons' would be to disregarded the stage-by-stage evolution of the statutory classification of the different kinds of taxpayers under s. 2(31) of the I.T. Act, 1961. The difference between an association and a body is too pronounced to be slurred over. While an 'association' might well connote an active element of combining or associating, a 'body' would include even a comparatively inert mass of people or institution. The only essential attributes of a body of individuals are that there should be a plurality of individuals and they must, in the gross, have a nexus to a source of income. This conception at once excludes the crucial

characteristics which we associate with an association of persons, such for instance, as a common intention and a common activity to produce taxable income. In other words, persons who do nothing but stand and wait may not be an association of persons; but, they may yet be a body of individuals, if they stand together, and wait for something to be shared between them. It would be a matter for the income-tax authorities as well as the Tribunal and the courts to consider the facts in each case to find out if any given group of people are to be regarded as a body of individuals or not. The individuals concerned may have something or other in common which brings them together with reference to an income or its source. It may be common intention; it may be common holding out; or it may be a sharing of common spoils. The list is not exhaustive. Nor is it necessary that all these characteristics must be present in every case. Much might depend on the relationship in the gross to the income or to the income-yielding asset in question. **N.P. Saraswathi Ammal - CIT, Mad.** *Sec.2(31)* **[138ITR19,]**

Book AP. Satyanarayana Publishing House - CIT

The word book is not defined in the Income-tax Act. Hence, the expression must be understood with reference to the connotation of the said word in common parlance; and in common parlance,the word book is comprehensive enough to include all kinds of books. Hence, books of all varieties except those which are specifically excluded from its purview under section 80QQ, such as magazines, brochures, newspapers, journals, etc., answer the description of books in section 80QQ. Guides which contain questions and answers and which are published for the benefit of students studying in schools and colleges, cannot be described as brochures or tract nor do they fall under the expression other publications of a similar nature by whatever name called. *Sec.80QQ.* **[118ITR519]**

Book assets AP. CIT - Warner Hindustan Ltd.

The expression book asset in rule 2 does not refer to a notional, fictitious and intangible asset. There is nothing to support this interpretation in the Explanation itself. The expression book asset should be understood in its normal sense as any asset appearing as such in the books of account of the company. The object of Explanation 1 to rule 2 is to prevent companies from bringing into existence or increasing paid-up share capital or reserve by revaluation or otherwise of any book asset. The Explanation does not apply where there has been a real increase in the value of the asset which is realised, as opposed to any unrealised increase by taking recourse to revaluation. <u>*Companies (Profits) Surtax Act*</u> **[46CTR34,158ITR51]**

Books **Bom. Sheraton Appearels - ACIT**

In traditional terms, books means a collection of sheets of papers bound together with the intention that such binding shall be permanent and the papers used are kept collectively in one volume. It may also be assumed that it connotes the intention that it should serve as a permanent record. <u>*Sec.271(1)(c).*</u>**[175CTR651,256 ITR20,123Taxman238]**

Books of ac/s. **Bom. Sheraton Appearels - ACIT**

If books of account is considered in isolation, then it may mean books in which merchants,traders and businessmen generally keep their accounts and which are maintained for recording (a) all receipts and expenses with matters relating thereto; (b) all sales and purchases; and (c) the assets and liabilities. They are the documents and ledgers which must be prepared and kept by the business entity including the profit and loss account and the balance-sheet. ....At the same time to account means to reckon, and it is difficult to conceive of any accounting which does not involve either additions or subtractions or both of these operations of arithmetic. A book which contains successive entries of items may be a good memorandum book; but until those entries are totalled or balanced, or both, as the case may be, there is no reckoning and no accounts. A book which merely contains entries of items of which no account is made at any time, is not a book of account in a commercial sense...Section 34 of the Evidence Act, 1872, refers to the words entries in books of account. The accounts under section 34 means accounts which are maintained in the regular course of business. Income-tax legislation has been using the term book or books of account right from its inception. But, these terms are defined in the Act for the first time by the Finance Act, 2001, with effect from June 1, 2001. Section 2(12A) defines the said term as including ledgers,day-books, cash books, account books and other books whether kept in the written form or as print-outs of data stored in a floppy, disc, tape or any other form of electromagnetic data storage device. The above definition appears to have been framed by the Legislature keeping in view the development of computer technology. If the newly inserted definition of books of account inserted in the Income-tax Act is examined in contrast to the definition given under section 34 of the Evidence Act, it will be clear that the stringent requirements of section 34 are not to be found in the said definition. Obviously, for the simple reason that the purpose of both the legislations are different. The term books of account referred to in sub-clause (1) of clause (a) of Explanation 5 to section 271(1)(c) means books of account which have been maintained for determining any source of income. *Sec.2(12A)r.w.s. 271(1)(c).***[175CTR651,256ITR20,123Taxman238]**

Books of the assessee P&H. CIT - Kartar Singh

The expression books of the assessee in the context in which it appears in section 10(2)(vii) of the Indian Income-tax Act, 1922, does not give any indication of the particular type of accounts which the assessee should maintain. That the accounts maintained by the assessee do not lead to the correct assessment of the income, profits and gains of the business has nothing to do with the allowance that can be granted under section 10(2)(vii). Therefore, where the assessee produced a memorandum book which showed the relevant entries regarding the purchase and sale of vehicles, and wherein the loss had been calculated and written off, the department cannot contend that such a book was not a book within the meaning of the section. *Sec.41.***[77ITR338]**

Borne SC CIT - Dalhousie Properties Ltd.

In the context of section 23(1), the expression borne should be construed as referring to the amount of tax which the owner is liable to discharge and not merely the actual sum paid by him in discharge of his liability. *Sec.23.***[42CTR142,149ITR708]**

Borne by AAR Dhv Consultants Bv

That the remuneration of the employees is borne by a permanent establishment, if the same is deductible in computing the permanent establishment's taxable profits in the source country. *Sec.44D.r.w.s.115A.***[197CTR105,277ITR97,147Taxman521]**

Borrowed and paid MP. Malwa Mills Karamchari Parasper Sahakari Sanstha Ltd. - CIT

Under s. 36(1)(iii) of the I.T. Act, 1961, deduction is allowed for the amount of interest paid in respect of the capital borrowed for the purposes of the business or profession. The words borrowed and paid clearly postulate two different entities, one which lends capital and the other which borrows and pays interest. The same entity cannot be its own lender and borrower nor interest can be paid to self. *Sec.36(1).*[140ITR379]

Borrowed money Cal. CEPT -Bhartia Electric Steel Co.Ltd.

The term borrowed money was used in Rule 5A of the First Schedule and Rule 2A of the Second Schedule to indicate and denote money actually borrowed as a loan in the ordinary sense and not also to indicate and denote money which, though obtained on some other basis and in some other kind of transaction, could be,on the failure of such basis and such transaction, made out to be borrowed money in essence and in law. *Excess Profits Tax Act.* **[25ITR192]**

Break- up Ass. Mahadeo Jalan - CWT

The expression break-up in section 7 of the Wealth-tax Act is inextricably connected with the situation where the company was reached the stage of winding up and that expression cannot be applied to a company which is a going concern and there is nothing ostensibly wrong with it. *Sec.7.WT.Act.*[69ITR170]

Broker dealer / profession Bom. CIT-Lallubhai Nagardas and Sons

...........From the definitions of broker, broker dealer and profession and the discussion on profession in the Corpus Juris Secundum,it is clear that a stock broker cannot be said to be engaged in the practice of a profession. The real job of a stock broker is to make arrangements for sale of shares or securities of others. Such activity clearly falls within the expression business and not profession .... There is no dispute that stocks and shares are commodities which are marketable. A person selling his own shares and securities has always been held to be carrying on business and not engaged in the practice of a profession. The distinction between a dealer and a broker in common parlance is that a dealer sells his own goods whereas a broker sells or arranges the sale of the goods of others though there may be cases where the same person does both the acts, that is, sells his own goods and also arranges for the sale of goods of others. In either event, he is engaged substantially in the sale of goods. The question is whether such a person can be held to be engaged in the practice of a profession. The answer to this question, in clear terms, is in the negative for the simple reason that the activity of a share broker is nothing but a business activity which falls within the expression business as defined in clause (13) of section 2 of the Act. It does not fall within the expression profession even if it is interpreted in the broader sense to include many of the activities which were earlier not included therein. A broker is not held to be engaged in the practice of a profession in the English speaking world as is evident from the discussion in the Corpus Juris Secundum. *Sec.2(13).***[114CTR58,204ITR93]**

Building SC CIT - Alps Theatre

The word building in clause (vi) of section 10(2) as much as in clauses (iv) and (v) means structures and does not include the site.

*Sec.32.***[65ITR377]**

2. Upon a plain construction of section 10(2)(vii) and considering that section 10(2) refers to allowances to be made in computing profits and gains of business, the word building in section 10(2)(vii) would include a part of the building. **[CIT-London Hotel, Bom.,** *Sec.32.***68ITR62]**

3. The word building in section 10(2)(vi) of the Indian Income-tax Act, 1922, means the superstructure and does not include the site on which it is erected.**[CIT-Kailash Motors,All.,***Sec.32.***82ITR353]**

4. The word building occurring in the Income-tax Act, 1961, has to be understood from the common sense point of view and its use in the Act has to be appreciated in the context in which the provision for depreciation on building has been made, treating the same as the capital asset of an assessee. The roads laid in the proximity of the building belonging to the assessee, a turf club, for providing access to the race course premises and other buildings within the compound, fall within the meaning of the word building as referred to in section 32(1)(ii) of the Act, read with Appendix I in rule 5 of the Income-tax Rules, 1962, and hence depreciation is allowable on such roads. **[CIT- Bangalore Turf Club Limited,Kar.,** *Sec.32(1).* **38CTR 162,150ITR23,15 Taxman 221]**

5. Building means structures and does not include site. **[CIT-Indian Oil Corporation Ltd., Bom.,***Sec.32.***134CTR416, 218 ITR511,87Taxman117]**

6. Roads laid within factory premises are necessary adjuncts to the factory building and are to be treated as building for purposes of depreciation under section 32 of the Income-tax Act,1961. **[CIT-Southern Petro Chemical Industries Corporation Ltd. (No.1),Mad.,*Sec.32.*151CTR188,233ITR391,149Taxation280,142 Taxman16]**

7. Building used as hotel and the tourist complex could not be treated as a plant within the meaning of section 32 of the Income-tax Act, 1961, and cognate provisions. **[CIT-Ajit Bhawan Hotel, Raj, *Sec.32.*186CTR138,274ITR66,135Taxman415]**

8. The expression building has not been defined under the Act and it would have to be understood in the ordinary sense in which the word is used. A flat in the building would be building. Merely because the flat in the building is in a co- partnership society that would be of no consequence as long as the flat was obtained for a consideration and the person allotted the flat is a member of the society.**[Bennett Coleman and Co. Ltd.-ACWT,Bom.,*Sec.2(ea)(i). WT.Act.*214CTR, 671,304ITR235]**

Building, flat Kar. Karnataka Bank Employees Association - CIT

The words building or flat have not been defined in the Income-tax Act. The meaning which is given in common parlance has to be ascertained. With the growth of civilization, and more particularly in recent years in the cities, instead of constructing an independent building to be used as a house, the concept of multi storeyed flats has developed. The structure intended for human habitation having its place and intended for the occupation of a family has been defined as a house. This purpose is achieved by an independent house as well as a flat. The vertical growth leads to multi storeyed construction of buildings where independent flats are constructed for human habitation. The word house or building covers a flat for the purposes of withdrawal from the recognised provident fund to meet expenditure on building a house or purchasing a site or a house or a house and a site under rules 68 and 69 of the Income-tax Rules, 1962. *Income-taxRules-68,69.***[151CTR487,233ITR628,102 Taxman107]**

Building constructed Kar. CWT - Giridhar G. Yadalam

.....building in the process of construction could not be understood as a building which had been constructed. Sec.2(ea)(i).WT.Act. **[211CTR43,325ITR223,163Taxman372]**

Building owned by the assessee **All.** **Addl. CIT - U.P. State Agro Industrial Corp. Ltd.**

The expression building owned by the assessee in section 32 of the 1961 Act has not been used in the sense of property, complete title in which vests in the assessee. The assessee will be considered to be an owner of the building under section 32, if he is in a position to exercise the rights of the owner not on behalf of the person in whom the title vests but in his own rights. *Sec.32*.**[20CTR141,127ITR97]**

2. Building owned by the assessee the expression as occurring in section 32(1) of the Income-tax Act means the person who having acquired possession over the building in his own right uses the same for the purposes of the business or profession though a legal title has not been conveyed to him consistently with the requirements of laws such as the Transfer of Property Act and the Registration Act, etc., but nevertheless is entitled to hold the property to the exclusion of all others. **[Mysore Minerals Ltd.CIT,AP.** *Sec.32.***AIR1999AP3185,156CTR 1,239ITR775,106 Taxman116, 152 Taxation 488]**

Business **All.** **Gayaprasad and Chotey Lal**

A return on the money invested by the assessee amounted to a business within the meaning of the Income-tax Act. Sec.2(13),28. **[3ITR177]**

2. Business must be some activity which has for its object the acquirement of some profit which can be claimed as of legal right .

[Mahammad Faruq,All.,*Sec.2.(13).***6ITR1]**

3. The term business as used in section 10 denotes an abstract and intangible thing, quite apart from any of these physical adjuncts, and quite apart also from such other elements as the goodwill,the business connections,the business reputation and so on. **[CIT-Bosotto Brothers Ltd., Mad.,***Sec.2(13).r.w.s.22, 32.***8ITR41]**

4. The term business is a word of very wide, though by no means determinate, scope. It is neither practicable nor desirable to make any attempt at delimiting the ambit of its connotation. Each case has to be determined with reference to the particular kind of activity and occupation of the person concerned. Though ordinarily business implies a continuous activity in carrying on a particular trade or avocation, it may also include an activity which may be called quiescent. **[CIT-Calcutta National Bank Ltd., SC,** *Excess Profits Tax Act* **37 ITR 171]**

5. The definition of business, being an inclusive definition and not being exhaustive, is indicative of extension and expansion and not restriction. **[P.Vadamalayan-CIT, Mad.,** *Sec.2(13).***74ITR94]**

6. The definition of business in section 2(4) was of wide amplitude and it could embrace within itself dealing in real property as also the activity of taking a property on lease, setting up a market thereon and letting out shops and stalls in the market. **[S.G.Mercantile Corporation P.Ltd.-CIT,SC,***Sec.2(13),22.***83ITR 700]**

7. Business connotes a continuous course of activities. All the activities which go to make up the business need not be started simultaneously in order that the business may commence. The business would commence when the activity which is first in point of time and which much necessarily preceded all other activities is started. **[CIT-Saurashtra Cement and Chemical Industries Ltd., Guj.,***Sec.2(13).***91ITR170]**

8. The word business has a wider connotation which comprehends within itself three important stages of the business of every industrial undertaking, viz., purchase, manufacture and sale. Therefore, when a particular asset is used in the capital work-in-progress that asset must be treated as an asset used in business. Similarly, if a piece of land is acquired or stores are put in transit they must be treated as having been used in business because user in business starts the moment the land or the store is purchased for the business of the undertaking and any attempt to limit the user in business to manufacture and subsequent sale would offend the clear and simple meaning of the word business, which has a wider significance. **[CIT-MohanMeakinBreweriesLtd.,HP.,***Sec.2(13). r.w.s.84.***11CTR52,122ITR203,2Taxman103]**

9. The expression business does not necessarily mean trade or manufacture only: it is being used as including within its scope professions, vocations and callings for a fairly long time.The word business is one of wide import and it means an activity carried on continuously and systematically by a person by the application of his labour and skill with a view to earning an income. **[BarendraPrasadRay-ITO,SC,** *Sec.2(13).r.w.s.5(2),9(1)(i).***129ITR 295,6 Taxman 19]**

10. A trading transaction very often results in loss to one party and profit to the other party but the transaction can not be viewed in one way for the party who makes a profit and in a different way for the party who suffers the loss.If the transaction is a trading transaction or an adventure in the nature of trade, it will amount to business whether it results in loss or profit. **[CIT-Bhikamchand Jankilal,MP.,** *Sec.2(13).r.w.s.28Expl-2,43(5),73(1).***131ITR554]**

11. The Income-tax Act does not deal merely with business of selling or purchasing. Under section 28 of the I.T. Act, 1961, the profits and gains of any business which was carried on by the assessee at any time during the previous year are assessable to tax. Under the term business, as defined in section 2(13) of the Act, it is not essential that there should be a series of transactions and even a single transaction may constitute business. **[CIT-Khairagarh Timber Traders,MP.,***Sec.2(13).r.w.s.28.***28CTR35,137ITR346]**

12. The word business connotes some substantial and systematic or organised course of activity or conduct with a set purpose. **[Shiv Naraîn Agarwal -CIT, All.,** *Sec.2(13).r.w.s.64(1)(i).***19CTR134, 139ITR999]**

13. The word business is a word of large and indefinite import. It is something which occupies the attention and labour of a person for the purpose of profit.It has a more extensive meaning than the word trade. An activity carried on continuously in an organised manner with a set purpose and with a view to earn profit is business. **[CIT-M.P.Bazaz,Ori.,***Sec.2(13).r.w.s.28.***200ITR131]**

14. The question whether a particular source of income is business must be decided according to ordinary notions of what a business is. The activity from which the income is derived must have a set purpose. The motive for the activity must be profit and not sport or pleasure. Even a single or isolated transaction can constitute business if it bears a clear indicia of trade, although the activity would normally be systematic and organised characterised by a course of dealings which are frequent, regular and continuous. **[CIT -R.M.Meenakshisundaram,Mad.,***Sec.2(13).***212ITR220]**

15. A single act of constructing a godown and letting it out can not be treated as a business.The expression business contemplates continuous activity from year to year. **[CIT-Veerabhadra Industries, AP.,***Sec.2(13).r.w.s.22,28.***240ITR 5,156 Taxation13, 106Taxman 370]**

16. ....the expression business though extensively used, is a word of indefinite import. In taxing statutes it is used in the sense of an occupation or profession which occupies the time, attention and labour of a person, normally with the object of making profit. To regard an activity as business, there must be a course of dealings, either actually continued or contemplated to be continued with a profit motive, and not for sport or pleasure. The carrying on of activity of agriculture which necessarily involves time, attention and labour, is to be regarded as business if it is done with a profit motive and not for sport or pleasure.**[CWT-K.Vijayakumar, Mad.,** *W.T.Act.Schedule.***166CTR246,243ITR271,107Taxman570]**

17. The word business has been defined to include every trade, occupation and profession. The question whether a particular letting amounts to business has to be decided in the circumstances of each case and each case has to be looked at from a businessman's point of view to find out whether the letting was the doing of a business or the exploitation of his property by an owner. It is the actual activity carried on by the assessee that has to be seen. **[CIT - Jai Bharat Theatre,All.,** *Sec.2(13).r.w.s.185* **167CTR274, 247ITR 295,110Taxman297]**

18. ... if the main activity is not business, then any transaction incidental or ancillary would not normally amount to business unless an independent intention to carry on business in the incidental or ancillary activity is established. In such cases, the onus of proof of an independent intention to carry on business in the incidental or ancillary activity of sales rests on the department. The inclusion of incidental or ancillary activity in the definition of business contained in section 2(11) pre-supposes the existence of trade or commerce, etc. **[Commissioner of Sales Tax- Sai Publication Fund, SC,** *Bombay Sales Tax Act.***177CTR1, 258ITR70]**

19. In order that an activity may constitute business, firstly, it must be a continuous course of dealing and, secondly, it must be carried on with a profit motive. **[Mrs. Kamala Muthia-CIT,Mad.,** *Sec.2 (13).r.w.s.28,45.***180CTR231,259ITR184,129Taxman803]**

20. The word business includes the whole process of the undertaking carried on by the assessee. It begins with the initial process to achieve the end product and ends with the realisation of the ultimate profit. Whatever means or mode is connected with such process are part of the business. The business may be of manufacturing or trading or otherwise. **[CIT-Birla Jute and Industries Ltd.,Cal.,** *Sec.2 (13).32.***180 CTR 339,260 ITR55, 133 Taxman 337]**

21. The term business is to be given a wide meaning and with the rapid advancement and growth in the field of science and technology even consultancy services offered would be covered under the term business. **[CIT-U.P. Electronic Corporation Ltd.,All.,** *Sec.2(13)r.w.s.35.***276ITR45,145Taxman 494]**

22. The word business has been defined under section 2(13) of the Income-tax Act, 1961, which includes any trade, commerce or manufacture or any adventure or concern in the nature of trade, commerce or manufacture. An isolated transaction or activity can also be part of business, but to consider the question of business, there must be regular activity of purchasing and selling. **[CIT-Suresh Chand Goyal, MP.,***Sec.2(13).28.***209CTR410,298 ITR277, 163Taxman 54]**

23. A business is nothing more than a continuous course of activities and for commencement of business all the activities which go to make up the business need not be started simultaneously. As soon as an activity which is the essential activity in the course of carrying on the business is started, the business must be said to have commenced. A finding regarding the date when a business was set up is a finding of fact. **[CIT-Espn Software India P. Ltd., Del.,***Sec.2(13).***301ITR368]**

24. The term business which prefixes generation of power in clause (iv) of the Explanation to section 115JA of the Income-tax Act, 1961, is not limited to one which is prosecuted only by engaging with an outside third party. The meaning of the word business as defined in section 2(13) includes any trade, commerce or manufacture or any adventure or concern in the nature of trade, commerce or manufacture. **[CIT-D C M Sriram Consolidated Ltd.,Del.,***Sec.2 (13).115J.***221CTR519,322ITR486,176Taxman49]**

25. Business as defined under section 2(13) of the Act includes any trade, commerce or manufacture or any adventure or concern in the nature of trade or manufacture which obviously means some effort, physical or intellectual. **[CIT-Manjoo and Co.,Ker.,** *Sec.2(13).* **236CTR356,335ITR527,195Taxman39]**

Business/speculation business **MP. CIT- Bhikamchand Jankilal**

A speculative transaction would amount to speculation business if it fulfils the definition of business under section 2(13) of the I.T. Act, 1961, or if it amounts to an adventure or concern in the nature of trade. Neither repetition nor continuity of similar transaction is necessary to constitute a transaction an adventure in the nature of trade and even a single transaction may constitute business...The object of Expln. 2 to section 28 was not to substitute a different definition of business from that contained in section 2(13). The object was merely to declare that speculation business will be treated as distinct and separate from any other business. The opening words say that the enacting provision in the Explanation will apply where speculative transactions carried on by an assessee are of such a nature as to constitute a business. But much significance can not be attached to the use of the expression speculative transaction in plural, for, as enacted in section 13(2) of the General Clauses Act, 1897, the words in the singular include the plural and vice versa. There is nothing in the context of Expln. 2 to show that a single speculative transaction, though it is an adventure in the nature of trade cannot amount to a speculation business. *Sec.2(13).r.w.s.28Expl-2,43(5),73(1).***[131ITR554]**

Business and profession **Mad.** **CIT - International Clearing & Shipping Agency**

The terms business and profession are defined in the Income-tax Act, 1961, in sections 2(13) and 2(36), respectively. Though the scope of the term business is wide, if the activity is properly to be characterised as profession, then that activity cannot also be regarded as business. The distinguishing feature of a profession is the possession by the practitioner of the profession of specialised knowledge involving intellectual skill and higher education and learning. The services rendered by a professional while practising the profession, are services for which he has been trained. The practice of a profession cannot be regarded as a commercial activity though the practice is not without compensation or profit. The compensation earned by the practitioner of a profession is by reason of the personal qualification possessed by him. *Sec.2(13),(36).* **[158CTR672,241ITR172,118Taxman730]**

2. The Legislature intended to have different scope for business and profession in section 32(1) of the Income-tax Act, 1961. From the scheme of the section it is discernible that the various clauses operate on further specific conditions laid down in each specific clause. In clause (iv) the Legislature has used the word business only. This means that the Legislature was conscious of the fact that business and profession are different and separate and they cannot be used interchangeably. Under clause (iv) the Legislature intended to restrict the benefit to the assessee carrying on business only and not to one carrying on a profession. **[G.K.Choksi and Co.-CIT,SC,***Sec.32(1)(iv).***295ITR376]**

Business and vocation **All.** **Upper India Chamber of Commerce-CIT**

Upon a proper construction of the words business and vocation in the context of the Indian Income-tax Act, there must be some real, substantive and systematic course of business or conduct before it can be said that a business or vocation exists the profits of which are taxable as such under the Act. In the case of a company,the words business and vocation are virtually synonymous, since there can be no room for a vocation, ...Business or vocation in ordinary parlance connotes activities in which a person is engaged with a set purpose, and the frequency or the repetition of the activity, though at times a decisive factor, is by no means an infallible test......The word vocation is a word of wider import than the word profession.

Sec.2(13).[**15ITR263**]

Business commences **Cal.** **Tetron Commercial Ltd. - CIT**

A business commences with the activities undertaken even at the preparatory stage for setting up of the business. *Sec.36(1)(iii).* **[182CTR124,261ITR422,133Taxman781]**

Business connection PC **CIT - Currimbhoy Ebrahim & Sons Ltd.**

The phrase business connection is different from, though doubtless not unrelated to, the word business of which there is a definition in the Act. *Sec.9(1),5(2)(b).*[**3ITR395**]

2. The word business does not merely qualify the word connection by describing the sort of connection. It has the significance indicated in section 2(4) of the Act and denotes an adventure or concern in the nature of trade, commerce or manufacture,and the word connection must be used in the sense of that with which one is connected; and the expression any business connection therefore, means any adventure or concern in the nature of trade,commerce or manufacture, being a business with which he (i.e., the person residing out of British India) is connected. **[CIT-P.V.R.M. Visalakshi Achi,Ran.,***Sec.9(1), 5(2) (b)* **5ITR448]**

3. The business connection contemplated by section 42 is something wider than mere doing business with India. All those cases in which the non-resident is merely doing business with India should be excluded from section 42. The language used by the Legislature in section 42(1) is very wide: any business connection is not necessary that business connection which is constituted by a permanent and exclusive agency. On the other hand, a mere casual connection, a connection which has no continuity, would also not be a business connection as contemplated by the Legislature under section 42. An isolated transaction through an agent, or even a connection for a short period, would not necessarily constitute business connection. It is impossible to give an interpretation to this expression which would be a good interpretation for all the innumerable cases which may arise. It is only possible to interpret the expression negatively rather than positively. In most cases it would depend upon the facts of the particular case as to whether the business connection has been established or not. **[Abdullabhai Abdul Kader-CIT, Bom.,** *Sec.9(1),5(2)(b).***22ITR241]**

4. The expression business connection in section 42(1) of the Indian Income-tax Act,1922, is an expression of wide and indefinite import and that expression is different from, though related to, the expression business as defined in the Act. Business connection of the nature contemplated in section 42(1) connotes some element of continuity between the person in the taxable territories who helped to make the profits and the person outside who receives or realises the profits, and an isolated transaction does not attract the operation of section 42. It is not the length of time during which the connection has subsisted but the nature of the connection which would determine whether a business connection within the meaning of section 42(1) has been established or not. A course of numerous dealings within a short time having an element of continuity about them would be sufficient to establish a business connection.

[Bikaner Textile Merchants Syndicate Ltd.-CIT, Raj., *Sec.9(1),5(2)(b).***58ITR169]**

5. To conform with the requirement of the expression business connection which have been fairly settled by judicial decisions, it is necessary that the common thread of mutual interest must run through the fabric of the trading activities carried on outside and inside the taxable territory and the same has been described as real and intimate connection. The commonnes of interest may be by way of management control or financial control or by way of sharing of profits. It may come into existence in some other manner but there must be something more than mere transaction of purchase and sale between principal and principal in order to bring the transaction within the purview of the expression business connection within the meaning of section 9(1)(i) and 163(1)(b) of the Income-tax Act. **[CIT-Hindustan Shipyard Ltd., AP.,***Sec.9(1)163(1).,* **109ITR158]**

6. The expression business connection used in section 9(1)(i) of the Act, 1961, undoubtedly means something more than business. A business connection involves a relation between a business carried on by a non-resident which yields profits or gains and some activity in the taxable territories which contributes directly or indirectly to the earning of those profits or gains. It predicates an element of continuity between the business of the non-resident and the activity in the taxable territories; a stray or isolated transaction is normally not to be regarded as a business connection. Business connection may take several forms; it may include carrying on a part of the main business or activity incidental to the main business of the non-resident through an agent,or it may merely be a relation between the business of the non-resident and the activity in the taxable territories which facilitates or assists the carrying on of that business. In each case, the question whether there is a business connection from or through which income, profits or gains arise or accrue to a non-resident must be determined upon the facts and circumstances of the case. **[Biyani and Sons(P.)Ltd.-CIT,P&H.,** *Sec.9(1),5(2)(b).* **120ITR887]**

7. In the context in which the expression business connection is used in section 9 (1), there is no warrant for giving a restricted meaning to it excluding professional connections from its scope. **[Barendra Prasad Ray- ITO, SC,***Sec.9(1),5(2)(b).***129ITR295,6 Taxman19]**

8. The expression business connection postulates a real and intimate relation between trading activity carried on outside the taxable territories and trading activity within the territories, the relation between the two contributing to the earning of income by the non-resident in his trading activity. In the case of an agency agreement, the crux of the matter is the agent's authority to accept offers or to bind the principal. If that element is absent, it cannot be said that there was any business connection as contemplated by section9.**[T.I.andM.SalesLtd.CIT,Cal.,***Sec.9(1),5(2)(b)***151ITR286]**

9. A business connection contemplated under section 9 of the Income-tax Act, 1961, involves a relation between a business carried on by a non-resident and some activity in the taxable territories which is attributable directly or indirectly to the earnings, profits or gains of such business. **[CIT - Atlas Steel Co. Ltd., Cal.,** *Sec.9(1),5(2)(b).***56CTR73, 164ITR401,28 Taxman462]**

10. ... expression business connection means something more than a business. It presupposes an element of continuity between the business of the non-resident and the activity in the taxable territory. A stray or isolated transaction would normally not be regarded as a business connection. Business connection may take several forms; it may include carrying on part of the main business or activity incidental to the non-resident through an agent or it might merely be a relation between the business of the non-resident and the activity in the taxable territory which facilitates or assists the carrying on of that business.A relation to be a business connection must be real and intimate and through or from which income must accrue or arise, whether directly or indirectly to the non-resident. Such a business connection could be spelt out on the terms of the agreements in question. **[AdvanceRulingP.No.8of 1995,AAR,**<u>*Sec.9(1)(i)r.w.s.245Q*</u>**136CTR451,223ITR416,135 Taxation269,90Taxman47]**

11....expression business connection has not been defined in the Act.In the absence of any definition, the ordinary meaning of the said expression must be taken. Rule 10(ii) of the Income-tax Rules, 1962, more or less uses the same expression as has been used in section 9. The words business connection for the purpose of sections 5 and 9 of the Act must be confined to profits arising out of business. When an income accruing or arising from business by reason of a legal fiction becomes assessable, it must be held that the same must be confined to receipts out of business and not out of the sale of capital assets.... Having regard to the purport and object there of, if the words business connection in India were wide enough to cover all transactions including transactions in capital assets, there was no reason for Parliament to specifically include income (a) through or from any property in India, (b) through or from any asset or source of income from India, and (c) through or from sale of a capital asset situate in India. From the very fact that in terms of section 9 of the Act, the transfer of a capital asset situate in India has been brought within the purview of the deemed income under section 9 of the Act, and rule 10(ii) of the Rules, it is clear that the intention of Parliament was not to bring within its purview any income derived out of sale or purchase of a capital asset effected outside India. **[CIT-QuantasAirwaysLtd.,Del.,** Sec.5,9.**175CTR 98, 256ITR84,122 Taxman 935]**

12. The essential features of business connection are : (a) a real and intimate relation must exist between the trading activities by a non-resident carried on outside India and the activities within India; (b) the relation contributes directly or indirectly to the earning of income by the non-resident in his business; and (c) a course of dealing or continuity of relationship and not a mere isolated or stray nexus between the business of the non-resident outside India and the activity in India,would furnish a strong indication of business connection. **[Sutron Corporation, AAR,** *Sec.9(1),5(2)(b).* **189CTR366,268ITR156,138Taxman87]**

13. The essential features of business connection are : (a) a real and intimate relation must exist between the trading activities carried on outside India by a non-resident and the activities within India; (b) such relation should contribute directly on indirectly to the earning of income by the non-resident in his business;(c) a course of dealing or continuity of relationship and not a mere isolated or stray nexus between the business of the non-resident outside India and the activity in India. **[ABC Ltd,AAR,***Sec.9(1),5(2)(b).r.w.s.195* **208CTR117,289ITR438,159Taxman334]**

Business expenditure/loss J & K. Chenab Forest Co.-CIT

There is a clear distinction between a business expenditure and a business loss; the former is indicative of a volition but a loss comes so to speak as ab extra. *Sec.28.***[96ITR568]**

Business income Cal. CIT - Ajmera Industries Private Ltd.

In order to be a business income there must be evidence of exploitation of a commercial asset....Where income is derived by the exploitation of the asset, and there is only a difference in the manner of exploitation that is to say, instead of user of the asset by the assessee himself there is a leasing out of the asset, the income derived must be considered to be business income. *Sec.28,56 (2)(iii).***[103ITR245]**

2. A commercial asset can be rightfully exploited by the owner thereof to the best advantage and the advantage can be materialised either by using it himself personally or by letting it out to somebody else. So long as the commercial asset is capable of being exploited as such, it has to be understood that its income is business income. **[CIT-Malabar and Pioneer Hosiery (P.) Ltd.,Ker.,***Sec.28. r.w.s.22.***132CTR193,221ITR117,86Taxman361]**

Business interest Del. Punjab Stainless Steel Inds. – CIT

The business interest of the assessee has to be distinguished from the personal interest of its directors or partners, as the case may be. In other words, there has to be a nexus between the advancing of funds and business interest of the assessee. The appropriate test in such a case would be as to whether a reasonable person stepping into the shoes of the directors/partners of the assessee and working solely in the interest of the assessee, would have extended such interest free advances. ... The test to be applied in such cases is not the source of the funds but the purpose for which the advances were extended. *Sec.36(1)(iii).*[324ITR396]

Business is set up Bom. CIT - L and T McNeil Ltd

When a business is established and is ready to commence business then it can be said of that business that it is set up. But before it is ready to commence business, it is not set up. *Sec.28.r.w.s.56.* **[112CTR405,202ITR662,117Taxation156]**

Business premises P&H. CWT - Jaidev Inder Singh

Where agricultural operations are carried on urban land, the land will fall within the definition of the expression business premises as provided in rule 1(i), Paragraph B, Part I, of the Schedule to the Wealth-tax Act, 1957, and would thus be entitled to exemption from the charge of additional wealth-tax on urban assets, under Paragraph A(2) of the Schedule. *Rule 1WT.Act.***[103 CTR173,195 ITR367,63Taxman25]**

Business purpose Cal. CIT - Kejriwal Enterprises

A business purpose is for the carrying on of the business and such purpose in capacity as a businessman; the expenses must not be of a private or domestic nature; there must be no dishonesty and no underhand attempts involved at motivated diversion of funds so as to get an otherwise unwarranted deduction of interest on borrowed funds. *Sec.36(i)(iii).***[181CTR305,260ITR341,133Taxman749]**

But does not include any tax chargeable with... SC T.S.Krishna- CIT

The words but does not include any tax chargeable with reference to the value of any particular asset of the business or profession in the last part of the Explanation to section 40(a)(ii-a) of the Income-tax Act, 1961, are related to the tax chargeable under a law in force in any country outside India and the exception contemplated by those words has no logical connection with wealth- tax chargeable under the Wealth- tax Act,1957. *Sec.40(a)(iia).***[87ITR429]**

Buyer **HP. Rudra and Co. - Union of India**

The Explanation to section 206C of the Income-tax Act, 1961, makes it clear as to who is a buyer. It says that the buyer is one who obtains in any sale, by way of auction, tender or any other mode, goods of the nature specified in the Table in sub-section (1) of section 206C. *Sec.206.***[152CTR27,233ITR66]**

2. Buyer means a person who, by virtue of the payment, gets a right to receive specific goods and not one who is merely allowed/ permitted to carry on business. **[Union of India -S.S.Om Prakash and Company,SC,***Sec.206 C.* **248 ITR105,115Taxman]**

By such person **Bom.** **Grasim Industries Ltd. - S.M. Mishra, CIT**

The expression by such person appearing in section 9(1)(vii)(b), refers to the recipient of the income and not to the person making the payment. This would be clear if one looks to the opening words of sub-section (1) of section 9 which reads the following income shall be deemed to accrue or arise in India. Section 9 (1) refers to the income which is deemed to have accrued or arisen in India by the recipient of the income. The expression such person appearing in sub-clause (b) of section 9(1)(vii) therefore, refers to the recipient because one has to consider whether the income received by him (the recipient) is deemed to have accrued or arisen in India. Section 9 does not contemplate taxing the payer but contemplates taxing the recipient for the income received by him. *Sec.9(1).***[332ITR276]**

C

Can not be recovered from him **Ker.** **R.K.-motors and Timbers (P.) Ltd. – CIT**

The words cannot be recovered from him in the proviso to section 26(2) of the Indian Income-tax Act, 1922, do not postulate the difficulties that may be experienced in collection or the time that may elapse or the impediments that may be put forward in the way of collection of tax from the person succeeded but of the availability of sufficient assets from which the tax can be collected. Where it is extremely doubtful whether there were assets available with the person succeeded, who transferred his business to the successor, it can not be said that proceedings against the successor under the proviso would be without justification. *Sec.187.* **[68ITR794]**

Capital **MP.** **CIT - Anand Bahri Steel and Wire Products**

The word capital in the business world means the net worth of an enterprise and excludes the borrowings. It is, therefore, wrong to assume that the expression capital employed is not open to the construction that it does not embrace moneys borrowed by the assessee and invested in the industrial undertaking. In some contexts, the expression capital employed may include the borrowed moneys or borrowed capital; but, in the context of section 80J, the expression does not include borrowed moneys and debts, ... The computation of capital employed has to be in the prescribed manner as is expressly provided in section 80J and, therefore, the rules can prescribe as to what should or should not be included in the computation. The provision for deduction of borrowed moneys and debts in rule 19A(3) is not such which changes the character of that which has to be computed under section 80J and is valid. *Sec.80J.* **[21CTR166,133ITR365]**

Capital asset **Kar.** **Syndicate Bank Ltd. - Addl. CIT**

The term capital asset as defined in section 2(14) of the Income-tax Act, 1961, has a wide meaning and includes every kind of property as generally understood except those that are expressly excluded in the definition. A business undertaking as a whole would constitute a capital asset within the meaning of section 2(14). *Sec.2(14).* **[45CTR68,155ITR681]**

2... till that point of time when the assessee has some right that could be termed to be property, it will not fall within the definition of capital asset as defined under section 2(14) of the Act. **[Guj. Patel Brass Works-CIT,Guj.,** *Sec.2(14).***205CTR139,286ITR598]**

3. Trade mark, brand, goodwill, technical know-how relating to the manufacture of goods would all qualify to be treated as capital assets within the meaning of section 2(14) of the Act. Section 55(2), which deals with the cost of acquisition of a capital asset, makes it clear that goodwill, trade mark or brand name associated with a business and other incorporeal rights mentioned therein, are treated as capital assets under the Act for the purpose of capital gains. **[Foster's Australia Ltd.AAR** *Sec.2(14)r.w.s.55(2).***217CTR 21,302ITR289,170Taxman 341]**

Capital borrowed **Del.** **CIT-Saraswati Chemicals and Allied Industries (P.) Ltd.**

The expression capital borrowed used in section 36(1)(iii) of the Income-tax Act,1961, in the context in which it is placed in the provision means money and not any other asset.........Interest payable on capital borrowed means interest,which actually becomes payable on an amount of money and not on any other asset. An amount due under a statute cannot be regarded as borrowed capital for the expression capital predicates the relationship of a borrower and a lender, which relationship has to be found as a matter of matter of fact. *Sec.36(1)(iii).***[167CTR150,249ITR235,114Taxman 564]**

Capital employed **Cal.** **CIT - Indian Oxygen Ltd.**

Section 84 of the Act refers to capital employed and not to assets used. Rule 19 is even more specific. It clarifies the expression capital employed and lays down that capital employed consists of certain classes of assets. It is clear from the rule that if an asset is acquired prior to the commencement of the accounting period, the question of its user or non-user is entirely immaterial. Whether such an asset is used or not, it will still be included in the capital employed in the business. *Sec.80J.***[113ITR109]**

2. The expression capital employed found in section 84(1) of the Act cannot be construed to mean capital used or utilised. Any asset purchased for the undertaking is capital employed in that undertaking. The capital which is employed in a particular undertaking should not be confused with the capital which is used in that undertaking. The whole capital which is employed by an entrepreneur in establishing a particular industry cannot and would not necessarily be utilised or used in manufacturing process. The employment of capital for the establishment of an industry has a very wide connotation.When an entrepreneur starts a new industry, for various commercial exigencies, he cannot utilise all the assets which are employed by him for starting that industry. Therefore,the value of the assets employed by that entrepreneur would always be more than the value of the assets which are actually utilised for bringing out production. **[CIT-Mohan Meakin Breweries Ltd., HP.,***Sec.84.***11CTR52,122ITR203,2Taxman103]**

3. There can be no doubt that the expression capital employed is susceptible of more than one interpretation and it may include long-term borrowings or it may not, depending on the context and the circumstances in which it is used. There is doubt even amongst lawyers and accountants whether short-term borrowings can be regarded as forming part of the capital employed. The expression capital employed is not a term of art nor is it an expression having a fixed connotation or meaning,but it is susceptible of varied meanings, including or excluding short-term borrowings or long-term borrowings, whether of all categories or of any particular category or categories,depending on its environmental context. Since the expression capital employed has a variable meaning, which in a given case may or may not include borrowed monies, the Board could,in exercise of its rule-making power, exclude borrowed monies in the computation of the capital employed and in doing so, it would not in any way be acting contrary to the mandate of the statute. **[Lohia Machines Ltd.-Union of India,SC,***Sec.80J.***44 CTR 328,152ITR308,20Taxman9]**

4. The moment capital is utilised for the purpose of acquiring any asset for a business, such capital becomes employed in the business. Whether the asset itself is actually used in the business or not, so far as the capital is concerned, it continues to be employed in the business. **[CIT-Century Spg. and Mfg. Co.Ltd.,Bom.,** *Sec.80J.***181ITR214]**

Capital expenditure **SC** **Gotan Lime Syndicate - CIT**

It is not the law in every case,that if an enduring advantage is obtained the expenditure for securing it must be treated as capital expenditure,for in the ordinary case, the cost of the material worked up in a manufactory is not a capital expenditure, it is a current expenditure, and does not become a capital expenditure merely because the material is provided by something like a forward contract, under which a person for the payment of a lump sum down secures a supply of the raw material for a period extending over several years. *Sec37.***[59ITR718]**

2. Payment of premium in consideration of the owner of property agreeing to grant a right to take and remove the stock-in-trade of the taxpayer is in the nature of capital expenditure. **[H. Dear and Co.(P.)Ltd.-CIT,SC,***Sec.37.***60 ITR546]**

3. The test to determine whether the amount spent is capital or revenue in nature is,whether by spending such amount the assessee has brought any new asset into existence with an enduring benefit. If the amount is spent for preserving and maintaining the present asset in existence, it cannot be said that the expenditure so incurred is capital in nature. **[Nathmal Bankatlal Parikh and Co. -CIT, AP.,***Sec.31.r.w.s.37.***122ITR168]**

4. Where a borrowing is made for the purpose of business,interest paid on such borrowing becomes eligible for deduction contemplated by section 36(1)(iii) of the Income-tax Act,1961, and this would be so, whether the capital is invested in order to acquire a revenue asset or a capital asset,because the act of borrowing capital is distinct from the act of investment of the capital to acquire an asset. However, if there is no existing business with reference to which the capital is borrowed and the borrowed capital is invested to purchase a new asset of enduring nature, then the interest paid on such borrowing till the asset so purchased goes into production increases the cost of installation of the said asset and hence should be treated as capital expenditure not covered by section 36 (1) (iii) of the Act. **[CIT-Shah Theatres (P.)Ltd.,Raj.***Sec.37.r.w.s.36(1).* **67CTR1,169 ITR499]**

5. Where the object of incurring an expenditure is to effect a capital structure as a result of which certain incidental advantage flows, the expenditure will be of capital nature.It is not the acquisition of a right of a permanent character alone the creation of which is a condition for the carrying on of the business that could be rightly treated as an expenditure on capital account.Capital expenditure can be incurred after a company is floated or after it starts its business. **[Vazir Sultan Tobacco Co. Ltd.-CIT,AP.** *Sec.37.r.w.s. 28.***74CTR1,174ITR689,41Taxman7]**

6. Ordinarily,the word capital expenditure refers to the expenditure which is of a permanent nature or for securing tangible or intangible property,corporeal or incorporeal right or enduring benefit to the business. An expenditure may be once and for all or in respect of fixed capital or for enduring benefit.Purchase of capital asset and stock-in-trade marks the distinction between capital and revenue expenditure. If the asset is acquired as stock-in-trade then it will be revenue expenditure. **[CIT- Jaipur Mineral Development Syndicate., Raj.,** *Sec.37.***27CTR253,216ITR469,82 Taxman 52]**

7. If an amount is spent for the purpose of bringing into existence a new asset or obtaining a new advantage,then such an expenditure would not be an expenditure of a revenue nature but it would be a capital expenditure. **[CIT - Bharat Aluminium Co. Ltd.Del.** *Sec.37.r.w.s.31.***[292ITR600,160Taxman388]**

Capital gain SC Commonwealth Trust Ltd.–CIT

In commercial parlance computation of capital gain would mean, the actual gain measured by the difference between the sale price and the cost of acquisition. *Sec.45,50(1).***AIR1997SC3580, 142 CTR91, 228ITR1,140 Taxation698,94Taxman236]**

2. Section 45 of the Income-tax Act,1961, brings to charge capital gains and its ingredients are (i) the existence of a capital asset owned by the assessee ;(ii) the transfer of such asset during the previous year ; (iii) arising of profits and gains from transfer of such assets, and (iv) such profits and gains must accrue and arise to the assessee. **[B.Raghurama Prabhu Estate-JCIT(Asstt.),Kar.,** *Sec.45* **[239CTR274,335ITR394]**

Capital loss Cal. Darjeeling Consolidated Tea Co. Ltd. - CIT

Capital loss is not defined anywhere in the Act. But it is clear that it must be a loss arising out of the same transaction as mentioned in section 45 of the Income-tax Act,1961, as giving rise to a capital gain. In the first place, there must be a transfer of a capital asset and in the second place,there must be a loss arising out of such transfer. *Sec.45.***[183ITR493]**

Capital of that concern **Guj.** **CWT-Lallubhai Gordhandas Charitable Trust**

.........the expression capital of that concern which occurs in the second proviso to section 21A is intended to take within its sweep all the types of concerns in which the trust could invest its funds. The word capital will therefore take colour from the type of concern to which it is applied.When the concern is a company registered under the Companies Act,1956, the word capital will have to be understood in the context of the provisions of that Act. As provided by section 13(4) of the Companies Act, in the case of a company having a share capital, the memorandum shall state the amount of share capital with which the company is to be registered and the division thereof into the shares of a fixed amount. The word capital is used in company law in various senses, but it is properly used to denote the share capital of a company. The nominal capital of a company sets the limit of capital available for issue and, therefore, the issued capital of a company cannot exceed that limit. The nominal capital is strictly speaking not capital at all, since it is only an authority by the shareholders to the directors to create new capital by the issue of shares. The issued capital is on the other hand a reality and not a mere authority to create new capital. The nominal capital must be stated in the memorandum of association and would be equal to nominal value of shares which the directors are authorised to issue. The alteration in the actual value of shares as contrasted with the nominal, does not affect the amount of issued capital and this would apply to the whole concept of share capital in whatever sense that term is used. The capital of a company would mean share capital in the context of the provisions of the second proviso to section 21A of the said Act and the meaning of the word capital cannot be made mercurial by attaching it to all the assets that the company may own, nor can it include the reserves of the company, which can at any subsequent time be distributed as dividend. The expression capital employed stands on a different footing in the context of the provisions of section 80J of the Income-tax Act read with rule 19A of the Rules framed there under and cannot be projected in the expression capital of that concern occurring in the second proviso to section 21A of the Act, the purpose underlying the provisions of section 80J being entirely different from the object sought to be achieved by the provisions of the second proviso to section 21A of the said Act, and for that matter even the provisions of section 13(4) of the Income-tax Act. Sec.21A.WT.Act. **[152CTR338,239ITR448,102Taxman174]**

Capital receipt **Bom.** **Bombay Burmah Trading Corporation Ltd. - CIT**

Cases of termination resulting in loss of employment or cessation of business must be distinguished from cases of cancellation of contracts which are of a trading nature or are entered into in the course of business.If a sum represents profits in a new form, then that is income. But, where the agreement relates to the structure of an assessee's profit-making apparatus and affects the conduct of the business,the money received for the cancellation or variation of such an agreement would be capital receipt. The question is a question of fact and must be decided by ascertaining the true nature and object of the transaction made between the parties. *Sec.41.*[**81ITR777]**

2.Income is something which flows from the property.Something received in place of the property will be a capital receipt. **[CIT-Udayan Chinubhai, SC,** *Misc.***141CTR387,222ITR456,140 Taxation244,98Taxman502]**

3. In case money is borrowed by a newly started company which is in the process of constructing and erecting its plant,the interest incurred before the commencement of production on such borrowed money can be capitalised and added to the cost of the fixed assets created as a result of such expenditure.By the same reasoning if the assessee receives any amounts which are inextricably linked with the process of setting up its plant and machinery, such receipts will go to reduce the cost of its assets. These are receipts of a capital nature and cannot be taxed as income. **[CIT-BokaroSteelLtd, SC,***Misc.***AIR1999SC387,151CTR276,236 ITR315,148 Taxation 515,102 Taxman94]**

Capital value of assets **SC** **Union of India -Harbhajan Singh Dhillon**

The expression capital value of assets does not take in either the general liabilities of the individual owning them or in particular the debts owed in respect of them. A valid tax on the capital value of assets including agricultural land cannot be imposed under the power under article 246(1) read with entry 86 in List I, as entry 86, which is the only entry authorising such a tax, restricts in express terms the power to impose a tax on the capital value of assets, exclusive of agricultural land, of individuals and companies. An interpretation of the content and scope of such power, however liberal, cannot be adopted to include within it anything which the entry in positive terms excludes or restricts.

Constitution of India.[**83ITR582**]

Capital /revenue account **SC** **CIT- Ponni Sugars and Chemicals Ltd.**

The character of the receipt of a subsidy in the hands of the assessee under a scheme has to be determined with respect to the purpose for which the subsidy is granted. In other words, one has to apply the purpose test.The point of time at which the subsidy is paid is not relevant.The source is immaterial. If the object of the subsidy is to enable the assessee to run the business more profitably then the receipt is on revenue account. On the other hand, if the object of the assistance under the subsidy scheme is to enable the assessee to set up a new unit or to expand an existing unit then the receipt of the subsidy would be on capital account.

Sec.80P(2)(a)(i).[**219CTR105,306ITR392,174Taxman87**]

Capital /revenue expenditure **Cal.** **CIT- Hindustan General Electrical Corporation Ltd.**

One of the primary rules for determining whether a particular expenditure is a revenue or capital expenditure,is that the court from the terms of the agreement between the parties and from the surrounding circumstances has to ascertain the purpose for which it is being incurred. If the expenditure is so related to the carrying on or the conduct of the business that it may be regarded as an integral part of the profit earning process,it should be held to be a revenue expenditure. Should, however,the purpose be the acquisition of an asset or a right of a permanent character the possession whereof is the condition precedent or the pre-requisite to the commencement or continuance of the business,the expenditure would be a capital expenditure. *Sec.37.*[**81ITR243**]

2. Acquisition of the goodwill of a business is, without doubt, acquisition of a capital asset,and,therefore, its purchase price would be capital expenditure. It would not make any difference whether it is paid in a lump sum at one time or in installments distributed over a definite period. Where, however, the transaction is not one for acquisition of the goodwill, but for the right to use it,the expenditure would be revenue expenditure. **[Devidas Vithaldas and Co.-CIT,SC,** *Sec37.* **1CTR28,84ITR277]**

3. In the case of mining leases,where minerals are part of the land and have to be won, extracted and brought to the surface, expenditure for acquiring the right over or in the land to win the minerals would be of a capital nature. But, where the mineral has already been gotten and is on the surface, expenditure incurred for obtaining the right to acquire the raw material, i.e.,the mineral, would be revenue expenditure laid out for the acquisition of stock-in-trade. **[R.B.Seth Moolchand Suganchand-CIT,SC,** *Sec37.* **1CTR28,86ITR647]**

4. Expenditure in the acquisition of a concern would be capital expenditure; expenditure in carrying on the concern would be revenue expenditure. An expenditure which has to be incurred by an assessee in the course of its trade to enable the assessee to carry on its trading operation should usually be considered to be revenue expenditure when other necessary conditions are satisfied.....An expenditure cannot be considered to be capital expenditure merely on the ground that the amount involved is large. The quantum of expenditure involved cannot alter the nature and character of the expenditure.**[Kalyanji Mavji and Co.-CIT,Cal.,***Sec.37.***87ITR228]**

5. An expenditure is incurred while the business is going on and is not incurred either for extension of the business or for the substantial replacement of its equipment, the aim and object of the expenditure would determine whether it is a capital expenditure or a revenue expenditure.The source or manner of the payment would then be of no consequence. If the expenditure is so related to the carrying on or the conduct of the business that it might be regarded as an integral part of the profit making process, it should be held to be revenue expenditure. If the purpose is the acquisition of an asset or a right of a permanent character,the possession whereof is a condition precedent to the commencement or continuance of the business,the expenditure would be of a capital nature. **[Acc-Vickers Babcock Ltd.-CIT,Bom.,***Sec.37.***103ITR321]**

6. The question whether a particular payment made by an assessee under the terms of an agreement forms a part of capital expenditure or revenue expenditure,would depend upon several factors, namely, whether the assessee obtained a completely new plan with a complete new process and completely new technology for manufacture of the product or the payment was made for the technical know-how which was for the betterment of the product in question which was already being produced;whether the improvisation made is part and parcel of the existing business or a new business was set up with the so-called technical know-how for which payments were made; whether on expiry of the period of agreement the assessee is required to give back the plans and designs which were obtained, but the assessee could manufacture the product in the factory that has been set up with the collaboration of the foreign firm; the cumulative effect on a construction of the various terms and conditions of the agreement; whether the assessee derived benefits coming to its capital for which the payment was made. **[Jonas Woodhead and Sons (India) Ltd.-CIT,SC,**<u>*Sec.37.*</u>**AIR1997 SC1105,138 CTR283,224ITR342,137Taxation120,91Taxman1]**

7. If the amount spent was for the purpose of bringing into existence a new asset or obtaining a new advantage, then obviously such an expenditure would not be an expenditure of a revenue nature but it would be a capital expenditure. ... If an amount is spent for the purpose of bringing into existence a new asset or obtaining a new advantage,then such expenditure would be capital in nature. Another yardstick,which is adopted is to find out whether the expenditure is incurred for obtaining an advantage of enduring benefit. If it is so,then the expenditure should normally be treated as capital in nature. **[Bharat Gears Ltd.-CIT,Del.,**<u>*Sec.37.r.w.s.31.*</u>**337ITR 368,201Taxman86]**

Capital /revenue receipt All. Seth Madan Gopal Bagla – CIT

The principle on the basis of which the question has to be decided is whether the document or the transaction embodied in the document is a transfer of any rights and the amount paid is the price of those rights or the transaction is merely the right or the liberty to use the property for a term of years and the amount paid is the price of the use. If it is the former,the receipt is a capital receipt; if the latter, the receipt is a revenue receipt. <u>*Sec.28(iii).*</u>**[49ITR546]**

2. Generally payments made in the settlement of rights under a trading contract are trading receipts and are assessable to revenue. Where the circumstances of any case indicate that compensation is paid to a person for cancellation of a contract which does not affect the trading structure of his business nor deprive him of what,in substance ,is his source of income, termination of the contract being a normal incident of the business,and such cancellation leaves him free to carry on his trade, the receipt is revenue. Where, however, by the cancellation of an agreement, the trading structure of the assessee is impaired or such cancellation results in loss of what may be regarded as the source of the assessee's income, the payment made to compensate for cancellation of the agreement is normally a capital receipt. **[CIT-Balaji Chitra Mandir,AP., *Sec.28(iii).*45CTR206,154ITR777,76Taxation104]**

3. If the object of a subsidy scheme is to enable the assessee to run the business more profitably the receipt is on revenue account. On the other hand, if the object of the assistance under the subsidy scheme is to enable the assessee to set up a new unit or to expand the existing unit,the receipt of the subsidy is on capital account. It is the quality of the payment that is decisive of the character of the payment and not the method of the payment or its measure. **[CIT - Rasoi Ltd.Cal.*Misc.*335ITR438]**

Capital/trading receipt Bom. Bombay Burmah Trading Corporation Ltd. - CIT

Compensation received for immobilisation,sterilisation, destruction or loss, total or partial,of a capital asset would be capital receipt. Where compensation is recovered for an injury inflicted on a man's trading,so to speak,a hole in his profits,the compensation would go to fill the hole and would be a trading receipt.On the other hand, where the injury is inflicted on the capital assets of the trade making,so to speak,a hole in them,the compensation recovered is meant to be used to fill that hole and is a capital receipt. *Sec.28(iii).* **[81ITR777]**

Carrying on business All Vijay Laxmi Sugar Mills Ltd. - CIT

When the liquidator of a company is engaged in merely realising the assets of the company, he cannot be said to be carrying on business. If, however, for the purpose of facilitating the winding up, the liquidator carries on the company's business and realises the assets of the company in such a way that what he does bears the characteristics of a continuing trading activity, it is possible to infer that the business of the company is being carried on. *Sec.28.* **[86ITR402]**

Carrying on business in good faith All. Saraya Sugar Mills (P.) Ltd.-CIT

Allowable business expense or loss is one which is incurred in carrying on the trade with a view to earn profits, bona fide and in good faith. Carrying on business in an unlawful manner, involving breach or infraction of the law, is not carrying it on in good faith or bona fide.Any expenditure due to infraction of law by the assessee is not a deductible item. *Sec.28(1).***[7CTR329,116ITR387]**

Carrying on business in India **Cal.** **Imperial Chemical Industries Ltd. - CWT**

To determine whether a foreign company is carrying on business in India or not, it has to be ascertained initially if transactions are being had or entered into to which the company is a party. If the participation of the company in the transactions is direct,e.g., where contracts are executed in its name and through its own officers and employees, then there is no difficulty in holding that the company itself is carrying on business in India. The position may not be so clear where the company is connected with the transactions indirectly, e.g., through a third party in India. In such a case, it has to be determined further whether the transactions are those of the third party on his own account or whether he is acting for or on behalf of the foreign company as an agent. In the latter case again, the transactions would be those of the foreign company and not those of the agent and it would follow that the foreign company is carrying on the business. To come to the conclusion that business is being carried on in India the transactions must be found to have some connection with this country.Even if some connection is established it may not follow that business is being carried on in India where goods are imported into India in the course of international trade. The supplier abroad may not in such cases be held to be carrying on business in India. The tests enunciated in decided cases to determine the situs where the business or trade is being carried on are as follows: (a) Where the goods involved in the transaction are brought, stored, or located in India, further dealings with the same in India may indicate that business is being carried on in India. (b) Where transactions are had pursuant to contracts entered into by and between parties in India, the business resulting from such transactions would be held to be carried on in India. (c) Where payments involved in the transactions are made and received in India it would be relevant evidence to show that the business is being carried on in India. (d) Similarly,where negotiations leading to the transactions and forming a crucial part of the transaction take place within India, then again it would be a piece of evidence to hold that business is being carried on in India.

*Sec.2(h).WT.Act.***[119ITR46,2Taxman127]**

Carrying out any work **Del.** **S.R.F.Finance Ltd. - CBDT**

It is most inappropriate to equate the rendering of a service with carrying out a work. That is why Parliament thought it expedient to expand the meaning of the word work by including in it the supply of labour. It is obvious that because the word workwould not include within its amplitude the supply of labour, Parliament added the same by including the latter in the former, thereby giving the word work an extended meaning. The extended meaning cannot travel beyond the actual extended area; Parliament had stretched the scope of the word to some extent only. The word work may have different and wider meanings but one has to find out the real meaning of the word in the context of its setting in section 194C. The meaning attributable should fit into the clause for carrying out any work. An architect is not engaged to carry out the work of drawing a sketch. A lawyer is not engaged to carry out the work of arguing a case; he is engaged, to argue a case or to conduct a case; he is paid a fee for the services rendered by him and not any price for the work done by him. **[*Sec.194C.*122CTR431,211ITR861,123 Taxation458,76Taxman 432]**

2. The words carrying out any work in section 194C are limited to any work which on being carried out culminates in a product or result. **[EastIndiaHotelsLtd.Bom.,*Sec.194C.*320ITR526, 179Taxman17]**

Case **All. Shyam and Company – CIT**

1. A perusal of the Explanation to section 127 of the Income-tax Act,1961, makes it clear that the expression case used in section 127 of the Act comprises not only a pending proceeding but also proceedings which may have been completed on or before the date of transfer. It also includes all proceedings which may be commenced after the order of transfer. *Sec.127.***[192ITR387]**

2. In section 127 of the Income-tax Act, 1961, the word case is used in a comprehensive sense for pending proceedings as well as for proceedings to be instituted in future.Consequently, an order of transfer can be validly made even if there is no proceeding pending and the purpose of transfer in such an event will simply be that all future proceedings have to take place before the officer to whom the case of the assessee is transferred. **[United Breweries Ltd.-DCIT (Assessment)Kar.*Sec.127.*211ITR256]**

3. Case is defined as any proceeding under the Income-tax Act for, or in connection with the assessment or reassessment of any person in respect of any year or years which is pending before the Income-tax authority on the date on which the application is made under section 245C of the Act. **[Mohanlal S. Doppa-CIT,Guj.,** *Sec.245C.***172CTR1,253ITR33,121Taxman671]**

4. A bare perusal of section 127 of the Income-tax Act,1961, makes it clear that the Director General or the Chief Commissioner or the Commissioner, as the case may be, is empowered to transfer a case from one jurisdiction to another on complying with the principles of natural justice. The Explanation to section 127(4) of the Act says what the word case means in relation to any person whose name is specified in any order or direction issued under section 127 of the Act. The Explanation says that case means all proceedings under the Act in respect of any year : (i) which may be pending on the date of the order or direction; (ii) which may have been completed on or before the date of the order or direction; (iii) including all proceedings which may be commenced after the date of the order or direction in respect of any year. In other words, the Explanation to section 127(4) of the Act talks of proceedings, past, present and future in respect of a person whose name is specified in the order or direction passed under section 127 of the Act and this would apply to any previous year. **[CIT-Sahara India Financial Corporation Ltd.,Del.,***Sec.127.***212CTR 178, 294ITR363,162Taxman357]**

Cash **Guj.** **CWT - Shri Sadiqali Samsuddin**

The word cash,whether in hand or in the bank, would mean moneys which are readily available and not any money without any implication of ready availability.An amount belonging to an Indian national and lying in a frozen fixed deposit account in a bank in a foreign country and subject to temporary restraint prohibiting repatriation of such amount and permitting repatriation, if at all, of the balance amount only after certain deductions would not amount to cash so as to enable the wealth-tax authorities to take the face value thereof for purposes of computation of the wealth-tax liability of the assessee. *Sec.7.WT.Act.***[41CTR282,152ITR190]**

Cash/mercantile system **All.** **Bishambhar Nath Swaroop Narain-CIT**

According to the cash system, a record is kept of actual receipts and actual payments, entries being made only when money is actually collected or disbursed, and if the profits of the business are accounted for in this way, the tax is payable on the difference between the receipts and the disbursements for the period in question. In the case of mercantile system,net profit or net loss is calculated after taking into consideration all the income and all the expenditure relating to the period, whether such income has actually been received or whether such expenditure had actually been paid or not. Sec.28(ii).**[119ITR681]**

Casual **Cal.** **Asiatic Oxygen Limited - CIT**

The word casual means (1) subject to or produced by change; accidental; fortuitous (2) coming at uncertain times; not to be calculated on, unsettled. A receipt which is foreseen, known, anticipated and provided for by agreement cannot be regarded as casual even if it is not likely to recur ever or at least for a considerable time. *Sec.10(3).***[189ITR483]**

Casual visit **Mad.** **A.M.M. Sayed Abdul Cader - CIT**

The test to determine whether a visit is casual or not is to find out whether it was of an accidental or uncertain nature and whether the intention of the person visiting the place was for a temporary stay with the intention of reverting back to the place of his abode. It is easy to give instances of casual visits rather than to define in precise terms the essence and the nature of a casual visit.....The question whether a visit is a casual or an occasional visit within the meaning of section 4A(a)(iii) depends not on mere presence in British India but on the quality of the presence in relation to the objects and intentions of the person sought to be charged to tax in a particular year. *Sec.6.***[18ITR310]**

Casual and non recurring **All.** **WG.CDR.K.P.K. Ghose - CIT**

The words casual and non-recurring have not been defined in the Act and they must, therefore, receive their plain and ordinary meaning. In the Oxford Universal Dictionary,the word casual has been defined as meaning: (i) subject to or produced by chance; accidental, fortuitous; (ii) coming at uncertain times; not to be calculated on; unsettled. In the context of the statute,unsettled seems to be the aptest meaning to be applied in such cases. *Sec.10(3).* **[191CTR32,268ITR260,140Taxman437]**

Catching fish **Bom.** **CIT - Fazalbhoy Ibrahim and Co. P. Ltd.**

Catching fish does not amount to manufacture or production of fish, because when a person catches fish, he does not manufacture or produce fish.Therefore,an assessee who is engaged in the business of fishing with a trawler is not an industrial undertaking manufacturing or producing articles and is not entitled to relief under section 80J of the Income-tax Act,1961. This position will not be affected even if the fish caught by the assessee in the deep seas are subjected to some process because the benefit of section 80J is restricted only to undertakings which manufacture or produce articles. *Sec.80J.* **[126CTR242214ITR239]**

Cause of action Ker. Kurumber Betta Estate -ITO

1. ..... Cause of action has not been defined in the Constitution or in the Code of Civil Procedure. A liberal interpretation of the expression is necessary. Cause of action means a bundle of essential facts which it is necessary for the party seeking relief to prove, if denied by the opposite party, in order to secure the relief prayed for. If one of such essential facts required to be proved to secure an order has arisen within the territory of a court it will have jurisdiction to entertain a writ petition presented in such territory. *Constitution of India.* **[197ITR499]**

2. The cause of action is a bundle of facts which would give the plaintiff a right to relief. Any and every fact, though relevant, would not necessarily be included in the bundle of essential facts constituting cause of action. If a cause of action arises within the jurisdiction of a High Court,the writ issued by it can extend and run beyond its territorial jurisdiction.

[Raj Kumar Mangla - Chairman, CBDT *Constitution of India.*
[154CTR228,234ITR113,102 Taxman110]

Ceasing to be a partner Mad. Kaithari Lungi Stores - CIT

A change in the constitution of a firm may arise by admission, retirement, expulsion, insolvency or death of one or more partners subject to the conditions referred to in sections 31 to 35 of the Partnership Act. In the case of death of a partner, there should be a contract express or implied between the partners that the firm shall not be dissolved by the death of a partner. Under the ordinary law, every change in the constitution of a firm amounts to a dissolution of the old firm and bringing into existence of a new firm. In law the firm also has no legal existence apart from its members and it is merely a compendious name to describe a collection of persons who are partners. But the mercantile usage recognises the firm as a distinct person. The Partnership Act struck a remedial note as between these two extreme propositions and recognised the continued existence of the firm in spite of a change in its constitution. The Income-tax Act went a little further and recognised the firm for the purpose of assessment as a unit independent of the partners constituting it. A change in the constitution of a firm is different from the dissolution of the firm. If there is a contract to the contrary against dissolution of a firm by the death of a partner, a change in the constitution of the firm also occurs by reason of the death of a partner provided there are at least two surviving partners. The words ceasing to be partners in section 187(2) of the Income-tax Act,1961, would also include a case of death of a partner when such death by reason of the contract to the contrary or by reason of any law did not bring about the dissolution of the partnership. However, where there is no contract to the contrary against the dissolution of the firm by death of a partner, it cannot be stated that death will amount to a change in the constitution of the firm within the meaning of section 187(2). Sec.187(2).**[104ITR160]**

Cessation of liability Cal. CIT - Sugauli Sugar Works P. Ltd.

There can be a cessation of liability of a debt by the bilateral acts of both the creditor and the debtor or by the refusal of the debtor to honour his liability when pressed for the dues or by the discharge of the debt by making a payment of the dues. In no case can a debtor bring the liability to an end on his own volition. *Sec.41.***[23CTR 286,140ITR286,7Taxman163]**

2. The cessation of the liability may accrue either by operation of law, i.e., on the liability becoming unenforceable in law by the creditor, provided the debtor unequivocally declares his intention not to own the liability even if demanded by the creditor. It may also accrue by way of a judicial pronouncement, absolving the assessee of the liability. It may accrue if there is a contract between the parties whereby the liability gets extinguished or it may come to an end by discharge of the debt. Some benefit however must accrue to the assessee by virtue of remission or cessation of the liability, as the case may be. Mere change of nomenclature in the books of account without anything more brings no benefit to the assessee and its liability to pay to the creditor does not get extinguished,merely by change of nomenclature or by change of the sub-head under which the liability is shown in the account books of the assessee. **[CIT Auto Kashyap IndiaPvt.Ltd., Del.,***Sec.41.***233CTR 166,330ITR435,190Taxman469]**

Cessation or remission of the liability **Cal.** **CIT - Agarpara Co. Ltd.**

The question whether a trading liability that was once incurred ceases to exist for the purpose of section 41(1) has to be decided in the light of the provisions of the Income-tax Act,1961, and the statute,if any, governing such liability. The assessee who maintains his accounts on the mercantile basis would be entitled to a deduction in respect of bonus in the year in which a liability arises under the statute, or the employees' claim for bonus is admitted by the assessee or is settled by an agreement between the parties or is adjudicated upon by an award.Under section 36(1)(ii) of the Act, payment of bonus to the employees is an allowable deduction. Under the Payment of Bonus Act,1965, liability to pay bonus has become a statutory obligation imposed upon an employer covered by the said Act. Under the Bonus Act, bonus is payable within a period of eight months from the close of the accounting year unless there is a dispute regarding such payment, in which case it is payable within a month from the date of the award becoming enforceable. Once bonus has been offered by the employer, but remains undrawn, it cannot be said that the liability subsists even after expiry of the time prescribed by the statute, particularly when there is no dispute pending regarding the payment of bonus. In the context of such facts and circumstances, it may be inferred that unclaimed or unpaid bonus is in excess of the liability of the assessee and, therefore, to that extent, in any event, the liability has ceased. Even assuming that there can be a cessation only on bilateral act by both the creditor and debtor, such act may be inferred from the conduct of the debtor and creditor. It need not be a positive act or a positive conduct.It can be inferred from the surrounding circumstances that there has been a cessation or remission of the liability of the assessee. *Sec.41(1).* **[158ITR78]**

Change in a constitution of a firm **MP.** **Girdharilal Nannelal - CIT**

In clause (a) of sub-section (2) of section 187 of the Income-tax Act, 1961, a clear provision is made to cover the situation ordinarily that of dissolution under the general law within the meaning of the expression change in the constitution of a firm for the purpose of this section. Resort to the general law and the concept of dissolution and change in the constitution of a firm according to the general law contained in the Indian Partnership Act, 1932, is, therefore, not warranted. The words and the case is not one covered by section 187 in section 188 also indicate that a situation which would be ordinarily one of succession of firm under the general law, can also be within the ambit of section 187 and, therefore, a case of that kind is excluded from the ambit of section 188 which provides for cases of succession of one firm by another. The definition of the expression a change in the constitution of a firm as contained in section 187(2) is a new definition.Such a definition did not exist in the 1922 Act. However, what was merely implicit in section 26(1) of the 1922 Act has been made explicit by the definition. There is no other significant difference between these corresponding provisions in the old and new Acts. Plainly construed,section 187(2) indicates that if one or more of the partners of the old firm continue to be partners in the new firm, it is a case of change as defined in section 187(2).If a firm is dissolved and succeeded by another firm which has, as its partners, one or more partners of the original firm, the case will be one covered by section 187, as it would be merely a change in the constitution of the firm as defined in sub-section (2) thereof. Sec.187(2).**147ITR529]**

Charge **Cal.** **CIT - State Bank of India**

The word charge in section 9(1)(iv) means payment and not mere security. What section 9(1)(iv) provides for is a deduction of certain sums out of the assessee's income, after the income has become his. The broad ground that where there is only an application of the income, there can be no claim to exemption from tax or deduction in the computation of the income is, therefore, not available as an argument against a literal construction of section 9. Sec.24. **[31ITR545]**

Charged **SC** **CIT - Kamalini Khatau**

Section 164 of the Income-tax Act,1961, does not create a charge on the income of a discretionary trust.The word charged in the context in which it is used in section 164 means only levied. Sec.164.**[119 CTR169,209ITR101,120Taxation422,74Taxman392]**

Charitable **All.** **Chamber of Commerce-CIT**

Before an institution can be held to be charitable there must be an element of altruism, that is to say, the beneficiaries must not be able to claim the benefit. Sec.2(15).**[4ITR 397]**

2. It cannot be said that a purpose would cease to be charitable even if public welfare is intended to be served. If the primary purpose and the predominant object are to promote the welfare of the general public the purpose would be a charitable purpose. **[CIT-Gujarat Maritime Board,SC,**<u>*Sec.2(15).*</u>**295ITR 561]**

Charitable object of general public utility,	**Mad.**	**CIT - Madras Stock Exchange Ltd.**

If the profit-making activity is the appointed means of achieving a charitable object of general public utility, the profit would be taxable. One cannot carry on a business and claim exemption on the income there from by merely saying that it is for a charitable purpose.There is a distinction between a business being held under trust whose profits feed a charity in which case the income is clearly exempt and the carrying on of a business in carrying out what is conceived as charitable purpose in which case the income may be taxable.The distinction is somewhat fine but it has to be kept in mind.The proposition that one must run the activity on a no profit no loss basis is applicable to a case where in the course of carrying out a charitable purpose there is an activity. In such a case if the aim was not to render the service on no profit no loss basis, the profit would be taxable.The test that the profit from a business which is taxable if carried on by a businessman does not cease to be taxable by a charity indulging in it is also applicable primarily to a case where the profit-making activity is embarked upon as the appointed means of achieving the purpose of the trust. <u>*Sec.2(15). r.w.s.11.*</u>**[105ITR546]**

Charitable organization	**Bom.**	**DIT(Exem.) Chembur Gymkhana**

The fact that the assessee provided service to its members did not detract from the position that it advanced a general public utility. The advancement of any object of benefit to the public or a section of the public as distinguished from a benefit to an individual or a group of individuals would be a charitable purpose. ....The assessee did not exist only for an individual or a group of individuals and fulfilled the definition of the expression charitable organization in section 2(15). <u>*Sec. 2(15),11.*</u>**[251CTR145,346ITR86]**

Charitable purpose	**CAL**	**CIT-Indian Chamber of Commerce**

The proper interpretation of the definition of charitable purpose in section 2(15) of the Income-tax Act, 1961, is to consider the expression not involving the carrying on of any activity for profits as qualifying the expression the advancement of any other object of general public utility and not the other classes of charitable purpose mentioned in that section like relief of the poor,education and medical relief. In other words, the advancement of any other object of general public utility would be a charitable purpose provided that its advancement does not involve the carrying on of any activity for profit. Parliament has thought it necessary to impose certain restrictions on the area of the object of general public utility and the area selected is that its advancement must not involve the carrying on of any activity for profit. *Sec.2(15)r.w.s.11.***81ITR417**

2. The last clause in the definition of charitable purpose in section 2(15) of the Income-tax Act,1961, namely, not involving the carrying on of any activity for profit qualifies only the object immediately preceding it, that is, any other object of general public utility. The first three objects in section 2(15), namely, relief of the poor, education and medical relief are by themselves considered as charitable purposes and there was no need to qualify them. Because Parliament was introducing into the definition of charitable purpose a new category of a general nature, namely,advancement of any other object of general public utility, it added the qualifying words appearing in the last clause to prevent adventures of a purely commercial nature claiming the benefit of exemption under the Act. **[Addl.CIT-Aroor Brothers Charitable Trust,Kar.,***Sec.2(15).***7_CTR 99,115 ITR418]**

3. Charitable purpose in section 2(15) of the Income-tax Act, 1961 includes relief to the poor, education, medical relief and the advancement of any other object of general public utility. The words public utility or general public utility are not capable of a precise meaning. The question whether service is public utility or not has to be determined in the context of different situations but it is well-settled that public utility means public purpose depending upon the context in which it is used in the statute or the Rules. Indeed, in some decisions, public utility is considered very similar public purpose. **CIT- Agricultural Market Committee, AP,** *Sec.2(15).* **[336ITR641]**

Cheritable purposes **PC** **All India Spinners' Association-CIT**

Section 4(3) of the Indian Income-tax Act gives a clear and succinct definition of charitable purposes which must be construed according to its actual language and meaning. Lord Macnaghten's definition of charity and English decisions on the law of charities have no binding authority on its construction and though they sometimes afford help or guidance, cannot relieve the Indian Courts from their responsibility of applying the language of the Act to the particular circumstances that emerge under conditions of Indian life.The difference in language, particularly the inclusion in the Indian Act of the word publicis of importance. *Sec.2(15).***12ITR482**

2. The words charitable purposes defined in section 2(15) of the Income-tax Act, 1961, include the advancement of any other object of general public utility. Prior to April 1, 1984, the words used in the section were advancement of any other object of general public utility not involving the carrying on of any activity for profit. The Finance Act, 1983, omitted the words not involving the carrying on of any activity for profit from the section with effect from April 1,1984. Thus, after April 1, 1984, even if there is some profit in the activity carried on by the trust/institution, so long as the dominant object is of general public utility, it cannot be said that the trust/ institution is not established for charitable purposes. **[CIT- Krishi Utpadan Mandi Samiti, All.** *Sec.2(15).***331ITR154]**

Charity **Bom.** **Chaturbhuj Vallabhdas - CIT**

The word charity if used generally or without qualifications or limitations denotes public charity and falls within the definition of charitable purpose in section 4(3) of the Indian Income-tax Act. *Sec.2(15).***[14ITR144]**

Chemical works **All.** **CIT - Babu Ram Ramesh Chand**

A higher rate of depreciation is allowable in the case of plant and machinery installed in a chemical works. The machinery and plant upon which the depreciation is claimed must be installed in a unit or plant which can be called chemical works. Merely because a certain chemical process is gone through, it cannot be said that it is a chemical works. For that matter, a chemical process is probably gone through in most of the manufacturing processes. To wit,even in a steel plant, some chemical process may be involved but on that account a steel plant cannot be called a chemical works. If the product is a chemical,the plant or factory producing the same can be called a chemical works. *Sec.32.r.w.r.5.***[95CTR232,190ITR535, 54Taxman348]**

Child **Gau.** **CIT - Saraswati Devi Singh**

Section 2(15A) of the Income-tax Act,1961, defined the term child as including a step-child. The object of the Explanation to section 64(1) is to determine in whose income, whether of the father or of the mother, the income of the child has to be included. This Explanation does not nullify the definition of child in section 2(15A). The step-child also would come within the purview of section 64(1)(iii) of the Act in view of the clear definition of child in section 2(15A). *Sec.2(1A).*<u></u>**[195ITR185]**

Chit fund scheme Mad. Bilahari Investments P. Ltd. - CIT

A chit fund scheme is a kind of savings scheme in which a specific number of individuals come together to pool a specific amount at periodic interval section.Usually, the number of individuals and the number of periods is the same.At the end of each period, there is an auction of the chit. The members of the chit participate in this auction for the pooled money during that interval. The bid amount is divided by the number of members thus determining the contribution per head during that period.When the chit is auctioned every month, the bidder takes the chit for an amount which is less than the face value of the chit.The difference between the face value and the auctioned value during every period is the gross dividend generated in that period.This amount of dividend gets distributed among all the members (subscribers) equally. The members (subscribers) need not pay the total monthly subscription and instead, they pay the monthly subscription after deducting the amount of dividend earned. Members who have bid for the chit in auction have the liability to keep the contribution to the chit till the end of the chit period and the prized members get dividend in future months also.Usually the discount, namely, the sum of money, which the prized subscriber is required to forgo, decreases over periods. The person getting money in the last period receives the full scheme amount. In the case of a debenture, it is a receipt, while in the case of a chit transaction, it is an amount paid followed by the encashment of the chit amount at an early date with the promise to pay the remaining installments. *Sec.5.r.w.s.145.* **[288ITR39]**

Circular Ker. CIT - Kerala Financial Corporation Ltd.

Unless copies of the circulars are addressed to the other authorities mentioned in section 116, the letter or document which is claimed to be a circular cannot be treated as such. The burden of proving that a particular document or letter is a circular is on the party who raises such a claim. *Sec.116.***[47TR297,155 ITR246]**

Clarificatory Bom Apar Industries Ltd.

An amendment which is intended to remove an ambiguity in the interpretation of a section must of necessity be regarded as clarificatory.If it is clarificatory in nature,it is expressive of a position in law which Parliament intended to hold the field at all material times and must consequently be regarded as operating with retrospective effect. *Sec.115JAA,234B.***[231CTR313,323ITR 411,190Taxman353]**

Clarificatory or declaratory AAR Rural Electrification Corporation Ltd.

It is well settled that an amending provision is regarded as clarificatory or declaratory when it is introduced to clear the doubts or ambiguity as regards its meaning in order to avoid unintended consequences. Moreover, in the absence of clear words indicating that the amendment was clarificatory, it would not be so construed when the pre-amended provision was clear and unambiguous. *Misc.***[223CTR105,312ITR122,180Taxman55]**

Class of income Bom. CIT- B.B.& C.I. Railway Co- op. Mutual Death Benefit Society

A class of income really means a category of income and it is a much wider expression than a head of income. *Sec.2(24).***[17ITR509]**

Closure A.P. Sadula Janardhan (HUF) - State Bank of Hyderabad

The word closure connotes an act or process of closing something.

Sec.54. **[206CTR117,286ITR291]**

Colourable device Del. Bhagat Construction Co. (P.) Ltd. - CIT

A colourable device is a colourable transaction which is seemingly valid,but a feigned or counterfeit transaction entered into for some ulterior purpose. *Misc.***[165CTR181,250ITR 291,114Taxman606]**

Commencement of business AP. CIT - Sponge Iron India Ltd.

The following are the principles applicable to determine whether a business has commenced:(i)whether a business has been commenced or not is a question of fact.However, what activities constitute commencement of business is a mixed question of law and fact and it has to be decided on the facts of each case; (ii) there is a distinction between setting up of business and commencement of business. A business is said to be set up when it is ready to commence; (iii) where the business consists of a continuous course of activities, for commencement of business all the activities which go to make up the business need not be started simultaneously. As soon as an activity which is the essential activity in the course of carrying on the business is started, the business must be said to have commenced. *Sec.37.*[**201ITR770]**

Commercial asset Mad. CIT-B.Nagi Reddy

1. There is no such thing as a natural commercial asset,because an asset becomes a commercial asset by the use to which it is put in business and not because of any inherent qualities.

[*Sec.22.***180 ITR 457,47Taxman95]**

2. A commercial asset does not cease to be a commercial asset by reason merely of an intervening lull in the business during which period the asset is indirectly exploited commercially by leasing it to a third party in the trade. **[CIT-Universal PlastLtd.Cal.** *Sec.28.,***197ITR1]**

3. A thing is not by its very nature a commercial asset.A commercial asset is only an asset used in a business and nothing else, and business may be carried on with practically all things. **[Sultan Brothers Pvt. Ltd.-CIT,SC,***Sec.56.* **51 ITR353]**

Commercial expediency MP. Addl.CIT - Kuber Singh Bhagwandas

The expression commercial expediency is not limited to an existing practice prevailing in any particular trade or business. Even if the incurring of a particular expenditure may not be supported by any prevailing practice, yet, if, at the time when the expenditure is incurred, commercial expediency justifies it, the expenditure would be taken to be for the purposes of the business. *Sec.37(1).* **[9CTR94,118ITR379]**

2. The expression commercial expediency is one of wide import and includes such expenditure as a prudent businessman incurs for the purpose of business. The expenditure may not have been incurred under any legal obligation,but yet it is allowable as business expenditure if it was incurred on grounds of commercial expediency.

[S. A. Builders Ltd. - CIT (Appeals), SC, *Sec.37(1).***[206CTR631,288ITR1,158Taxman74]**

3. The commercial expediency would include such purpose as is expected by the assessee to advance its business interest and may include measures taken for preservation, protection or advancement of its business interests. **[Punjab Stainless Steel Inds.- CIT,Del.,***Sec.37(1).***[324ITR396]**

Commercial vehicles Mad. Sundaram Industries Ltd. - CIT

The term commercial vehicles used in the Ninth Schedule is not to be understood as being applicable only to a commercial vehicle, which is complete in all respects. Any such construction would render the manufacturer of a chassis ineligible to be regarded as a manufacturer of commercial vehicles. Without the chassis there can be no commercial vehicle. It is the essential base and it is that chassis with the other things included therein, viz., the engine, the wheels, the steering mechanism, etc.,that provides the mobility required in order to qualify it for being regarded as vehicle.The body built thereon is essential in order to make the vehicle usable for the purpose for which the vehicle is intended to be used as a commercial vehicle. *Sec.32A.Sch.IX.***[176CTR229, 258ITR38,131 Taxman93]**

Commission Bom. Harihar Cotton Pressing Factory -CIT

The expression commission has no technical meaning but both in legal and commercial acceptation of the firm it has definite signification and is understood as an allowance for service or labour in discharging certain duties such as for instance of an agent, factor, broker or any other person who manages the affairs or undertakes to do some work or renders some service to another. Mostly it is a percentage on price or value or upon the amount of money involved in any transaction of sale or service or the quantum of work involved in a transaction. It can be for a variety of services and is of the nature of recompense or reward for such services.

<u>*Sec.37.*</u>**[39ITR594]**

Commission or brokerage **Ker.** **Kerala State Stamp Vendors Association - Office of A.G.**

It is clear from section 194H of the Income-tax Act, 1961, that what is subject to deduction of Income-tax at source is only commission or brokerage. It is clear from the definition of commission or brokerage as contained in the Explanation to section 194H that the commission or brokerage that attracts TDS is the one paid for services rendered in the course of sale which obviously can be services rendered by a third party like a broker or an agent and cannot be by the buyer as the buyer is not rendering any service except buying.A discount given on price by the seller to the purchaser cannot be termed as commission or brokerage for services rendered in the course of buying and selling of goods as the act of buying does not constitute rendering of any service. Wherever the Legislature wanted to levy tax on trade discount, the legislature specifically provided for the same which is clear from the provisions of section 194H of the Act which provides for deduction of tax on discount paid to the lottery dealers in the form of commission.

<u>*Sec.194H.*</u> **[202CTR231,282ITR7,152Taxman398]**

2. Commission or brokerage defined under Explanation (i) to section 194H of the Income-tax Act,1961, has a wide meaning and it covers any payment received or receivable directly or indirectly by a person acting on behalf of another person for services rendered. It is clear from section 194H that payment includes credit of such sum to the account of the payee or at the time of payment of such income in cash or by the issue of cheque or draft or by any other mode. **[CIT-Director,Prasar Bharti. Ker.** <u>*Sec.194H.*</u>**230 CTR 277, 325 ITR 205, 189 Taxman315]**

Commodity **Bom.** **CIT - Bharat R. Ruia (HUF)**

Since the expression commodity is not defined under the Act the expression commodity in section 43(5) has to be given meaning as understood in common parlance.The expression commodity means an article of trade or commerce which is tangible in nature. .. Futures contracts being articles of trade and commerce which are legally permitted to be traded on the stock exchange, the transactions in futures would be transactions in a commodity as contemplated under section 43(5)........ The expression commodity would cover all articles of trade including stocks and shares. Even under section 43(5), the expression commodity is not expanded to include stocks and shares. In fact, the use of a comma between the word commodity and the words including stocks and shares in section 43(5) make it clear that transactions for purchase of any commodity would include transaction for purchase or sale of stocks and shares. *Sec43(5).*[**241CTR1,337ITR452,199 Taxman87]**

Commutation of the value of an annuity **Bom.** **CWT- Hirji Cowasji Jehangir**

The commutation of the value of an annuity necessarily implies that the annuitant gives up his right to receive the annuity. Commutation is a bilateral transaction in which the grantee of the annuity gives up his right to receive a sum annually in return for a lump sum and the grantor gets rid of his recurring liability to pay an annuity annually. The question whether in the case of an annuity,the value thereof could not be included in the net wealth of an assessee under section 2(e)(1)(iv) of the Wealth-tax Act,1957, does not necessarily get answered in favour of the revenue merely because the terms and conditions of the grant of the annuity do not contain any positive provision which prevents a commutation of the annuity into a lump sum. It is when there is no express provision which prevents commutation of annuity into a lump sum grant that a further enquiry becomes necessary in that direction and such an enquiry must necessarily be directed at finding out whether the preclusion of such a right can be inferred from the terms and conditions of the grant. *Sec.2(e)(1).WT.Act.*[**17CTR36,129ITR642]**

Company carrying on an industrial undertaking in India, **Pat.** **Harsco Corporation - CWT**

The phrase any company established with the object of carrying on an industrial undertaking in India etc.,in section 45(d) of the Wealth-tax Act, 1957, must mean in the context the formation and registration of a company or its incorporation either in India or outside in any foreign country. It cannot mean the establishment of a place of business in India. *Sec.45(d).WT.Act.*[**60ITR715]**

Company dealing in or holding investments **SC** **Nawn Estates (P.) Ltd. - CIT**

The expression company whose business consists wholly or mainly in the dealing in or holding of investments in section 23A and Explanation 2(i) thereto of the Indian Income-tax Act,1922, is not limited to companies whose principal business is the acquisition and holding of shares, debentures, stocks or other securities,but covers companies whose primary or principal source of income is house property or capital gains as well. *Sec.104,109.*<u></u>**[106 ITR45]**

Company whose business consists wholly .. holding **Mad.** **CIT - Emcete and Sons P. Ltd.**

The expression company whose business consists wholly or mainly in the dealing in or holding of investments consists of two parts, viz., (1) a company whose business consists wholly or mainly in the dealing in investments, and (2) a company whose business consists wholly or mainly in the holding of investments.*Sec.104.***[215ITR817]**

Compensation **Del.** **CIT - D.R.Sondhi**

The primary significance of the word compensation is equivalence and the secondary or more common meaning is something given or obtained as an equivalent. The large number of ways in which this expression compensation has been interpreted has one common factor running through them all, that is, compensation is regarded as an equivalent or recompense which makes good the lack of variation of something else. Flowing from the concept enlargement of the meaning of this expression takes in that which compensates for loss or provision, amends, remunerates or recompenses. *Sec.28 (ii)(a).***[164CTR560,248ITR695,114Taxman259]**

2. ...The expression compensation is a well known term.It means a payment to the holder of an office as compensation for being deprived of profits to which as between himself and his employer he would, but for an act of deprivation by his employer or some third party such as the Legislature, have been entitled.

[CIT - P. Surendra Prabhu, Kar., *Sec.10.r.w.s.17.*
[198CTR209,279ITR402,149Taxman82]

Compensatory Kar. Dr. S. Reddappa -Union of India

In the fiscal statute,the word compensatory would mean to make good the loss suffered by the Revenue on account of acts of commission and omission attributable to the assessee. Creating circumstances for compelling the assessee to discharge his statutory obligation cannot be termed to be a penalty. The collection of tax being an act of the State for providing protection, security and other amenities to the society, cannot, in all circumstances, be termed to be either a penalty or a punishment. The failure on the part of the assessee to abide by the provisions of the Income-tax Act has been made a basis for forcing him to compensate society by paying interest in terms of sections 234A, 234B and 234C of the Act. The amount on which interest is levied is an amount which can legitimately be said to be public revenue though payable by the assessee, but not paid by him, despite his knowledge of the position of law. Levy of interest on such amount which is utilised by the assessee for his own purposes has rightly been directed to be compensated by means of directing him to pay the interest at the rates specified under the sections. The Legislature in its wisdom thought it appropriate to replace penal provisions by incorporating compensatory provisions. It is not possible to hold that the provis-ions of sections 234A, 234B and 234C are provisions of a penal nature simply because in actual application of these provisions there may be situations where an assessee may render himself liable to payment of interest under each one of these provisions simultaneously for the same period nor can the compensatory nature of the provisions be deemed to have been lost simply because in a given situation the provisions may on account of their simult-aneous application to an assessee raise the liability to pay interest for the overlapping period to a rate higher than 2% per month. So long as the basic character of the levy remains compensatory, the rate of interest which is levied either by the provision itself or on account of its dual effect in a given situation will be wholly immaterial. The argument that the charge of interest for a period in excess of the period for which the assessee withheld the amount of tax payable by him amounted to penalty,is also without any substance. Such a provision appears to have been made with a view to facilitate the computation of liability of the assessee, without imposing any penalty upon him. The safeguards provided in the statute itself unambiguously show that the sections cannot be termed to be penal in character and thereby unconstitutional on the ground of violating the principle of audi alteram partem *Sec.234.* **[149CTR521,232ITR62]**

Compensatory tax SC Jindal Stainless Ltd. - State of Haryana

From the point of view of the Government, a compensatory tax is a charge for offering trading facilities. It adds to the value of trade and commerce which does not happen in the case of a tax as such.A tax may be progressive or proportional to income, property, expenditure or any other test of ability or capacity (principle of ability). Taxes may be progressive rather than proportional. Compensatory taxes, like fees, are always proportional to benefits. They are based on the principle of equivalence. However, a compensatory tax is levied on an individual as a member of a class, whereas a fee is levied on an individual as such. The theory of compensatory tax is that it rests upon the principle that if the Government by some positive action confers upon individual(s), a particular measurable advantage, it is only fair to the community at large that the beneficiary shall pay for it. For a tax to be compensatory, there must be some link between the quantum of tax and the facility/ services. Every benefit is measured in terms of cost which has to be reimbursed by compensatory tax or in the form of compensatory tax. In other words, compensatory tax is a recompense/ reimbursement. In the context of article 301, therefore, compensatory tax is a compulsory contribution levied broadly in proportion to the special benefits derived to defray the costs of regulation or to meet the outlay incurred for some special advantage to trade,commerce and intercourse. Constitution of India.**[283ITR1]**

Complete Del. Chhotey Lal Bharany - CIT

......The word complete in the context meant free from deficiency, entire or perfect. To attract the provision contained in section 145, the return had to be correct and complete,which meant that the disclosure made therein must be full. Sec.145.**[46CTR317, 161ITR552]**

Completed P &H. CIT - Smt. Santosh Mahey

The word completed has to be given its full meaning in a case where in appeal, the assessment order, during the course of which the penalty proceedings are initiated, is set aside and the case is remanded for fresh assessment. The assessment proceedings cannot be held to be completed till such time the order is passed in the remand proceedings. *Sec.275(a)(i).***[210CTR,259,293ITR573, 161Taxman340]**

Completed assessment Cal. Hansraj Dhingra -Union of India

Completed assessment in section 155(1) means a positive act of completion which should be clearly distinguished from lapse of authority to assess on the expiry of the period of limitation prescribed. It cannot be held that on the lapse of the authority to assess on the expiry of the period of limitation, the provisional assessment became final and complete in this case. *Sec.155(1).* **[98ITR397]**

2. Section 155 of the Income-tax Act, 1961, gives a limited jurisdiction to amend an order of assessment with a view to include the proper share in the assessment or correction thereof as may be found necessary as a consequence of the assessment or reassessment of the firm or any reduction or enhancement of the income thereof.The words completed assessment employed in section 155(1) mean a positive act of completion. **[CIT - S.Subramaniam, Mad.** *Sec.155.***291 ITR63]**

Complex MP. Sewaram Takhtani - ACIT

It is the subjective satisfaction of the authority concerned to decide on the basis of the material on record,as to whether the accounts maintained by the assessee are complex in nature or not. The word complex is not defined in the Act and hence it has to be given a wide and liberal meaning. *Sec.142(2A).***[188CTR441,270ITR509, 138Taxman290]**

Complexity SC Rajesh Kumar - DCIT

The expression complexity in section 142(2A) would mean the state or quality of being intricate or complex or that it is difficult to understand. Difficulty in understanding would, however, not lead to the conclusion that the accounts are complex in nature. The formation of opinion under section 142(2A) that the accounts of the assessee require an expert audit should indisputably be based on objective considerations. No order can be passed on whims or caprice. All that is difficult to understand should not be regarded as complex.What is complex to one may appear simple to another. It depends upon one's level of understanding or comprehension. Sometimes, what appears to be complex on the face of it, may not be really so if one tries to understand it carefully. Therefore, special audit should not be directed on a cursory look at the accounts. There should be an honest attempt to understand the accounts of the assessee. <u>*Sec.142(2A).*</u>**[206CTR175,287ITR91,157Taxman168]**

Computation Cal. CIT - Ashim Krishna Mondal

The word computation has a narrower meaning than the word assessment. It is also completely distinct and different from estimate. In the context of the Income-tax Act, computation is a calculation, a method of determination by reckoning through calculation. It involves some methodical process with some amount of approximate mathematical precision based on the calculable data available. <u>*Sec.158BB.*</u>**[192CTR336,270ITR160,144Taxman365]**

Computer Gau. CIT - Technotive Eastern (Pvt.) Ltd.

Computer means one that computes; specifically a programmable electronic device that can store, retrieve, and process data. <u>*Sec.80HH, 80-I.*</u>**[176CTR422,255ITR253,124Taxman769]**

Concealment All. Mohammad Ibrahim Azimulla

The dictionary meaning of the word concealment is to hide, to keep secret. The Explanation to section 271(1)(c) does not alter or extend this meaning. It only assumes the concealment to exist if the assessee fails to prove that the non disclosure was not due to fraud or wilful conduct.Between fraud and innocence, carefulness and gross or wilful conduct, there may be numerous stages. One may not be innocent yet he may not be a fraud. An action may not be careful, but for that reason only, it cannot be considered gross or wilful. The assessee may lead evidence in penalty proceedings or may rely on circumstances on record to show that the non-disclosure was not due to fraud or wilful conduct, and the Tribunal may find that the assessee was not careful. If non-disclosure is due to not being careful, it is not due to fraud or wilful conduct and the assessee cannot be said to be guilty of concealment. Even after the rejection of his explanation the authorities may find that the conduct of the assessee was not fraudulent or wilful. Disclosure under threat or pressure implies knowledge, and cannot be equated with disclosure which, with more alertness, an assessee could have made. The one assumes knowledge,the other negatives it. Proof that the disclosure was not due to care would establish that it was not due to fraud or gross or wilful conduct. <u>*Sec.271(1)(c).*</u>**[131ITR 680]**

2. Concealment can be said to be in law the intentional suppression of truth or fact known to the injury or prejudice of another. Where, in order to make good an omission in the originally filed return, the assessee voluntarily furnishes a revised return inclusive of the income so omitted, a question arises whether the filing of a revised return will not expatiate the contumacious conduct, if any, on the part of the assessee in not having disclosed the true income in the originally filed return. Blame worthiness attached to the assessee with reference to the original return cannot be avoided by filing a fresh return after concealment was detected by the assessing authority. .... that a return in response to a notice under section 148 was not to be treated at par with or compared to a revised return. That being the position, the filing of the return including the agreed concealed income did not constitute a mitigating circumstance and penalty had been rightly levied. **[P.C. Joseph and Brothers-CIT,Ker.,** *Sec.271.(1) (c).***[158CTR104,243ITR 818,108Taxman253]**

3. Concealment inherently carries with it the element of mens rea. The fact that some figure or some particulars have been disclosed, even if it takes out the case from non-disclosure, would not by itself take the case out of the purview of furnishing inaccurate particulars. Mere omission from the return of an item of receipt amounts neither to concealment nor to deliberate furnishing of inaccurate particulars of income,unless and until there is some evidence to show or circumstances are found from which it can be gathered that the omission was attributable to an intention or desire on the part of the assessee to hide or conceal the income so as to avoid imposition of tax thereon. **[K.C.Builders-ACIT,SC,** *Sec.2717(1)(c).***186CTR721,265ITR562,135Taxman461]**

4. Concealment means an attempt to hide an item of income or a portion thereof from the knowledge of the income-tax authorities. **[CIT-Mahabit Prasad Bajaj,Jhar.,***Sec.271.***298ITR109]**

Concealment of income SC Dilip N. Shroff - JCIT

Concealment of income and furnishing inaccurate particulars are different.Both concealment and furnishing of inaccurate particulars refer to deliberate acts on the part of the assessee. A mere omission or negligence would not constitute a deliberate act of suppressio veri or suggestio falsi. *Sec.271.***[291ITR519,161Taxman218]**

Concern Bom. Dr. J.M.Mokashi - CIT

...the expression concern appearing in section 64(1)(ii) is a word of wide import and takes within its sweep and ambit all organisations or establishments engaged in business or profession, whether owned by a company, partnership or individual or any other entity. Sec.64(1).**[115CTR73,207ITR252,72Taxman98]**

Concession SC Arun Kumar - Union of India

Section 17(2)(ii) declares that the value of any concession in the matter of rent respecting any accommodation provided to the employee by his employer would be a perquisite. Nevertheless, it must be a concession in the matter of rent respecting any accommodation provided by the employer to his employee. It is, therefore, clear that before section 17(2)(ii) can be invoked or pressed into service and before calculation of concession as per rule 3, even after its amendment in, is made, the authority exercising power must come to a positive conclusion that it is a concession. The concession is, thus, a foundational, fundamental or jurisdictional fact. Sec.17(2).**[205CTR193,286ITR89,155Taxman659]**

Conclusive proof SC P. R. Metrani - CIT

Conclusive proof gives an artificial probative effect by the law to certain facts. No evidence is allowed to be produced with a view to combatting that effect. In this sense, this is an irrebuttable presumption. Misc.**[203CTR290,287ITR209,155Taxman186]**

Consideration Ker. CGT - C.K.Nirmala,

1. One of the essential ingredients constituting a gift under section 2(xii) of the Gift-tax Act, 1958, is that the transfer of property by one person to another must be without consideration in money or money's worth. However, the word consideration is not defined in the Act and, therefore, it must carry the meaning assigned to it in section 2(d) of the Indian Contract Act, 1872. Section 2(d) of the Indian Contract Act provides that when at the desire of the promisor, the promisee or any other person had done or abstained from doing, or does or abstains from doing, or promises to do or to abstain from doing,something, such act or abstinence or promise is called a consideration for the promise. Under the Contract Act, the adequacy or inadequacy of the consideration is immaterial but under the Gift-tax Act an agreement to transfer property otherwise than for adequate consideration gives rise to a gift to the extent of the inadequacy. There is nothing to show in the definition of the term consideration that the benefit of any act or abstinence must enure directly to the promisor. *Sec.2(xii)G.T.Act.* **[215ITR165]**

2. The term consideration has not been defined under the Gift-tax Act. The term would carry the meaning assigned to it in section 2(d) of the Indian Contract Act, 1872. Consideration is that which creates a contractual relationship between the promisor and promisee in regard to the performance of promise and in regard to which the parties to the agreement or contract get related to each other. Consideration may be relating to a party other than the promisor and promise. **[CGT-Smt. K.Nagammal, Ker.** *Indian Contract Act.* **135CTR17, 226 ITR598]**

3. The word consideration in section 35AB of the Income-tax Act,1961, is to be understood in the sense in which it has been used in the Indian Contract Act, 1872.Therefore the word consideration would include the entire obligation of the assessee, without which the assessee would not be able to acquire the know-how. **[Tata Yodogawa Ltd.-CIT,Jhar.***Sec.35AB.***196CTR353,335ITR53]**

Constituted under an instrument **Pun.** **Padam Parshad Rattan Chand-CIT**

The words constituted under an instrument in section 26A of the Indian Income-tax Act, 1922, mean created or formed by a formal deed. But the fact that the partners of a firm who jointly execute a partnership deed choose to allege therein that, under an oral agreement, they have previously been partners for some time on the same terms as those embodied in the deed does not debar the firm from registration under section 26A. The deed cannot have retrospective effect as regards the Income-tax assessment of the firm but there is no objection to the firm being treated as constituted under the instrument as from the date of the instrument itself. *Sec.187.* **[25ITR335]**

Constituted under an instrument of partnership **SC** **R.C.Mitter &Sons - CIT**

The words constituted under an instrument of partnership in section 26A of the Income-tax Act include not only firms which have been created by an instrument of partnership but also those which may have been created by word of mouth but have been subsequently clothed in legal form by reducing the terms and conditions of the partnership to writing. Firms which were created by word of mouth but the constitution of which has subsequently been reduced to writing can also, therefore, be registered under section 26A.... . It is however necessary that the instrument of partnership should have been in existence in the accounting year in respect of which assessment is being made. <u>*Sec.187.*</u>**[36ITR194]**

Contingent/accrued liability **Mad.** **Associated Printers (Madras) Private Ltd. - CIT**

A claim by workmen for bonus, to whatever period it relates, is at best a contingent liability of the employer at the stage when the claim is preferred. It becomes an accrued liability if the claim is admitted by the employer. If the claim is denied and the workmen do not pursue the claim it will never accrue as a liability. If the claim is denied by the employer and it is referred as an industrial dispute, no liability accrues, if the industrial Tribunal negatives it. If, however, the claim is upheld by the industrial Tribunal after adjudication, it becomes an accrued liability when the award becomes enforceable. If the claim for bonus is settled by agreement between the employer and the workmen, it becomes an accrued liability on the date of the agreement. <u>*Sec.37.*</u>**[43ITR281]**

Contingent liabilities **Bom.** **India United Mills Ltd. - CIT**

The gratuity is payable when a workmen dies, retires, resigns or is removed from service. These events no doubt take place in the future but they cannot be said to be uncertain. The services of every workman are bound to come to an end on account of one or other of the causes stated above. Under the award, every employer is bound to pay gratuity to his workman for his past and future services. In the circumstances,every businessman would make provision every year for his liability under the award. Under the mercantile system of accounting, an expenditure is admissible not when it is actually paid but when the liability for the expenditure is incurred. It is legitimate in such a scheme of gratuity to estimate the liability by actuarial valuation and deduct such estimated liability in the profit and loss account while working out the net profits. While working out the net profits, a trader can also provide for his liability to pay a certain sum for every year of service which he receives from his employees. This can be done if such liability is property ascertainable and it is possible to arrive at a proper discounted present value. Even if the liability is a contingent liability, provided its discounted present value is ascertainable, it can be taken into account.Contingent liabilities discounted and valued as necessary can be taken into account as trading expenses if they are sufficiently certain to be capable of valuation and if profits cannot be properly estimated without taking them into account. Sec.37. **[98ITR426]**

2. It is settled law that till such time that an arbitrator's award is not made a rule of the court,and a decree obtained on its basis, it cannot be enforced against the party against whom it had been made and remains at best a contingent liability. Even in a case where the mercantile system of accounting is followed, the liability for which a provision is made must have accrued in the previous year, and should not partake of the nature of a contingent liability. **[A.P.S.Cold Storage and Ice Factory-CIT,All.,***Misc.***119ITR709]**

3. Where a statutory liability has accrued for the first time during the relevant year and an assessee deducts or seeks to deduct the estimated value of that liability for computing the profit for that year and the contingent part is so insignificant or so irrelevant that it cannot be considered contingent in the real sense of the term. **[CIT-Eastern Spinning Mills Ltd.Cal.,***Sec.36(1)(v).***19CTR94, 126 ITR686]**

Contingent liability not provided for **MP.** **CWT - Gwalior Rayon Silk Mfg. (Wvg.) Co. Ltd. (No. 1)**

A debt which does not exist on the valuation date could not be taken into consideration for determining the net wealth for purposes of the Wealth-tax Act, 1957, and such a debt must exist on or before the valuation date. There is no statutory liability on an assessee to pay dividends compulsorily on cumulative preference shares, whether profits are earned or not, and the amounts of arrears of dividend on cumulative preference shares which are shown in the balance-sheet under the head contingent liability not provided for are not deductible in computing the net wealth as the liability is not a debt owed under section 2(m) of the Wealth-tax Act,1957, where the general body of the company has not declared the dividend before the valuation date. *Sec.2(m).r.w.s.7.WT.Act.***[131ITR140]**

Continue **Bom.** **Hiralal Jeramdas - CIT**

Though it is true that one of the shades of the meaning of the word continue is to resume, and one of the shares of the meaning of the word resume is to take up a thing after an interruption, having regard to the different shades of meaning of both the words, the main idea conveyed is continuation of a thing without break. Even assuming that the word resume means taking up a thing after having given it up, having regard to the scheme underlying the relevant provisions of section 24 the words provided that the business, profession or vocation in which the loss was originally sustained continued to be carried on by him in that year are not capable of bearing a construction so as to include restarting a business, in which loss was suffered after its discontinuance. *Sec.73, 74.***[58ITR1]**

Contract becoming frustrated AP. CIT - M.D. Manohar Rao

The doctrine of frustration applies to contracts of sale. In a contract of sale simpliciter, where the parties do not expressly provide for the contingency of acquisition or what is to happen in such an eventuality,or where it cannot be said that the acquisition of the land was within the contemplation of the parties, the agreement of sale becomes frustrated if the land is acquired by the State before the execution of the sale deed. In cases, however, where the parties expressly provide for such contingency, or even in cases where there is no such express stipulation but, in the facts and circumstances of that case, it can be said that the acquisition of land was within the contemplation of the parties at the time of entering into the agreement of sale, the contract cannot be said to have become frustrated. *Sec.4(1)r.w.s.54.***[48CTR13,155ITR696,22Taxman429]**

Contract settled Del CIT - Hans Machoo and Co.

A contract can be said to be settled if instead of effecting the delivery or transfer of the commodity envisaged by the contract, the promisee, in terms of section 63 of the Indian Contract Act, 1872, accepts any satisfaction which he thinks fit. ...The word settled or settlement in connection with the contract has not been defined in the Income-tax Act or in the Contract Act or in the Sale of Goods Act,or in any other statute.The proper meaning to be given to the words contract settled in the definition clause would be a contract determined or concluded or disposed of. By the use of the expression settled,what is intended to be dealt with is a case of performance of contract and not non-performance. *Sec.28Expl.2.r.w.s.43(5).***[164 CTR 93,247ITR79,113Taxman427]**

Contractor AAR A.A.R. No. 542 of

The word contractor is defined in Concise Oxford Dictionary to mean a person who undertakes a contract to provide material or labour for a job. Contractor is a person who, in the pursuit of an independent business, undertakes to do specific jobs of work for other persons, without submitting himself to their control in respect to the detail of the work. *Sec.9(1)(i)Expl.1.*[**193CTR328,274ITR 501,143Taxman71]**

Control and management **Mad.** **CIT Excess Profits Tax - Erin Estate**

Control and management signifies the controlling and directive power,the head and brain as it is sometimes called, and situated implies the functioning of such power at a particular place with some degree of permanence while wholly would seem to recognise the possibility of the seat of such power being divided between two distinct and separate places. The control and management of a firm's business is situated at the place where the central management and control actually abides. Control of a business does not necessarily mean the carrying on of the business. The place of control may be different from the place where the physical operations of the business are carried on and the place where trading activities are carried on need not necessarily be the place of control or management........Control and management means de facto control and management. The existence of a mere right or power of control and management is not the criterion. When Section 4A(b) refers to the control and management of its affairs by the firm it means the affairs which have some relation to the business whose income, profits and gains are sought to be taxed and not the private or domestic affairs of the individual partners having no relation to the firm's business. The control and management of a firm's affairs may be divided and may abide in more than one place. *ExcessProfit tax Act.*[**20I TR412]**

2. The expression control and management meant de facto control and management and not merely the right or power to control and manage. **[CIT - Nandlal Gandalal,SC,***Sec.6.***40ITR1]**

3. Under the provisions of section 6(3) of the I.T.Act, 1961, a non-Indian company is said to be resident in India in any previous year if during that year the control and management of its affairs is situated wholly in India. The expression control and management-means de facto control and management and not merely the right or power to control and manage.

[CIT - Bank of China, Cal., *Sec.6(3).*[**154ITR617]**

4. The expression control and managementmeant de facto control and management and not merely the right or power to control and manage. Since the de facto control and management of the affairs of the firm would be with the resident partner the firm could not be said to be a non-resident. **[Ms.Meenu Sahi Mamik,AAR,**<u>*Sec.6(2).*</u> **287ITR514]**

Controlling interest Cal. CEPT - Jeewanlal Ltd.

The phrase controlling interest occurring in section 2(21) of the Excess Profits Tax Act, 1940, refers to the director's power of controlling by votes the decisions binding on the company in the shape of resolutions passed at a general meeting. The fact that the beneficial interest in the shares is in a third party or that a vote-carrying share is vested in a director as trustee is, therefore, immaterial. No distinction can be made between the case when the director trustee has, and when he has not, a beneficial interest in the shares. <u>*Excess Profits Tax Act.*</u> **[20 ITR39]**

2. In common parlance a person is said to have a controlling interest in a company when such a person acquires, by purchase or otherwise,the majority of the vote-carrying shares in that company, for the control of the company resides in the voting powers of its shareholders. In this sense, the directors of a company may well be regarded as having a controlling interest in the company when they hold and are entered in the share register as holders of the majority of the shares which,under the articles of association of the company, carry the right to vote. It is not, however, necessary that in order to have a controlling interest the person or persons who hold the majority of the vote-carrying shares must have a beneficial interest in the shares held by them. They may hold the shares as trustees and may even be accountable to their beneficiaries and may be brought to book for exercising their votes in breach of trust, nevertheless, as between them as shareholders and the company, they are the shareholders, and, as such, have a controlling interest in the company. When a shareholder holding the majority of shares authorises an agent to vote for him in respect of the shares so held by him, the agent acquires no interest, legal or beneficial, in the shares.The shares being always subject to the will and ordering of the shareholder, the controlling interest which the holder of the majority of shares has never passes to the agent. **[CIT - Jeewanlal Ltd., SC,** <u>*Excess Profits Tax Act.*</u> **24 ITR475]**

Cooperation Kar. Mahalakshmi Rice Mills –CIT

The expression cooperation in any enquiry relating to the assessment, in section 273A should be held to mean that the assessee did not resort to litigation, obstruction or evasive tactics in concluding the assessment and no more. *Sec.273A.* **[19CTR177,129ITR53]**

2. The word cooperation employed in the amnesty scheme was required to be read and interpreted in the context of the amnesty scheme. Once the income was denied and alternatively claimed as business loss,the element of co-operation evaporates. **[CIT Bimal kumar Damani,Cal.,***Amnestyscheme.***180CTR452,261ITR87,129 Taxman564]**

Copyright AAR Dassault Systems K. K

That passing of a right to use and facilitating the use of a product for which the owner had a copyright was not the same thing as transferring or assigning rights in relation to the copyright. Where the purpose of the licence or the transaction was only to establish access to the copyrighted product for internal business purpose, it was not legally correct to say that the copyright itself had been transferred to any extent. Merely authorizing or enabling a customer to have the benefit of data or instructions contained therein without any further right to deal with them independently did not amount to transfer of rights in relation to copyright or conferment of the right of using the copyright. *Sec.9(1)(vi).***[229 CTR105,322 ITR125,188Taxman223]**

Corrosive chemicals P&H. CIT - Saraswati Industrial Syndicate Ltd.

The words corrosive chemicals employed in entry(ii) B(7) of Para III of Part I of Appendix I to the Income-tax Rules, 1962, contemplates only free chemicals and does not include non- free chemicals of corrosive effect. *Sec.32.ITRules,Appendix I.***[178ITR419,45 Taxman11]**

Cost of acquisition Cal. CIT - Octavious Steel and Co. Ltd.

The cost of acquisition of an asset, be it a capital asset or any other asset must be understood in its common sense, that is, it must represent the expenditure incurred in acquiring the asset. *Sec.45.* **[137CTR257,221ITR810,82Taxman9]**

Cost to the business **Guj.** **CIT- Kaira District Co-operative Milk Producers' Union Ltd.**

Where an assessee converts his capital assets into stock-in-trade and starts dealing in them, the taxable profit on the sale must be determined by deducting from the sale proceeds the market value on the date of their conversion into stock-in-trade, since this would be the cost to the business and not the original cost to the assessee. *Sec.43(1).*[165CTR57,247ITR314,114Taxman215]

Cottage industry **All.** **District Co-operative Federation Ltd.-CIT**

The expression cottage industry occurring in section 14(3)(i)(b) contemplates an industrial activity of which a well-recognised feature is that it is commonly located in the cottages or homes of the artisans. It is carried out on a small scale with a small amount of capital and a small number of workers and has a turnover which is correspondingly limited. *Sec.80P.*[87ITR639]

2. Primarily a cottage industry is carried on by the families in their dwelling houses but when the term cottage industry is applied to co-operative societies,the idea of a family unit does not fit in. A co-operative society can in a way be likened to a family constituted of its members.Where the members of a cooperative society are engaged in the manufacture of goods in their cottages or dwelling houses it can be said that the family constituted of its members is engaged in a cottage industry.Before it can be said that a co-operative society is engaged in an industry it is necessary that there must be an activity relating to an industry. An industry implies manufacture of certain articles and cannot embrace a business of mere purchase and sale of goods. **[Addl.CIT - Indian Co-operativeUnionLtd.,Del.,** *Sec. 80P.***134ITR108]**

3. A cottage industry is one carried on by persons in their homes. But the phrase cottage industry used in section 80P of the I. T. Act, 1961,has to be understood in the context of the Act and in the background of the section.The concept of cottage industry in the context of section 80P cannot be limited to industries carried on by families in their own houses. **[CIT-TaxTextile Industries Co-operative SocietyLtd.Ker.,** *Sec.80P.***66CTR103,170ITR465,35 Taxman271]**

Course SC CIT - East West Import and Exports P. Ltd.

Course ordinarily conveys the meaning of a continuous progress from one point to the next in time or space and conveys the idea of a period of time;duration and not a fixed point of time. In the course of such previous year in the Explanation to section 23A(1) would therefore, refer to the period commencing with the beginning of the previous year and terminating with the end of the previous year. Therefore,it would necessarily mean that free transferability of the shares by the holders to other members of the public should be present throughout the previous year. *Sec.104,109.* **[76CTR9, 176ITR155,43Taxman26]**

Crane Guj. Gujco Carriers - CIT

The word crane when used for an inanimate object means a machine for moving heavy objects usually by suspending them from a projecting arm or beam. Crane is any of a diverse group of machines that not only lift heavy objects but also shift them horizontally. Movable cranes are mounted on railway, cars, motor trucks or chassis equipped with caterpillar treads and the hoisting machinery is mounted so as to counterpoise part of the load on the boom and thereby,prevent the entire crane from overturning while carrying the load. A mobile crane mounted on a truck constitutes a single unit known as truck crane which is adapted for use on roads for special services.The truck on which the crane is mounted is constructed and adapted specially to carry the crane. *Sec.32. Rule.5,Appex I.***[174CTR324,256ITR50,122Taxman206]**

Credit society Mad. CIT - Coral Mills Workers Co- op. Stores Ltd.

A credit society within the meaning of section 14(3)(i) of the Indian Income-tax Act, 1922, and section 81(i) of the Income-tax Act, 1961, can only mean a society which provides credit by way of loans of money to its members and not a society which sells goods on credit.

*Sec.80P.***[106ITR868]**

Criminal contempt SC ITAT - V.K.Agarwal

....the definition of criminal contempt under section 2(c) of the Contempt of Courts Act, 1971, refers not merely to publication by words, signs, etc.,but includes the doing of any act whatsoever which scandalises or tends to scandalise or lowers or tends to lower the authority of any court [section 2(c)(i)] or interferes or tends to interfere with, or obstructs or tends to obstruct, the administration of justice in any manner [section 2(c)(iii)]. Therefore, any act which tends to interfere with the administration of justice or tends to lower the authority of any court can be punished for contempt. It is not necessary that there should be an actual interference with the course of administration of justice. It is enough if the offending act or publication,tends in any way to so interfere. If there are insinuations made which are derogatory to the dignity of the court and are calculated to undermine the confidence of the people in the integrity of the judges, the conduct would amount to contempt.... Contempt of Courts Act.**[150CTR513,235ITR175,101Taxman382]**

Current income Bom CIT - Kotak Mahindra Finance Ltd.

Under section 207 of the Income-tax Act, 1961, advance tax is payable during any financial year in respect of the current income. The words current income are very crucial. The words current income refer to computation of total income under the provisions of the Income-tax Act including section 115J. Under section 207 of the Income-tax Act, the words total income have been equated to the expression current income. Sec.207.**[183CTR491,265ITR119, 130Taxman730]**

Current repairs Bom. New Shorrock Spinning & Manu. Co.Ltd.-CIT

The word current in the expression current repairs does not mean petty but denotes such repairs which are attended to when the need for them arises and are not allowed to fall into arrears or to be accumulated. If the assessee, although the need has arisen, does not attend to that need and allows the repairs to get accumulated, it could not be said that when he is expending money on these repairs he is expending them on current repairs. But if the need for repairs had not arisen before they were actually effected, even repairs done long after the acquisition of the asset would be current repairs.

Sec.31.**[30ITR338]**

2. The expression current repairs in section 10(2)(v) of the Indian Income-tax Act,1922, indicates expenditure incurred by an assessee in relation to a building or machinery necessitated during the time when his business was conducted and by reason of the business itself, such as an account of wear and tear and the like. There can be no expenditure incurred on account of current repairs either before the business started or after the business has ceased. **[A.Y.S. ParisuthaNadar-CIT,Mad.**<u>*Sec.30.*</u>**[46ITR1041]**

3. Current repairs are necessary repairs which are needed for the maintenance of building and machinery, etc. They are not luxury repairs,the element of need being implicit in the expression. Whether a particular repair carried out was an essential repair or not should be judged from the viewpoint of commercial expediency and not by academic or theoretical standards. Primarily, it is for the businessman to decide when his building or machinery, etc., require repairs.However, if, by carrying out the repairs,a new asset or an advantage of enduring nature is created,the expenditure on such repair cannot be regarded as an expenditure on current repairs.

[Guntur Merchants Cotton Press Co. Ltd. - ITO, AP. <u>*Sec.30(a)(ii).*</u>**108ITR620]**

4. The Legislature has used different language in the provisions of section 30(a) (i)and section 30(a)(ii) of the I.T. Act, 1961.A tenant is entitled to deduction of the amount spent on account of the cost of repairs to the premises when he has undertaken to bear the cost of the repairs.If the amount is spent by the assessee otherwise than as a tenant, the amount paid by him on account of only current repairs is allowable. The latter provision applies to the assessee occupying the premises otherwise than as a tenant, as an owner or mortgagee in possession, and in those cases the deduction is restricted in respect of the current repairs to the premises. So far as a tenant is concerned, it is the repairs to the premises... **[Installment Supply P. Ltd.-CIT,Del.,**<u>*Sec.30(a)(ii).*</u>**40CTR313,149ITR52,17Taxman172]**

5. One of the ingredients of an amount being allowed as a deduction under section 30(a)(ii) of the Income-tax Act, 1961, is that the amount must be spent for purposes of carrying out current repairs. An amount spent in carrying out repairs which were long overdue cannot be said to be spent on current repairs. Section 30(a)(ii) is not concerned with the question as to whether the nature of the repairs is capital or revenue. As long as the repairs which are carried out fall under the category of current repairs, then, irrespective of the fact that the repairs have been carried out to a capital asset and may otherwise have been regarded as capital expenditure,the section specifically allows deduction. Current repairs must necessarily mean repairs which are required to be carried out from time to time as and when a defect arises. If there has been wear and tear on an item, like the floors of a building, over a number of years and ultimately they are replaced, then such replacement cannot be regarded as current repairs. **[Modi Spinning and Weaving Mills Co. Ltd.- CIT, Del.,** <u>*Sec. 30(a)(ii).*</u>**[200ITR544]**

6. The expression current repairs has not been defined in the Act. It has, therefore, to be taken in its popular or commercial sense. In commercial parlance, it means repairs which are undertaken in the normal course of user for the purpose of preservation, maintenance or proper utilisation. It does not mean petty repairs or repairs necessitated by wear and tear during the particular year. Payments on account of current repairs must be understood in contradistinction to payments for additions or improvement. The object of the expenditure should not be to bring a new asset into existence or to obtain a new or different advantage. The quantum of expenditure incurred on the repairs is not relevant for determining whether it is an expenditure on current repairs or not, because the extent of repairs and the amount spent would depend upon various factors. **[CIT-ChowguleandCo.Pvt.Ltd.,Bom.**<u>*Sec.31.*</u>**[125CTR642,214 ITR 523,128Taxation3,81Taxman384]**

7. The expression current repairs in section 31 of the Income-tax Act, 1961, means expenditure on building, machinery, plant which is not for renewal or restoration. It is only for preserving or maintaining an already existing asset which does not bring a new asset into existence or does not give to the assessee a new or different advantage and they must be such repairs as are attended to as and when the need for them arises and the question as to when a building, machinery, plant or furniture requires repairs and when the need arises must be decided not by any academic or theoretical test but must be decided by the test of commercial expediency.But,if the amount spent was for the purpose of bringing into existence a new asset or obtaining a new advantage,then such an expenditure would not be an expenditure of a revenue nature but it would be a capital expenditure. **[CIT -Volga Restaurant, Del.**<u>*Sec.31.*</u>**170CTR206, 253ITR 405,119Taxman757]**

8. .... In order to constitute current repairs the expenditure must have been incurred to preserve and maintain an already existing asset, and the object of the expenditure must not be to bring a new asset into existence or for obtaining a new advantage. It is difficult to accept that for the purpose of determining the allowability of expenditure under the head Repairs the entire productive apparatus of a manufacturing company be treated as one single asset and wholesale replacement of complete identifiable and distinct parts be regarded as a repair effected to the production facility as a whole. While judging a claim for deduction under the Income-tax Act the scope of the term repair cannot be stretched beyond all recognition. **[CIT-Madras Cements Ltd., Mad.** *Sec.31.***175CTR 551,255 ITR243,123Taxman412]**

9. ..to decide the applicability of section 31(i) the test was .....whether the expenditure was current repairs. The basic test was to find out whether expenditure was incurred to preserve and maintain an already existing asset, and the expenditure must not be to bring a new asset into existence or to obtain new advantage.

[CIT- Saravana Spinning Mills P.Ltd.,SC,*Sec.31(i).***293ITR201]**

Current repairs to machinery **Pat.** **CIT - Darbhanga Sugar Co. Ltd.**

The expression current repairs to machinery in section 10(2)(v) of the Indian Income-tax Act must be interpreted to mean repairs to machinery in the current accounting year. There is nothing in that section to suggest that the expenditure on repairs cannot be allowed as a proper deduction if the repairs are not petty. Current cannot be interpreted to mean petty. The section does not say anything about the magnitude of the expenditure. *Sec.31***[29ITR21]**

D

Data processing **Gau.** **CIT - Technotive Eastern (Pvt.) Ltd.**

Data processing means the converting of raw data to machine-readable form and its subsequent processing (as storing, updating, combining re-arranging or printing out) by a computer. <u>*Sec.80HH, 80-I.*</u>**[176CTR422,255ITR253,124Taxman769]**

Dealer **SC** **CIT - Sai Publication Fund**

The dealer under section 2(11) in respect of the goods sold or purchased by him unless he carried on the business of buying and selling of such goods . . .. on a combined reading of sections 3,2(5A) and 2(11), the tax under the Act was leviable on the sales or purchases of goods by a dealer and not every person. . . . The definition of dealer in section 2(11) clearly indicates that in order to hold a person to be a dealer,he must carry on business and then only he may also be deemed to be carrying on business in respect of transactions incidental or ancillary thereto. <u>*Bombay Sales Tax Act.*</u>**[177CTR1,258 ITR70]**

Debatable issue **Cal.** **Hindustan Lever Ltd. - Joint CIT**

The question whether interpretation of a provision of law is right or wrong is per se a debatable issue. <u>*Sec.154.*</u>**[204CTR125,284ITR42]**

Debenture **Raj.** **CIT -Shree Rajasthan Syntex Ltd.**

The term debenture is not a technical term nor a term of art,but in its ordinary sense denotes one of the modes for borrowing money by any company in exercise of its borrowing powers....However,in the ordinary business sense,a debenture is generally understood to be a document usually but not necessarily under seal, acknowledging a debt and securing repayment thereof by mortgage or charge on the company's property or undertaking and providing that until repayment, interest will be paid thereon at a fixed rate payable usually either half- yearly on fixed dates. *Sec.37.r.w.s.117. Companies Act.* **[186CTR59,269ITR461,134Taxman577]**

2. In the absence of any definition of debenture in the Income-tax Act,1961, reliance could be placed upon the definition given in section 2(12) of the Companies Act, and also the common parlance in which this term is understood. Sub-section (5) of section 11 of the Act, deals with the investment in bonds of financial corporation, State Government or Central Government. Though a specific provision is made in respect of investment in these particular kinds of bonds, that would not mean that one has to give restrictive meaning to the term debenture more particularly when the term is not defined under the Act. The principle of interpretation is that in the absence of any definition given to a particular term in a statute, the meaning which is to be given to the term is the meaning in which it is understood in common parlance. **[DIT- Shree Visheshwar Nath Memorial Public Charitable Trust,Del.,** *Sec.11(5),13 (1)(d).r.w.s(12)Companies Act.* **333 ITR 248, 194Taxman280]**

Debt Cal. Ramesh Behari Ghose - Union of India

The word debt means a sum payable in respect of a liquidated money demand recoverable by action. The word is a generic name for various kind of money payable, for example, a judgment debt or a public demand under section 3(6) of the Public Demands Recovery Act or a private demand for money had and received. *Sec.220.* **[45ITR 622]**

2. Debt means a certain sum due from one person to another by record under a speciality or deed or under simple contract by writing or oral. **[CWT-Ahmed Tea Co. Ass.***Sec.2(m).WT.Act.***48 ITR943]**

3. Broadly stated, debt is a liquidated money obligation for the recovery of which an action will lie. It is an ascertained liquidated quantified obligation enforceable in presenti or in futuro. A future contingent liability is not a debt due and owing. An inchoate liability with a fair prospect of maturity into a debt in future and still in its embryo stage would not answer the description of a debt. **[CWT-Pierce Leglie And Co.Ltd.,Mad.,***Sec2(m).WT.Act.***AIR1963 Mad. 356,48ITR1005]**

4. It cannot, as an invariable rule, be said that for an existing obligation to pay money to be a debt, it must necessarily be for an ascertained sum. The word debt has been used in some enactments in the sense of an existing obligation to pay an ascertained sum of money and in some other enactments to include an existing obligation to pay a sum of money, though its ascertainment may come at a later stage. The word debt in section 2(m) of the Wealth-tax Act is capable of being understood to include an existing obligation to pay an ascertained sum of money as well as an existing obligation to pay money, though the amount may not have been yet ascertained. **[CWT-StandardMillsCo.Ltd.,Bom.,**<u>*Sec.2(m).WT.Act.*</u>**50ITR267]**

5. A debt is one which can be ascertained or readily calculable amount. **[CWT-Harison and Crossfield Ltd.,Ker.**<u>*Sec.2(m).WT.Act.*</u> **[AIR1965,Ker.209,54ITR 587]**

6. Consideration which remains upaid in respect of assets purchased by assessee from another company is a debt due from assessee. **[CWT-M/s.Spencer&Co.Ltd.,SC.**<u>*Sec2(m)(ii).WT.Act.*</u>**AIR1973,SC, 2376,88ITR429]**

7. A debt is a present obligation to pay an ascertainable sum of money either immediately or in future and the provision for payment of Income-tax is a debt owed within the meaning of section 2(m) of the Wealth-tax Act, 1957, and as such deductible in computing the net wealth of an assessee. **[Bhagwandas Jain-Addl. CWT, MP.**<u>*Sec2(m).WT.Act.*</u>**7CTR240, 116ITR347]**

8. The word debt is synonymous with liability. It may be payable now or it may become payable in future but the obligation must be a present one and should be in respect of an ascertainable sum of money. The liability to pay income-tax is embedded in the accrual or earning of the income itself. It is ascertainable also because the rates are provided in the Finance Act relating to the concerned years. Its actual quantification may be delayed. That will not, however, change the nature of the liability. Therefore, all kinds and categories of debts are deductible irrespective of the purpose or use thereof. Therefore, the liability to Income-tax on the concealed income in respect of an assessment year, which though quantified later on as a result of a settlement between the assessee and the department, was a debt owed within the meaning of section 2(m) of the Wealth-tax Act, 1957, and was deductible in computing the net wealth of the assessee for that assessment year. It would not be a liability falling within clause (iii) of section 2(m) as, on the relevant valuation dates, the liability had not been quantified by way of an assessment nor was the assessee claiming in appeal, revision or other proceeding that the amount was not payable by him nor was it outstanding for a period of more than twelve months on the relevant valuation dates. **[CWT-Dina Nath, All.** *Sec.2(m)(iii). WT.Act.* **136ITR499]**

9.A loan of money undoubtedly results in a debt,but every debt does not involve a loan. Liability to pay a debt may arise from diverse sources, and a loan is only one of such sources. Every creditor who is entitled to receive a debt cannot be regarded as a lender. The facility of drawing hundis and late payment of price of goods purchased by the assessee,cannot be equated with borrowing of capital. **[CIT-Orient Trading Co.Guj.***Sec.36(1)(iii).***208ITR216]**

10. An Income-tax liability is a debt within the meaning of section 2(m) of the Wealth-tax Act,1957. **[CWT-Late Rao Karan Singh, Raj.***Sec.2(m).WT.Act.***171CTR477,252ITR447,118Taxman84]**

Debt has become bad Guj Kamla Cotton Co. - CIT

The requirement that a debt has become bad or irrecoverable does not mean that the Department can insist upon demonstrative and infallible proof that the debt had become bad. It is not compulsory for the assessee to take legal proceedings against the debtor for recovery of the claim before writing it off as a bad debt. When a creditor bona fide writes off the debt because there appears no chance of its recovery in the foreseeable future or where the recovery proceedings would be so cumbersome and expensive as to outweigh any advantage of instituting any recovery proceedings, the assessee discharges the onus and would be entitled to claim deduction of the bad debt under clause (vii) of section 36(1) of the Income-tax Act, 1961. *Sec.36(1)(vii).***[138CTR336,226ITR605,141 Taxation83]**

Debt must be both...... Guj. CIT - Ahmedabad Electricity Co. Ltd.

A debt must be both bad and actually written off before any deduction can be claimed under section 36(1)(vii) of the Income-tax Act, 1961, read with section 36(2). Whether or not a debt is bad is a question to be determined objectively and not by subjective opinion of any particular individual. The objective test in deciding whether a debt is bad or not is whether a reasonable prudent business person would conclude that there is no reasonable likelihood that the debt will be paid by the debtor or someone else either on behalf of the debtor or otherwise. The debt in question must be bad to be written off as irrecoverable in terms of section 36(2)(i)(b) of the Act. This is essentially a question of fact. However, a debtor does not need to be insolvent for the debt to be bad. It is sufficient,if on a bona fide assessment, the debtor is unlikely to make the payment of the debt. The objective test is that, on the balance of probabilities, the circumstances must indicate to a reasonable and prudent business person that the debt is unlikely to be recovered. Merely because legal steps for recovery are not taken, it cannot in such a case be inferred that the debt is not bad. The relevant factors that can enable forming of the view that there is no reasonable likelihood that the debt will be paid are: (i) the length of time the debt is outstanding-the longer the debt is outstanding, the more likely it is to be considered a bad debt by a prudent business person; (ii) the efforts that a creditor has taken to collect a debt-the greater the extent to which a person has unsuccessfully tried to collect the debt, the more likely is it to be considered a bad by a prudent business person; and (iii) information about the debtor-a creditor may have obtained reliable information about the debtor, such as financial difficulties and defaults of the debtor towards other customers or insolvency, that would lead a prudent business person to conclude that a debt is bad...In order to support a claim for deduction of a bad debt written off for tax purposes, there should be sufficient evidence to show that reasonable steps based on sound commercial considerations were taken to recover the debt. Such steps would include one or more of the following steps: (a) issuing reminder notices; (b) restructuring of debt; (c) rescheduling of debt settlement; (d) negotiation or arbitration of a disputed debt; and (e) legal action. The decision not to take legal action should be considered to be reasonable if it can be shown that the anticipated cost of any legal action is prohibitive in relation to the amount of the debt.

Sec.36(1)(vii).,36(2). **[181CTR222,262ITR97,129Taxman190]**

Debt owed **SC** **Kesoram Industries and Cotton Mills Ltd.-CWT**

....debt owed within the meaning of section 2(m) of the Wealth-tax Act, 1957, could be defined as the liability to pay in praesenti or in futuro an ascertainable sum of money. A debt involves a present obligation incurred by the debtor and the liability to pay a sum of money in present or in future. The liability must however, be to pay a sum of money, i.e., to pay an amount which is determined or determinable in the light of factors existing on the date when the nature of the liability has to be ascertained. On the terms used in section 3 of the Indian Income-tax Act, 1922, liability to be taxed becomes effective not later than the last day of the year of account. That liability does not give rise to any obligation to pay a sum of money either determined or determinable in the light of the factors existing on that date. The liability to pay tax however, arises only when the Finance Act becomes operative on the first day of April of the assessment year... *Sec2(m).WT. Act.***59ITR767**

2. Income-tax liability amounts to debt owed. **[CWT-Sahu Jain Ltd.Pat.***Sec.2(m).WT.Act.***AIR 1966, Pat.281,60 ITR592]**

3. Liability to pay Income-tax on income earned before the valuation date is a debt owned at the time of the valuation date and a provision made for such income-tax is deductible from the gross wealth for determining the net wealth upon which wealth-tax is levied. **[CIT-AmcoBatteries(P.)Ltd.,Mys.***Sec.2(m).WT.Act.* **52ITR370]**

4. The mere fact that some adjudication had taken place, the nature of which was not fully disclosed,would not affect the position that the amounts set apart were on the same basis on which the payments were made during the previous year. If the employer did not dispute his liability to pay bonus, the amounts so set apart represented an ascertained present liability in respect of the payment of bonus. Therefore, in computing the net wealth of the assessee, the provision for bonus was a debt owned under section 2(m) and was a permissible deduction. **[CWT-Associated Cement Co. Ltd., Bom.,** *Sec.2(m).WT.Act.***128 ITR347]**

Debts owed **Mys. CWT-D.C. Basappa,**

The Expression Debts owed includes debts payable at once and those payable in future and debts ascertained and those not ascertained. *Sec.2(m).WT.Act.***[AIR1964Mys.204,51ITR790]**

2. An obligation to pay a sum of money at a future date beyond the valuation date is not covered by the expression debts owed. **[CWT-Raipur Manufacturing Company, Guj.***Sec2(m).WT.Act.***AIR 1964Guj154,52ITR482]**

Declaratory Act **AP. CIT - Agriculture Market Committee**

... a declaratory Act is intended to remove doubts regarding common law which are to be construed according to common law; (ii) declaratory Acts are also made to rectify or clarify a gross mistake, or the omission in the former statute, in which event the latter statute relates back to the time when the former Act is made; (iii) the purpose of a declaratory Act is to remove a doubt as to the meaning of an existing law or to correct a construction considered erroneous by the Legislature. If a declaratory Act is by way of an explanatory Act, one should see whether it is intended to supply an obvious omission or clear up doubts as to the meaning of the previous Act. In the absence of clear words indicating that the amending Act is declaratory, it would not be so construed when the pre-amended provision was clear and unambiguous; (iv) if a statute is curative,or merely declarative, retrospective operation is generally intended; and (v) in determining the nature of the Act, substance is more important than the form. If the provision is clear and unambiguous, the question of treating the amending Act as declaratory would not arise, even if the amending Act uses the expression for the removal of doubts which itself is not conclusive as to an amendment being clarificatory or declaratory in nature.

<u>*Misc.*</u>**[337ITR299]**

Declared **SC** **K.P.Varghese - ITO**

The word declared in sub-section (2) of section 52 is very eloquent and revealing. It clearly indicates that the focus of sub-section (2) is on the consideration declared or disclosed by the assessee as distinguished from the consideration actually received by him and it contemplates a case where the consideration received by the assessee in respect of the transaction is not truly declared or disclosed by him but is shown at a different figure. <u>*Sec.52(2).*</u> **[131ITR597]**

Deemed **Cal.** **CIT - Oriental Co. Ltd.**

Income is assessable in the hands of an assessee who is resident in India only if it falls under section 5(1)(a) or (b) or (c) of the I.T. Act, 1961. Income which is deemed to be received in a foreign country is not assessable. Sometimes the expression deemed is included to give a comprehensive description that includes what is obvious, what is uncertain and what is in the ordinary sense impossible. A deeming provision, however, is never used by the Legislature to mean what is obvious or to include the natural meaning of the expression. <u>*Sec.5(1).*</u>**[25CTR75,137ITR777]**

Deemed dividend **SC** **CIT - Mukundray K. Shah**

That the concept of deemed dividend under section 2(22)(e) postulated two factors : whether the payment was a loan and whether on the date of payment there existed accumulated profits. These two factors had to be correlated. <u>*Sec.2(22)(e).*</u>**[209CTR97, 290ITR433, 160Taxman276]**

Deemed to accrue or arise in India **Kar.** **Vdo Tachometer Werke, West Germany - CIT**

In order that the income of a non-resident can be deemed to accrue or arise in India under section 9 of the I.T. Act, 1961, two factors must be established, namely, (1) the said income must have accrued or arisen whether directly or indirectly through or from any business connection in India or from any other source mentioned in section 9(1)(i), and (2) in the case of a business, some operations must have been carried on by the non-resident in India. It follows that if no operations are carried out in India, no income can be deemed to accrue or arise in India even though there may be business connection in India. Merely because certain goods are manufactured in India by an Indian company with the aid of technical know-how obtained by it abroad from a non-resident company, it cannot be said that the non-resident company has been carrying on business operations in India, in the absence of any other material which points to the contrary. *Sec.9.***[117ITR804]**

Deemed to accrus SC CIT - Bhogilal Laherchand

The term deemed in section 4 brings within the act of chargeability income not actually accruing but which is supposed notionally to have accrued. *Sec.5.***[AIR1954SC155,25ITR50]**

Deemed to have been effected Del. CIT- Madhsy Films P. Ltd.

Section 27 of the General Clauses Act, 1897, provides that service by post is deemed to have been effected by properly addressing, prepaying and sending by registered post, a letter containing the notice required to be served. Unless the contrary is proved the service is deemed to have been effected at the time when the letter would be delivered in the ordinary course of post. *Sec.282.***[216CTR145, 301ITR69,175Taxman347]**

Deemed to have been served PH. Ramesh Khosla - ITO

The words issue and service are not synonymous, yet in the light of the provisions of section 114 of the Indian Evidence Act, 1872, and section 27 of the General Clauses Act, 1897, a notice is deemed to have been served on the addressee if it has been sent through prepaid registered post or letter. Even in cases where the addressee has refused to accept the registered envelope tendered to him by the postman, he is to be imputed with the knowledge of the contents thereof and in spite of that refusal, the notice or the contents of the envelope are to be deemed to have been duly served on the addressee. <u>*Sec.282.*</u>**[41CTR138,154ITR556,73Taxation74,17 Taxman317]**

Deeming **Bom.** **Habib and Sons - CIT**

Deeming a past event or thing to be something other than what it was or is, may be an interference with the course of nature, since it creates artificial data in the place of existing fact. But it is not a retrospective changing of the statute law. <u>*Misc.*</u>**[15ITR132]**

Defective return **Cal** **National Insurance Co. Ltd. - CIT**

A return of income has to be regarded as defective only if it contains any of the defects referred to in the Explanation to section 139(9). In other words, the provisions of section 139(9) will not be applicable in the case of returns which do not contain any of the specified defects. <u>*Sec.139.*</u>**[213ITR862]**

Deferred benefit **AP.** **CIT - T.Ponnaiah**

....the words deferred benefit occurring in section 64(1)(vii) must be construed as being a benefit deferred to a year subsequent to the accounting year in which the income was taxable, so long as it is not deferred beyond the minority of the child. <u>*Sec.64(1)(vii).*</u>**[69CTR 150,172ITR269,36Taxman18]**

Definite information **All.** **Badar Shoe Stores**

The words definite information are placed in section 34 to protect the subject against an assault by an Income-tax Officer based upon mere suspicion. The definite information, which is something more than mere gossip or rumour, must lead to the discovery or belief. Provided the information is definite and does lead to that belief, it need not necessarily be information of fact, though in ninety-nine cases out of a hundred it would inevitably be information of fact. Still less need it be information of actual escape from assessment or under assessment. It may well be information of circumstances not themselves amounting to under assessment or escape from assessment, but leading to the belief of under assessment or escape from assessment.In short, it may be circumstantial evidence.

<u>*Sec.147.*</u>**[14ITR431]**

2. The words definite information are placed in the section to protect the subject against any assault by the Income-tax Officer based upon mere suspicion. Definite information, which is something more than mere gossip or rumour, must lead to the discovery or belief. A mere change of opinion based on the same facts and figures does not amount to discovery. **[Kedar Nath-CIT, All.** *Sec.147.***15ITR224]**

Definite Portions SC Udayam Chimnbhai - CIT

The expressions Definite portions indicate a physical division in which a member takes a particular item of property,and not an undivided share in property. *Misc.***[AIR1967SC762,63ITR416]**

Demolished or destroyed Cal. CIT- Bengal Assam Steamship Co.Ltd.

The expression demolished or destroyed in section 41(2) referred in physical demolition or destruction. *Sec.41(2).***[161ITR576]**

2. The expression demolished or destroyed in section 41(2) of the Income-tax Act,1961,refers to physical demolition or destruction. **[Kerala Shipping Corporation Ltd.-CIT,Ker.,***Sec.41(2).***185CTR 364,265ITR13,132Taxman 763]**

Demurrage Kar. CBDT - Chowgule and Co. Ltd.

Demurrage has several meanings depending on the context. It may be an extended freight when the person responsible for payment of freight has to pay, in addition thereto,demurrage for the delay caused by him. It may mean compensation or damages, when demurrage is payable for the delay in taking delivery of goods from a warehouse. It may be an ingredient of hire charges payable by the charterer of the ship, for the delay in sailing the ship and delivering it in time; sometimes, it is treated as in part penalty. *Sec.172.* **[98CTR124,192ITR40]**

Denying his liability to be assessed under this Act Oudh. Rani Anand Kunwar - CIT

The words denying his liability to be assessed under this Act... are not confined to the denial of liability before the Income-tax Officer but include his denial in the appeal filed by him. There is nothing in the Indian Income-tax Act to justify the view that it is incumbent upon the assessee to deny his liability to assessment before the Income-tax Officer to invest him with a right of appeal under Section 30 of the Act. *Sec.246.***[8ITR126]**

Dependent AP. His Highness Prince Azam Jah. -ETO

. ... the term dependant includes in the case of an individual his wife or husband and minor child even if they were not dependent on him or her. *Expenditure-tax Act.***[55ITR 230]**

2. It is only a spouse or minor child who is wholly or mainly dependent on the assessee for support and maintenance that falls within the definition of the term dependent. **[Rajkumarsinghji-CET,MP.,***Expenditure-tax Act***.78ITR405]**

3. The word dependant is not a term of art in Taxation and should bear its natural meaning which may not include one who is independent and who does not require and get the assistance of another for support and maintenance. **[CET-Krishna,Mad.,** *Expenditure-tax Act.***78ITR541]**

Deposit SC CIT-Bazpur Co-operative Sugar Factory Ltd.

The essence of a deposit is that there must be a liability to return it to the party by whom or on whose behalf it is made on the fulfilment of certain conditions. *Misc.***[70CTR 94,172ITR321,38 Taxman195]**

2. From the definition of deposit in Explanation (b) to section 40A (8), it is manifest that deposit could mean any deposit of money with the company including any amount borrowed by the company. Borrowing is a generic term. It includes deposits as well as loans and the distinction between deposits and loans is thus obliterated and furnishes no saving from the mischief of the provisions of section 40A(8). **[CIT-Suman Tea and Plywood Industries (P.) Ltd.,Cal.,***Sec.40A(8).***115CTR156,204ITR719,71Taxman622]**

3. The specific definition of the word deposit in clause (b) of the Explanation to section 40A(8) does not in any manner contemplate which persons, and/or the nature or character of the persons, from whom the assessee-company may have received such amounts. Thus,the fact that the amount was received by the company from directors, friends or relatives, etc., is totally irrelevant for the purposes of this provision.Clause (b) of the Explanation to section 40A(8), while defining the word deposit,has made no such exclusion in favour of directors, friends, relatives, etc., and, therefore, deposits by such persons in their current accounts cannot be excluded. Therefore, under section 40A(8), the interest paid by the assessee-company on amounts received from friends and relatives of directors and directors and shareholders can be subjected to a disallowance of 15 per cent. as contemplated by the section. **[Agew Steel Manufacturers Pvt. Ltd.-CIT,Guj.**<u>*Sec.40A(8).*</u>**209ITR77]**

4. The definition of deposit is very wide which is evident from the word any that precedes deposit. The term deposit has been defined in clause (b) of the Explanation to mean any deposit of money with a company. The nature of the deposit, viz., fixed deposit, call deposit, deposit in current account or savings account, is immaterial and irrelevant for determining whether it is a deposit within the meaning of section 40A(8) of the Act or not. The decisive factor is the true relationship between the assessee-company and its credi-tors. A deposit would not cease to be so merely because of the fact that it is kept in a current account if it does not fall in any of the exclusions contained in clause (b) of the Explanation to section 40A(8).**[CIT-Jhaveri Bros. and Co. Pvt. Ltd. Bom.**<u>*Sec.40A(8).*</u>**214ITR374]**

5. The word deposit has been explained to mean any deposit of money with, and includes any money borrowed by a company. **[Daga and Co.(P.)Ltd.-CIT,Cal.,**<u>*Sec.40A(8).*</u>**,227ITR480]**

6. The term deposit as defined by the Explanation to section 40A(8) also means any deposit of money and includes any money borrowed by the company.There is therefore, no occasion for drawing a difference between deposits made by directors and shareholders and deposits made by public,other than directors and shareholders of the company.The Company Law Board circular deals with deposits dealt with by the Companies Act and not for the purpose of the Income-tax Act. **[CIT-Bhandari Machinery Co. (P.) Ltd.,Del.,** <u>*Sec.40A(8).*</u> **151CTR334,231ITR294]**

7. ....from the definition of the word deposit as contained in Explanation (b) to sub-section (8) of section 40A of the Act, it is amply clear that unless interest liability is incurred on the deposit of money which includes money borrowed by a company, no disallowance as envisaged by sub-section (8) of section 40A could be made. **[CIT-Commonwealth Trust India Ltd.,Ker.,** <u>*Sec.40A (8) .Expl. (b).*</u>**153CTR485,240ITR758,106Taxman39]**

8. The word deposit has not been left to take its colour from the general law but has to be construed in the context of the provision and the special meaning assigned to it, though even alluding to the general meaning of deposit in the context of the provision of sub-section (8) of section 40A, the current account of directors, etc., with the company would not make difference. The word deposit has been defined to mean any deposit of money with, and includes any money borrowed by a company. This is wide enough to obliterate the distinction between a deposit and a loan that is generally kept in view in the context of determining when the same becomes payable, that is particularly with reference to finding out when the payment was due and computing the period of limitation for filing a suit. **[CIT-Navjivan Roller and Pules Mills,Guj.***Sec.40A(8).Expln.(b).* **251ITR661,121Taxman200]**

Deposits Mad CIT- Khivaraj Motors Ltd.

The expression deposits has been defined in Explanation (b) to section 40A(8) of the Income-tax Act, 1961, and it means any deposit of money, which includes any money borrowed by a company, but does not include any amount received by the company mentioned in sub-clauses (i) to (ix). The deposits made by directors in a company in their current account do not form part of any of the exempted categories. Deposits which are understood in the business of a bank may be in the current account, savings bank account and fixed deposit account.Payment into the current account cannot be excluded from the nature of deposits which are made in the bank. The only distinction between a fixed deposit and this deposit is that the term for which the payment has been made in the case of a fixed deposit is a fixed one whereas, in the case of a current account, no time is fixed therein and this distinction will not take the amount outside the purview of deposit used in the clause. Any assistance from the Companies (Acceptance of Deposit) Rules, 1975, cannot be taken because the said rules came into force on February 3, 1975, and clause (ix) was specifically added in the definition of deposit in the said Rules with effect from September 18, 1975, which excluded deposits by the directors from the term deposit. This specific exclusion by the amendment in the Rules makes it more clear that deposits by the directors were included in the term deposit and it is by way of a specific provision that the same has been excluded. The words deposit by the directors' which were excluded by the insertion of clause (ix) of the definition in the Rules, refers to all deposits whether they are for a fixed period or in their current accounts. In Explanation (b) to section 40A(8) of the Income-tax Act,by defining the word deposit, no such exclusion has been made and, therefore,deposits by the directors in their current accounts cannot be excluded. On a correct interpretation of the provisions of section 40A(8) the payments which are made by a director to the company in the current account of the said director on which the company is paying interest will be considered as a deposit. Sec.40A(8). **146CTR700,227ITR473]**

2. A perusal of section 40A(8) of the Income-tax Act, 1961, makes it absolutely clear that the word deposits has not been qualified by any phrase limiting it to public or private deposits. The word deposit is in respect of any deposits received by the company. Any deposits would definitely cover within its ambit all the deposits whether received from the public or shareholders or from the directors. **[CIT-Nath Roller Flour Mills (P.)Ltd.,All.,** Sec.40A(8).**192CTR328 , 273 ITR432,142Taxman 578]**

Depreciation Guj. CIT - Daudayal Hotels P. Ltd.

The term depreciation as understood in commercial circles and those dealing with the assessment of income from business, as well as within the meaning of accountancy practice, means wear and tear of the assets used for the purposes of earning revenue on user of the assets. In other words, one cannot deduce the correct income without taking into account the wear and tear which an asset undergoes while being used for the purpose of generating receipts, which on finalisation of accounts, result in taxable profits. The concept of depreciation is that any asset, on account of normal wear and tear, is required to be replaced at a point of time in future. Therefore, to enable a business to meet the cost of such replacement, the wear and tear is permitted to be calculated at a notional rate of percentage of the cost/written down value of the assets. Sec.32. **[199CTR556,282ITR132,152Taxman389]**

Depreciation actually allowed Bom. Allied Publishers Pvt. Ltd.- CIT

The words depreciation actually allowed in section 10(5)(b), mean depreciation of which the assessee has received effective advantage or benefit and not merely depreciation which is notionally allowed or which is allowable. Sec.43(6)(b).**[68ITR546]**

Derive income from land Mad. CIT - Maddi Venkatasubbayya

Agricultural income cannot be said to accrue to every person into whose hands the produce of the land passes.It is only the owner, landlord or ryot, or persons having a derivative interest in the land from these persons that can be said to derive income from the land by the performance of agricultural operations in it. Sec.2(1). **[20ITR151]**

Derived **Bom. Hindustan Lever Ltd.-CIT**

As far as Income-tax law is concerned,the word derived has been given a narrow meaning-a strict meaning by courts, and has been understood in the restricted sense of a direct derivation and not understood in the broad sense as equivalent to derived directly or indirectly. In other words,only the proximate source has to be considered and not the source to which it may ultimately be referable. *Misc.***[12CTR55,121ITR951]**

2. The word derived in section 80J of the Income-tax Act,1961, is not a term of art. Its use in the definition indeed demands an enquiry into the genealogy of the product. But the enquiry should stop as soon as the effective source is discovered. Profit or gain can be said to have been derived from an activity carried on by a person only if the said activity is the immediate and effective source of the said profit or gain. There must be a direct nexus between the activity and the earning of the profit or gain. The income, profit or gain cannot be said to have been derived from an activity merely by reason of the fact that the said activity may have helped to earn the said income or profit in an indirect or remote manner. The expression derived from the business in section 80J should receive a restricted meaning and if it is an income directly relatable to the business activities of the assessee, it will be deemed to be derived from the business of the assessee. **[CIT-Eastern Seafoods Exports (P.) Ltd.,Mad.,***Sec. 80J.***215ITR64]**

3. ....cash assistance, duty drawback and import entitlements though undoubtedly attributable to business carried on by the assessee as the assessee would not have been in a position to receive any of those benefits, had the assessee not been carrying on business, it cannot be said that such income was derived from business as to qualify for deduction under sections 80J and 80HH of the Act. .... the expression derived from is narrower than the expression attributable to.... the word derived is usually followed by the word from, and it means : get or trace from a source; arise from, originate in; show the origin or formation of. **[CIT-Viswanathan and Co., Mad.,***Sec.80J.80HH.***181CTR335,261ITR 737,133Taxman 476]**

4. The word derived would mean derived from the source.The source has to be in section 14. **[CIT-Williamson Fin.Services,SC.,** *Sec.80HHC.r.w.r.8(1).***213CTR,612,297ITR17,165Taxman638]**

Derived from **Ker. Sterling Foods – CIT**

The expression derived from has a definite but narrow meaning and it cannot receive a flexible or wider concept. *Sec.80HH.* **[150ITR292]**

2. The expression, derived from should be given a restricted meaning and whenever the Legislature wants to give a wider expression, the Legislature employs the expression, attributable to and the use of the expression, derived from indicates that the profit or gain should be derived from the conduct of the business. There is no justification to give the expression,derived from a wider meaning to cover every receipt connected with the industrial undertaking. It is not all business receipts that would qualify for the deduction and the Legislature has apparently not intended to give the benefit of deduction to all business income.If the intention of the Legislature was to grant relief to all business income, it could have used the expression, profits and gains of industrial undertaking. **[CIT-Pandian Chemicals Ltd.Mad.***Sec.80HH.***147CTR5,233ITR497]**

3. The meaning of the expression derived from has got only a limited import and therefore, the expression derived from as used in section 80HHC must be understood as profit directly arising from the export of the goods and not incidental to the export. **[K.RavindranathanNair-DCIT(Asst.),Ker.,***Sec.80HHC.***181CTR 310,262ITR669]**

4. The words derived from in section 80HH of the Income-tax Act, 1961, must be understood as something which has a direct or immediate nexus with the assessee's industrial undertaking. Although electricity may be required for the purposes of the industrial undertaking, the deposit required for its supply is a step removed from the business of the industrial undertaking. **[Pandian Chemicals Ltd.- CIT, SC,** *Sec.80HH.* **[183CTR 99,262ITR278, 129 Taxman 539]**

5. There must be, for the application of the words derived from, a direct nexus between the profits and gains and the industrial undertaking. **[CIT-Ritesh Industries Ltd. Del.,***Sec.80I.***192CTR 81, 274ITR324,142 Taxman551]**

6. The connotation of the words derived from is narrower as compared to that of the words attributable to.By using the expression derived from Parliament intended to cover sources not beyond the first degree. **[LibertyIndia,SC,***Sec. 80IB.***225CTR,233,317ITR218, 183Taxman349]**

7. There was a clear distinction between the terms derived from and attributable to. The expression derived from had a very strict interpretation relatable only to income derived from the business and not to any other source of deposits made. The interest earned by the assessee from and out of fixed deposits could not be brought under the expression derived from the industrial undertaking. **[CIT –Tribalogy India Ltd.,Mad.,*Sec.80I.*238CTR192,335ITR12]**

8. ...the expression derived from occurring in section 80HHD of the Act meant that such an income should be the result of services provided to foreign tourists. The source of income should be generated directly from the foreign tourists. **[CIT-Le Passage to India Tour and Travels P.Ltd.,Del.,** *Sec.80 HHD.* **241CTR535, 335ITR69]**

Derived from / attributable Mad. CIT - B.Desraj

The expression used in section 80HHC is derived from.Whenever the said expression is used, the Legislature wanted to give a restricted meaning. If the expression attributable to is used, it means the Legislature wanted to give wider meaning. *Sec.80HHC.* **[293ITR214]**

Derived from an industrial undertaking MP. CIT- Paras Oil Extraction Ltd.

The words derived from an industrial undertaking mean that the income has been derived from industrial activity which the industry is undertaking and it does not mean any industrial activity undertaken by the assessee. The words industrial undertaking have to be construed narrowly and cannot be given a wide meaning. What is to be seen is as to what is the activity of the industrial undertaking of the assessee. *Sec.80HH.80I.* **[230ITR266,96Taxman234]**

Derived from exports Guj. Ahmedabad Mfg. and Calico Printing Co. Ltd. - CIT

The words derived from exports cannot be interpreted as meaning referable to exports. Profits and gains can be said to have been derived from an activity carried on by a person only if the activity is the immediate or effective source of the profits or gains. There must be a direct nexus between the activity and the earning of profits or gains. *Income-tax (Determination of Export Profits)Rules.***[25CTR 263,137ITR616]**

Destroyed Bom. Otis Elevator Co. (India) Ltd.-CIT

The expression destroyed used in the section has a wider connotation than mere physical destruction. It also applies to a case where the use of the asset is denied to the assessee in its business for an indefinitely long duration, even though there may be no evidence of its physical destruction.Where the asset, of the nature described in the section,becomes unavailable to the assessee for an indefinitely long period, albeit on account of an act of a stranger,the asset must be said to have been destroyed as far as the assessee is concerned. ...Moreover, though sale, discarding or demolition required a voluntary act on the part of the assessee, destruction can be either voluntary or involuntary, there being more likelihood of its being involuntary. *Sec.32(1)(iii).*[**195ITR682,60Taxman215]**

Detected **Guj.** **Manilal Gafoorbhai Shah - CIT**

Considering the scheme of section 24, the detection, which is stipulated by this sub-section, is the detection of the declared amount as the income of the declarant himself and of none else. This idea is that if the declared amount is already detected by the Income-tax authorities as the income of the declarant, or is deemed to have been so detected prior to the date of declaration, the declarant should not be allowed to take the special and peculiar benefits which the Act contemplates. The obvious anxiety of the Legislature was, therefore, to bring out the income which was not already detected or deemed to have been detected in the hands of an assessee. *VDIS.*[**95ITR624]**

Detection **Cal.** **CIT - Bimal Kumar Damani**

....The process undertaken for issuing notice under section 132A for requisitioning the materials from the customs authority were an attempt to obtain the documents. Until those documents were requisitioned, scrutinised and investigated upon there was no question of detection. *Sec.132A.*[**180CTR452, 261ITR87,129 Taxman564]**

Detriment **Del.** **Sony India Ltd. - CIT**

The dictionary meaning of the word detriment is any loss or harm suffered in person or property. It could also be understood that the promisee has, in return of a promise, forborne some legal right which he otherwise would have been entitled to exercise, or that he has given up something which he had a right to keep. Legal detriment again refers to giving up something which immediately prior thereto the person had the privilege to retain. (Black's Law Dictionary, 6th Edition).This expression has been used simply to imply that there should be no loss or injury to the Revenue.Loss of revenue would normally refer to non- recovery of its dues or that a demand of tax may not become irrecoverable.The assessee by his acts and deeds may not be able to frustrate the recovery of tax. Intention on the part of the assessee to cause loss or injury to the Revenue should be evident from the reasons recorded by the Officer in support of his belief that grant of full period would be detrimental to the interest of the Revenue. *Sec.220(1).r.w.s.156.* **[196CTR81,276ITR278,146Taxman98]**

Development SC Gujarat Industrial Development Corp. - CIT

The word development in section 10(20A) of the Income-tax Act, 1961, should be understood in a wide sense. There is no warrant to exclude all development programmes relating to any industry from the purview of the word development in the said clause. There is no indication in the Act that development envisaged therein should be confined to non- industrial activities. Development of a place can be accelerated through varieties of schemes and establishment of industries is one of the modes of developing an area. *Sec.10(20A).* **[227ITR414,94Taxman64]**

Devolution Mad. CIT - S. Krishnamurthy

The word devolution in section 49(1)(iii)(a) of the Income-tax Act, does not denote merely succession on death but includes also a change of ownership from one living being or body,whether incorporated or not, to another living being or body, whether incorporated or not. Consequently, the act of throwing the individual property into the family hotchpot could be considered to be an act of devolution within the meaning of the said provision. *Sec.49(1).***[38CTR71,152ITR669]**

Dharmadaya SC CIT - Bijli Cotton Mills (P.) Ltd.

Dharmada or Dharmadaya,in common parlance,means anything given in charity or for religious or charitable purposes.Among the trading or commercial community in various parts of this country a gift or payment for Dharmada is by custom invariably regarded as a gift for charitable purposes. *Misc.***[116ITR60]**

Direct and intimate connection **SC** **Indian Aluminium Co. Ltd.- CIT**

...to be a permissible deduction,there must be a direct and intimate connection between the expenditure and the business, i.e., between the expenditure and the character of the assessee as a trader, and not as owner of assets, even if they are assets of the business, needs to be qualified by stating that if the expenditure is laid out by the assessee as owner-cum-trader,and the expenditure is really incidental to the carrying on of his business, it must be treated to have been laid out by him as a trader and as incidental to his business. *Sec.37.*[**1CTR51,84ITR735]**

Direct costs **Bom** **CIT - King Metal Works**

The expression direct costs is defined to mean costs directly attributable to the trading goods exported out of India including the purchase price of such goods.Direct costs, therefore, cover the purchase price of the trading goods exported and costs which may be directly attributable to the trading goods. Freight and insurance cannot be regarded as costs directly attributable to the trading goods within the meaning of clause (b) of the Explanation to sub-section (3).The plain meaning of the section must be given effect to by the court in advancing the legislative intent. Freight and insurance attributable to the transportation of goods beyond the customs station does not constitute a part of the direct costs which are defined to mean costs directly attributable to the trading goods exported out of India. The words, exported out of India are used in a descriptive sense. In order that the costs can be regarded as direct costs within the meaning of Explanation(b),they must be attributable to the trading goods which are eventually exported out of India. *Sec.80HHC.*[**235CTR234,329ITR426,192Taxman294]**

Direction **All.** **Gangadhar Baijnath - CIT**

The expression direction in the proviso could only refer to the directions which the Appellate Assistant Commissioner or other tribunals can issue under the powers conferred on him or them under the respective sections.Therefore, the expression finding as well as the expression direction can be given full meaning, namely, that the finding is a finding necessary for giving relief in respect of the assessment of the year in question and the direction is a direction which the appellate or revisional authority, as the case may be, is empowered to give under the sections mentioned therein.

Sec.153.**[69ITR500]**

2. The finding referred to in section 34(3) must be one which is necessary for the disposal of the appeal.The finding should not be an incidental finding and the direction must be one which the authorities who, give it, must be entitled to give in the exercise of their powers as either appellate or revisional authorities. **[CIT-Homi Mehta and Sons P. Ltd.,Bom.,**Sec.153(3)(ii). **27CTR 238,137 ITR 217]**

Directions and findings Pat. CIT - Joharmal Parsuram (Firm)

The expression direction could only refer to the directions which the Appellate Assistant Commissioner or other Tribunals can issue under the powers conferred on him or them under the respective sections.Therefore,the expression finding as well as the expression direction can be given full meaning, namely,that the finding is a finding necessary for giving relief in respect of the assessment of the year in question and the direction is a direction which the appellate or revisional authority, as the case may be, is empowered to give under the sections mentioned therein.The words in consequence of or to give effect to have to be collated with, and cannot enlarge, the scope of the finding or direction under the proviso.

Sec.153(3)(ii).**[62ITR729]**

Directors may veto any transfer Cal. Star Company Ltd. - CIT

A provision in the articles of association of a company that the directors may veto any transfer without assigning any reason operates as a restriction on transfer and where the articles contain such a provision the shares of the company are not freely transferable, even though the directors might not have in fact objected to the transfer of any shares. Misc.**[58ITR149]**

Discarded Guj. CIT-Elecon Engineering Co. Ltd.

The word discarded has been understood to cover the case of obsolescence.Know-how in whatever form it may be is capable of diminishing in value over the years by obsolescence.It would, therefore, be included within the meaning of the word plant in section 32. Sec.32.**[96 ITR672]**

Discarding/demolishing or destroying **Mad.** **B.Nagi Reddy - CIT**

Conversion of the individual property into assets of a partnership in which the assessee is also a partner,would not constitute a sale nor could it be regarded as amounting to discarding,demolishing or destroying the building, machinery, plant or furniture or any of the assets. *Sec.34(2)(ii).* **[199ITR451]**

Disclosure **Cal.** **CIT - Bimal Kumar Damani**

The word disclosure means to disclose,reveal, unravel or bring to notice.Disclosure in Part III satisfied the above test of disclosure. ......The meaning of the phrase in good faith and true disclosure had to be construed in a wider amplitude.It was not the disclosure of the amount alone but also the nature of the amount of the concealed income of the assessee.If one claims that this was not a concealed income of the assessee and the income of someone else, then it could not be said to be a true disclosure of concealed income.The disclosure could not be treated to be bona fide or to have been made in good faith and a true disclosure nor could it be treated to be a situation within the meaning of clause (c) of the amnesty scheme.

Amnesty scheme. **[180CTR452,261ITR87,129Taxman564]**

Discontinuance of business **Pat.** **Hanutram Bhuramal - CIT**

The phrase discontinuance of business is apt to be used in an ambiguous sense.Where a business changes hands or a partner ceases to be a partner there is no discontinuance of business,but only a change in the ownership of the business. *Misc.* **[6ITR290]**

Discover **Bom.** **CIT- Sir Mahomed Yusuf Ismail**

The word discover in section 34 does not mean a mere change of opinion on the same facts or on a question of law,or the mere discovery of a mistake of law. *Sec.147.* **[12ITR8]**

Discovers **All.** **Badar Shoe Stores**

So far as the word discovers in the amended section 34 of the Indian Income-tax Act is concerned, it requires that the Income-tax Officer should have formed an honest and reasonable belief upon material which could reasonably support such belief. In the nature of things, it cannot amount to a conclusion of certainty. *Sec.147.*[14ITR431]

Discovers Pun. CIT-Shree Jagan Nath Maheshwary

The word discovers has been interpreted by English Courts to mean, comes to the conclusion from the examination the inspector makes,and from any information he may choose to receive or has reason to believe, or finds or satisfies himself, or honestly comes to the conclusion from information before him.The courts in India have adopted the same interpretation of the word.

Sec.147.[AIR1957Pun226,32ITR418]

Discretion Del. VLS Finance Ltd. - CIT

Discretion must be a sound one governed by law and guided by rule, not by humour. There is nothing like unfettered discretion immune from judicial reviewability.Courts stand between the executive and the subject, alert, to see that discretionary power is not exceeded or misused. Discretion is a science of understanding to discern between right or wrong, between shadow and substance,between equity and colourable glosses and pretence and not to do according to one's will and private affections.The action of the State,an instrumentality, any public authority or person whose actions bear the insignia of public law element or public character are amenable to judicial review and the validity of such action would be tested on the anvil of article 14 of the Constitution. *Sec.281B.*[163CTR343, 246ITR707,112Taxman295]

Dismissed Mad. M.R.M.Periannan Chettiar - CIT

The term dismissed has a definite legal connotation,implying a final disposal by the Tribunal rejecting the case of the suitor.

Misc.[39ITR159]

Disposition AP. Kancharla Kesava Rao-CED

The term disposition covers every mode by which property can pass by act of parties and it should be liberally construed.

Estate Duty Act.[74ITR428]

2. The word disposition is not a term of law.It has no precise meaning.Its meaning has to be gathered from the context in which it is used.In the context in which it is used in section 2(xxiv), it cannot mean to dispose of. In that sub-section,it is used along with the words conveyance, assignment, settlement, delivery,payment or other alienation of property. Hence, it is clear from the context that the word disposition therein refers to a bilateral or a multilateral act. It does not refer to a unilateral act. **[Goli Eswariah-CGT,SC,** *Sec.2(xii),(xxiv) (d).G.T.Act.***76ITR675]**

3 The term disposition is not a term of art nor legalese but a plain English word of wide import. What is more, the word has acquired, beyond its normal ambit, an extended meaning on account of the special definition in section 2(15) with Explanation 2 super added. Explanation 2 is deliberately designed to take into its embrace what otherwise may not be disposition or conform to its traditional concept. **[CED-Kantilal Trikamlal,SC,***EstateDutyAct.***105ITR92]**

Distributed MP. Central India Industrial Corporation Ltd. - CIT

For the purpose of section 23A of the Income-tax Act, 1922, no real distinction can be drawn between declaration and distribution of dividend and if dividends are declared by a company, then for the purpose of that provision there is a distribution of dividends.To read the word distributed as meaning actual payment would be contrary to the object of the provision and would lead to a manifest absurdity amounting to an inconsistency with the provisions of section 16(2). *Sec.104,109.***[48ITR543]**

Distribution SC Punjab Distilling Industries Ltd. - CIT

The word distribution in the context of section 2(6A)(d) means allotment or apportionment of the surplus amongst the share-holders; the allotment takes place and each shareholder gets a vested right to his portion of the surplus as soon as the capital stands reduced. *Sec.2(22).***[57ITR1]**

Diversion of income by an overriding title SC CIT-Imperial Chemical Industries (India) (P.) Ltd.

An obligation to apply the income in a particular way before it is received by the assessee or before it has accrued or arisen to the assessee results in the diversion of income. An obligation to apply income which has accrued or arisen or has been received amounts merely to the apportionment of income and the income so applied is not deductible.The true test for the application of the rule of diversion of income by an overriding title is whether the amount sought to be deducted in truth never reached the assessee as his income. *Misc.*[74ITR17]

Diversion of income by an overriding title **Raj** **Smt. Savita Mohan Nagpal - CIT**

In order to determine whether there has been a diversion of income by overriding title, the true test is whether the amount sought to be deducted, in truth, never reached the assessee, as his income. Obligations, no doubt, there are in every case, but it is nature of the obligation that is the decisive factor. Where by the obligation income is diverted before it reaches the assessee, it is deductible; but where the income is required to be applied to discharge an obligation after such income reaches the assessee, the same consequence in law does not follow. *Misc.***42CTR 211,154 ITR 449**

2. In determining whether there has been diversion of income by overriding title, it is the nature of the obligation which is the decisive fact. There is a difference between an amount which a person is obliged to apply out of his income and an amount which by the nature of the obligation cannot be said to be a part of the income of the assessee. Where by an obligation, income is diverted before it reaches the assessee, it is deductible; but where the income is required to be applied to discharge an obligation after such income reaches the assessee, the same consequence, in law, does not follow. It is the first kind of payment which can truly be excused and not the second. The second payment is merely an obligation to pay another a portion of one's income, which has been received and is since applied. The first is a case in which the income never reaches the assessee, who even if he were to collect it, does so, not as part his income, but for and on behalf of the person to whom it is payable. **[K.C. Bose and Co.-CIT,Cal.** *Misc.***156ITR701]**

3. In deciding whether there has been a diversion of income by overriding title,the true test is whether the amount sought to be deducted,in truth, never reached the assessee as his income. It is the nature of the obligation which is the decisive factor. There is a difference between an amount which a person is obliged to apply out of his income and an amount which by the nature of the obligation cannot be said to be a part of the income of the assessee. Where,by the obligation, income is diverted before it reaches the assessee, it is deductible; but where the income is required to be applied to discharge an obligation after such income reaches the assessee, the same consequence, in law,does not follow. It is the first type of obligation which diverts income by reason of overriding title and not the second. The second type of obligation is merely an obligation to pay another a portion of one's own income, which has been received and is since applied. **[CIT-Sh.& Smt. Ganesh G.K. Azrenkar, Bom.*Misc.*123CTR158, 217ITR148,130 Taxation 442,79Taxman141]**

4. There is a difference between an amount which a person is obliged to apply out of his income and an amount which by the nature of the obligation cannot be said to be a part of the income of the assessee. Where by obligation income is diverted before it reaches the assessee, it is deductible; but where the income is required to be applied to discharge an obligation (self-imposed and gratuitous) after such income reaches the assessee the same consequence, in law,does not follow. Diversion of income by overriding title can be created even by a contract between the two parties before the income reaches the hands of the assessee. **[CIT-MadrasRaceClub,Mad.*Misc.*219ITR39,135Taxation500]**

5. In order to determine whether there has been a diversion of income by overriding title the true test is whether the amount sought to be deducted, in truth, never reached the assessee as his income. Obligations, no doubt, there are in every case, but it is the nature of the obligation which is the decisive fact. There is a difference between an amount which a person is obliged to apply out of his income and an amount which by the nature of the obligation cannot be said to be a part of the income of the assessee. Where by the obligation income is diverted before it reaches the assessee, it is deductible; but where the income is required to be applied to discharge an obligation after such income reaches the assessee, the same consequence, in law, does not follow. It is the first kind of payment which can truly be excused and not the second. The second payment is merely an obligation to pay another a portion of one's own income, which has been received and is since applied. The first is a case in which the income never reaches the assessee,who even if he were to collect it, does so, not as part of his income, but for and on behalf of the person to whom it is payable. **[Smt. K. Sarala Devi-CIT,Ker. *Misc.*132CTR464,222ITR211,136 Taxation219,88 Taxman18]**

6. In order to decide whether there has been a diversion of income by overriding title, the true test is whether the amount sought to be deducted, in truth, never reached the assessee as his income.

Obligations, no doubt, there are in every case, but it is the nature of the obligation which is the decisive fact. There is a difference between an amount which a person is obliged to apply out of his income and an amount which by the nature of the obligation cannot be said to be a part of the income of the assessee. Where by the obligation, income is diverted before it reaches the assessee, it is deductible,but where the income is required to be applied to discharge an obligation after such income reaches the assessee, the same consequence, in law, does not follow. It is the first kind of payment which can truly be excused and not the second. The second payment is merely an obligation to pay another a portion of one's own income, which has been received and is since applied. **[CIT-Madras Race Club, Mad.**<u>*Misc.*</u>**175CTR264,255ITR 98,164 Taxation704,126Taxman6]**

7. When a third person becomes entitled to receive the amount in question under an obligation of the assessee even before the assessee could lay a claim to receive it as his income, there would be a diversion of income by overriding title ; but when after receipt of the income by the assessee, the same is passed on to a third person in discharge of the obligation of the assessee, it will be a case of application by the assessee and not of diversion of income by over riding title. **[CIT-Sunil J. Kinariwala,SC.**<u>*Sec.60.*</u>**AIR 2003 SC668,179CTR15,259 ITR 10,126Taxman161]**

8. The fundamental principle is that an application of income is an allocation of one's own income after it accrues or has arisen, although such application may be under a contract or obligation, whereas diversion of income is that which diverts away or deflects before it accrues or reaches the assessee and it is received by him only for the benefit of the person who is entitled to the income under an overriding charge or title. There is a distinction between obligation to spend money in a particular manner attached to an income, and a similar obligation attaching to the source of the income. If the obligation is on the source of the income it is a case of diversion of income by overriding title, but if the obligation is to spend the money in a particular manner it is only an application of the income. **[Jit and Pal X-ray Pvt. Ltd.- CIT, All.**<u>*Misc.*</u>**186CTR541,267ITR 370,134Taxman62]**

9. The principles relating to diversion of income by overriding title are (i) if a third person becomes entitled to receive an amount under an obligation of an assessee even before he could claim to receive it as his income, there would be a diversion of income by overriding title but when after receipt of the income by the assessee, the same is passed on to a third person in discharge of the obligation of the assessee, it will be a case of application of income by the assessee and not of diversion of income by overriding title ; (ii) if income does not result at all, there cannot be a tax,even though in book-keeping, an entry is made about the hypothetical income which does not materialise; (iii) the existence or absence of entries in books of account cannot be decisive or conclusive in the matter ;(iv) the concept of real income must be applied in appropriate cases but with circumspection and must not be called in aid to defeat the fundamental principle of law of Income-tax as developed. **[U.P.Bhumi Sudhar Nigam-CIT,All.**<u>*Misc.*</u>**195CTR450,280ITR 197,144Taxman 91]**

Diversion of income by reason of overriding title **Bom.** **Colaba Central Co-operative Consumers' Wholesale and Retail Stores Ltd - CIT**

The doctrine of diversion of income by reason of overriding title applies only in cases where the income never reaches the assessee as his income. The mere fact that the assessee has an obligation to apply a certain amount out of its income for a particular purpose cannot make it a case of diversion of income by overriding title. An obligation to apply the income accrued, arisen or received amounts merely to the apportionment of income and the income so applied is not deductible. There is a difference between an amount which a person is obliged to apply out of his income and an amount which, by the nature of the obligation, cannot be said to be a part of his income. Where by the obligation income is diverted before it reaches the assessee, it is deductible; but where the income is required to be applied to discharge an obligation after such income reaches the assessee, the same consequence, in law, does not follow. The obligation to apportion the income in a particular manner cannot be termed diversion of income by overriding title. <u>*Misc.*</u> **[142CTR394,229ITR209,97Taxman1]**

Dividend **Bom.** **Vidyutrai Y. Desai - CIT**

Dividend under section 2(6A)(c) of the Income-tax Act does not connote the actual income of the assessee but what is deemed income for the purposes of Income-tax law. The assessee may have incurred an actual loss by subscribing to the capital of the company and may have, in the distribution of assets in winding up, received much less than the capital subscribed; but in whatever form the distribution may have been made, to the extent to which that distribution is out of accumulated profits,it is a dividend and attracts tax. <u>*Sec.2(22).*</u>**[33ITR510]**

2. Dividend need not be distributed in money; it may be distributed by delivery of property or right having monetary value.

[Kantilal Manilal - CIT,SC.*Sec. 2(22).*41ITR275]

3. Dividend, in its ordinary connotation, means the sum paid to or received by a shareholder proportionate to his shareholding in a company out of the total sum distributed. Dividend distributed by a company being a share of its profits declared as distributable among the shareholders, is not impressed with the character of the profits from which it reaches the hands of the shareholders.

[CIT - Nalin Behari Lall Singha,SC.*Sec.2(22).*74ITR849]

4. Since the Act has not provided for any other definition of the word dividend except the ones enumerated in section 2(22) of the Income-tax Act, 1961, it should be construed that this definition would be applicable to all provisions which contain the term dividend in the Act. As per section 104 an Income-tax Officer, if satisfied that a company in respect of any previous year has not distributed as required by the statute, dividends from out of its profits and gains, shall make an order in writing that the company shall,apart from the sum determined as payable by it on the basis of the assessment under section 143 or 144, be also liable to pay Income-tax at the rate provided in that section. The object of the Legislature in enacting section 2(22)(e) and section 104 of the Income-tax Act,1961, is one and the same, namely, to prevent the escapement of tax by some shareholders and/or companies.While under section 2(22)(e) of the Act, by a deeming provision,the Legislature has made payment of any advance or loan to a shareholder a deemed dividend so as to subject such payments to the levy of tax in the hands of the receiver of the said amount, section 104 of the Act provides for levy of tax on companies which attempt to avoid payment of tax by their shareholders by not distributing their surplus profits and income. In either case, the object of the Act is to see that evasion of tax is prevented. Thus it is clear that the Act did not contemplate the levying of tax twice, namely, once in the hands of the shareholder who has received it as a deemed dividend and again in the hands of the company which, according to the assessing authority, has failed to declare the dividend. The two sections have used two different verbs but that by itself would not take away the effect of the deeming provision found in the definition clause. If actually the Legislature wanted the deeming clause not to be made applicable to the provisions of section 104 of the Act then it would have said so in categorical terms in the statute, in the absence of which the statutory definition given under section 2(22)(e) of the Act will have to be applied to the word dividend as found in section 104 also. **[CIT-Mysodet(P.) Ltd.,SC.**<u>*Sec.2(22).*</u>**152CTR531,237ITR35,103 Taxman336]**

Documents evidencing the creation of trust MP. Laxminarayan Maharaj -CIT

That the words documents evidencing the creation of the trust embrace all evidential documents, i.e.,all documents which afforded a logical basis for inferring the creation of the trust. .... documents, though not directly evidencing the creation of the trust, afforded a logical basis for inferring the creation of the trust... <u>*Sec.12A.*</u>**[150 ITR 465,17Taxman80]**

Donee Guj. Bhavna Nalinkant Nanavati-CGT

The definition of donee specifically states that where a gift is made to a trustee for the benefit of another person it includes both the trustee and the beneficiary, meaning thereby that the trustee does not receive the property for his absolute use but receives the same for the benefit of another person and such other person when read in the context of a trust and as distinguished from the trustee shall have to be the beneficiary. <u>*Sec.2(viii).G.T.Act*</u>.**[174CTR152,255 ITR529]**

Drawn up Mad. CWT - T.R. Kannan

The expression drawn up is not defined. A fair reading of Explanation I shows that if a Balance-sheet was drawn up on the valuation date, then that balance- sheet should be taken into account for the purpose of rule 1D. The second contingency provided in the rules is that if the balance-sheet was not drawn up on the valuation date,then the balance-sheet drawn up on a date immediately preceding the valuation date should be taken into account.The third situation that is contemplated in the rule is that in the absence of both, the balance-sheet drawn up on the date immediately after the valuation date should be taken into account.

<u>*Sec.7WT.Act.r.w.r.1D.*</u>**[172CTR422,252ITR382,105Taxman669]**

Due Cal. Smt.Usharani Roy Choudhurani

The word due in section 7(1) is not limited to salary due in respect of the accounting year. <u>*Sec.15.*</u>**[10ITR199]**

2. The word due is normally understood as conveying the meaning imminent. Money payable becomes due for payment on the date when it is to be paid; such a date is usually called the due date. Unless the money payable is due, its recovery cannot be enforced; a creditor cannot demand payment, unless it is due. **[CIT - H.S. Shivarudrappa,Kar.**<u>*Sec.41(2).*</u>**200ITR1]**

3. ......the word payable is used with reference to the payer. Due means as a noun, an existing obligation or a simple indebtedness without reference to the time of payment ; or a debt ascertained and fixed though payable in future. As an adjective, due means capable of being justly demanded, payable or owing and unpaid. The expression due is defined by Webster to mean that which is owed ; that which custom, statute or law requires to be paid.

[CIT -Upnishad Investment P. Ltd.,Guj.<u>*Sec.18.*</u>**260ITR532]**

4. The word due has several shades of meaning. The word is used to refer to debt or obligation which has become immediately payable. It is also used to refer to simple indebtedness without reference to the time of payment. Though the amount becomes payable when the asset is transferred in the absence of any agreement to the contrary, and the amount also becomes due in the sense that that amount is owed, the determination of the amount if not made at the time of the transfer, and is made subsequently, the amount can be said to become due at that subsequent point of time when the amount payable is determined.Section 41(2) of the Income-tax Act, 1961,does not equate payable with due.The word due in section 41(2) of the Act should be understood as the time at which the person entitled to the monies could enforce the payment of the same which had by then been determined. **[CIT-Southern Roadways Ltd.,Mad.,**<u>*Sec.41(2).*</u>**88 CTR53,266 ITR135]**

Due and payable Guj. Baroda Board and Paper Mills Ltd. - ITO

Under section 530(1)(a) of the Companies Act,1956, in a winding up, there shall be paid in priority to all other debts,inter alia, all revenues, taxes, cesses and rates due from the company to the Central or a State Government or to a local authority at the relevant date as defined in section 530(8)(c) and having become due and payable within the twelve months next before that date. The only meaning that could be attached to the word due occurring in the section is that it must be presently due and the words,due and payable mean the same thing,namely, that it must be presently payable. Therefore, so far as section 530(8)(a) is concerned, the revenue, tax, cess or rate due from the company to the Central or State Government or to a local authority must be presently payable, that is, that the liability could be enforced as at the relevant date and, secondly,it must have to become presently payable within twelve months immediately preceding the relevant date.

Companies Act.[**102ITR153]**

Due date **Ker. CIT - G.T.N.Textiles Ltd.**

The expression due date means the time stipulated for payment. As per the Explanation to clause (va) of section 36(1) of the Income-tax Act, 1961, for the purpose of the clause due date means the date by which the assessee is required as an employer to credit an employee's contribution to the employee's account in the relevant fund. *Sec.36(1).*[**269ITR282]**

2. As per the Explanation to clause (va) of section 36(1) for the purpose of the clause due date means the date by which the assessee is required as an employer to credit an employee's contribution to the employee's account in the relevant fund. The amount is deductible only if the assessee credits the amount to the employee's account in the relevant fund on or before the date by which he is legally or contractually required to do so. The right to deduction would be lost if the sum is credited after the due date. **[CIT- Jairam and Sons, Ker.** *Sec36(1).* **187CTR99,269ITR285, 134Taxman503]**

Due time **SC Prakash Nath Khanna - CIT**

Section 276CC refers to due time in relation to sub-sections (1) and (2) of section 139 and not sub-section (4). It cannot be said that the Legislature without any purpose or intent specified only sub-sections (1) and (2) and the conspicuous omission of sub-section (4) has no meaning or purpose behind it. Sub-section (4) of section 139 cannot control the operation of sub-section (1) wherein a fixed period for furnishing the return is stipulated. The mere fact that for the purposes of assessment and carry forward and set off of losses the return filed under sub-section (4) is treated as one filed within sub-section (1) or (2) would not amount to the return having being filed within due time. *Sec.276CC.*[**187CTR97,266ITR1,135 Taxman327]**

During a period of not less than seven years. SC CIT - Braithwaite and Co. Ltd.

The only interpretation which can be given to the expression during a period of not less than seven years in the proviso is that the period should go beyond seven years. The period of seven years would not be complete till the last minute or even the last second is counted. In other words, till the last minute of the seven- year period is completed, the period remains less than seven years. *Companies (Profits) Surtax Act.***[110CTR290,201ITR343,67 Taxman155]**

During the pendency of any proceeding Del. Nimitya Properties Ltd.-CIT

Sub-section(1) of section 281B of the Income-tax Act, 1961, authorises the Assessing Officer to pass a provisional order of attachment of property belonging to the assessee during the pendency of any proceeding for the assessment of any income or for the assessment or reassessment of any income which has escaped assessment. .... The Explanation to sub-section (1) gives the scope of proceeding and categorically provides that proceedings under sub-section (5) of section 132 shall be deemed to be proceedings for the assessment of any income, etc. The ambit and scope of the word proceeding is expanded by this Explanation to include not only those proceedings,which are pending on the date of attachment of the property, when the search is authorized under section 132, but also proceedings which may have been completed on or before such date and includes all proceedings under this Act, such as those under sections 153A or 153C which may be commenced after such date in respect of any year, as a result of search. *Sec.281B .r.w.s.153.***[231CTR179,322ITR668]**

Dwelling place Bom. CIT - Fulabhai Khodabhai Patel

The connotation of a dwelling place is quite different from a mere residence or a mere house in which one finds oneself for a temporary or short period. A dwelling place connotes a sense of permanency, a sense of attachment, a sense of surroundings, which would permit a person to say that this house is his home. A man may have more than one home; he may have a home at different places; but with regard to each one of these he must be able to say that it is something more than a mere house or a mere residence.It is not sufficient in order to satisfy this test that there is a dwelling place in the taxable territories in which the assessee goes and lives. What is necessary and what is essential is that the dwelling place must be maintained for him. In other words, there must be a house or a building or a part of the building which must be set apart and made available for him, in which he could live if he so desired as a home. There must also be in him a right to live in such a dwelling place maintained for him, because without that right, it could not be said that he has either maintained a dwelling place or a dwelling place has been maintained for him. *Sec.6.* **[31ITR771]**

E

Earned Guj. CIT - S.G.Pgnatale

The word earned has two meanings. One is the narrow meaning of rendering services, etc. The word earned is also used in the wide sense treating income as earned only if the assessee has contributed to its accrual or arising by rendering services and in respect of which a debt is created in his favour. Unless there is a debt in favour of the assessee by reason of his rendering services it cannot be said to be income earned in the wide sense. There is a difference in the language used in clause (ii) and clause (iii) of section 9(1) of the I.T. Act, 1961. In cl. (ii), the Legislature has referred to income which falls under the head Salaries if it is earned in India. In cl. (iii), income chargeable under the head Salaries payable by the Government to a citizen of India for service outside India has been referred to. Therefore, if the Legislature wanted to indicate that the word earned, in the context of salaries, was to have the narrower meaning, it would have specifically said so by referring in cl. (ii) to the rendering of services at a particular situations. In view of the clear indication given by the Legislature itself by using a different phraseology in cl. (iii) as compared with cl. (ii), the words earned in India occurring in cl. (ii) must be interpreted as arising or accruing in India and not from service rendered in India.So long as the liability to pay the amount under the head Salaries arises in India, cl. (ii) can be invoked. If the liability to pay arises outside India and the amount is payable outside India, cl. (ii) cannot be invoked.

Sec.9.r.w.s.17. **[16CTR337,124ITR391]**

Earned in India Ker. CIT - S.R.Patton

Prior to April 1,1979, the words earned in India occurring in clause (ii) must be interpreted as arising or accruing in India and not from service rendered in India. So long as the liability to pay the amount under the head Salaries arises in India, clause (ii) can be invoked. If the liability to pay arises out of India and the amount is payable outside India, clause (ii) cannot be invoked. <u>*Sec.9(1)(ii).*</u>**[92CTR 197,193ITR49]**

Education Raj. CIT-Maharaja Sawai Mansinghji Museum Trust

Education connotes the process of training and developing knowledge of students by normal schooling.A museum cannot be taken to be an educational institution existing solely for educational purposes.It is not entitled to exemption under section 10(22) of the Income-tax Act, 1961. <u>*Sec.10(22).*</u>**[65CTR46,169ITR379, 33Taxman279]**

2. Advancement of knowledge brings within its fold suitable methods of its dissemination and though the primary method of sitting in a class-room may remain ideal for most of the initial education it may become necessary to have a different outlook for further education.It is not necessary to nail down the concept of education to a particular formula.Its progress lies in acceptance of new ideas and development of appropriate means to reach them to the recipients. **[Gujarat State Co-operative Union-CIT,Guj.** <u>*Sec.10(22), 11.*</u>**103CTR206,195ITR279]**

3. The word education connotes the process of training and developing the knowledge, skill, mind and character of students by normal schooling and has not been used in the wide and extensive sense according to which every acquisition of further knowledge constitutes education. **[CIT-Oxford University Press, Bom.** <u>*Sec.10(22).*</u>**[135CTR163,221ITR77,89Taxman353]**

Effective connection AAR Worley Parsons Services Pte. Ltd.

The effective connection should be between the royalty generating services and the permanent establishment. It is not enough that there is a permanent establishment of the non-resident in the source country carrying out some activities in connection with the project or the work....The terminology effective connection denotes a real and intimate connection. Clear correlation between the services which give rise to royalty income and the permanent establishment is a key factor... <u>*Sec.5(1)(2)(b)r.w.s.9(1)(i),(vi).*</u>
[223CTR,46,312ITR273,179Taxman347]

Either before or after the institution of .. MP. Laxmandas Pranchand - Union of India

The Income-tax Act does not provide that before instituting a criminal proceeding under the Act, the accused has to be given a notice in order to afford him an opportunity to compound the offence.The court cannot insist upon issuance of prior notice as a condition precedent for criminal prosecution...The composition of an offence under Chapter XXII of the Income-tax Act, 1961, is permissible either before or after the institution of proceedings. It is for the offender to take steps towards composition and not for the authorities to ask him whether he is intending to compound and save himself from prosecution. This can be done even after the proceedings in court. That being so, it would be improper to invalidate the proceedings for want of notice before institution of proceedings. Firstly, there is no condition of notice in the Act. Secondly, the occasion to compound is not frustrated even after the institution of the proceeding.The law must, therefore, sustain the validity of prosecution even when there is no prior notice. <u>*Sec.278B. r.w.s.279(2).*</u>**[154CTR315,234ITR261,98Taxman203]**

Electrical machinery All. CIT-Saran Khandsari Udyog

If a machinery has in-built motor and operates electrically, then the same would be electrical machinery. The generator generates electricity, but no in-built motor is fitted into it and it is not operated by electricity and, therefore, the same cannot be said to be an electrical machinery. <u>*Sec.32.*</u>**114CTR410,204ITR447, 73Taxman 303]**

2. For machinery to come within the scope of the term electrical machinery it is not necessary that the machinery should produce, transmit or store electricity. It would not also include any machinery which is run by electricity or whose motive power is electricity. What it comprehended is that the machinery is such that inbuilt into it is the electric motor which forms a vital and inseparable part of the machinery. **[CIT-S.R.P.Tools Ltd.,Mad.,** <u>*Income-tax Rules-r.5.*</u>**242 ITR 636]**

Employed Bom. CIT - Tata Engineering and Locomotive Co. Ltd.

The word employed used in section 32(1)(iv) is a very wide expression and takes within its ambit not only persons who are contractually employed in the strict sense of the term,but all persons engaged in the business of the assessee. It has no relation to the capacity in which the assessee contracts to employ them. What is required is that such persons should be so engaged in the business that one can say that they are employed. That being so, an apprentice will clearly fall within the expression persons employed in the business. Sec.32(1).**[112CTR328,201ITR1036]**

Employee Ker. CIT - Travancore Tea Estates Co. Ltd.

The expression employee in section 40(a)(v) is qualified by the word such and refers to the type of employees in the earlier part of the section.The earlier part of the section makes it clear that the employee must have had some benefit or amenity arising out of the expenditure. *Sec.40(a)(v).***[122ITR557]**

2. Therefore, the word employee will include only such workers who are directly employed by the assessee. **[R and P Exports-CIT,All.,** *Sec.80HH(2).***196CTR45,279ITR536,146Taxman404]**

Employer SC CBDT - Aditya V Birla

Employer means the use of services of any person: it comprehends whole time servant or part-time engagee. *Sec.10(6)(vii)r.w.s.80RRA.* **[67CTR165,170ITR137,36Taxman9]**

Employment Ker. CIT - O. Abdul Razak

Explanation (a) to section 6(1)(c) of the Income-tax Act,1961 is an exception, under which 60 days residence referred to in clause (c) is substituted by 182 days if the assessee went abroad in the previous year for the purpose of employment.No technical meaning is intended for the word employment used in the Explanation. Going abroad for the purpose of employment only means that the visit and stay abroad should not be for other purposes such as a tourism, or for medical treatment or for studies or the like.Going abroad for the purpose of employment therefore means going abroad to take up employment or any avocation as referred to in Circular No. 346, dated June 30,1982,which takes in self-employment like business or profession.The taking up of his own business by the assessee abroad satisfies the condition of going abroad for the purpose of employment covered by Explanation (a) to section 6(1)(c). <u>*Sec.6.*</u> **[241CTR485,337ITR267,198Taxman1]**

Employs Guj. CIT - Prithviraj Bhoorchand

The expression used in the statute is employs twenty or more workers. The plain dictionary meaning of the said term employ is to use the services of a person in return for payment. As long as the industrial undertaking manufactures articles or things, and where the manufacturing process is carried on without the aid of power, it employs twenty or more workers, the requirements of the provision are fulfilled. When the provision is clear and unambiguous, there is no need to read anything more into the same. <u>*Sec.80-I.*</u>**[200CTR82, 280ITR94]**

Encumbrance Bom. Jethamal Mohanlal Khivansara - Union of India

The word encumbrance means interests in or burden or charge upon property.The estate (encumbrances) which may be carved out of ownership, may be classified as (1) securities; (2) leases; (3) servitudes and (4) trusts. All such estates may be called encumbrances. The encumbrance in a given case may include in its fold claim by adverse possession which again creates interests in property. Thus, it would be clear that the word encumbrances can only mean interest in the property or land which is a subject-matter of agreement to sale...encroachment, i.e., an unlawful gaining upon the possession of a neighbour without having any right or interest in the land can hardly be said to be an encumbrance on the property. <u>*Sec.269.*</u>**[188CTR546,272ITR143,137Taxman545]**

End of the previous year SC CIT - K. Srinivasan and K.Gopalan

The expression end of the previous year in sub-section (3) and (4) of section 25 of the Indian Income- tax Act, 1922, in the context of those sub-sections means the end of an accounting year (a period of full 12 months)expiring immediately preceding the date of discontinuance or succession. <u>*Sec.3.*</u>**[23ITR 87]**

Enduring All. Girdhari Dass and Sons - CIT

The word enduring means enduring in the way that fixed capital endures and it does not connote a benefit that endures in the sense that for a good number of years it relieves the assessee of a revenue payment. *Sec.37,30(a)(i).***[105ITR339]**

Enduring benefit SC CIT - Coal Shipments P. Ltd.

Although an enduring benefit need not be of an everlasting character it should not be so transitory and ephemeral that it can be terminated at any time at the volition of any of the parties.

*Sec.30(a)(i)r.w.s.37.***[82ITR902]**

Engaged Del. National Projects Construction Corp.Ltd. - CWT

The word engage may have a variety of meanings depending on the context and setting in which it is used. Ordinarily the expression connotes doing of more than one act or one transaction. Continuity of action is implicit in the meaning of the word. It has also been used in the sense of being busy or conducting or devoting attention or effect or employing oneself. The words engaged in the manufacture,production, etc., should normally,therefore, mean continuously occupied in the manufacture as a principal business as distinguished from an occasional participation or single act or casual employment or a mere supervision without physical participation. The extent of activity would be a relevant factor and if such activity is at an extended scale it may be suggestive of being engaged in manufacturing activity. *Sec.45(d).WT.Act.***[74ITR465]**

Entertainment Guj. CIT - Patel Brothers and Co. Ltd.

The dictionary meaning of the term entertainment is to receive and treat with hospitality, which broadly means receiving and entertaining guests in a friendly, generous and liberal way. The term entertainment in the context of section 37 (2A) and(2B) of the Income-tax Act,1961,on its true construction and meaning, would include the acts or practice of receiving and entertaining strangers and friends in a friendly, generous and liberal way. These acts may consist of providing, inter alia, a formal or elegant meal, a banquet and being hospitable in providing for the wants of a guest in a liberal and generous manner.If the act of entertaining is on a lavish and a grand scale involving wasteful expenditure,it would, no doubt, amount to entertainment. On the other hand, if the acts or practice of being hospitable in the sense of providing meals, drinks or other wants of the persons entertained, whether they may be employees, workmen or officers,servants or agents in the service of an assessee, are an express or implied condition of service, they would not amount to acts of entertainment. Similarly, if the acts or practice of being hospitable in the sense of providing meals, drinks or satisfying any other wants of guests,whether they are friends, strangers or customers, are a part and parcel of the express or implied terms and conditions of business, trade or profession, or on account of long standing custom in such trade, business or profession, they would not amount to acts of entertainment.In the area lying between these two termini the broad dictionary meaning of the term 'entertainment' indicated above should be adopted.No doubt, entertainment is hospitable treatment of guests and every act of entertainment includes hospitality.But that would not warrant the converse position that every act of hospitality would constitute entertainment. Hospitality shown on account of obligation of business arising as a result of express or implied contract or arising on account of long standing custom of a trade, business or profession cannot amount to entertainment, and acts done in discharge of such obligation cannot be included and covered in the term entertainment. Though ultimately in each case it would be a question of fact depending on many factors, the following broad tests will provide a guideline to determine the nature of entertainment expenses: (a) If the provision of food, drinks or any amusement to a client, constituent or customer is on a lavish and extravagant scale, or is of wasteful nature, it is entertainment per se (b) If the provision of food, or drinks to a client, constituent or customer is in the nature of bare necessity, or by way of ordinary courtesy, or as an express or implied term of the contract of employment spelled out from long-standing practice or custom of trade or business, it will not amount to entertainment.(c) If the provision of food or drinks to a client, customer or constituent is in a liberal and friendly way, it may amount to entertainment having regard to the place, item and cost of such provision.(d) The provision of amusement to a client, customer or constituent by way of hospitality or otherwise will always be entertainment. *Sec.37.* **[106ITR424]**

2. Entertainment is hospitable treatment of guests and every act of entertainment includes hospitality. But that would not warrant the converse proposition to be correct and every hospitality would not constitute entertainment.Hospitality shown on account of obligation in the business arising as a result of an express or implied contract or arising on account of a long-standing custom of a trade, busi ness or profession cannot amount to entertainment and acts done in discharge of such obligation cannot be included and covered in the term entertainment in section 37(2A) of the Income-tax Act, 1961. **[CIT-Nadh Shah Kapur and Sons,P&H.** *Sec.37(2A).* **14 CTR 77, 122ITR972]**

3. The term entertainment in the context of section 37(2B) of the Income-tax Act, 1961, on its true construction and meaning,would include the practice of receiving and entertaining strangers and friends; but if the practice of being hospitable in the sense of providing meals, drinks or other wants of guests are a part and parcel of the express or implied terms and conditions of a business, trade or profession or on account of longstanding custom in such trade,business or profession, they would not amount to acts of entertainment.Thus, hospitality shown on account of obligation of business arising as a result of an express or implied contract or arising on account of the longstanding custom of a trade cannot amount to entertainment. **[CIT-Rajkumar Mills Ltd.,MP.** *Sec.37(2B).* **135ITR811]**

4. Generally, entertainment expenditure is an expression of wide import. However, in the context of disallowance of entertainment expenditure as a business expenditure by virtue of sub-section (2A) of section 37 of the Income-tax Act, 1961, the word entertainment must be construed strictly and not expansively. Ordinarily entertainment connotes something which may be beneficial for mental or physical well-being but is not essential or indispensable for human existence. A bare necessity, like an ordinary meal, is essential or indispensable and, therefore, is not entertainment. Where such a bare necessity is offered, it is hospitality not entertainment .Unless the definition of entertainment includes hospitality, the ordinary meaning of entertainment cannot include hospitality. For this reason, the expenditure incurred in extending customary hospitality by offering ordinary meals as a bare necessity, would not be entertainment expenditure. **[CIT-Patel Brothers and Co.Ltd., SC,** *Sec.37(2A).***AIR1995SC 1895,126 CTR133,215ITR165,126 Taxation393,81Taxman156]**

5. The meaning of the term entertainment is to receive and treat with hospitality or entertaining guests in a friendly, generous way. Generally entertainment expenditure is an expression of wide import. However, in the context of disallowance of entertainment expenditure it must be construed strictly and not expansively. **[CIT-AssamAsbestosLtd.,Gau.,***Sec.37(2A).***158CTR252,240ITR 297,108Taxman438]**

Entertainment expenditure Mad. CIT - K.P.V.Shaik Mohamed Rowther & Company

The very essence of entertainment expenditure is the purveying of food,refreshment, etc.,to the entertainers, especially in clubs and other public places.If, the whole scheme of this expenditure is to further the business of the entertainer, the intention behind the expenditure as well as the nature of the expenditure combine to stamp the outgoings as entertainment expenditure in the manner comprehended by the provisions of section 37(2A) of the Act.

Sec.37(2A).[**145ITR37,14Taxman440]**

Entries in books of account Bom. Sheraton Appearels - ACIT

Section 34 of the Evidence Act, 1872, refers to the words entries in books of account.The accounts under section 34 means accounts which are maintained in the regular course of business.

Evidence Act.[**175CTR651,256ITR20,123Taxman238]**

Erroneous Del. Gee Vee Enterprises - Addl.CIT

The word erroneous in Sec. 263 includes the failure to make such an enquiry. The order becomes erroneous because such an inquiry has not been made and not because there is anything wrong with the order if all the facts stated therein are assumed to be correct.

Sec.263.[**99ITR375]**

2. It is incumbent on the Income-tax Officer to further investigate the facts stated in the return, when circumstances would make such an inquiry prudent and the word erroneous in section 263 of the Income-tax Act,1961, includes the failure to make such an enquiry. The order becomes erroneous because such an inquiry has not been made and not because there is anything wrong with the order if all the facts stated therein are assumed to be correct.

[Duggal and Co.-CIT,Del. *Sec.263.***122CTR171,220ITR456,71 Taxman331]**

3. The term erroneous could be looked into only from the facts and circumstances which were placed before the assessing authority at the time of the assessment. There is no scope for reopening the assessment on a subsequent event or on any new material.

[Jai Kumar Kankaria-CIT,Cal.*Sec.263.***171CTR483,251ITR707, 120Taxman810]**

Erroneous and prejudicial to the revenue **SC** **Tara Devi Aggarwal - CIT**

Where an assessee is assessed on an income voluntarily returned, it is not prejudicial to the interests of the revenue only if it is found that the assessment was made on the basis that the income had been earned by the assessee which was assessable. Where an income has not been earned and is not assessable, merely because the assessee wants it to be assessed in his or her hands in order to assist someone else who would have been assessed to a larger amount, an assessment so made will be erroneous and prejudicial to the revenue and the Commissioner has jurisdiction under section 33B of the Indian Income-tax Act, 1922, to cancel the assessment and proceedings for assessment may be initiated under the provisions of the Act against some other assessee who according to the income-tax authorities would be liable for the income thereof. *Sec.263.*[**88ITR323**]

2. A bare reading of section 263 of the Income-tax Act, 1961, makes it clear that the prerequisite for the exercise of jurisdiction by the Commissioner suo motu under it, is that the order of the Income-tax Officer is erroneous in so far as it is prejudicial to the interests of the Revenue. The Commissioner has to be satisfied of twin conditions, namely, (i) the order of the Assessing Officer sought to be revised is erroneous; and (ii) it is prejudicial to the interests of the Revenue. If one of them is absent if the order of the Income-tax Officer is erroneous but is not prejudicial to the Revenue or if it is not erroneous but is prejudicial to the Revenue recourse cannot be had to section 263(1) of the Act. The provision cannot be invoked to correct each and every type of mistake or error committed by the Assessing Officer, it is only when an order is erroneous that the section will be attracted. An incorrect assumption of facts or an incorrect application of law will satisfy the requirement of the order being erroneous. In the same category fall orders passed without applying the principles of natural justice or without application of mind. The phrase prejudicial to the interests of the Revenue is not an expression of art and is not defined in the Act. Understood in its ordinary meaning it is of wide import and is not confined to loss of tax. The scheme of the Act is to levy and collect tax in accordance with the provisions of the Act and this task is entrusted to the Revenue. If due to an erroneous order of the Income-tax Officer, the Revenue is losing tax lawfully payable by a person, it will certainly be prejudicial to the interests of the Revenue. The phrase prejudicial to the interests of the Revenue has to be read in conjunction with an erroneous order passed by the Assessing Officer.Every loss of revenue as a consequence of an order of the Assessing Officer, cannot be treated as prejudicial to the interests of the Revenue, for example, when an Income-tax Officer adopted one of the courses permissible in law and it has resulted in loss of revenue, or where two views are possible and the Income-tax Officer has taken one view with which the Commissioner does not agree, it cannot be treated as an erroneous order prejudicial to the interests of the Revenue unless the view taken by the Income-tax Officer is unsustainable in law. **[Malabar Industrial Co. Ltd.-CIT, SC,** *Sec.263.* **159CTR1,243ITR83,109Taxman66]**

Erroneous Judgment **Gau. B & A Plantation and Industries Ltd. - CIT**

The expression erroneous has not been defined in the Act. Erroneous assessment refers to an assessment that deviates from the law and, hence, is invalid. The erroneous assessment pertains to a defect, which is jurisdictional in nature. It does not refer to the judgment of the AO. in fixing the amount or valuation of property. Erroneous judgment means one rendered according to course and practice of court; but contrary to law, upon a mistaken view of law or upon an erroneous application of legal principles.The two elements namely, that the order is erroneous and it is prejudicial to the interests of the Revenue must co-exist.When an AO. has several choices and he adopts one of the choices, it cannot be interfered with unless it is shown that the choice of exercise by the AO. is without application of mind or wholly contrary to the law. *Sec.263.* **[290ITR395]**

Erroneous order

An erroneous order resulting in prejudice to the interests of the Revenue need not be a motivated or mala fide one. Decisions/ orders of subordinate authorities which are erroneous mean not only when erroneous in point of law but erroneous in any sense and if that error results in causing prejudice to the interests of the Revenue it is susceptible to be corrected by the CIT in exercise of his jurisdict-ion under sub-Sec. (2) of Sec. 25. **[CWT-N.T.RamaRao, AP.** *Sec. 25.WT.Act.***[182CTR625,261ITR611, 129Taxman430,TaxLR912]**

2. An erroneous order does not mean a wrong order; it does not mean an order with which the CIT is unable to agree. An erroneous order would be an order which suffers from a patent lack of jurisdiction; the error must be with reference to jurisdiction. **[Smt.LilaChoudhury-CIT,Gau.**<u>*Sec.263.*</u>**209CTR266,289ITR226]**

3. An order cannot be termed erroneous unless it is not in accordance with law.If an Income-tax Officer acting in accordance with law makes a certain assessment, it cannot be branded as erroneous by the Commissioner simply because, according to him, the order should have been written more elaborately. **[CIT-Sunbeam Auto Ltd.,Del.,**<u>*Sec.263.*</u>**227CTR133,332ITR167, 189 Taxman 436]**

Error apparent from the record All. Devendra Prakash - ITO

There is no invariable rule of law that,to constitute an error apparent from the record, the error must be apparent from the record of the proceedings against the assessee himself. <u>*Sec.264.*</u> **[72ITR151]**

Error apparent on the face of the record Ker. Upasana Hospital and Nursing Home - CIT

The error apparent on the face of the record cannot be said to be the record of one particular assessment, but the entire record of the assessee relating to all the assessment years. <u>*Sec.154.*</u>**[173CTR54, 253ITR507,120Taxman545]**

Escape Lah. Amir Singh Sher Singh - CIT

There is no justification for confining the meaning of the word escape in section 34 of the Indian Income-tax Act to those cases only which have not come to the notice of the Income-tax Officer at all and excluding those cases where he has applied his mind but on account of an error of judgment has set any part of the income free from assessment. The section is wide enough to include any case of non-assessment, to whatever cause it may be due.The section, therefore, empowers of Income-tax Officer to revise an assessment already made and to assess a sum which had not been assessed by his predecessor on account of a wrong application of the Act. <u>*Sec.147.*</u>**[3ITR171]**

Escaped Lah. Diwan Kishen Kishore – CIT

The expression escaped as used in section 34, Income-tax Act, connotes failure by the taxing authority to tax the income owing to accidental or deliberate omission by the assessee to declare it or to some similar circumstances. It does not include cases where the income is known or disclosed to the Income-tax authority and has been the subject of assessment which has however, been set aside by a superior authority owing to some mistake in procedure or to the income being treated in a wrong category... Sec.147.**[1ITR143]**

2. The word escaped in section 34 should not be read in the widest sense that that word is capable of bearing. It means eluded notice in the course of assessment and does not mean had avoided being assessed and the expression had escaped assessment is not equivalent to had not been assessed. In order to escape assessment the income must avoid both calculation and charging and there is no distinction between mistakes of fact and mistakes of law on the part of the Income-tax Officer. **[Madan Mohan Lal -CIT,Lah.** Sec.147.**3ITR438]**

Escaped assessment Sindh. CIT- Lokumal Bhojumal

Where an assessee makes a false return and by misleading the Income-tax authorities evades income-tax, the income-tax authorities have power, on finding out that income had been concealed, to take proceedings under section 34 of the Indian Income-tax Act for fresh assessment.The words escaped assessment in section 34 are wide enough to cover such cases. Sec.147.**[6ITR51]**

2. The burden of proving that income has escaped assessment within the meaning of section 34 is on the Income-tax authorities. Section 34 is not a charging section but a machinery section; and a machinery section should be so construed as to effectuate a charging section,and in interpreting such a provision,the rule is that that construction should be preferred which makes the machinery workable.In the ordinary course, an order of assessment made after investigation by a particular Officer should not at his sweet will and pleasure be allowed to be revised. There must exist something, either suppressed by the assessee, or a fact or a point of law which was inadvertently or otherwise omitted to be considered by the Income-tax Officer,before he can proceed to act under section 34.A mere change of opinion on the same facts and law is not covered by that section. Under section 34, the Income-tax Officer cannot institute a fishing investigation or enquiry merely with the object of finding out facts which would entitle him to re- open a past year's assessment. But to enable the Income-tax Officer to initiate proceedings under section 34, it is enough that he, on the information which he has with him and in good faith, considers that he has a good ground for believing that the assessee's profit has for some reason escaped assessment or been assessed at too low a rate, so that a notice can be served if the Income-tax Officer is bona fide of opinion that the income has escaped. Action can be taken with reference to events which happened subsequently, these events having relation to the facts on which the original assessment had been made. In order to hold that income may have escaped assessment, there must have been either some fresh facts brought to the notice of the Income-tax authorities, or some change in the law, which were in existence during the chargeable accounting period, but which were not brought to the notice of, or taken notice of by, the Income-tax authorities during the assessment year, but which arose subsequent to assessment, having relation to the facts on which the original assessment had been made. Under section 34(1), the belief of the Income-tax Officer that income has escaped assessment or has been under-assessed must be that of an honest and reasonable person, based upon reasonable grounds; the Income-tax Officer may act under this section on direct or circumstantial evidence,but not on mere suspicion, gossip or rumour. The powers under the present section are wide, but they are not plenary. The words of the section are reason to believe, and not reason to suspect.

[Bhimraj Panna Lal-CIT,Pat., *Sec.147.***32ITR289]**

3. When the liability to pay tax is evaded by one method or other there is an escapement of assessment.The term escaped assessment includes both non-assessment as well as underassessment. When a person is not assessed to tax though he is liable to be taxed the tax escapes assessment.

[TaxOfficer-cum-Regional Transport Officer-Durg Transport Co.P.L.,SC*M.P.MotorVehicles(Taxation of Passengers)Act.***85ITR 156]**

4. The power to make assessment or reassessment within four years of the end of the relevant assessment year would be attracted even in cases where there has been a complete disclosure of all relevant facts upon which a correct assessment might have been based in the first instance, and whether it is an error of fact or law that has been discovered or found out justifying the belief required to initiate the proceedings.The words escaped assessment,where the return is filed,cover the case of discovery of a mistake in the assessment caused by either an erroneous construction of the transaction or due to its non-consideration, or caused by a mistake of law applicable to such transfer or transaction even where there has been a complete disclosure of all relevant facts upon which a correct assessment could have been based...In cases where the Assessing Officer had over-looked something at the first assessment, there can be no question of any change of opinion, when the income which was chargeable to tax is actually taxed as it ought to have been under the law, but was not, due to an error committed at the first assessment. **[Praful Chunilal Patel-M.J. Makwana,ACIT,Guj.,*Sec.147.*148 CTR62,236ITR832]**

5. In section 147 of the Income-tax Act,1961, provision is made for both assessment and reassessment in a case where any income chargeable to tax has escaped assessment for any assessment year. The proviso treats at par the assessment under section 143(3) and under section 147 and makes no distinction whether the escapement of income is by reason of the failure on the part of the assessee to make a return under section 139 or in response to a notice issued under sub-section (1) of section 142 or section 148. Under clauses (a) and (b) of Explanation 2 to section 147, cases where no return has been furnished by the assessee and where a return of income has been furnished by the assessee but no assessment has been made, have both been included in the expression escaped assessment. **[K.Govindan and Sons-CIT, SC.*Sec.147.*164CTR490, 247ITR192,114Taxman94]**

Escaping assessment SC CIT - M/s. Narsee Nagsee and Co.

The words escaping assessment apply equally to cases where a notice was received by the assessee but resulted in no assessment at all and to cases where due to any reason no notice was issued to the assessed and therefore there was no assessment of his income.

*Sec.34.***[AIR1960SC123240ITR307]**

Estimate and underestimate All. CIT - Elgin Mills Co. Ltd.

The words estimate and underestimate have not been defined in the Income-tax Act. The dictionary meaning of the word estimate is the action of valuing or appraising;and approximate calculation based on probabilities.The word underestimate signifies an estimate which is below the truth or which is at too low a rate. Both for making an estimate and an underestimate an application of mind is required on the part of the maker. The question whether there was justification for the estimate or whether it was an underestimate has to be examined with reference to the time when the estimate was filed. It is not correct to say that while making an estimate the maker should project himself into the future. It at the time when the estimate is filed there is proper basis and justification shown for it, then it cannot be said that it is an underestimate. Sec.216. **[123ITR712,3Taxman529]**

Etc. **Ker. CIT - Maulana Tea Co.**

The word etc.does not share the character of an inclusive definition. Usually, after reciting the initiatory words of a set formula,or a clause given in full, etc. is added as an abbreviation for the sake of convenience...the word etc. cannot be given any restrictive meaning. *Sec.271B,AmnestyScheme.*[**160CTR291,244ITR589,110 Taxman303]**

Every order **Mad Visvas Promoters (P) Ltd. - ITAT**

The term,every order referred to in section 260A of the Income-tax Act, 1961, apart from being qualified as one involving substantial question of law, would mean an order which finally disposes of the rights of the parties in controversy. In cases where an application for rectification is made and an order passed under section 254(2) of the Act by merely rejecting such application, it does not decide the substantial issue involved between the parties, since the issue has already been decided under section 254(1) of the Act by the Tribunal and if the application is filed subsequently under section 254(2) by way of miscellaneous petition for rectification of mistake within the period of limitation, viz., four years, and it is rejected, it would only make the original order passed by the Income-tax Appellate Tribunal final and if at all there is any substantial question of law that may arise or any right of the parties is finally decided by the Income-tax Appellate Tribunal, that can be only under the orders passed under section 254(1) of the Act. Therefore, the order passed under section 254(2) cannot be construed as any order referred to under section 260A of the Act to enable the aggrieved party to file appeal before the High Court. Sec.254. **[323ITR114,185Taxman145]**

Evidence **Lah. Paras Dass Munna Lal - CIT**

The word evidence as used in sub-section (2) of section 23 of the Income-tax Act,is not confined to direct evidence,but is comprehensive enough to cover circumstantial evidence.Sec.143(2)**[5ITR523]**

Examining the record **Guj. CIT - Arunaben Sumankumar**

... in view of the clear words used in clause (b) of the Explanation to section 263(1), it has to be held that while calling for and examining the record of any proceeding under section 263(1), it is and it was open to the Commissioner not only to consider the record of that proceeding but also the record relating to that proceeding avilable to him at the time of examination. <u>*Sec.263.*</u>**[177CTR470,259ITR 386,124Taxman57]**

Exceptional and not systematic **MP.** **CIT -Vallabh Leasing and Finance Co. Pvt. Ltd.**

... the juxtaposition of the words exceptional and not systematic shows that the Legislature was using the word systematic in contradistinction to exceptional. The avoidance of tax must be exceptional, that is, by way of exception to the normal practice of the assessee and it should not be systematic, that is, part of a regular reprehensible practice carried on by the assessee. <u>*Sec.94(3).*</u> **[187CTR20,265ITR1,134Taxman770]**

Exchange **Bom.** **CIT - Rasiklal Maneklal (HUF)**

1. The word exchange as defined in the Transfer of Property Act or as understood in ordinary parlance connotes the following: Existence of different properties owned by different persons; as a result of the transaction of exchange both properties continue to exist; the properties continue to be owned by two different parties but the ownership of one is transferred to the owner of the other and vice versa. <u>*Sec.45.*</u>**[95ITR656]**

2. An exchange involves the transfer of property by one person to another and reciprocally the transfer of property by that other to the first person.There must be a mutual transfer of ownership of one thing for the ownership of another. **[CIT-Rasiklal Maneklal (HUF),SC,**<u>*Sec.45.*</u> **[AIR1989SC1333,77CTR31,177ITR198]**

Exclusive use **Mad.** **CWT - Smt. Muthu Zulaikha**

A mere perusal of section 5(1)(iv) of the Wealth-tax Act,1957, clearly shows that it will apply only in the case of individual assessments of co-owners and not to the total value of the property jointly owned by the co-owners. Section 5(1)(iv) reads thus : one house or part of a house belonging to the assessee and exclusively used by him for residential purposes.The purpose should be residential,meaning thereby that it should not be used for commercial or non-residential purposes and it should be used exclusively for residential purposes. The expression cannot be read as solely for residential purposes. What is to be seen is whether the intention of the assessee is to live in the house. The expression exclusive use should be read to mean that the house should be used for residential purposes meaning thereby that it should not be let out for rent or given on license or used for commercial purposes. The requirement of exclusive use of the building for residential purposes must, therefore, be construed in a practical and pragmatic way rather than in a pedantic sense. *Sec.5(1)(iv).WT.Act.* **[164CTR613,245ITR800,115Taxman85]**

Exclusively used by him for residential purposes throughout **Mad.** **CWT - W.Doraisamy**

The expression exclusively used by him for residential purposes throughout the period of twelve months immediately preceding the valuation date can mean nothing but that the intention or animus manendi of the assessee was to live in the house and that he had not created any interest in the said property in favour of any other person, that is to say, there is no element of right in the property of any other person in the house. In such a case, the house belonging to the assessee is exclusively used by him for residential purposes throughout the period he has kept it available to himself for occupation whenever he desired. Any trespass or permissive possession cannot rob the owner, who has chosen to keep the house vacant,of his right of residence and so long as such a right of residence is not disturbed, it is he who is in occupation and thus it is he alone who is using it exclusively and no one else. The position is completely different, however, of a person, who has admitted a lessee upon the land on rent or even licenses for use to another on receiving a fee. In the case of a lease,there is an element of transfer of property.In the case of licence,there is a withdrawal by the owner on conditions of payment of a fee,from occupation and the use is not exclusively by him. *Sec.7.WT.Act.* **[130CTR157,215ITR853, 87Taxman260]**

Executor **Guj.** **CIT - Navnitlal Sakarlal**

1. The word executor as defined in the Explanation to section 168 of the Income-tax Act,1961, means an executor as known under the Indian Succession Act, 1925, as well as an administrator known under the Indian Succession Act and, what is more, any other

person administering the estate of a deceased person is also included in this special definition of executor, though, under the Indian Succession Act, such other person administering the estate of the deceased would never be referred to either as an executor or as an administrator. Reading the provisions of section 2(11) of the CPC in the light of the provisions of section 2(29) and section 159 of the Income-tax Act, 1961, it is clear that an executor, who certainly represents the estate of a deceased person in law, and an administrator,who also similarly represents the estate of a deceased person in law,and also any person who intermeddles with the estate of the deceased, all fall within the definition of legal representative for the purposes of the Income-tax Act. *Sec.168.***[7CTR258, 67ITR125]**

2. The term executor in section 168 of the Income-tax Act,1961, is not to be understood in any restricted sense as the Explanation to the section has given an extended meaning to the word executor so as to include an administrator or any other person administering the estate of the deceased person, that is, including one who is in de facto management of the property of the deceased person.

[CITP.Visalakshi,Mad.*Sec.168.***130CTR488,217ITR282,132Tax man281]**

Existing solely Mad. Valliammai Society - DG (Inv.)

Section 10(23C)(vi) of the Income-tax Act,1961, grants exemption to any educational institution existing solely for educational purposes and not for purposes of profit. The Legislature by using the word solely makes it very clear that the sole purpose of the institution shall be for the purpose of perpetrating the object of the institution. Again,the word existing is to be construed in the sense that the existence of the institution in so far as it relates to section 10(23C)(vi) is only for educational purpose and not commercial or profit-making purpose. *Sec.10(23C).***[232CTR121,327ITR337]**

Expenditure SC CIT - Nainital Bank Ltd.

In its normal meaning, the expression expenditure denotes spending or paying out or away, i.e., something that goes out of the coffers of the assessee. A mere liability to satisfy an obligation by an assessee is undoubtedly not expenditure: it is only when he satisfies the obligation by delivery of cash or property or by the settlement of accounts that there is expenditure. But expenditure does not necessarily involve actual delivery of or parting with money or property.If there are cross claims one by the assessee against a stranger and the other by the stranger against the assessee and as a result of the accounting the balance due only is paid, the amount which is debited against the assessee in the settlement of the accounts may appropriately be termed expenditure within the meaning of section 10(2)(xv). *Sec.37.*[**62ITR638**]

2. Expenditure is equal to expense and expense is money laid out by calculation and intention. **[South India Insurance Co.Ltd.-CIT,Bom.***Sec.44.***106ITR 969]**

3. The term expenditure used in section 40A(3) of the Act has a wide meaning and includes anything that is paid out or away, including payments for stock-in-trade. **[Ratan Udyog -ITO,All.** *Sec.40A(3).***6CTR134,109ITR1]**

4. The expression expenditure is not defined in the Income-tax Act, 1961,and must be understood in the context in which it is used. The Legislature has used the expression allowances and depreciation in several sections in the scheme in Chap. IV within which section 44A appears. Section 37 of the Income-tax Act, 1961,enjoins that any expenditure not being expenditure of the nature described in sections 30 to 36 laid out or expended wholly and exclusively for the purpose of the business or profession should be allowed in computing the income chargeable under the head Profits and gains of business or profession.In sections 30 to 36, the expression expenses incurred as well as allowances and deprec-iation has been used. For example, depreciation and allowances have been dealt with in section 32. Therefore, the Legislature was using the expression any expenditure in section 37 to cover both. It has used both the expressions allowances contemplated under section 32 as well as the actual expenditure as contemplated under section 30. Section 44A appears in Chap. IV,which comprises of sections 14 to 44B, and, under that chapter,under sub-heading D, section 44A appears.Sub-heading D deals with profits and gains of business or profession.Though,normally,an association like a trade, professional or similar association would have been entitled to provide for depreciation if it was using those assets for the purpose of its business, yet, as it cannot carry on any business as such, it would not be entitled to any depreciation. Therefore, it cannot be said that it does not fall within the expression expenditure deductible,which describes the quality and nature of the expenditure under the provisions of any Act. Though an exemption must be strictly construed and the provisions in fiscal statutes must receive strict construction, yet section 44A is a special provision for deduction in the case of certain computations. In that sense it is a beneficial provision. It is also a canon of construction that even in a statute which requires to be strictly construed, if there is any beneficial provision it should be liberally construed. The provisions of section 44A being intended for the benefit of a trade, professional or similar association, in case of doubt, it should be construed liberally and in favour of the taxpayer.**[CIT-Indian Jute Mills Association,Cal.** *Sec.44A.***23CTR198, 134ITR68]**

5. The word expenditure in section 40A(3) is of wide import and includes expenses which are taken into account while determining the gross profit. It includes the price paid for stock-in-trade and/or raw materials, etc. Spending in the sense of paying out of away of money is the primary meaning of expenditure. **[Fakri Automobiles-CIT,Raj.***Sec.40A(3).***160ITR504]**

6. Spending in the sense of paying out of away of money is the primary meaning of expenditure. Thus expenditure is what is paid out or away and is something which is gone irretrievably. The word expenditure in section 40A(3) of the Income-tax Act,1961,should not be given a restricted meaning because section 40A(3) is designed to check tax evasion by claims of cash expenditure which are difficult to be properly investigated. The term expenditure in section 40A(3) includes within its ambit payments made for the purchase of goods. **[Kejriwal Iron Stores-CIT,Raj.***Sec.40A(3).***62CTR227,169ITR 12,31Taxma311]**

7. Spending in the sense of paying out or away of money is the primary meaning of expenditure. Expenditure is something which is gone irretrievably. **[CIT - Hardware Exchange, Gau.,** *Sec.40A(3).***95CTR183,190ITR61]**

8. The word expenditure in section 40A(3) has not been defined in the Act. It is a word of wide import. Section 40A(3) refers to the expenditure incurred by the assessee in respect of which payment is made.It means that all outgoings are brought under the word expenditure for the purpose of the section. The expenditure for purchasing stock-in-trade is one of such outgoings. Rule 6DD of the Income-tax Rules,1962, also contemplates payments made for stock-in-trade and raw materials.This rule is in accordance with the terms of section 40A(3). Section 40A(3) is, therefore, attracted to payments made for acquiring stock-in-trade and other materials.

[Attar Singh Gurmukh Singh–ITO,SC,*Sec.40A(3).***AIR1991SC 2109,97CTR251,191ITR667]**

9. The object underlying section 40A (3)of the Income-tax Act, 1961, is to prevent the use of unaccounted money in carrying on business. Hence,the meaning of the expression expenditure cannot be restricted to deductions provided by sections 28 to 43A alone. The emphasis is upon the word expenditure and not upon the word deduction,as would be evident from a reading of the sub-section. The amounts spent by the assessee in purchasing goods for the purpose of resale is expenditure within the meaning of section 40A(3). **[JanataMetalSupplyCo.–CIT,All.***Sec.40A(3).***100CTR52, 193ITR 646]**

Expenditure for the purpose of earning any income or profits **SC** **Bharat Commerce and Industries Ltd. - CIT**

When interest is paid for committing a default in respect of the statutory liability to pay advance tax, the amount paid and the expenditure incurred in that connection is in no way connected with preserving or promoting the business of the assessee.This is not expenditure which is incurred and which has to be taken into account before the profits of the business are calculated.The liability in the case of payment of income-tax and interest for delayed payment of income-tax or advance tax arises on the computation of the profits and gains of business.The tax which is payable is on the assessee's income after the income is determined. This cannot,therefore,be considered as anexpenditure for the purpose of earning any income or profits. *Sec.37***[AIR1998SC,145 CTR340,230 ITR733,144Taxation496,98Taxman 151]**

Expenditure in the nature of entertainment **Mad.** **CIT-Karuppuswamy Nadar & Sons**

There is no universal formula for finding out what is the expenditure in the nature of entertainment or what is not such an expenditure. The very concept of entertainment would rule out cases of business or commercial courtesy extended to clients who visit the businessmen for doing business.Entertainment postulates that it is some seeking of pleasure and cannot comprehend customary hospitalities shown to persons who come to do business. This running after pleasure may be indicated in cases where a lavish party is given in a posh hotel even to clientele. But the ordinary, elementary extension of business courtesy or civility to visiting clients by allowing them to quench their thirst cannot be understood as expenditure in the nature of entertainment.

*Sec.37.***[120ITR140]**

2. In order that an expenditure should constitute expenditure in the nature of entertainment within the meaning of section 37(2A) of the Income-tax Act,1961, it should have been expended for entertaining persons, whether customers or guests, either by way of providing food, drinks,refreshment or in providing any type of pleasure or amusement or an expenditure of a like nature incurred by the assessee for the purpose of entertainment, pleasure or amusement for himself or for his employees. Whatever be the case, the dominant purpose of incurring the expenditure for the purpose of food, drinks, refreshment and amusement, must be for deriving or providing pleasure.

[Addl.CIT-Bangalore Turf Club Ltd., Kar.,
*Sec.37(2A).***[19CTR172,126ITR430]**

Expenditure incurred solely and exclusively **Bom.** **CIT - Elphinstone Spinning and Weaving Mills Co. Ltd.**

The amount expended by a company for making alterations in its memorandum and articles of association in order to bring them into accord with changes brought about in the law relating to companies is expenditure incurred by the assessee solely and exclusively for the purpose of its business.It is incurred in order that the company should continue to function in accordance with law.

*Sec.37.***[100ITR139]**

Expenditure is laid out **HP.** **Mohan Meakin Breweries Ltd. -CIT (No.1)**

Where an expenditure is laid out by an assessee by way of tax on the property used for his business as owner-cum-trader and the expenditure is incidental to the carrying on of his business, it must be treated to have been laid out by him as a trader and as incidental to his business for the purpose of section 37(1) of the I.T. Act, 1961. *Sec.37(1).***[117ITR501]**

Expenditure laid out wholly and exclusively **Bom.** **CIT-Rajaram Bandekar**

The true test of an expenditure laid out wholly and exclusively for the purposes of trade or business is that it is incurred by the assessee as incidental to his trade for the purpose of keeping the trade going and of making it pay and not in any other capacity than that of a trader. The question should be decided on ordinary principles of commercial trading and commercial expediency. The mere fact that to some extent, the expenditure may enure to a third- party's benefit cannot, in law, defeat the effect of the finding as to the purpose of the expenditure.

*Sec.37(1).***[121CTR233,208ITR503,76Taxman156]**

Expenditure laid out wholly and .... **Cal.** **British Electrical and Pumps (P.) Ltd. - CIT**

Expenditure incurred not with a view to the direct and immediate benefit for purposes of commercial expediency and in order indirectly to facilitate the carrying on of the business is expenditure laid out wholly and exclusively for the purposes of the trade, within the meaning of section 37 of the Income-tax Act,1961. It is not correct to say that the primary motive in incurring the expenditure admissible to deduction under section 37 must be directly to earn income thereby. *Sec37.***[106ITR620]**

Expenditure made wholly and exclusively for the purpose of the... **Ker.** **CIT - Standard Furniture Co. Ltd. (in liquidation)**

It is not necessary that there should be any direct co- relation in point of time between an expenditure and the earning of profits in order that it may constitute business expenditure. A sum of money may be spent not because of necessity or with a view to direct and immediate benefit to business, but if the expenditure is made on the ground of commercial expediency and indirectly to facilitate the carrying on of the business it may still be expenditure made wholly and exclusively for the purpose of the business. The liability to pay gratuity is a contingent liability but an estimate of the liability is ascertainable with substantial accuracy and can be taken into account for arriving at the true profits and gains of an assessee. The incurring of expenditure for gratuity much ahead of the actual time of payment of the gratuity would amount to an expenditure incurred wholly and exclusively for the purpose of the business.

Sec.37(1). **[8CTR54,116ITR751]**

Expenditure of capital nature **HP** **Mohan Meakin Breweries Ltd.- CIT(No.2)**

An expenditure which is made for an initial outlay or for extension of a business at a subsequent stage or for acquiring or bringing into existence an asset or advantage for the enduring benefit of a business should be treated as an expenditure of capital nature. The advantage of enduring nature need not necessarily be tangible in character. *Sec.37(1).***[117ITR505]**

Expenditure wholly and exclusively incurred for purposes of business **Mad.** **M.Seshadri Iyengar (M.) & Sons - CIT**

In order to be a permissible allowance, the expenditure must be for the purpose of carrying on the business, and where the business is closed and as a result of the closure of the business the liability to pay retrenchment compensation has arisen, the liability to pay retrenchment compensation cannot be said to be a liability which arose at the time when the business was carried on and cannot be termed as expenditure wholly and exclusively incurred for purposes of business. *Sec.37* **[152ITR734,76Taxation59]**

Expenses for the purpose of business **Cal.** **Royal Calcutta Turf Club - CIT**

It is not merely direct benefit to the trade which can be the objective of expenditure made wholly and exclusively for the purpose of the trade. Expenditure incurred for securing ends which indirectly facilitate the carrying on of business can also be expenditure laid out wholly and exclusively for the purpose of the business. *Sec.37.***[AIR1998Cal.166,33ITR616]**

Explanation **Ker.** **CIT- S. R. Patton**

The mere use of the label Explanation is not decisive of the true meaning and scope of the provision. Ordinarily, the purpose of an Explanation in a statute is to clarify or explain or settle any doubt or ambiguity or controversy. It may even widen the scope of the main provision in rare cases. *Misc.***[92CTR197,193ITR49]**

2. An Explanation added to a statutory provision is not a substantive provision in any sense of the term but as the plain meaning of the word itself shows it is merely meant to explain or clarify certain ambiguities which may have crept in the statutory provision. **[K.A. Ramaswamy Chettiar - CIT,Mad.*Misc.*220ITR657,88Taxman 526]**

3. An Explanation brought on the statute book is ordinarily clarificatory in nature and has retrospective effect, as the Explanation so brought to a provision in the statute simply explains the law as it has always been in the main provision. **[Laxmi Industries Ltd.-ITO,Raj.*Misc.*149CTR349,231ITR514]**

4. The object of an Explanation to a statutory provision is : (a) to explain the meaning and intendment of the Act itself; (b) where there is any obscurity or vagueness in the main enactment, to clarify the same so as to make it consistent with the dominant object which it seems to subserve; (c) to provide an additional support to the dominant object of the Act in order to make it meaningful and purposeful; (d) An Explanation cannot in any way interfere with or change the enactment or any part thereof ; but where some gap is left which is relevant for the purpose,the Explanation, in order to suppress the mischief and advance the object of the Act, can help or assist the court in interpreting the true purpose and intendment of the enactment, and the right with which any person under a statute has been clothed. It cannot set at naught the working of an Act by becoming a hindrance in the interpretation of the same. **[CIT-OrissaCement Ltd.,Del. *Misc.*173CTR371,254ITR24,122Taxman 353]**

5. The label Explanation is not decisive of the true meaning and scope of the provision.Ordinarily the purpose of an Explanation in a statute is to clarify or explain or settle any doubt or ambiguity or controversy.It may even widen the scope of the main provision in rare cases.The words used alone can reflect the true intent and they should be construed on their own terms. **[CIT- Kerala ElectricLamp WorksLtd.,Ker.,*Misc.*[183CTR182, 261ITR721, 129Taxman549]**

6. An Explanation to a statutory provision is intended to explain its contents and cannot be construed to override it or render the substance and core nugatory.**[CIT-Jet Airways (I) Ltd.,Bom., *Misc.* 239CTR183,331ITR236,195 Taxman117]**

7. Although ordinarily the object of an Explanation is not to enlarge the scope of the original section that it is supposed to explain, if on a true reading of an Explanation it appears that it has widened the scope of the main section, effect should be given to the legislative intent notwithstanding the fact that the Legislature named that provision as an Explanation.**[R. P. G. Industries Ltd. – CIT,Cal.,** *Misc.***241CTR19,338ITR313,198Taxman349]**

Exploitation of a commercial asset Cal. CIT-Ajmera Industries Private Ltd.

Exploitation of a commercial asset does not necessarily mean exploitation by the assessee himself at all material times. The assessee may temporarily cause it to be exploited by another person against payment of consideration and for this purpose may execute a lease for a fixed period even with option to renew. *Sec.28.* **[114ITR769]**

Export turnover SC CIT - K. Ravindranathan Nair

Profit incentives like rent, commission, brokerage charges, etc., though they formed part of the gross total income, had to be excluded as they were independent incomes which had no element of export turnover.All the four variables in the section were required to be kept in mind. If all the four variables are kept in mind, it becomes clear that every receipt is not income and every income would not necessarily include the element of export turnover. *Sec. 80HHC.***295ITR228]**

2. The definition of export turnover excludes freight and insurance. Since export turnover has been defined by Parliament and there is a specific exclusion of freight and insurance, the expression export turnover cannot have a different meaning when it forms a constituent part of the total turnover for the purposes of the application of the formula. ....Freight and insurance do not have an element of turnover. For this reason in addition,these two items would have to be excluded from the total turnover. **[CIT-Gem Plus Jewellery India Ltd. Bom.***Sec.10A.***233CTR248, 330ITR175,194 Taxman192]**

Extinguishment of rights Bom. Bharat Forge Co. Ltd. - CIT

The phrase extinguishment of rights takes colour from the associated words and expressions and will have to be restricted to the sense analogous to them. Hence, the expression extinguishment of any rights therein will have to be confined to the extinguishment of rights on account of transfer and cannot be extended to mean any extinguishment of right independent of or otherwise than on account of transfer. *Sec.2(47).***[205ITR339]**

F

Factory **Pun.** **CIT - Sarveshwar Nath Nigam**

The word factory is not defined in the Income-tax Act, and in its popular sense, kohlus, which are used in connection with the manufacture of gur, must be deemed a factory. <u>*Misc.*</u>**[48ITR853]**

Fails **All.** **Swadeshi Polytex Ltd. - ITO**

...the word fails occurring in section 144(b) must be interpreted to mean omits or does not. The word fails is wide enough to include an act of inevitable necessity. It does not involve any want of care, neglect or default on the part of the assessee but places on him an absolute duty to comply with the directions issued under section 142(2A) to have the accounts audited. It is significant that while in contingencies contemplated by clauses (a), (b) and (c) of section 144 an assessee can apply on sufficient cause being shown for cancellation of the assessment,he has no such rights in a case where he fails to comply with a direction issued under sub-section (2A) of section 142. This clearly indicates that the reasons which result in the omission on the part of an assessee to comply with the direction issued under section 142(2A) are wholly irrelevant. The failure thus may be on account of reasons within or without his control. In either case he incurs the liability to be assessed under section 144(b) of the Act. <u>*Sec.144(b).*</u>**[127ITR287,4Taxman490]**

Fails to furnish the return **AP.** **MullapudiVenkatarayudu-Unionof India**

The argument for the petitioner that under section 271(1)(a) penalty can be levied only in cases where an assessee fails to file the return under section 139(1)(a) and not in cases where there was a mere omission on the part of the assessee to do so,.. .... under section 139(1) as obligation has been cast on an assessee to file a return within the period prescribed in that section, and, there being an obligation non-furnishing of the return would be a failure to furnish the return within the ambit of section 271(1) of the Act.

<u>*Sec.271(1).*</u>**[99ITR448]**

Failure **Bom.** **Pannalal Nandlal Bhandari - CIT**

According to 34(1)(a) of the Income-tax Act failure connotes that there is an obligation which has not been carried out and if there was no obligation upon the assessee to make a return then it would not be a failure on his part to carry out that obligation. <u>*Sec.147.*</u> **[30ITR57]**

Failure on the part of the assessee to disclose material facts **Bom.** **ICICI Bank Ltd. - K.J. Rao**

Having furnished all material facts even if an assessee erroneously claims higher depreciation, it will not be a case of failure to disclose fully and truly all material facts. At what rate the depreciation is to be claimed is a matter of legal inference to be drawn from the material facts.If the legal inference drawn from the material facts is erroneous it cannot be said that there is failure on the part of the assessee to disclose material acts. <u>*Sec.147.*</u>**[188CTR380, 268ITR203,136Taxman669]**

Family **SC** **C. Krishna Prasad - CIT**

Family always signifies a group. Plurality of persons is an essential attribute of a family. A single person, male or female, does not constitute a family. A family consisting of a single individual is a contradiction in terms. Section 2(31) of the Income-tax Act,1961, treats a Hindu undivided family as an entity distinct and different from an individual. Assessment in the status of a Hindu undivided family can be made only when there are two or more members of the Hindu undivided family. <u>*Sec.2(31).*</u>**[97ITR493]**

Family settlement **Gau.** **CGT-S.N.Zaman and S.M.Elahi**

A family settlement is made just to avoid disputes to maintain the honour and dignity of a family. It is neither a partition nor an exchange.Dispute not only means existing dispute, but also a dispute which is possible or is likely to occur in future. *Sec.2(xxiv). G.T.Act.***[157CTR385,221ITR842]**

Fee P& H. CIT - Dr. Mrs. Usha Verma

According to Corpus Juris Secundum fee in a generic sense, implies compensation or salary; but if used in its narrow, distinctive sense it signifies the compensation for particular acts or services rendered in the line of official duties. It has been defined as a charge fixed by law for the services of a public officer, or for the use of a privilege under the control of the Government; a charge for services; a charge or emolument . . .This meaning conforms to the provisions of section 17 of the Act and it is in consonance with the broad concept of salary as compensation for services rendered. *Sec.17.* **[172CTR98,254ITR404,120Taxman738]**

2.....fee or a compensatory tax has to be broadly proportional and not progressive. In the principle of equivalence,which is the foundation of a compensatory tax as well as a fee, the value of the quantifiable benefit is represented by the costs incurred in procuring the facility/services,which costs in turn become the basis of reimbursement/recompense for the provider of the services/ facilities.Compensatory tax is based on the principle of pay for the value. It is a sub-class of a fee. **[Jindal Stainless Ltd.-State of Haryana,SC.***Constitution of India.***283ITR1]**

Fee paid for technical services AP. Clouth Gummiwerke Aktiengesellschaft - CIT

From Explanation 2 below section 9(1)(vii) of the Income-tax Act, 1961, it is clear that any fee paid for technical services is income within the meaning of section 9 of the Act and, therefore, it is taxable.Under Explanation 2 fee offered for technical services means any consideration paid for technical services and excludes consideration paid for any construction, assembling, mining or like project undertaken by the recipient. In other words, any fee paid for technical services is income, if it does not include any consideration paid for any construction undertaken by the recipient. *Sec.9(1).* **[238ITR861,152Taxation127,103Taxman280]**

Fees deemed to accrue or arise in India AP. Elkem Technology-DCIT

Section 9(1)(vii)(b) of the Income-tax Act, 1961, uses the expression fees for services utilised in India and not fees for services rendered in India. Accordingly, if the fees are paid for services utilised by the Indian company in its business carried on by it in India, irrespective of the place where the services were rendered, the amount of the fees should be deemed to accrue or arise in India.

Sec.9(1).[169CTR49,250ITR164,117Taxman382]

Fees for technical services **Mad.** **Skycell Communications Ltd.-DCIT**

Fees for technical services is not defined in section 194J of the Income-tax Act,1961. Explanation (b) in that section provides that that expression shall have the same meaning as in Explanation 2 to clause (vii) of sub-section (1) of section 9. This definition shows that consideration paid for the rendering of any managerial, technical or consultancy service, as also the consideration paid for the provision of services of technical or other personnel,would be regarded as fees paid for technical service. The definition excludes from its ambit consideration paid for construction, assembly, or mining or like project undertaken by the recipient, as also consideration which would constitute income of the recipient chargeable under the head Salaries. Thus while stating that technical service would include managerial and consultancy services, the Legislature has not set out with precision as to what would constitute technical service to render it technical service. Having regard to the fact that the term is to required to be understood in the context in which it is used, fees for technical services could only be meant to cover such things technical as are capable of being provided by way of service for a fee. The popular meaning associated with technical is involving or concerning applied and industrial science.Technical service referred to in section 9(1) contemplates rendering of a service to the payer of the fee. Mere collection of a fee for use of a standard facility provided to all those willing to pay for it does not amount to the fee having been received for technical services.

Sec.9(1)(vii).[170CTR238,251ITR53,119Taxman496]

2.The term fees for technical services as used in Explanation 2 to section 9(1)(vii) was at par with the term fees for included services in article 12(4)(a) of the DTAA.

[Imt Labs (India) P. Ltd., AAR, *(India-US DTAA),Sec.9(1).*
206CTR 304,287ITR450,157 Taxman213]

3. The expression fees for technical services in section 194J of the Income-tax Act,1961, has the same meaning as given to the expression in Explanation 2 to section 9(1)(vii) of the Act. In the

Explanation the expression fees for technical services means any consideration for rendering of any managerial, technical or consultancy services. The word technical is preceded by the word managerial and succeeded by the word consultancy. Since the expression technical services is in doubt and is unclear, the rule of noscitur a sociis is clearly applicable.This would mean that the word technical would take colour from the words managerial and consultancy, between which it is sandwiched. Both the words managerial and consultancy involve a human element. And, both, managerial service and consultancy service, are provided by humans. Consequently, applying the rule of noscitur a sociis, the word technical in Explanation 2 to section 9(1)(vii) would also have to be construed as involving a human element.

[Bharti Cellular Ltd., Del., *Sec.194J.r.w.s.9(1)(vii).* **220CTR258, 319ITR139,175Tax- man 573]**

Financial company **Guj.** **Barkha Investment and Trading Co. - CIT**

Under clause (c) of the Explanation to sub-section (8) of section 40A of the Act a financial company means (i) a hire-purchase finance company; (ii) an investment company; (iii) a housing finance company; (iv)a loan company; (v) a mutual benefit finance company; and (vi) a miscellaneous finance company. *Sec.40A(8).* **[200CTR 342,281ITR316,150Taxman528]**

Finding

The word finding in the proviso to section 34(3) must be given a wide significance so as to include not only findings necessary for the disposal of the appeal but also findings which are incidental to it. It will also apply to cases where it is held that the income in question was not received during the year with which the appellate authority was concerned. If, in pursuance of such a finding, the Income-tax Officer proceeds to investigate afresh as to in which year the income was received, the action of the Income-tax Officer would still be the result of or the logical consequence of the finding arrived at for the purpose of the disposal of the appeal and the proviso to section 34(3) would apply to such a case. **A.S.Khader Ismail -ITO, P.Sahadeva Mudaliar,Mad.***Sec.153(3).***[47ITR16]**

2.The expression finding in the second proviso to section 34(3) of the Indian Income-tax Act,1922, means a finding necessary for giving relief in respect of the assessment for the year in question and the said proviso did not save the time-limit prescribed under section 34(1) in respect of an escaped assessment of a year other than that which was the subject-matter of the appeal or revision, as the case might be. **[Sabita Bhagwandas Shah-K.P.Majumdar,2nd ITO, Bom.,**Sec.153.**59ITR652]**

Finding and direction SC ITO-Murlidhar Bhagwan Das

That the expressions finding and direction, in the second proviso to section 34(3), meant respectively, a finding necessary for giving relief in respect of the assessment for the year in question, and a direction which the appellate or revision authority, as the case may be, was empowered to give under the sections mentioned in that proviso. A finding, therefore, could only be that which was necessary for the disposal of an appeal in respect of an assessment of a particular year. The Appellate Assistant Commissioner might hold, on the evidence, that the income shown by the assessee was not the income for the relevant year and thereby exclude that income from the assessment of the year under appeal. The finding in that context was that the income did not belong to the relevant year. He might incidentally find that the income belonged to another year, but that was not a finding necessary for the disposal of an appeal in respect of the year of assessment in question.

Sec.147.**[52ITR335]**

The expressions finding and direction in Sec. 153(3) of the Act, are limited in meaning. A finding given in an appeal, revision or reference arising out of an assessment must be a finding necessary for the disposal of the particular case, that is to say, in respect of the particular assessee and in relation to the particular assessment year. **[Raj Kishore Prasad-ITO,All.,**Sec.153.**88CTR152, 195ITR438,102Taxation224]**

Firm Gau. CIT - Gaya Prasad Khemani

A firm, in a strict sense,has no legal identity of its own, independent of the partners, though for the purpose of income- tax, it has been treated as a taxable entity, independent of and distinct from its partners. Misc.**[196ITR389]**

First put to use Kar. Biyar Rubber P. Ltd. - CIT

The words used in the section (32 A) first put to use in the immediate preceding year would indicate that not only should the machinery or plant have been put in place but it should be capable of production and should have been put to use. *Sec.32A.*[**210CTR 191,292ITR251,160Taxman377]**

Fixed/circulating capital **Bom.** **Bombay Burmah Trading Corporation Ltd. - CIT**

Fixed capital is what the owner turns to profit by keeping it in his own possession; circulating capital is what he makes profit of by parting with it and letting it change masters. Circulating capital is capital which is turned over, and in the process of being turned over,yields profit or loss. What is capital asset in the hands of the one person may be a trading asset in the hands of another. The determining factor is the nature of the trade in which the asset is employed. *Misc.*[**81ITR777]**

For all purposes **SC** **ITO -Tata Engineering and Locomotive Co. Ltd.**

The vicarious liability imposed upon a person by his appointment as the statutory agent of a non-resident under section 43 of the Indian Income-tax Act,1922,extends to the liability for the assessment year for which the appointment is made and cannot extend to the liability for any other assessment year. expression for all purposes in section 43 only indicates that when an appointment is made for a particular assessment year it is good for all purposes as far as that assessment is concerned, i.e., for all purposes for imposing tax liability,determining the quantum of the liability and for recovering it. *Sec.163.*[**71ITR457]**

For any year and that year **SC** **First Addl.ITO - H.N.S. Iyengar**

The period of limitation for the service of a notice under section 34(1)(a) of the Indian Income-tax Act,1922, for the assessment or reassessment of income escaping assessment, is eight years from the end of the relevant assessment year and not the accounting year. The words for any year and that year in section 34(1)(a) have reference to the assessment year and not the accounting year, for the assessment is for the assessment year although of the income of the previous year. *Sec.149.*[**44ITR437]**

For every month during which the default continued **SC** **CWT - Suresh Seth**

The words for every month during which the default continued indicate only the multiplier to be adopted in determining the quantum of penalty and do not have the effect of making the default in question a continuing one.Nor do they make the amended provisions modifying the penalty applicable to earlier defaults in the absence of necessary provisions in the amending Acts. *Sec.18 (1)(a).WT.Act.*<u></u>**[21CTR349,129ITR328,6Taxman35]**

For purposes of business of **PH** **Dalmia Biscuits Ltd. - CIT**

The crucial words in section 32(1)(vi) of the Income-tax Act,1961, are new machinery or plant installed for purposes of business of generation or distribution of electricity or any other form of power. The use of the words here for purposes of business of admit of only one interpretation,namely,that the machinery or plant installed must be for the purposes of the business of generation or distribution of electricity or any other form of power.In other words, it is only where the business is of generation or distribution of electricity or any other form of power that the installation of new machinery or plant qualifies for initial depreciation. *Sec.32(1).* **[98CTR182,194ITR749,104Taxation148]**

For the benefit of the minor child **Mad** **CIT - K. J. Ramaswamy**

..for the benefit of the minor child means that the minor should have the benefit of the income.In other words, the income must be readily available for the use of the minor. If the income is only to be credited to the minor's account in the trust and to remain there till the minor attains majority, then, it is not for the immediate use of the minor,which means that the minor does not get any benefit at all for the present. He is not entitled to receive the money till he attains majority. So long as it is shown that the beneficiary/minor child of the individual has to receive his share of income only on attaining majority, clause (iii) to section 64(1) of the Act would not be attracted. *Sec.64(1).*<u></u>**[205CTR352 ,286 ITR77,157Taxman 2]**

For the business of the assessee **MP.** **CIT - Malwa Vanaspati and Chemical Co. Ltd.**

No expense which is paid by way of penalty for a breach of the law can be said to be for the business of the assessee. The question as to whether the breach of the law involves moral turpitude or only a violation of some technical provision is not decisive of the question as to whether the amount is an allowable expenditure under section 37(1) of the I.T. Act, 1961.In the case of expenses incurred on account of penalty levied for a breach of the law, the penalty is imposed on the assessee personally and the expenses incurred in that behalf cannot be held to be laid out wholly and exclusively for the purpose of the business of the assessee. *Sec.37(1).***[135ITR221]**

For the purpose of business Bom. Ebrahim Aboobaker - CIT

The expenditure incurred for the preservation of the entire business as an entity and for defending against a claim of hostile title or against nationalisation must be construed to be an expenditure incurred for the purposes of business which is deductible under section 10(2)(xv) of the Indian Income-tax Act,1922.Deductible expenditure would not include merely expenditure incurred for protecting individual asset or assets, but would include expenditure incurred in defending a challenge to the title of the trader to the entire business as an entity. *Sec.37(1).***[81ITR664]**

2. For the purpose of business includes expenditure voluntarily incurred for commercial expediency, and it is immaterial if a third party also benefits thereby. **[S.A. Builders Ltd.-CIT(Appeals), SC,** *Sec.37(1).***206CTR631,288 ITR 1,158Taxman4]**

For the purpose of its business Guj CIT - Pinnacle Finance Ltd.

Where the business of the assessee consists of hiring out machinery and/or where the income derived by the assessee from the hiring of such machinery is business income,the assessee must be considered as having used the machinery for the purpose of its business.

*Sec.32.***[188CTR446,268ITR395,139Taxman328]**

For the purpose of making or earning such income SC Vijaya Laxmi Sugar Mills Ltd. - CIT

The requirement under section 57(iii) that the expenditure should have been incurred for the purpose of making or earning such income shows that the object of spending or the end or aim or the intention of such spending was for earning the interest income.

Sec.57(iii).[**97CTR 257,191 ITR 641]**

For the purpose of protecting.. AP. Society for Integrated Devlopement in Urban & Rural Areas - CIT

The expression for the purpose of protecting the interests of the Revenue in section 281B is very wide in its meaning.The provisional attachment provided in section 281B is more like an attachment before judgment under the Code of Civil Procedure and it is a liability on the property. It should neither be used as a tool to harass the assessee nor should it be used in a manner which may have an irreversible detrimental effect on the business of the assessee. Not to misuse this power a number safeguards have been provided in the section itself. *Sec.281B.*[**172CTR77,252 ITR642,119Taxman289]**

For the purposes of the business MP. Addl. CIT - Badrinarayan Shrinarayan Akodiya

The phrase for the purposes of the business in section 37(1) of the Income-tax Act,1961,has a particular meaning, namely,either the expenditure must be necessary for the purposes of the business or it may have to be incurred on account of the practice prevailing in any particular trade or business.It will not include voluntary donation made by an assessee for the purpose of charity.What is necessary is that direct connection should be established between the expenditure and the character of the assessee as a trader.

Sec.37(1).[**101ITR817]**

2.A trader carries on the business for the purposes of earning profits and not for the purposes of paying the income-tax. Though the earning of profits and the payment of taxes are not isolated and independent activities of a business, yet the expenditure incurred or laid out for the purpose of the payment of Income-tax would not fall within the scope of the expression for the purpose of the business.

[East India Pharmaceutical Works Ltd.-CIT,Cal.,*Sec.37.***114 ITR591]**

3.The expression for the purpose of the business is wider in scope than the expression for the purpose of earning profits. Its range is wide : it may take in not only the day to day running of a business but also the rationalisation of its administration and modernization of its machinery; it may include measures for the preservation of the business and for the protection of its assets and property from expropriation, coercive process or assertion of hostile title; it may also comprehend payment of statutory dues and taxes imposed as a pre-condition to commence or for the carrying on of a business; it may comprehend many other acts incidental to the carrying on of a business. The only limitation is that it should be for the purpose of the business, that is to say, the expenditure incurred should be for the carrying on of business and the assessee should incur it in his capacity as a person carrying on the business. It cannot include sums spent for purposes unconnected with the business.**[Krishna Sahakari Sakhar Karkhana Ltd.-CIT,Bom.***Sec.37.***229 ITR577]**

4.The expression for the purpose of the business as it occurs in the section is wider in scope than the expression for the purposes of earning profits. It may take in not only the day-to-day running of a business but also many other acts incidental to the carrying on of a business. So, to decide whether a payment of money or incurring of expenditure is for the purposes of the business and an allowable expenditure,the test applied is of commercial expediency and principles of ordinary commercial trading.If the payment or expenditure is incurred to facilitate the carrying on of the business of the assessee and is supported by commercial expediency, it does not matter that the payment is voluntary or that it also enures to the benefit of a third party. If the sole object is business promotion, the expenditure would be one incurred wholly and exclusively for the purposes of the assessee's business even though some other object necessarily results, being inherent in the nature and quality of the expenditure. **[Addl.CIT-Kuber Singh Bhagwandas,MP.,** *Sec.37.***9CTR94,118 ITR379]**

5.The phrase for the purposes of the business used in section 37 of the Income-tax Act,1961, has a wide import. It has to be assigned a meaning according to the circumstances of each case and it cannot be assigned a limited meaning.The things done to cut down losses, when the business is still running, are for the purpose of the business.**[Ambala Cantt. Electric Supply Corporation Ltd. - CIT,PH.,***Sec.37.***26CTR361,133ITR343]**

Foreign enterprises **SC** **Petron Engg.Construction (Pvt.) Ltd. - CBDT**

A branch, unit or establishment of an Indian company doing business in a foreign country cannot be said to be a foreign enterprise within the meaning of section 80-Oof the Income-tax Act, 1961.A foreign enterprise contemplated by section 80-O is an enterprise situate in a foreign country having been created or registered in accordance with the laws of that country. This view is supported by the setting in which the expression has been placed and the circumstances in which the law came to be passed.

<u>*Sec.80-O.*</u>**[75CTR20,175ITR523,41Taxman294]**

Forward **AP. CIT - Shahzadi Begum**

The word forward is a word of common usage and has to be understood in its usual meaning. From the dictionary meaning of the word forward the connotation of the expression is abundantly clear. It denotes sending forward or sending towards the place of destination or transmission. In all these situations the thing forwarded moves from the place of its origin towards its destination. That would show that it has left the place from which it was forwarded. It follows that on the date when a letter is dispatched in the usual course, either through special messenger, or by handing it over to a responsible person for posting in the normal course or when a letter is handed over at the post office for delivery to the addressee, it can be said that the letter is forwarded on that date. The draft assessment order would be deemed to have been forwarded on the date on which it was given for purposes of service on the assessee,to any process server of the Income-tax Department or when it was delivered at the post office or when it was handed over to any person, for posting the same through the media of the post office or when it is handed over to the representative of the assessee. <u>*Sec.144B.r.w.s.153(1)(a).*</u>**[142CTR471,225ITR963, 94Taxman336]**

Founder of the institution **SC** **DIT-Bharat Diamond Bourse**

The expression founder of the institution in section 13(3)(a) means that the person concerned should be the originator of the institution or at least one of the persons responsible for the coming into existence of the institution. Contribution of money is not an inexorable test of a person being a founder, though it might happen often that the person who originates an institution may often also fund it. <u>*Sec.13(3).*</u>**[179CTR225,259ITR280,126Taxman365]**

Four assessment years immed-iately succeeding the initial **All.** **CIT - Laxmi Metal Ind.**

The expression and each of the four assessment years immediately succeeding the initial assessment year occurring in sub-section (2) of section 80J is to be construed in its plain and ordinary sense. When so construed, there is no room for ambiguity on a plain reading of those words. The failure to make a claim for relief under section 80J in the initial assessment year or in any of the succeeding assessment years in no way creates a hurdle for making a legitimate claim for the remaining period of exemption provided in sub-section (2) of section 80J. The assessee can legitimately make a claim for relief under section 80J within the period of exemption of five years by showing that it fulfils all the requirements for the grant of relief contemplated under section 80J of the Act. There is nothing in the statutory provisions which imposes a limitation that the relief must necessarily be claimed for all the five assessment years. *Sec.80J(2).*[**146CTR722,236ITR130,100Taxman619]**

Fraud or willful default **Guj** **Dahod Sahakari Kharid Vechan Sangh Ltd. - CIT**

Absence of proof acceptable to the Department cannot be equated with fraud or wilful default. *Sec.271.*[**200CTR265,282ITR321, 149Taxman456]**

'From' ..... 'to' Gau. **CIT** **Highway Construction Co. (P.) Ltd. (No. 1)**

Where a statutory period runs 'from' a named date 'to' another, or the statute prescribes some period of days or weeks or months or years within which some act has to be done, although the computation of the period must in every case depend on the intention of Parliament as gathered from the statute, generally the first day of the period will be excluded from the reckoning, and consequently the last day will be included. *Misc.*[**138CTR43,223 ITR32,135Taxation517 , 92Taxman365]**

Full and true disclosure **P&H.** **R.P.Handa - ITO**

The words full and true disclosurehas not been defined in the Act. It must be seen whether, on the facts of a case, the assessee had made a full and true disclosure of his income. *Sec.273A.*[**198ITR 54**]

Full and true disclosure of income **MP.** **Sureshchandra Babulal Mittal - ACIT (Inv.)**

The Explanation appended to the section defines the meaning of the words full and true disclosure of income.By a deeming fiction, if it is found that the excess of income assessed over the income returned is of such a nature as not to attract the provisions of clause (c) of section 271(1) then it is deemed that the assessee had made a full and true disclosure of his income or particulars relating thereto. By this deeming fiction, the Legislature has given certain benefits to the assessee which if proved would result in reduction or waiver of penalty. *Sec.273A.*[**169CTR29,249ITR603,118Taxman65**]

Full disclosure **Bom.** **Siemens Information Systems Limited - ACIT**

A full disclosure is a disclosure which provides a complete statement of all material facts. A true disclosure is a disclosure which does not suppress material facts from the Assessing Officer and does not contain any taint of falsehood. *Sec.10A,147.* [**343ITR188**]

Full value of consideration **Bom.** **CIT - Shakuntala Kantilal**

The expression full value of consideration contemplates both additions to, as well as deductions from, the apparent value. What it means is the real and effective consideration. So far as clause (i) of section 48 is concerned, the expression used by the Legislature in its wisdom is the expenditure incurred wholly and exclusively in connection with such transfer. The expression in connection with such transfer is wider than the expression for the transfer. Any amount the payment of which is absolutely necessary to effect the transfer will be an expenditure covered by this clause. In other words, if, without removing any encumbrance, sale or transfer could not be effected, the amount paid for removing that encumbrance will fall under clause (i). *Sec.48.r.w.s.55.*[**190ITR56**]

2. When a sale of property takes place, the capital gains arising out of such a transfer has to be computed by taking into account the full value of the consideration received or accruing as a result of such transfer.From the full value of the consideration, the amount of expenditure incurred wholly and exclusively in connection with such transfer as also the cost of acquisition of the asset and the cost of any improvement thereto have to be deducted.The expression full value of consideration cannot be construed as having reference to the market value of the asset transferred but only means the full value of consideration received by the transferor in exchange of the capital asset transferred by him. In the case of a sale the full value of consideration is the full sale price actually paid. Full value means the whole price without any deduction, whatsoever, and it cannot refer to the adequacy or inadequacy of the price bargained for. Nor does it have any necessary reference to the market value of the capital asset which is the subject-matter of the transfer. The full value of consideration does not have any reference to the market value but only to the consideration referred to in the sale deeds as the sale price of the assets which have been transferred.

[CIT-Smt.NiloferI.Singh,Del._Sec.45,48._**221CTR277,309ITR233, 176Taxman]**

Full value of the conside-ration ..of the capital asset	**SC**	**CIT- George Henderson and Co. Ltd.**

The expression full value of the consideration for which the sale, exchange or transfer of the capital asset is made,appearing in section 12B(2) of the Indian Income-tax Act,1922,does not mean the market value of the asset transferred, but the price bargained for by the parties to the sale, etc.The consideration for the transfer of a capital asset is what the transferor receives in lieu of the asset he parts with, viz., money or money's worth, and therefore the very asset transferred or parted with cannot be the consideration for the transfer. The expression full consideration in the main part of section 12B(2) cannot be construed as having a reference to the market value of the asset transferred but the expression only means the full value of the thing received by the transferor in exchange for the capital asset transferred by him. The main part of section 12B(2) provides that the amount of capital gain shall be computed after making certain deductions from the full value of the consideration for which the sale, exchange or transfer of the capital asset is made. In the case of a sale, the full value of the consideration is the full sale price actually paid. The legislature had to use the words full value of the consideration because it was dealing not merely with sale but with other types of transfers, such as exchange, where the consideration would be other than money.

The expression full value means the whole price without any deduction whatsoever and it cannot refer to the adequacy or inadequacy of the price bargained for. Nor has it any necessary reference to the market value of the capital asset which is the subject-matter of the transfer. <u>*Sec.45,48.*</u>**[66ITR622]**

Fully and truly **Bom-(Panaji)** **Sociedade De Formento Industrial P. Ltd.-ACIT**

The duty of the assessee is not merely to disclose all material facts but to disclose them fully and truly. The term truly prima facie indicates that a full disclosure is not sufficient and that the assessee must also disclose the material truly. The ambit of the expression fully and truly is wider than the term fully . There may, therefore, be a case where a full disclosure may have been made by the assessee but the disclosure may not have been made truly. It is neither possible nor appropriate to categorise or limit the type of the cases which fall within the mischief of that expression. <u>*Sec.147.*</u> **[232CTR335 ,339ITR595]**

Fund **Cal.** **Duncan Brothers and Co. Ltd. – CIT**

1. The expression fund used in clause (ii) of rule 2 of Sch. II of the Companies (Profits) Surtax Act,1964, has been construed by the Board in Circular No. I. P. (XV-5) of 1968, dated 23rd January, 1968,giving it its ordinary meaning as understood in common parlance. A fund as contemplated under rule 2(ii) need not be a free fund available for any purpose and not earmarked for a particular purpose.A provision for Taxation would constitute a fund within the meaning of that rule. <u>*Companies (Profits) Surtax Act.*</u>**[128ITR302]**

2. In construing the provisions of section 13(2)(h) of the Income-tax Act, 1961, the expression funds has to be understood in the context of the provisions of section 13(2)(h) of the Income-tax Act,1961, and not only with reference to dictionaries or to commercial parlance or to the principles of accountancy. The expression used is funds and not fund.Funds means money in hand or cash according to some dictionaries. This would be the proper meaning to be attributed to the expression funds as appearing in the provision. The fundamental requirement of section 13(2)(h) is that there must be investment of funds of a trust. If any expanded meaning is given to include assets other than money in hand or cash or credit balance in a bank account, it is evident that they are not capable of being invested as such.Other assets of the trust apart from money in hand or cash or balance in bank will have to be converted into money or cash before the same can be invested. The expression invest connotes a positive act on the part of the trust whereby the funds of the trust are laid out or committed in any particular property or business or transaction with the object of earning a profit or financial advantage or return. What is contemplated is the trust having assets in the form of money or cash or balance in a bank or any other form capable of being invested or by a positive act which pursuant to a decision of the trust is laid out or committed in a concern of a nature specified before it can be held that such an investment comes within the mischief of section 13(2)(h).

[CIT-Sir Sobha Singh Public CharitableTrust,Del._Sec.13(2)(h)._ **167CTR358,250ITR475]**

3. In construing section 13(2)(h) of the Income-tax Act,1961, the meaning of the expression fund will have to be determined not only with reference to dictionaries or to commercial parlance or to the principles of accountancy but also in the context of the said section itself. The expression in the said section is funds and not fund. One of the meanings of the said expression funds in the dictionaries is money in hand or cash. In the context of this provision, this is the meaning which should be attributed to the expression funds in construing the said section.The section contemplates that there will be investment of funds of a trust.If any other meaning is given to the expression funds,the same will not be available for investment or capable of being invested. If the funds of the trust are construed to include assets other than money in hand or cash or a credit balance in a bank account, the same are not capable of being invested as such. Other assets of the trust apart from money in hand or cash will have to be converted into money or cash before the same can be invested. **[CIT - Birla Charity Trust, Cal.,** _Sec.13(2)(h)._**66CTR172,170ITR150,34Taxman504]**

4. In construing the provisions of section 13(2)(h) of the Income-tax Act,1961, the expression funds has to be understood in the context of the provisions of section 13(2)(h) of the Income-tax Act,1961, and not only with reference to dictionaries or to commercial parlance or to the principles of accountancy. The expression used is funds and not fund.Funds means money in hand or cash according to some dictionaries. This would be the proper meaning to be attributed to the expression funds as appearing in the provision.

[CIT- Shri Ram Foundation, Del._Sec.13(2)(h)._**167CTR349,250 ITR 55,116Taxman113]**

Furnishing inaccurate particulars **SC** **Reliance Petroproducts Pvt. Ltd.**

Where there is no finding that any details supplied by the assessee in its return are found to be incorrect or erroneous or false there is no question of inviting the penalty under section 271(1)(c). A mere making of a claim, which is not sustainable in law, by itself, will not amount to furnishing inaccurate particulars regarding the income of the assessee. Such a claim made in the return cannot amount to furnishing inaccurate particulars. Sec.271(1)(c).**230CTR320, 322ITR158]**

Furnishing of inaccurate particulars of income **Ker.** **CIT -Gates Foam and Rubber Co.**

Placing of the bogus debit as genuine constituted furnishing of inaccurate particulars of income. Sec. 271(1)(c).**[91ITR467]**

G

Galvanisation and manufacturing **Cal.** **CIT - Hindusthan Metal Refining Works (P.) Ltd.**

Galvanization is an act or process of galvanizing or coating by which iron or steel is coated with zinc to protect it from rust. The process of galvanizing does not result in the manufacture or production of new goods as such. <u>*Sec.84.*</u>**[128ITR472]**

Garnishee **Bom.** **Bank of Rajasthan Ltd.- Union of India**

When a person or an institution or corporation is holding some amount which is receivable by the assessee though that amount is connected with a commercial transaction, the Revenue is entitled to treat such person, institution or corporation as a garnishee and resultantly a deemed assessee.

<u>*Sec.226(3).*</u>**[176CTR51,259ITR586,171Taxation379, 122Taxman608]**

General and public utility **Guj.** **CIT -Ahmedabad Rana Caste Association**

A purpose would be charitable if it is for the advancement of any other object of general public utility.General means pertaining to a whole class. Public means the body of people at large including any class of the public and utility means usefulness. Therefore, the advancement of any object beneficial to the public or a section of the public as distinguished from an individual or group of individuals would be a charitable purpose. <u>*Sec.2(15).*</u>**[88ITR354]**

General deposits **Raj.** **CIT - Gandhi Metals Mills (P.) Ltd.**

In legal parlance, a general deposit is where the money deposited itself is not returned but equivalent to the money (i.e., a like sum) is to be returned.A perusal of section 40A(8) of the Income-tax Act,

1961,shows that the word deposit mentioned therein has not clearly been defined. In the Explanation to section 40A(8), an inclusive definition of deposit has been given which states that the word deposit means any deposit of money with, and includes any money borrowed by, a company. This Explanation in clause (b) has defined the word deposit in a wider sense and besides any deposit of money with a company, the money borrowed by the company is also taken within the ambit of the word deposit. A distinction has been drawn in this definition with regard to the deposit of money and money borrowed.In the said Explanation, certain exceptions have been provided which would not include the amount received by the company as deposit. Admittedly, the amount deposited by a director in the company in its current account has not been excluded there from. The deposits which are understood in the business of a bank may be in the current account, savings bank account and fixed deposit account. The payment in the current account cannot be excluded from the nature of deposits which are made in the bank. The only distinction between a fixed deposit and this deposit is that the term for which the payment has been made in the case of a fixed deposit is a fixed one whereas, in the case of a current account, no time is fixed therein and this distinction will not take the amount outside the purview of deposit used in the clause. ... In the Explanation to section 40A(8) referred to above, by defining the word deposit,no such exclusion has been made and, therefore, deposits by the directors in their current accounts cannot be excluded. On a correct interpretation of the provisions of section 40 A(8), the payments which are made by a director to the company in the current account of the said director on which the company is paying interest will be considered as a deposit. *Sec.40A(8).* **[200ITR252]**

General public utility

1. The words general public utility in section 4(3) are very wide words. They would exclude the object of private gain, such as an undertaking for commercial profit though all the same it would subserve general public utility. **All India Spinners' Association - CIT,PC.***Sec.2(15).***[12ITR482]**

2. An object beneficial to a section of the public is an object of general public utility. To serve as a charitable purpose, it is not necessary that the object must be to serve the whole of mankind or all persons living in a country or province. **Hiralal Bhagwati-CIT,Guj.***Sec.2(15).***161CTR401,246ITR188]**

Genuine and valid hedging contract AAR Soprropha S.A

In regard to the scope of hedging contracts the following position emerges: (a) Hedging contracts can be both for purchase or sale. (b) In order to be a genuine and valid hedging contract of sales, the total of such transactions should not exceed the total stock of the raw material or merchandise on hand. (c) In order to be a genuine and valid hedging contract of purchase, there should be an existing forward contract of sale by actual delivery. (d) The hedging contract need not necessarily be in the same variety of the commodity : they could be in connected commodities, e.g., one type of cotton against another type of cotton.

*Sec.43(5).***[189CTR457,268ITR37,138Taxman75]**

Genuine Hardship Ker. Commonwealth Trust (India) Ltd

1. It should be noted that the expression hardship is qualified and stipulated as genuine and not severe or grave etc. Even if the hardship is really severe or grave, it need not be genuine. The stress is more on the background and conduct and not on the quantum.

[*Sec.220 (2A).***200CTR45,280ITR70,149Taxman648]**

2. A genuine hardship would, inter alia, mean a genuine difficulty. That per se would not lead to the conclusion that a person having large assets would never be in difficulty as he can sell those assets and pay the amount of interest levied.

[B.M. Malani,SC,*Sec.220(2A).***219CTR,313,306ITR196, 174Taxman363]**

Gift Del. Shiv Shankar Lal – CIT

The definition of the word gift in the Gift-tax Act, 1958, rather than that in the Transfer of Property Act,was applicable to the term used in section 47(iii) of the Income-tax Act,1961.

*Sec.2(14)(iii)r.w.s.47(iii)and 2(xii).G.T.Act.***[94ITR433]**

2. It is quite evident from the language of section 3 of the Gift-tax Act,1958, that it is only a gift as defined in this Act which is subjected to gift- tax. Having regard to the definition of the word gift and the expression transfer of property it is clear that unless the transfer in question is a transfer of property within the definition given in section 2(xxiv), it is not capable of being a gift within the meaning of the Act.**[CGT-Mrs. Jer Mavis Lubimoff,Bom.** *Sec.2(xxiv).G.T.Act.***7CTR28,114ITR90]**

3. The expression gift in section 47(iii) of the Income-tax Act,1961, should be construed as meaning gift under general law. If the expression gift is confined to the gifts contemplated by the Gift-tax Act,1958, then section 52 of the Income-tax Act would become otiose for the purpose of application of section 45. The definition of gift in section 2(xii) of the Gift-tax Act is only for the purpose of that Act and not for the purpose of any other Act. The definition clause says so clearly. Section 4(1) of the Gift-tax Act of creates a legal fiction. The fiction is limited for the purposes for which it is created. The term transfer, as defined in section 2(47) of the Income-tax Act, 1961,is an indication that the transfer contemplated in section 45 and onwards cannot be confined to transfers contemplated by the Gift-tax Act. **[CIT-Sham Narayan Mehrotra,Cal.** *Sec.47(iii)r.w.s.45.***122ITR313]**

4. Where two legislations are pari materia in that they deal with the same subject- matter and are mean to enforce each other, it is permissible to assign the meaning, given to a word in an earlier enactment, to the same word used in a later enactment. The Gift-tax Act was enacted in the year and the Income-tax Act was enacted in the year 1961 by Parliament. Both the enactments impose direct taxes and are administered by the same administrative agency. The Legislature gave an enlarged definition to the word gift in Gift-tax Act,1958, so as to include every sale for inadequate consideration so as to prevent evasion of Gift-tax by clothing a gift in the form of sale for a notional consideration. It is for the reason that every sale for inadequate consideration had been defined as a gift and the provisions of Gift-tax Act were made applicable, the legislature deliberately excepted the gifts out of the provisions of section 45, by section 47 (iii).If every sale of capital asset for inadequate consideration also falls within the purview of section 52(2) of the Act,when the fair market value as on the date of sale exceeds 15 per cent. of the consideration actually received, section 47(iii) would be rendered meaningless, which is not a permissible rule of interpretation. By giving the word gift used in section 47(iii) the same meaning as given in the Gift-tax Act, there would be complete harmony between section 47 and 52 of the Income-tax Act. Every sale for inadequate consideration falls within the meaning of the word gift and consequently it would attract the levy of Gift-tax and be outside the purview of capital gains tax and every sale in which there is under statement of the value falls within the mischief of section 52 and attracts capital gains tax and not the provisions of the Gift-tax Act. The provisions would be complementary to, and enforcing, each other. Hence, the word gift used in section 47(iii) of the Income-tax Act, should be given the same meaning as given to it in the Gift-tax Act, and consequently neither, section 45 nor section 52 of the Income-tax Act is attracted to a case of transfer of capital asset for inadequate consideration. **[SanjivV.Kudva-CIT,Kar.** *Sec.45,47,52.r.w.s.G.T.Act.4.***20CTR1,127ITR354]**

5. Section 2(xii) of the Gift-tax Act, 1958, read with section 2(xxiv)(d) shows that before a transaction can be termed a gift here must be a transfer by one person to another of an existing movable or immovable property made voluntarily and without consideration. The first part of the definition in section 2(xii) requires at least two persons. The transaction must be bilateral or multi lateral. A unilateral transaction by which one party releases his interest in a given property is not covered by this definition. **[CGT-Ansuya Sarabhai,Guj.***Sec.2(xii)G.T.Act.***22CTR201,133ITR108]**

6. A perusal of sections 2 and 4 of the Gift-tax Act, 1958, shows that Parliament has sought to give an extended meaning to the expression transfer of property and consequently to the expression gift, very transfer of existing movable or immovable property, made without consideration, constitutes gift. **[CIT-Jagtram Ahuja,AP.** *Sec.2,4.G.T.Act.***[172ITR632]**

7. Section 2(xii) of the Gift-tax Act, 1958, defines the term gift. According to the definition a transaction to be understood as a gift has to be a transfer by one person to another of any existing movable or immovable property. Secondly, the transaction has necessarily to be without consideration. **[CGT- Smt. K.Nagammal, Ker.***Sec.2(xii).***[135CTR17,226ITR598]**

8. Gift has been defined in section 2(xii) of the Gift-tax Act, 1958, as the transfer by one person to another of any existing movable or immovable property made voluntarily and without consideration in money or money's worth. The expression transfer of property has been defined in section 2(xxiv). Under sub-clause (d) of section 2(xxiv) transfer includes any transaction entered into by any person with intent thereby to diminish directly or indirectly the value of his own property and to increase the value of the property of any other person. The propositions established are (1) the transaction referred to takes colour from the main clause, each of which deals with one or the other mode of transfer. It must be a transfer of property in some way; (2)the transaction must be with some other person and it cannot be a unilateral act; (3) the required intent must be shown to be existing.The main ingredient of the clause is intent.Intent means the main or substantial object of the transaction. **[M.A. Ismail -CGT,Ker.**<u>*Sec. 2(xii).G.T.Act.*</u> **157CTR 557,240ITR539]**

9. The definition of gift makes it clear that there has to be a transfer by one person to another of movable or immovable property ; such transfer has to be voluntary and without consideration in money or money's worth. What is, therefore, absolutely essential for the purpose of a gift is a transfer of property. Transfer of property is defined for the purpose of the Gift-tax Act as any disposition or conveyance, or assignment or settlement or delivery or payment or other alienation of property. **[CGT-T.M.Louiz,SC,** <u>*Sec.2(xii),(xxiv)G.T. Act.*</u> **163CTR359,245ITR831,112Taxman622]**

10. A gift for the purposes of the Gift-tax Act,1958, means transfer of any property made voluntarily and without consideration by one person to another.The definition of transfer of property takes within its sweep the creation of a trust in property.The donor is a person who makes the gift and the donee is a person who acquires any property under a gift and includes both the trustee and the beneficiary where a gift is made to a trustee for the benefit of another person. **[Bhavna Nalinkant Nanavati-CGT,Guj.,** <u>*Sec.2(viii),(ix)(xii)(xxiv). G.T. Act.*</u>**[174CTR152,255ITR529]**

Godowns or warehouses **Guj.** **CIT - Ahmedabad Maskati Cloth Dealers Co-op. Warehouse Soc.**

The Legislature advisedly used the words godowns or warehouses because the intention was to encourage co-operative societies to construct godowns and warehouses which would prove useful for the rural economy. Literal interpretation of the words godowns or warehouses will not lead to any absurdity or produce any manifestly unjust result.The dictionary meanings of godowns and warehouses show that both these terms are synonymous and interchangeable. The common parlance meaning which can be attributed to godowns or warehouses is that it must be used for the purpose of storage of goods even for a temporary period. The expression facilitating the marketing of commodities would suggest a stage anterior to the actual sale of the commodity. The words facilitating the marketing of commodities would not lend colour to the words godowns or warehouses so as to enlarge their meaning. Rental income derived from letting of shops used for business cannot be said to be income derived from the letting of godowns or warehouses and exemption under section 80P(2)(e) will not be available in respect of such income. <u>*Sec.80P(2).*</u>**[50CTR283, 162ITR142]**

2. Section 80P(2)(e) of the Income-tax Act,1961,provides for deduction of income derived by a co-operative society from the letting of godowns or warehouses for storage,processing or facilitating the marketing of commodities. It will appear from the dictionary meanings of godowns and warehouses that both these terms are synonymous and interchangeable. The common parlance meaning which can be attributed to godowns or warehouses is that they must be used for the purpose of storage of goods even for a temporary period.**[CIT- District Co- operative Federation,All.** *Sec.80(P)(2).***193 CTR 992, 271ITR 22, 144Taxman333]**

Good faith Cal. Pragodas Mathuradas - ITO

A state of mind indicating honesty and lawfulness of purpose ..If it was proved that the Government servant knew at the time he did the particular act complained of that he had no authority under the Act to do it,and was intentionally doing it in contravention of the Act,he is not doing the act bonafide and the immunity of the section 293 does not come into play.If on the other hand he does that act under the bonafide belief that he had the power to do it under the Act, although actually there may not be any such power, his act will still enjoy the protection of section 293. *Sec.293.***18ITR757**

2. The good faith which is required to be established for invoking the provisions of sections 273A(1)(ii) is that, in making the disclosure, the petitioner must have acted honestly. In other words, he should not have been guilty of having acted dishonestly in making the disclosure. The fact that before making the disclosure his conduct had been dishonest or that he did not act in good faith is irrelevant for the purpose of applying these provisions. The disclosure under this section is made by an assessee for the purpose of getting the benefits provided therein. The fact that in the past the assessee did not make a full disclosure of his income and concealed the same is immaterial. **[Jakhodia Brothers - CIT, All.** *Sec.273A(ii).***[7CTR400,115ITR61]**

3. The requirement of section 18(2A) (now section 18B) of the Wealth-tax Act,1957,is that a full and true disclosure of the net wealth must be in good faith. If it has been made honestly, it should be considered as having been done in good faith whether or not it has been done negligently, i.e.,even in a case where a disclosure has been made negligently,it can still be considered as having been done in good faith if in fact it has been done honestly. **[Radhakrishna - CWT, Kar.***Sec.18B.WT.Act.***121ITR722]**

4. The expressions good faith and full and true disclosure of the income mean that the assessee in the circumstances must have felt that he has filed the return voluntarily and in good faith and according to him has made a full and true disclosure of his income. **[Seetha Mahalakshmi Rice and Groundnut Oil Mill Contractors Co.-CIT, AP.**<u>*Sec.273A.*</u>**127ITR579]**

5. The expression good faith is not defined in the Act. Under clause (22) of section 3 of the General Clauses Act, 1897, a thing shall be deemed to be done in good faith where it is in fact done honestly whether it is done negligently or not. Thus if the element of honesty is present, the requirement of good faith is satisfied. The good faith which is required to be established for invoking section 273A is that in making the disclosure the petitioner must have acted honestly. **[K. Ramuluand Bros.- CIT, AP.**<u>*Sec.273A.*</u>**185ITR517,51Taxman 57]**

6. The term good faith is not defined under the Act but is defined under section 2(22) of the General Clauses Act.Either under the General Clauses Act or in ordinary parlance, an act done in good faith means an act done honestly even if it is tainted with negligence or mistake.All that is required is that the disclosure of income must be full and true according to the honest belief of the assessee.Voluntary means without compulsion. **[Rohitkumar and Co.-CIT,Bom.**<u>*Sec.271(1),273A.*</u>**92CTR260,190ITR93]**

7. The expression good faith used in section 18B(1)(b), has not been defined in the Act. However, under the General Clauses Act, a thing shall be deemed to be done in good faith when it is, in fact, done, honestly, whether it is done negligently or not. Good faith has relevance to the frame of mind of the person at the relevant time when he furnished voluntarily the return making the disclosure of his net wealth. In a given case,in deciding the question of good faith what is relevant to be kept in consideration is the intention of honesty and absence of bad faith or mala fides. In a case where the disclosure has been made negligently, it may still be considered as having been made in good faith if, in fact, it was made honestly. A mistake committed unintentionally in making a full and true disclosure by oversight or negligence without anything more would not render the disclosure outside the purview of good faith.**[Sardar GurIqbal Singh- CWT,All.**<u>*Sec18B(1)(b).WT.Act.*</u>**[197ITR269]**

8. The expression good faith means an act done honestly even if the same be tainted with negligence or mistake. Section 2(22) of the General Clauses Act, 1897, lends a similar meaning to the said expression.In order that a disclosure is termed as having been made in good faith,the same must be demonstrably honest. A disclosure which is made under the compulsion of a possible penalty or other proceedings cannot be termed honest or one made in good faith, the underlying object of any such disclosure being not to come clean on the subject but to avoid the adverse consequences that may follow a non-disclosure. **[K.L.Swamy-CIT,Kar.**<u>*Sec.271(1)(c),273A.*</u>**57CTR 489,239ITR386,102Taxman491]**

9. Good faith entails honest intent free from taint of fraud or fraudulent design.An erroneous decision or erroneous assumption of jurisdiction, if done bona fide,cannot be labelled as an act not done in good faith. **S.P.Goyal-Director of Income-tax(Inv.),Del.** <u>*Sec.293.*</u>**202CTR169,288ITR595,152 Taxman481]**

Goods

The term goods, for the purposes of sales tax, cannot be given a narrow meaning.Properties which are capable of being abstracted, consumed and used and/or transmitted, transferred, delivered, stored or possessed, etc., are goods for the purpose of sales tax.The test to ascertain whether a property is goods for the purposes of sales tax is not whether the property is tangible or intangible or incorporeal.The test is whether the concerned item is capable of abstraction, consumption and use and whether it can be transmitted, transferred, delivered, stored, possessed, etc.In the case of software,both canned and uncanned, all of these are possible. Intellectual property when it is put on a media becomes goods. **[Tata Consultancy Services-State of A.P.,SC.,**<u>*Constitution of India.*</u>**192CTR257,271ITR401,141Taxman132]**

2. Goods do not include electromagnetic waves or radio frequencies for the purpose of article 366(29A)(d) of the Constitution of India. The goods in telecommunication are limited to the handsets supplied by the service provider. There are two reasons : (i) Electromagnetic waves are neither abstracted nor consumed in the sense that they are not extinguished by their user. They are not delivered, stored or possessed. Nor are they marketable. They are the medium of communication. What is transmitted is not an electromagnetic wave but the signal through such means. The signals are generated by the subscribers themselves. In telecommunication what is transmitted is the message itself by means of the telegraph. No part of the telegraph itself is transferable or deliverable to the subscriber. (ii) The second reason is more basic : a subscriber to a telephone service cannot reasonably be taken to have intended to purchase or obtain any right to use electromagnetic waves or radio frequencies when a telephone connection is given. Nor does the subscriber intend to use any portion of the wiring, the cable, the satellite, the telephone exchange, etc. At the most the concept of sale in the subscriber's mind would be limited to the handset that may be purchased. As far as the subscriber is concerned no right to the use of any other goods, incorporeal or corporeal, is given to him or her with the telephone connection. Electromagnetic wave (or radio frequencies) do not fulfil the parameters applied for determining whether they are goods, the right to use of which would be sale for the purposes of article 366(29A)(d). The essence of the right under article 366(29A)(d) is that it relates to the user of goods. It may be that the actual delivery of the goods is not necessary for effecting the transfer of the right to use the goods but the goods must be available at the time of transfer, must be deliverable and delivered at some stage. If the goods or what are claimed to be goods are not deliverable at all by the service providers to the subscribers, the question of the right to use those goods would not arise. **[Bharat Sanchar Nigam Ltd.-Union of India,SC,** *Constitution of India.* **201 CTR346,282ITR273,152Taxman135]**

3. The provisions of the Income-tax Act do not define the word goods or merchandise. Hence, a reference to the meaning of goods and merchandiseas can be inferred from the law settled so far would be necessary. The Supreme Court in Tata Consultancy Services v. State of A.P.(271 ITR 401), while testing whether the property involved in a transaction is goodsfor the purposes of sales tax, held that the term goods includes all types of movable properties, whether those properties be tangible or intangible.The test is whether the item concerned is capable of abstraction, consumption and use and whether it can be transmitted, transferred, delivered, stored,possessed, etc. This view was adopted in BSNL v. Union of India (282 ITR 273). Goods may be tangible property or intangible. A thing would become goods provided it has the attributes thereof having regard to (a) its utility; (b) capability of being bought and sold; and (c) capability of being transmitted, transferred, delivered, stored and possessed. If the above attributes are satisfied the same would be goods. **[CIT - Superstar Music, Mad.** *Sec.80HHC.HHF.* **210CTR128,291ITR8]**

Goods carriage **Guj.** **Gujco Carriers - CIT**

The truck on which the crane is mounted is constructed and adapted specially to carry the crane. Goods carriage as defined in section 2(14) of the Motor Vehicles Act, 1988, means any motor vehicle constructed or adapted for use solely for the carriage of goods, or any motor vehicle not so constructed or adapted when used for the carriage of goods. IT Rules- r.5.Appex I. **[174CTR324,256ITR50,122Taxman206]**

Goods or merchandise **Bom. Abdulgafar A. Nadiadwala - ACIT**

The Income-tax Act does not define the words goods or merchandise.It is well-settled that in the absence of there being anything contrary to the context the language of a statute should be interpreted according to the plain dictionary meaning of the terms used therein.It would be clear that the word goods has been understood to mean all items of merchandise, supply of raw materials, finished goods, all things specifically manufactured which are movable at the time of identifying for trade and sale, other than money. *Sec.80 HHC,HHF.***[188CTR232,267ITR488,137Taxma112]**

Goodwill Del CIT - Hindustan Coco Cola Beverages P. Ltd.

Goodwill conveys a positive reputation built by a person/company/ business concern over a period of time.... basically intangible assets were identifiable non-monetary assets that could not be seen or touched or physical measures which were created through time and/ or effort and that were identifiable as a separate asset. They could be in the form of copyrights, patents, trademarks, goodwill, trade secrets, customer lists, marketing rights, franchises,etc. which either arise on acquisition or were internally generated. *Sec.32.* **[238CTR1, 331ITR192,198Taxman104]**

Government AAR DLJMB Mauritius Investment Company - CIT

Though the tax treaty does not define the expression Government, under article 3(2), any term not defined therein will,unless the context otherwise requires, have the meaning which it has under the laws in force in India.Both under the Constitution of India and, on general principles, the Reserve Bank of India, which is a Government agency completely owned by the Government of India and the FIPB which is only a Committee of the Government of India,can be appropriately described as Government for the purposes of this clause. *Sec.245Q.***[141CTR413,228ITR268, 94Taxman218]**

Government security **Bom.** **British Bank of The Middle East - CIT**

Government security has been defined in clause (2) of section (2) of the Public Debt Act, 1944, to mean (a) a security created and issued by the Government for the purpose of raising a public loan and having one of the following forms, namely, (i) stock transferable by registration in the books of the bank; or (ii) a promissory note payable to order; or (iii) a bearer bond payable to bearer; or (iv) a form prescribed in this behalf; (b) any other security created and issued by the Government in such form and for such of the purposes of the Act as may be prescribed...Government security has been defined in clause (a) of section 2 of the Indian Securities Act, 1920, to mean promissory notes (including Treasury Bills), stock certificates, bearer bonds and all other securities issued by the Central Government, etc... Therefore, it is clear from the definition of Government security contained in the above two enactments that Treasury Bills are Government securities. *Public Debt Act.* **[149CTR169,233ITR251]**

Gross total income **Bom.** **CIT - Nagpur Zilla Krishi Audyogik Sahakari Sangh Ltd.**

.....the words gross total income referred to in section 80P(1) must be given the defined meaning which means total income computed in accordance with the provisions of the Act, but before making any deduction under Chapter VI-A or section 280-O.Computation in accordance with the provisions of the Act must mean computation in accordance with section 29. It would be consistent and reasonable to hold that the expression the amount of profits and gain used in sub-section (2) of section 80P cannot be understood in a different sense.The expression must mean income as computed under section 29. *Sec.80P(1).***[209ITR481]**

2. The effect of clause (5) of section 80B of the Income-tax Act,1961, is that gross total income will be arrived at after making the computation as follows : (i) making deductions under the appropriate computation provisions;(ii) including the incomes,if any, under sections 60 to 64 in the total income of the individual; (iii) adjusting intra-head and ITR or inter-head losses; and (iv) setting off brought forward unabsorbed losses and unabsorbed depreciation, etc. Only if the gross total income so determined is positive the question of allowing the deductions under Chapter VI-A would arise, not otherwise. **[Synco Ind. Ltd.-AO,SC,**<u>*Sec.80A,80AB,80B(5).*</u> **215CTR385,299ITR444,168Taxma224]**

Growing Crops Mad. M.Rangaswamy - CWT

A perusal of the Explanatory Notes on Clauses to the Finance Act, 1969,shows that what was intended to be exempted was growing crops, fruits on trees and grass in agricultural land. Even applying the common sense concept of crops, the same can by no stretch of imagination be extended to the tree or plant concerned and merely because the word used is growing crops it will not make anything which grows, including the tree or plant, also a crop merely because the crop is a yield from the tree or plant concerned. Tea or coffee bushes or plants, apart from the two leaves and the bud of the tea plant or the berries of the coffee plant, cannot be considered as growing crops within the meaning of section 5(1)(viiia) of the Act.

<u>*Sec.5(1).WT.Act.*</u>**[136CTR76,221ITR39]**

Guest house Ker. CIT-Commonwealth Trust Ltd.,

1. Section 37(3) of the Income-tax Act, 1961, treats a guest house somewhat differently from a residential accommodation in the general sense of the term. A residential accommodation is such, for instance, as is provided for by a company, by way of quarters for residence of its officers or others,whereas a guest house is what is provided for the periodical stay of those concerned with the company. <u>*Sec.37(3).*</u>**8CTR62,120ITR491]**

2. Guest house as defined by sub-rule (4) of rule 6C of the Income-tax Rules,1962, includes accommodation hired or reserved by the assessee in a hotel for a period exceeding 182 days during the previous year.This provision also indicates that expenses towards food and other amenities are included in the accommodation provided by an assessee in the nature of a guest house. Therefore, the expenses of a kitchen attached to the guest house for providing food and other amenities fall within the scope of section 37(3) of the Income-tax Act,1961. **[Saraswati Industrial Syndicate Ltd.-CIT,P&H.**<u>*Sec.37(3).*</u> **24CTR246,136ITR361]**

3. A guest house is a guest house whatever may be the category of the guests to which it caters and whatever may be the standard of comforts and amenities provided and no matter whether the intention of the party is merely to provide them with shelter and food or to entertain and amuse or gratify the visitors. **[Kesoram Industries and Cotton MillsLtd.-CIT,Cal.*Sec.37(3).*191ITR518]**

4. Sub-section (5) of section 37 of the Income-tax Act, 1961, which has been inserted by the Finance Act, 1983,with retrospective effect from April 1, 1979, has clarified that accommodation maintained by an assessee to provide lodging or boarding and lodging to any person including any employee or a director or the holder of any office in the assessee-company would be in the nature of a guest house within the meaning of sub-section (4) of section 37. **[CIT-Ocean Carriers Pvt. Ltd. Bom.]**

H

Hardship Ker. Commonwealth Trust (India) Ltd. - DCIT(Asst.)

...the expression hardship is qualified and stipulated as genuine and not severe or grave, etc. Even if the hardship is really severe or grave, it need not be genuine. The stress is more on the background and conduct and not on the quantum. Whether the non-payment of the tax is due to circumstances beyond the control of the assessee and whether the payment of interest would cause genuine hardship have to be analysed and appreciated taking into account various other factors also. *Sec.220(2A)*[**200CTR245,280ITR70, 149Taxman648]**

Harijan All. CIT - Harijan Evam Nirbal Varg Avas Nigam

...the word Harijan should be interpreted in accordance with its meaning in common parlance. Commonly understood, the term Harijan refers to the members of the Scheduled Castes only.

Sec.10(26B).[**131CTR169,226ITR696]**

Has been succeeded All. Shiva Shakti Saran Raghubir Saran - CIT

The words has been succeeded in section 26(2) are governed by the words in such capacity which must mean the capacity in which the persons succeeded was carrying on the business. *Sec.170(3).* **[19ITR314]**

Has escaped assessment PC Rajendranath Mukherjee - CIT

The words has escaped assessment in section 34 cannot be read as equivalent to has not been assessed, and the word assessment is not confined to the definite act of making an order of assessment. *Sec.147.*[2ITR71]

Has reason to believe Ori. Vishnu Borewell - ITO

The expression has reason to believe in section 147 of the Income-tax Act,1961,is stronger than the words is satisfied. The belief must be based on reasons which are relevant and material. The court can examine whether the reasons are relevant and have a bearing on the matters with regard to which he is required to entertain the belief before he can issue notice under section 148. If there is no rational and intelligible nexus between the reasons and the belief, so that, on such reasons, no one properly instructed on facts and law could reasonably entertain the belief, the inevitable conclusion is that the Assessing Officer could not have reason to believe that any part of the income of the assessee had escaped assessment.

Sec.147.[178CTR409,257ITR512,125Taxman696]

Having become due and payable within the twelve months Bom. Kewalramani Bros. - CIT

The words having become due and payable within the twelve months next before the relevant date in section 530(1)(a) have to be understood as putting a restriction on or cordoning off the amount for which priority is claimable and not in respect of each and every debt on account of taxes, rates and cesses, etc., which may be outstanding at that time and payable. *Companies Act.*[189ITR90]

Having regard to SC Rajesh Kumar - DCIT

The expression having regard to in section 142(2A) is significant. An opinion must be formed strictly in terms of the factors enumerated therein. The expression indicates that in exercising the power regard must be had also to the factors enumerated therein together with all factors relevant for the exercise of the power. The factors enumerated in section 142(2A) are not exhaustive......If the assessee is put to notice, he could show that the nature of the accounts is not such as would require appointment of special auditors. The assessee could further show that what the Assessing Officer considers complex is in fact not so. It would also be open to him to show that the same would not be in the interest of the Revenue. *Sec.142(2A).*[206CTR175,287ITR91,157Taxman168]

He has maintained for him a dwelling place **SC** **CIT-K.S.Ratnaswamy**

The expression he has maintained for him a dwelling place in section 4A(a)(ii) meant he causes to be maintained for him a dwelling place. In either of these expressions the volition on the part of the assessee in the maintenance of a dwelling place emerged very clearly; whether he maintained it or he caused it to be maintained, the maintenance of the dwelling place had to be at his instance, behest or request and when it was maintained by someone else other than the assessee, it had to be for the assessee or for his benefit. *Sec.6.*[**14CTR377,122ITR217,3Taxman7]**

Heavy chemical **Bom.** **Colour Chem Limited-CIT**

.....that heavy chemical is described as heavy chemical on the basis of the tonnage whereas fine chemical is defined on the basis of smaller quantity. In other words, the parameter of tonnage applies in order to ascertain whether the chemical is a heavy chemical or a fine chemical. *Sec.32A.*[**155CTR23,238ITR171,105Taxman28]**

Hedging contract **Cal.** **CWT - Gourepore Co. Ltd.**

The hedging contract is so called because it enables the persons dealing with the actual commodity to hedge themselves, i.e., to insure themselves against adverse price fluctuations. A dealer or a merchant enters into a hedging contract when he sells or purchases a commodity in the forward market for delivery at a future date. His transaction in the forward market may correspond to a previous purchase or sale in the ready market or he may propose to cover it later by a corresponding transaction in the ready market, or he may offset it by reverse transaction on the forward market itself. Hedging contracts, in order to be out of speculative transactions, must be only in respect of raw materials so far as the manufacturer is concerned, though these contracts may be both with regard to sales and purchases. Hedging contracts need not succeed the contracts for sale and actual delivery of goods manufactured, but the latter may be subsequently entered into, provided they are within reasonable time not exceeding generally the assessment year. In order to be genuine and valid hedging contracts of sales, the total of such transactions should not exceed the total stocks of the raw materials or the merchandise on hand which would include existing stocks as well as the stocks acquired under the firm contracts of purchase. Sec.43(5).[135ITR606]

Hedging transactions Guj. CIT - Mohanlal Ranchhoddas

The Legislature has, after carving out a subordinate source of business income styled as speculative business, enjoined that any loss under this head cannot be set-off against profits and gains from any non-speculative transaction, though it may be part of the larger head of income of business or profession. However, clauses (a), (b) and (c) of the proviso to section 43(5) of the Income-tax Act, 1961, by legal fiction take out of the purview of speculative transactions, hedging transactions.Hedging transactions are genuine transactions entered into for the purpose of insuring against adverse price fluctuations.In hedging transactions, neither delivery nor transfer is contemplated, and yet, they cannot be considered as speculative transactions in commercial parlance.By hedging transactions, a trader by corresponding contract of sale and purchase in the forward market tries to offset the likely loss which may arise in the ready market due to adverse price fluctuations. A dealer in stocks and shares can enter into a contract to guard against loss in his holdings of stocks and shares which may arise due to adverse price fluctuations.However, the only condition which should be satisfied before he can claim that a contract entered into by him should not be considered as a speculative transaction is that he must have entered into such a contract to guard against the loss due to adverse price fluctuations of shares in respect of which he might have entered into contract of sale by actual delivery. Sec.43(5). **[203ITR304]**

Held by the assessee Raj. Sushila Devi - CWT

To interpret the meaning of the term held by the assessee appearing in section 5(3) of the Wealth-tax Act, 1957, it would be profitable to have an idea of the object and purpose of enacting clause (xxvi) of section 5(1) of the Act. In order to widen the area of investment in financial assets qualifying for exemption from levy of wealth-tax, six types of investments enumerated in clauses (xxii) to (xxvii) were inserted in section 5(1).One type of investment amongst them such as deposits in banks falls in clause (xxvi) of section 5(1) of the Act.In order to enable the assessee to have the benefit of clause (xxvi), it is required that the property, i.e., the deposit in the bank,remains in his/her name till its maturity. Sec.5(3)WT.Act. **[61CTR75,172ITR74,34Taxman1]**

Hindu Undivided Family PC. Kalyanji Vithaldas-CIT

The expression Hindu undivided family is used in the Indian Income-tax Act,1922, with reference not to one school only of Hindu Law, but to all schools, and it is a mistake to read it as equivalent to the narrower expression Hindu coparcenary.It cannot be said that no female can be a member of a joint Hindu family. *Misc.***[5ITR90]**

2. The expression Hindu undivided family is much wider than a Hindu coparcenery. **[CIT-RaniRudhKumari,Oud.,***Misc.* **8ITR607]**

3. The expression Hindu undivided family in section 3 of the Wealth-tax Act,1957, includes a Jain undivided family. The word Hindu preceding the words undivided family signifies that the undivided family should be of those to whom Hindu law applies. Even though Jains may not be Hindus by religion they are governed by the same laws as Hindus.In this view the expression Hindu undivided family will include the Jain undivided family. The words Hindu undivided family are used in Taxation statutes with reference not to one school of Hindu law only but to all schools. **[CWT- Smt. Champa Kumari Singhi, SC,** *Sec.3.WT.Act.***83 ITR 720]**

4. The expression Hindu undivided family for the purpose of the Income-tax Act, or the Wealth-tax Act, or the Estate Duty Act, is not co-terminous with a Hindu coparcenary. The expression Hindu undivided family is used in the sense in which it is understood in the personal laws of Hindus. That means, a joint family may consist of a single male member,his wife and daughters, and there is no embargo either under the Income-tax Act, or the Wealth-tax Act, or the Estate Duty Act, to suggest that unless there is more than one male member, there cannot be an assessable unit of the Hindu undivided family. **[DulariDevi-CED,Ori.***Estate Duty Act.* **211ITR524]**

5. The expression Hindu undivided family in the Income-tax Act, 1961, is used in the sense in which a Hindu joint family is understood under the personal law of Hindus. **[CIT- Arun Kumar Jhunjhunwalla &Sons, Gau.** *Misc.***[223ITR45]**

Hire **Gau.** **A.B.C. India Ltd.-CIT**

The expressions hire and transportation are not defined in the Act. However, the dictionary meaning of hire is compensation for the use of a thing, or for labour or services. It also includes the other meaning to purchase the temporary use of a thing, or arrange for the labour or services of another for a stipulated compensation.

<u>*Sec.32.*</u>**[145CTR158,226ITR914]**

2. The word hire used in entry No. III(2)(ii) of Appendix I to the Income-tax Rules,1962, is only meant to denote that the use of the vehicle is not by the owner himself for his own purpose, but it is given to another for use for a limited period, of that other for a consideration. For this purpose, there is no qualitative difference between lease of the vehicle for a specified period for consideration and letting the vehicle on hire for short duration on payment of hire charges. **[CIT-Madan and Co.,Mad.,** <u>*Sec.32.r.w.ITRulesAppex.I.*</u> **174CTR172,254ITR445,128Taxman116]**

Hire/lease **Bom** **Kotak Mahindra Finance Ltd. - DCIT**

Entry III of Appendix I to the Income-tax Rules,1962, read as a whole,states that the assessee must run the vehicles on hire or the assessee must carry on the business of running the vehicles on hire. There is a basic difference between lease and hire. This difference is borne out by the basic difference in the meaning of the expression property and the expression possession. A transaction of hire is essentially a contract of bailment of a vehicle. In the case of a hire, only a licence is given to the hirer to use the vehicle for a temporary period the vehicle so hired. In the case of hire, the hirer has an option to buy the equipment which is one of the main distinguishing features between the words hire and lease. <u>*Sec.32.r.w.ITRules Appex.I.*</u>**[184CTR527,265ITR114,130Taxman442]**

His income **SC** **CIT - P.K. Kochammu Amma**

The total income of the assessee chargeable to tax would include the amounts representing the shares of the spouse and minor child in the profits of the firm. If this be the correct legal position, there can be no doubt that the assessee must disclose in the return submitted by him all amounts representing the shares of the spouse and minor child in the profits of the firm in which he is a partner, since they form part of his total income chargeable to tax. The words his income in section 139(1) must include every item of income which goes to make up his total income assessable under the Act. *Sec.139 (1).***[125ITR624,4Taxman11]**

Hold P&H. CIT - Ved Parkash and Sons (HUF)

The word hold according to the dictionary, means to possess, be the owner, holder or tenant of (property, stock, land....) Thus, a person can be said to be holding the property as an owner, as a lessee, as a mortgagee or on account of part performance of an agreement,etc. *Sec.2(42A)r.w.s.45.***[207ITR148]**

Hospitality Guj. CIT - Gujarat Carbon Ltd.

The term hospitality normally cannot be included in the ordinary meaning of the term entertainment, but falls with the enlarged meaning given to the words by Explanation 2 to section 37(2A) of the Income-tax Act, 1961. *Sec.37(2A).***[196CTR614,277ITR349]**

House Ori. CWT - K.B.Pradhan

The word house has no statutory definition and, therefore, has to be given the common parlance meaning.The dictionary meaning of the word is building for dwelling in, a building in general, a dwelling place. It also has the meaning of abode, habitation, etc. Though the concept of residence has been omitted from section 5(1)(iv), proviso, of the Wealth-tax Act, 1957,by the amendment made by the Finance (No. 2) Act, 1971, with effect from April 1, 1972, house or part of a house cannot cover a situation where the premises is not habitable.If a house was once habitable and became uninhabitable on account of want of repairs, the exemption provided by section 5(1)(iv), proviso, may operate.Where, however, the house is in the process of construction and, on account of the fact that it is not complete,has not reached a habitable stage, the concept of a house cannot be extended to cover such an incomplete construction. Parliament has also exempted a part of the house where the assessee's interest extends to a part of it. The concept of habitability is inherent in the word house. *Sec.5(1)(iv).WT.Act.***[130ITR393]**

2.Under section 5(1)(iv) of the Wealth-tax Act,1957, one house or part of a house belonging to the assessee is exempt from tax. The Act has not defined house.The word house would include any building irrespective of its use, that is to say, it may be used for business purposes or as a residence or as a school, and it hould not be restricted to a dwelling house.Section 5(1)(iv) does not contemplate that every room or a single self- contained unit of a house occupied by one person, although not in itself a division, is a separate house in law. **[CWT-Mahal Chand Pandia,Gau.,** *Sec.5(1)(iv).WT.Act.***134CTR120, 219 ITR 733,73Taxman534]**

3. CBDT Circular No. 317/8/73/WT/dated July 9,1973,lays down that a building used for purposes other than residential would also qualify for exemption under section 5(1)(iv) of the Wealth-tax Act, 1957.According to the Board's circular, the term house in section 5 (1)(iv) of the Act would refer not only to a building used for residential purposes,but for other purposes as well. **[CWT-M. Appuswamy,Mad.,***Sec.5(1)(iv).WT.Act.***233ITR460]**

4. The word house in section 5(1)(iv) of the Wealth-tax Act,1957, is not limited to a structure designed for human habitation and it includes any building or shed intended or used as a habitation or shelter for animals of any kind, a building in the ordinary sense or any building, edifice or structure enclosed with walls and covered, regardless of the fact of human habitation. **[CWT-N.Thavamani, Mad.,***Sec.5(1)(iv).WT.Act.***237ITR152]**

5. Section 5(1)(iv) of the Wealth-tax Act,1957, at the relevant time stated that the following asset shall not be included in the net wealth of the assessee : One house or part of the house belonging to the assessee. The word house has not been defined in the Wealth-tax Act nor in the General Clauses Act, 1897. However, the word building has been used in section 5(1)(iii) of the Wealth-tax Act and in section 5(1)(i) the word property has been used. Thus,the Wealth-tax Act has used the words house,building and property in different places, and hence different meanings should ordinarily be ascribed to these words in accordance with the settled principles of interpretation.All buildings cannot be regarded as houses. In common parlance a house means a place where people live. Of course a residential building can also be given for commercial purpose and yet it will remain a house.However, by no stretch of imagination can a cinema hall be regarded as a house. No one ever calls a cinema hall a house. **[CIT - Jai Kishan Gupta, All.** *Sec.5(1)(iv).WT.Act.* **185CTR340,264 ITR 482, 137Taxman388]**

6. House is a building where people live and reside. It is mainly for residential purposes. Cinema building cannot be treated as a house.

[CWT- Smt. Angoori Devi,All.,*Sec.5(1)(iv).WT.Act.***195CTR381, 273ITR500,144Taxman643]**

7. The word house has neither been defined under the Wealth-tax Act nor under the General Clauses Act. The word building has been used in section 5(1)(iii) and the word property has been used in section 5(1)(i) of the Act. In common parlance,house means a dwelling place where people live.However, a residential building or house can also be used for commercial purposes. **[Hiro J. Nagpal, Raj.,**<u>*Sec.5(1)(iv).WT.Act.*</u>**205CTR1,313ITR28,156Taxman286]**

House property Del. Addl.CIT - Vidya Prakash Talwar

...House property for the purposes of section 54 of the Income-tax Act, 1961, has the same meaning as the concept of house property under sections 22 to 27, which make it clear that the expression house property takes into account an independent residential unit. When section 54 talks of house property it does not mean an independent and complete house: it takes into account all independent residential units, particularly in these days when multi-storeyed flats are becoming the order of the day. <u>*Sec.54.*</u>

[132ITR661]

2. A house property for the purposes of section 54 of the Income-tax Act,1961,has the same meaning as the concept of house property under sections 22 to 27 which takes into account an independent residential unit and does not mean an independent and complete house; it takes into account all residential units, particularly in these days when multi storeyed flats are becoming the order of the day. **[CIT-Kodandas Chanchlomal,Guj.,**<u>*Sec.54.*</u>**155ITR273]**

3. Section 22 of the Income-tax Act, 1961, does not refer to house property despite its caption. The language employed in the section shows that the income referred to therein is not necessarily income from houses. It is income from property consisting of any building or lands appurtenant thereto of which the assessee is the owner. The word building is not confined in its scope only to dwelling houses.The word house in association with other words also has many other meanings. But, a commercial building is not regarded as a house. That, however, would not take the income from such buildings out of the ambit of section 22. Though it is not clear from the context why the Act describes income from property as income from house property, the substantive provision of law which creates the charge and obligates the person who receives such income to have it assessed under that head does not confine its application only to house property, but extends to all buildings whether such building is used as a dwelling house or for other purposes. **[CIT-Chennai Properties and Investments Ltd.,Mad.,** <u>*Sec.22.*</u> **186CTR409,266ITR685,136Taxman202]**

4. A plain reading of sections 22 to 27 of the Income-tax Act,1961, shows that house property means property constructed for dwelling or residential purpose, a portion of which or whole of the property or its appurtenant land is let out on rent to earn profit. It does not include a property or building constructed or purchased under planned business activity. **[CIT- Goel Builders,All.** Sec.27. **235CTR472,331ITR344,192Taxman28]**

Hundi transaction AP. CIT - Dexan Pharmaceuticals Pvt. Ltd.

The main characteristics of hundi transactions are as follows: (1) There are always three parties to such transaction. They are the drawer, the drawee and the payee. The drawer cannot himself also be the drawee. If the transaction is bilateral it is a very strong indication to show that it is not a hundi transaction. (2) A hundi is payable to satisfy a person or order but negotiable without endorsement by the payee. (3) The holder of a hundi is entitled to sue on its basis without any endorsement in his favour. (4) A hundi, once accepted by the donee, can be negotiated without endorsement. (5) In the case of loss of a hundi, the owner can claim duplicate or triplicate from the drawer and present the same to the drawee for claiming payment.(6) A hundi is normally in oriental language as per the mercantile custom. The above characteristics emanate from the long-standing custom of hundi transactions. Sec.69D.**[126CTR 57,214CTR576,82Taxman620]**

Hybrid systems of accounting Bom. CIT - CITI bank N.A.

Though the cash system and mercantile system of accounting are the two most common systems of accounting prevalent in the country,there are also innumerable other systems of accounting besides these two systems.Such systems are commonly known as hybrid systems of accounting. In such a system, there is a certain element of both cash and mercantile systems. An assessee following such a system may employ one method of accounting for one class of business or one class of customers or transactions and a different method for another class. Sec.145r.w.s.148 **[119CTR383,208ITR930,124Taxation19,75Taxman433]**

I

If he considers **SC** **CIT - Sunderlal**

.... the words if he considers used in the sub section postulates a scrutiny by the CIT of all the relevant facts for holding that the order is prejudicial to the interest of the revenue. An order passed by the CIT under Sec. 33B without giving any reasons for it,is,therefore, vitiated in law. *Sec.263.***[96ITR310]**

2.The words if he considers used in the Section postulates a scrutiny by the CIT of all the relevant facts for holding that the order is prejudicial to the interest of the revenue. **[CGT-O.E.Arumugha Mudaliar, Mad.,***Sec.24.G.T.Act.* **237ITR789154, Taxation 467]**

Illegal **Del.** **Remfry & Sons - CIT**

The expression illegal has been defined as being contrary to law and is normally applicable to everything which is an offence or which is prohibited by law. This word has an extensive meaning including anything and everything which is prohibited by law which constitutes an offence or which furnishes the basis for a civil suit. The expression in a wider meaning has been equated to unlawful but it may still not be void. This expression is distinguishable from the expression irregularity which is defined as a want of adherence to some prescribed rule or mode of proceedings and primarily consists of omitting to do what is necessary for due and orderly conduct of the proceedings.The word illegality on the other hand denotes a complete defect in jurisdiction or proceedings.

*Misc.***[195CTR66,276ITR1,145Taxman22]**

Immediate or deferred **Bom.** **Yogindraprasad N. Mafatlal - CIT**

The words immediate or deferred were added in section 64(v) in order to get over the reasoning adopted by the Bombay High Court in Manilal Dhanji's case .... the word minor in the expression minor child occurring in clause (v) of section 64 will have to be given a proper meaning and the word minor in that expression cannot be regarded as merely a word descriptive of the child. This expression used in section 64(v) requires that the income should be for the benefit of a minor child even when it is deferred. Looked at from this point of view the expression immediate or deferred cannot be regarded as having been added with the intention of covering a case where the benefit had been deferred beyond the minority of the child or children concerned. *Sec.64(v).***[6CTR439,109ITR602]**

Immediately before Mad. Thangam Textiles - First ITO

Sections 187 and 189(1) and (3) of the Income-tax Act, 1961, make it clear that the persons who are liable to bear the tax are the partners of the firm at the time of its dissolution. Hence, the notice under section 283(2) has to be issued only to such of those persons who were partners immediately before the dissolution of the firm. The phrase immediately before in section 283(2) means preceding the date of dissolution and has no reference to the year of assessment. *Sec.283(2).***[90ITR412]**

Implied decision Guj. CIT - Steel Cast Corporation

What is meant by implied decision is that though a point might have been raised before the Appellate Assistant Commissioner, in his final order the Appellate Assistant Commissioner might not have dealt with that point and thereby impliedly rejected it. A party may be aggrieved by an express decision of the Appellate Assistant Commissioner or by an implied decision of the Appellate Assistant Commissioner. *Sec.253,254.***[107ITR683]**

Impound any document or thing Bom. Avinash Bhosale

Supreme Court held that the phrase impound any document or thing does not include a passport. After considering the relevant provisions in relation to search and seizure, the Supreme Court has recorded a clear finding that the term document would not include a passport. The Supreme Court was dealing with power of a court to impound a document and, in that context, held that document does not include a passport. *Sec.131(3).***[322ITR381]**

Improve **Mad.** **CIT - Ramaswamy Mudaliar**

The cost of improvement of a capital asset can be taken into account in computing capital gains. The word improve has various shades of meaning and it includes everything by doing which there is an enhancement in the value of the asset or there is a rise in its price or the asset is made to grow better. *Sec.45,48.*[**196ITR939**]

in accordance.......... Parts II and III of Schedule VI to **Mad.** **CIT - Rajanikant Schnelder and Associates P. Ltd.**

The use of the words in accordance with the provisions of Parts II and III of Schedule VI to the Companies Act in section 115JA of the Income-tax Act,1961,is made for a limited purpose of empowering the Assessing Officer to rely upon the authentic statement of accounts of the company.While so looking into the accounts of the company, he has to accept the authenticity of the accounts with reference to the provisions of the Companies Act which obligates the company to maintain its accounts in a manner provided by that Act and the accounts to be scrutinised and certified by the statutory auditors and approved by the company in the general meeting and, thereafter, to be filed before the Registrar of Companies who has a statutory obligation to examine and be satisfied that the accounts of the company are maintained in accordance with the requirements of the Companies Act. Sub-section (1A) of section 115JA does not empower the Assessing Officer to embark upon a fresh enquiry in regard to the entries made in the books of account of the company.

Sec.115JA.[**302ITR22**]

2. The use of the words in accordance with the provisions of Parts II and III of Schedule VI to the Companies Act in section 115J was made for the limited purpose of empowering the Assessing Officer to rely upon the authentic statement of accounts of the company. While so looking into the accounts of the company, the Assessing Officer has to accept the authenticity of the accounts with reference to the provisions of the Companies Act, which obligate the company to maintain its accounts in a manner provided by that Act and the same to be scrutinised and certified by statutory auditors and approved by the company in general meeting and thereafter to be filed before the Registrar of Companies who has a statutory obligation also to examine and be satisfied that the accounts of the company are maintained in accordance with the requirements of the Companies Act.Sub-section (1A) of section 115J does not empower the Assessing Officer to embark upon a fresh enquiry in regard to the entries made in the books of account of the company. **[Apollo Tyres Ltd.-CIT** *Sec.115JA.***174CTR521,255ITR273, 122Taxman562]**

In any year **Cal.** **Burn & Co.**

.......the expression in any year in section 34 must be read as meaning the year during which the proceedings in assessment in respect of that very year should have been initiated, that is, the actual twelve months in which an assessment would normally and properly be made.... *Sec.147.***[2ITR30]**

In computing the total income of any individual **SC** **CIT-Shri Om Prakash**

Sub-section (1) of section 64 of the Income-tax Act,1961, opens with the words in computing the total income of any individual. Then it proceeds to say that in the total income of such individual shall be included the income of his spouse arising from the membership of such spouse in the partnership firm in which such individual is the partner.It further says that the income arising to the minor children of such individual who are admitted to the benefits of partnership wherein such individual is a partner shall also be included in the total income of such individual. An individual can be a partner in a partnership firm in his individual capacity or in the capacity of the karta of a Hindu undivided family or, for that matter, in any other capacity, e.g., as a trustee. There may be a firm comprising an individual and his wife, to which their minor children are admitted. There can also be a firm comprising two or more individuals wherein the wife/wives of one or more of the partners are also partners. The minor children of one or more of the partners may also have been admitted to the benefits of the partnership firm. So far as other partners in the partnership firm are concerned, they are not really concerned in what capacity a particular person is a partner, i.e., whether as an individual, as a karta, as a trustee or otherwise. To them, he is an individual, a person.This aspect, however, becomes relevant as between the partner and those whom he represents in the partnership firm. To wit,where a person is a partner as the karta of a Hindu undivided family,the capacity in which he is a partner in the partnership firm is relevant as between him and the other members of the Hindu undivided family. For, the income the karta receives as a partner is not his individual income; it is the income of the Hindu undivided family and he receives it on behalf of the Hindu undivided family. It is for this reason that the income of the wife and minor children arising from their membership/admission to the benefits of the partnership firm, is held not includible in the income of the Hindu undivided family since the total income of the Hindu undivided family is not the total income of the individual (husband or father, as the case may be). For section 64(1) to get attracted, it is necessary that the husband/father should be a partner in a partnership firm as an individual, i.e., in his individual capacity. It is not attracted where he is a partner as the karta of the Hindu undivided family to which such wife and/or minor children belong.

Sec.64(1).[**217ITR785,84Taxman156]**

In connection with **Kar.** **Stumpp, Schuele and Somappa Ltd. - CIT**

The expression in connection with in section 37(3) includes matters occurring prior to as well as subsequent to or consequent upon, so long as they are related to the principal thing.In other words, whatever has nexus to the travel undertaken in connection with the work outside the headquarters resulting in the stay, such stay, whether actual work in connection with company affairs was carried out or not, would be relatable to the travel undertaken which was indisputably in connection with the work of the company and, therefore, the only logical inference to be drawn is that the stay also was in connection with the work as it is intimately connected with the travel undertaken. The entire expenses of such travel would be covered by section 37(3). *Sec.37(3).*[**190ITR152,61 Taxman278]**

In connection with such transfer **Ker.** **V.A. Vasumathi - CIT**

In computing capital gains,expenditure incurred wholly and exclusively in connection with the transfer of a capital asset has to be deducted under section 48 of the Income-tax Act, 1961.The words in connection with such transfer occurring in the section mean intrinsically related to the transfer. Only such expenditure as is wholly and exclusively related in an intrinsic manner to the transfer is a deductible expenditure. *Sec.48.*[**123ITR94]**

In good faith made full disclosure of its net wealth **All.** **Hasan Ahmad Khan - CWT**

The question whether the assessee made full disclosure of his net wealth or not, has not to be looked into with reference to the net wealth as ultimately evaluated by the Wealth-tax Officer, but from the point of view of the assessee, that is, whether or not, while disclosing his net wealth fully, he acted honestly. The expression in good faith made full disclosure of his net wealth merely means that the assessee should have honestly described all his assets and liabilities which go to constitute his net wealth, along with their estimated value. *Sec.18(2A)WT.Act.***[99ITR414]**

2. The expression in good faith, made full disclosure of his net wealth occurring in section 18(2A) of the Wealth-tax Act,1957, merely means that once the assessee lays open his assets and liabilities which go to constitute his net wealth and makes an honest estimate of the same, the requirement of the provision would be satisfied. The word disclosure indicates that something is exposed, brought to light, uncovered, laid open or revealed, which was earlier kept hidden. Therefore, if an assessee honestly makes a full disclosure of his net wealth, although the disclosure made by him is ultimately found to be inaccurate, it cannot be said that the assessee did not act in good faith. **[Sardar Joginder Singh-CWT,Gau.,** *Sec.18(2A)WT.Act.***134ITR636]**

In part **Guj.** **Arvindkumar J. Saheba - CIT**

The expression in part in section 11(1)(b) of the Income-tax Act, 1961, does not refer to an aliquate part. If half a house is held in trust wholly for religious or charitable purposes, the subject-matter of the trust is only the said half of the house and that half is held wholly for religious or charitable purposes. There may be instances where,though there is a trust, it involves only a partial dedication of the property held under trust in the sense that only a part of the income of that property is utilised for religious or charitable purposes. The dichotomy between the two expressions wholly and in part is not based upon the dedication of the whole or fractional part of the property but between the dedication of the property wholly for religious or charitable purposes or in part for such purposes.

*Sec.11(1)(b).***[22CTR245,131ITR86]**

In the course of marketing of business Ker. V.O. Markose- CIT

The expression in the course of carrying on of business in section 5(1)(xiv) means that the gift should have some relationship with the carrying on of the business. It is not sufficient that the gift is made during the carrying on of a business. It must further be established that there was some integral connection or relation between the making of the gift and the carrying on of the business. *Sec.5(1) (xiv).G.T.Act.* **[98ITR504]**

In the like manner.. Cal. CWT - Official Trustee of W.B. For Trust Murshidabad Estate

A trustee is assessed under the scheme of the Wealth-tax Act,1957, only as a representative assessee although for certain purposes of the Act a trustee is an assessee. Section 21 of the Wealth-tax Act makes it clear that wealth-tax shall be levied upon a trustee and is recoverable from the trustee in the like manner and to the same extent as it would be leviable upon and recoverable from the person on whose behalf or for whose benefit the assets are held. Section 21(2) also provides that the Wealth-tax Officer can make a direct assessment on the beneficiary and recover the tax from the beneficiary in respect of the assets held under trust. The Wealth-tax Officer,therefore, has an option either to assess the beneficiary directly or to assess the trustee in respect of the assets held under trust. The Act does not envisage that when an assessment is made on a trustee the incidence of tax will be heavier. This is clearly provided by the expression in the like manner, and to the same extent used in section 21 of the Act. *Sec.21(2).WT.Act.* **[29CTR50, 136ITR162,10Taxman67]**

In the nature of Pat. CIT - Alkem Laboratories P. Ltd.

Section 37(1) of the Act is a provision of residuary nature and once a provision is found to be residuary in nature, the mere fact that a claim does not fall under any of sections 30 to 36 of the Act,will not automatically, make the claim unsustainable under section 37 of the Act.Section 37 of the Act is a general section for grant of deduction on certain accounts not enunciated in sections 30 to 36 of the Act.The expression in the nature of used in section 37(1) of the Act does not intend to stultify a legitimate claim in accordance with the principles of accountancy and well established commercial practice.If the expenses are not deductible under sections 30 to 36 of the Act but the conditions prescribed under section 37 are satisfied, then such expenses are required to be deducted while computing the income.Therefore, certain expenses which may not admissible under section 30(a)(i) of the Act could be admissible under section 37(1) of the Act. <u>*Sec.37(1).*</u>**[314ITR329]**

In the nature of entertainment expenditure **P & H.** **CIT - Khem Chand Bahadur Chand**

In construing the provisions of sub-sections (2) and (2A) of section 37 of the Income-tax Act, 1961, what first meets the eye and deserves highlighting is the designed and considered use of the phraseology by Parliament.The phrase deliberately employed is in the nature of entertainment expenditure.It is not merely entertainment expenditure or business entertainment simpliciter. ......whilst using the larger and compendious expression in the nature of entertainment expenditure, Parliament had an obvious purpose behind it. The expression is much wider in its connotation, inasmuch as it would take within its ambit not merely what may stricto sensu be regarded as entertainment expenditure proper but also all other expenditure of allied nature partaking of some, if not all, of the characteristics of entertainment expenditure. The phrase has a wide amplitude and its use leaves little doubt that the intention of the Legislature in employing it was to cast the net sufficiently wide so as to bring within its field all types of hospitality. Any lavish hospitality expended for business purposes would amount to entertainment expenditure.If that is so, all hospitality, even though frugal,would be well within the ambit of the larger phrase in the nature of entertainment expenditure. <u>*Sec.37(2A).*</u>**[23CTR319, 131ITR336]**

2. The expression in the nature of entertainment expenditure has not been defined in the Income-tax Act. Entertainment involves hospitality of any kind, which an assessee extends to a customer, client or a constituent for furtherance of business or profession. It must be wholly and exclusively necessary for the purpose of business or profession. The primary motive behind laying out or incurring entertainment expenditure is commercial or professional expediency. The phrase in the nature of entertainment expenditure encompasses within its ambit entertainment expenditure proper as well as expenditure akin to it partaking of some, if not all, of the characteristics of entertainment expenditure. Even lavish or frugul hospitality is none the less hospitality. The Income-tax authorities have to minutely scrutinise the accounts in each case item wise and see whether the expenditure by the assessee is wholly and exclusively for the purpose of the business or profession or whether it is in the nature of entertainment expenditure. Even if it comes under sections 37(1), 37(2) and 37 (2A), it has still to be investigated whether it is reasonable expenditure. The reason is obvious. By a fastidious or lavish person, the act of hospitality with luxurious dishes and costly hot drinks may be regarded as an ordinary lunch or dinner, but, to a common man, it would be a lavish meal or staggering wasteful expenditure. The test whether the hospitality is appropriate and befitting is an elusive and ambiguous one. What is appropriate and befitting is always a big question mark affording a carte blanche right to an assessee to expend lavishly exhibiting ostentation with the belief that the public exchequer would defray the wasteful expenditure. If this consideration is allowed to prevail, sub-sections(2), (2A) and (2B) of section 37 would be rendered otiose and ineffectual, defeating the legislative animation. Therefore, the fact-finding authorities have to carefully scrutinise the records and give the necessary allowance as per law. **[CIT-Navabharat Enterprises (P.)Ltd.AP.**<u>*Sec.37(2A)*</u>**63CTR187,170ITR332]**

In the performance of his duties Ker. A.K.Venkiteswaran - CIT

Section 16(v) of the Income-tax Act,1961, allows a deduction from salary of any amount actually expended by the assessee which, by the conditions of his service, he is required to spend out of his remuneration wholly,necessarily and exclusively in the performance of his duties.It is not sufficient if the expenditure claimed is wholly met in the performance of the assessee's duties or has necessarily to be so met nor is it sufficient if it is shown that it was met exclusively in the performance of the duties of the assessee. The requirements are cumulative and, therefore, it must be shown that the assessee was required to spend the amount out of his remuneration not only wholly but also necessarily and exclusively in the performance of his duties. More important is the requirement that it is not for the performance of his duties that he has to spend it but in the performance of his duties. The term in the performance is much narrower in scope than the term for the purpose of the performance.Many items of expenditure may be incurred by an assessee to enable to perform his duties properly.These are not expenses incurred in the performance of his duties as these are limited to expenses incurred during the process of the performance of the duties. The expenses must be incurred wholly in the discharge of the duties and exclusively in the discharge of the duties. Only such expenditure as is incurred after the process of performance of duties has commenced is deductible. Sec.16(v). [92ITR233]

In the performance of official duties SC Matajog Dobey-H.C.Bhari

In order that an act done by an officer may be an act done in the performance of his official duties,there must be a reasonable connection between the act and the discharge of official duty; the act must bear such relation to the duty that the officer could lay a reasonable,but not a pretended or fanciful claim, that he did it in the course of the performance of his duty. *Criminal Procedure Code* **[28ITR941]**

In the prescribed manner Mad. M.CT.Muthiah Chettiar Family Trust - Fourth ITO

The words in the prescribed manner in section 11(2)(a) of the Income-tax Act,1961,do not confer power on the rule-making authority to prescribe a time limit for making an application for exemption under the section. Therefore, paragraphs 2 and 4 in Form No.10 issued in pursuance of rule 17 of the Income-tax Rules, 1962, are ultra vires in that the rule-making authority has exceeded its limit in including in the Form the said two paragraphs.

*Sec.11(2)(a).***[86ITR282]**

In this behalf Kar. Southern Veneers and Wood Works Ltd. - CIT

The income that is to be excluded under section 10(15) of the Income-tax Act,1961,is the interest payable on any moneys borrowed in respect of the purchase of raw materials to the extent to which such interest does not exceed the amount of interest calculated at the rate approved by the Central Government in this behalf.The term in this behalf is indicative of the idea that the approval of the Central Government should be on behalf of or pertaining to the subject-matter referred to therein. In other words, the Central Government has to approve the rate of interest for the purpose of section 10(15)(iv)(c). The term in this behalf is a term frequently found in several provisions of the Income-tax Act. In fact, in the very section 10 of the Act, some of the clauses use this term. The idea behind section 10 and the requirement of the approval of the Government of India is that the exemption is available only when the Government of India deems it necessary and fit to grant exemption.It is because of this that, wherever specific approval or a notification is necessary, the term in this behalf had been used. For example,under clause (10B), the particular notification has to specify the amount in this behalf. Under clause (10C), a similar terminology is used.Several sub-clauses under section 88(2) which require issuance of a notification by the Government also refer to the specification in the notification in this behalf. Hence, the approval under section 10(15)(iv)(c) by the Central Government will have to be a specific approval concerning the subject stated therein and not a general approval for any other purpose. *Sec.10(15)(iv).* **[196ITR718]**

Inaccurate **SC** **Dilip N. Shroff - JCIT**

The word inaccurate signifies a deliberate act or omission on the part of the assessee.Such deliberate act must be either for the purpose of concealment of income or furnishing inaccurate particulars.The term inaccurate particulars is not defined.

Sec.271(1)(c).**291ITR519,161Taxman218]**

2. The word inaccurate in the context of levying penalty under section 271(1)(c) signifies a deliberate omission on the part of the assessee. Such deliberate act must be either for the purpose of concealment of income or furnishing of inaccurate particulars.

[T. Ashok Pai- CIT, SC.,*Sec.271(1).***210CTR259, 292ITR11, 161 Taxman 340]**

Incharge **SC** **State of Karnataka -Pratap Chand**

A person in charge must mean that the person should be in overall control of the day to day business of the firm. *Drugs and Cosmetics Act.***[128ITR573]**

Incidental to the carrying on of the business **Pat.** **CIT- Parmanand Makhan Lal**

The loss sustained as a result of misappropriation by an agent is one which is incidental to the carrying on of the business by the principal and is deductible in computing the profits and gains under section 10(1) of the 1922 Act corresponding to section 28 of the 1961 Act. *Sec.28.***[142ITR800]**

Includes Bom. CIT - Tata Hydro Electric Power Supply

The word includesis often used in an interpretation clause in order to enlarge the meaning of the words or phrases occurring in the body of the statute. When it is so used, these words and phrases must be construed as comprehending not only such things as they signify, according to their nature and import, but also those things, which the interpretation clause declares, that they shall include. *Sec.33,43(3).* **[112ITR288]**

Includes any profits SC Synco Ind. Ltd. - A.O.(Income-tax)

The words includes any profits in section 80-I(1) are important and indicate that the gross total income of an assessee shall include profits from a priority undertaking. While computing the quantum of deduction under section 80-I(6) the Assessing Officer, no doubt, has to treat the profits derived from an industrial undertaking as the only source of income in order to arrive at the deduction under Chapter VI-A. However, the non obstante clause in section 80-I(6)is applicable only to the quantum of deduction, whereas, the gross total income under section 80B(5) which is also referred to in section 80-I(1) is required to be computed in the manner provided under the Act which presupposes that the gross total income shall be arrived at after adjusting the losses of the other division against the profits derived from an industrial undertaking. *Sec.80-I.*

[215CTR,385,299ITR444,168Taxman224]

Inclusive definition Mad. DIT (Exemp.)- Agri.Horticultural Society

An inclusive definition widens the etymological meaning of the expression or term including therein that which would ordinarily not be comprehended therein.The words used in an inclusive definition denote extension and cannot be treated as restricted in any sense.While dealing with an inclusive definition it would be inappropriate to put a restrictive interpretation upon terms of wider denotation. *Misc.***[273ITR198]**

2. When the Legislature wants to enlarge the natural meaning of a word or phrase, it uses the word includes and, in such a context, an inclusive definition means that, over and above the natural meaning of the words, the specially provided meaning of the word will also have to be attributed for the purpose of interpretation of that particular statute or that particular chapter. **CIT-Premanand Industrial Co-operative Service Society Ltd., Guj.** *Misc.* **[16CTR6,124ITR772]**

Income Ran. CIT -Bengalee Urban Co-op. Credit Society Ltd.

Income as contrasted with profits or gains in the Income-tax Act means a periodical monetary return coming in and accruing to the assessee independently and not as the net proceeds of a business carried on by the assessee. *Sec.2(24).***[2ITR121]**

2. The word income is not limited by the words profits and gains and anything which can properly be described as income is taxable under the Act unless expressly exempted. **[Maharajkumar Gopal Saran Narain Singh-CIT.,PC.,** *Sec.2(24).***3ITR237]**

3. The element of periodical receipt or regularity or expected regularity of monetary return is an essential ingredient of income under the Indian Income-tax Act. **[Major A.U. John.,All.,** *Sec.2(24).***6ITR434]**

4. The word Income is not used in section 28 of the Indian Income-tax Act in the popular meaning of money received,but is used in a much wider sense and connotes the assessable figure arrived at after accounting for all the legitimate deductions and exemptions.

[Nagin Chand Shiv Sahai -CIT.,Lah.*Sec.2(24).***6ITR534]**

5. Income is generally though not necessarily a recurrent return from a definite source.Receipts from capital which is exhausted in the process of realisation may none the less be income. The fact that the receipts are from a source which is a wasting asset like that of a mine is an irrelevant consideration and anything which can properly be described as income would be taxable unless expressly exempted from Taxation.

[Jyotirendra Narayan Sinha Choudhury - State of Assam., Ass., *Assam Agricultural Income-tax Act.***19ITR379]**

6. Interest on securities only becomes income when it is actually received and not when it is due or capable of being received by the assessee. **[SethLalbhaiDalpatbhai-CIT.,Bom.***Sec.18,2(24).***22ITR 13]**

7. The word income........should be given its widest connotation in view of the fact that it occurs in a legislative head conferring legislative power and it includes a capital gain. It will be wrong to interpret the word in the light of any supposed English legislative practice. **[NavinchandraMafatlal-CIT,SC,**<u>*Sec.2(24).*</u>**26ITR758]**

8. The income by sale of forest trees is income and taxable.

[Sir Kamesiwar Singh-CIT, Pat.,<u>*Sec.2(24).*</u> **AIR1954Pat.146,26 ITR121]**

9.Income means-that which comes in as the periodical produce of one's work,business,lands or investments; annual or periodical receipts accruing to a person or corporation.**[V.RamawamiNaidu-CIT,Mad.,**<u>*Sec.2(24).*</u>**AIR1959Mad.126 5ITR33]**

10. The word income as used in the Income-tax Act,1961,is wide and vague in its scope. It is a word of elastic import and its extent is not controlled and is not governed by the words profits and gains in section 10 of the Indian Income-tax Act, 1922.Every receipt generally may be described as income unless it is expressly exempt. There is nothing to indicate that the source must be one which is recognised under the law.Even income derived from an illegal business could be liable to tax. **[CIT - Shanti Meattle.,All.,** <u>*Sec.2(24).*</u>**90ITR385]**

11. The deemed dividend contemplated by Section 2(6-A) cannot be considered as income under Section 44F. **[CIT-VadilalLalubhai., SC,**<u>*Sec.2(24).*</u>**AIR1972 SC1016,86ITR2]**

12. The word income in the Constitution cannot be restricted to the meaning of that term in a fiscal statute like the Income-tax Act which seeks to tax some kinds of income leaving out others. The word income in entry 82 of the Union List has been used in a very wide sense to include every kind of receipt and gain.Even under the Income-tax Act,the sources of income mentioned therein are not exhaustive and they can be expanded and indeed they have been expanded from time to time,as,for instance,by the inclusion of capital gains in section 6 of the Income-tax Act,1961,by the Amending Act No. 22 of 1967. **[Bhola Nath Kesari-Director of State Lotteries,Mad.,**<u>*Sec.6.r.w.s.80TT.*</u>**95ITR171]**

13. The definition of income in section 2(24) of the Income-tax Act, 1961, is an inclusive definition. It enlarges the meaning of income and cannot be a valid ground for contending that because rents of self-occupied properties are not mentioned in it, the notional annual value of such properties cannot be subjected to tax. ... Section 14 enumerates the heads of income under which income of an assessee falls to be charged. .... Income from property which is made liable to tax is not its actual income in money but an artificial or statutory income as defined in section 23 and that artificial or statutory income is the annual value of the property. The contention that the properties are self-occupied and,therefore, the owner does not receive anything in cash or in money's worth is entirely beside the point. The notional letting value of self-occupied property would constitute income ... **Sakarlal Balabhai - ITO, Guj..** *Sec.2(24).***[100ITR97]**

14. Section 2(6C) of the Indian Income-tax Act, 1922, and section 2(24) of the Income-tax Act, 1961, give inclusive definitions of the word income.Whenever a term is given such a definition in a statute, it means not only the things mentioned therein but also includes in its ambit the meaning of the term as generally understood. The word income has thus to be given a very wide meaning.

[Raja Ragavendra Singh - State of Punjab, P & H.*Sec.2(24). r.w.s.14.***102ITR40]**

15. Income is a monetary return expected by the assessee for the labour and/or skill bestowed, and/or capital invested by him; coming in from a definite source, which need not be a legal source, in the sense that the failure to pay the same need not be enforceable in a court of law; and excluding a receipt in the nature of a mere windfall, which would mean a windfall in regard to its very nature and not in regard to its extent or quantum. **[Mehboob Productions Pvt. Ltd.-CIT,Bom.,** *Sec.2(24).***106ITR758]**

16. It is well known in the world of commerce that the income of a businessman is the surplus by which the gross receipts from his business exceed the expenditure incurred on land, labour, capital and managerial skill and all other tangible and intangible assets used for the purpose of the business. **[Nippon Electronics (P.) Ltd.-CIT,Kar.,***Sec.2(24).***116ITR231]**

17. The definition of the word income in section 2(24) is inclusive and not exhaustive and includes whatever is income construed in its natural or ordinary sense to which is added certain artificial categories also. Barring these artificial categories, the natural concept would determine the quality of income and it is in this sense the word profits is understood in the natural and proper sense, that is in a sense which no commercial man would misunderstand.

[CIT- Express Newspapers Ltd.,Mad.,*Sec.2(24).* **15CTR383,124 ITR117]**

18 The expression income, according to the dictionary, means a thing that comes in. Income may also be defined as the gain derived from land, capital or labour or any two or more of them. Even in its ordinary economic sense, the expression income includes not merely what is received or what comes in by exploiting the use of a property but also what one saves by using it one self. That which can be converted into income can be reasonably regarded as giving rise to income. **[Bhagwan Dass Jain - Union of India,SC,** *Sec.2(24).***128ITR315]**

19. The word income in section 16(3) would include loss **[CIT - J.H. Gotla. SC.***Sec.64.r.w.s.2(24).***48CTR363,156ITR323]**

20. The Legislature has advisedly refrained from defining the term income.Income is not limited by the words profits and gains and anything which can properly be described as income is taxable under the Act unless expressly exempted.The element of return is not an essential ingredient of income.

CIT-SahneySteeland Press Works Ltd.. AP.*Sec.2(24).* **[44C TR 243,152ITR39,17Taxman403]**

21. The definition of income in section 2(24) of the Income-tax Act, 1961, is an inclusive definition and is wide enough to include the receipts earned from the business carried on by the official liquidator after the winding up of a company.

[Golecha Properties (P.) Ltd.– CIT, Raj.,*Sec.2(24).* **36CTR227, 171ITR47,66 Taxman246]**

22. ....the expression income as used in section 69A of the Income-tax Act,1961,had a wide meaning which meant anything which came in or resulted in gain. **[Chuharmal-CIT,SC,** *Sec.69A.***70CTR88,172ITR250]**

23. ....the word income is an inclusive definition. It is not exhaustive. It has a wide import. It has got a legal concept. The scheme of section 2(24) of the Act read with sections 4 and 10 of the Act is that, given its ordinary and natural meaning, the word income will take in any monetary return coming in. It will take in voluntary and gratuitous payments which are connected or linked with any office, vocation or occupation. Any amount received by an assessee by virtue of his profession, vocation or occupation will constitute his income. Section 10(3)(ii) of the Act takes within its fold any receipts arising from the exercise of a profession or occupation in the total income. The receipts may be casual or of a non-recurring nature. **[Father Epharam-CIT,Ker.,** *Sec.2(24) r. w.s.10 (3).***[75CTR142,176ITR78]**

24. The expression income in entry 82,List I, cannot be subjected by implication, to any restriction by the way in which that term might have been deployed in a fiscal statute. A particular statute enacted under the entry might, as a matter of fiscal policy, seek to tax some species of income alone. The definition would, therefore, be limited by the consideration of the fiscal policy of a particular statute. But the expression income in the legislative entry has always been understood in a wide and comprehensive connotation to embrace within it every kind of receipt or gain either of a capital nature or of a revenue nature.The taxable receipts as defined in the statute cannot be held to fall outside such a wider connotation of income in the wider constitutional meaning and sense of the term as understood in entry 82, List I........ The word income is of elastic import. In interpreting expressions in the legislative lists, a very wide meaning should be given to the entries. In understanding the scope and amplitude of the expression income in entry 82, List I, any meaning which fails to accord with the plenitude of the concept of income in all its width and comprehensiveness should be avoided.

[Elel Hotels and Investments Ltd.-Union of India, SC, *Constitution of India.***77CTR168,178ITR140,44Taxman304]**

25. The interest paid on compensation awarded for compulsory acquisition of land under section 28 of the Land Acquisition Act, 1894, is of the nature of income and not capital. **[K.S.Krishna Rao-CIT,SC,***Land Acquisition Act.***181 ITR408]**

26. Income under section 2(24) includes any capital gains chargeable under section 45.The words average rate of Income-tax and income, for the purposes of section 85A,have the same meanings as set out in sections 2(10) and 2(24), respectively. **[CIT-Central Bank ofIndiaLtd.,Bom.,***Sec.2(24).***185ITR6]**

27. The Income-tax Act defines the expression income in clause (24) of section 2, but that definition cannot be read back into entry 82 of List I of the Seventh Schedule to the Constitution. Even the said definition is an inclusive one and has been expanding from time to time. Several items have been brought within the definition from time to time by various amending Acts.The said definition cannot, therefore, be read as exhaustive of the meaning of the expression income occurring in entry 82 of List I in the Seventh Schedule. This, of course, does not mean that an amount which can, by no stretch of imagination, be called income can be treated as income and taxed as such by Parliament. It must have some characteristics of income as broadly understood. So long as the amount taxed as income can rationally be called income as generally understood, it is competent for Parliament to call it income and levy tax thereon.

[Sat Pal and Co.- Excise and Taxation Commissioner,P & H., *Sec.44AC.r.w.s. 206C.***185 ITR375]**

28. The definition of income in section 2(24) was inclusive, the purpose of the definition was not to limit the meaning of income but to widen its net, and the several clauses therein were not exhaustive of the meaning of income; even if a receipt did not fall within the ambit of any of those clauses, it might still be income if it partook of the nature of income ...The word income is of the widest amplitude and it must be given its natural and grammatical meaning. The prize money in a motor rally received by the winner is income.

[CIT-G.R.Karthikeyan,SC,*Sec.2(24).***[AIR1993SC1671,112CTR 302,201ITR866,68Taxman145]**

29. The definition of income in section 2(24) is an inclusive definition.It adds several artificial categories to the concept of income but, on that account, the expression incomedoes not lose its natural connotation.Anything which can be properly described as income is taxable under the Act unless it is exempted under one or the other provisions of the Act. **[Emil Webber-CIT,SC,***Sec.2 (24).***200ITR483]**

30. Income is something which flows from the property.

[Udayan Chinubhai- CIT, SC,*Misc.***AIR1996SC315,141CTR387, 222ITR456,140Taxation244,98 Taxman 502]**

31. Although the system of accounting adopted by the assessee may be a relevant factor, even in the mercantile system of account only such amounts can be assessed which the assessee had a right to receive or which had accrued.A mere claim or a mere demand without anything else is not income within the meaning of section 5 of the Income-tax Act,1961.

[Hope (India) Ltd. (Now Poddar UdyogLtd.)-CIT,Cal.,*Sec.5.* **155 CTR561,238ITR740,155Taxation 174, 05Taxman723]**

32. The word income is of the widest amplitude and it must be given its natural and grammatical meaning.The principles culled out from the decision are as follows :(i) since the definition of income in section 2(24) of the Income-tax Act,1961,was an inclusive one,its ambit should be the same as that of the word income in entry 86 of List I of Schedule VII to the Constitution of India; (ii) the words other games of any sort in section 2(24)(ix) were of wide amplitude and their meaning was not confined to games of a gambling nature alone,and therefore, section 2(24)(ix) was not confined to mere gambling or betting activities; (iii) assuming that the expression winnings had acquired a particular meaning, viz., receipts from activities of a gambling or betting nature only, it did not follow that monies received from non-gambling or non- betting activities were not included within the ambit of income; (iv) the definition of income in section 2(24) was inclusive, the purpose of the definition was not to limit the meaning of income but to widen its net, and the several clauses therein were not exhaustive of the meaning of income; even if a receipt did not fall within the ambit of any of those clauses,it might still be income if it partook of the nature of income. **[CIT-Avinash Pasricha,Del.,** *Sec.2(24).***170CTR287,251ITR360]**

33. The expression income in terms of section 2(24) of the Income-tax Act,1961, would include profits and gains and any capital gains chargeable under section 45 of the Act. ...In the case of a non-resident, it would include all income from whatever source derived, which is received or is deemed to be received in India in such year by or on behalf of such person or accrues or arises or is deemed to accrue or arise to him in India during such year.

[CIT-Quantas Airways Ltd., Del.,*Sec.2(24).r.w.s.45.***175CTR98, 256ITR84,122Taxman935]**

34. The definition of income is inclusive.The purpose of the inclusive definition is not to limit the ordinary meaning of the word income but to widen its net, and the several clauses therein are not exhaustive of the meaning of the word income. It is the true nature and quality of the receipt and not the head under which it is entered in the account books,that would prove decisive.

[CIT-Motor and General Sales(P.)Ltd.,All.*Misc.***188CTR42,266 ITR261,135 Taxman426]**

35. The inclusive definition of the word incomein section 2(24) of the Income-tax Act, 1961 adds several artificial categories to the concept of income, but on that account, the expression does not lose its natural connotation. It has to be construed as comprehending only such things which are income according to the natural import of the term......It is the income which has really accrued or arisen to the assessee that is taxable.Whether the income has really accrued or arisen to the assessee must be judged in the light of the reality of the situation.When in reality there is neither accrual nor receipt of income by the assessee, even though an entry to that effect might in certain circumstances have been made in the books of account, it would not constitute income for the purpose of levy of tax.A rebate obtained by the purchaser or remission of debt by a creditor would not result in the creation of income in the hands of the purchaser or debtor. **CIT-Industrial Credit and Dev. SyndicateLtd.,Kar.,***Sec.2(24).* **203 CTR 413,285ITR310]**

Income accrues **Guj. Anup Engineering Ltd. - CIT**

Income accrues only when the assessee gets a right to receive the same and if the right to receive is not established, no income would accrue or arise to the assessee. Unless and until a debt is created in favour of the assessee, which is due by somebody, it cannot be said that the assessee has acquired a right to receive the income or that the income has accrued to him. Sec.5.**[165CTR21,247ITR457,114 Taxman584]**

Income and Capital **Pun.** **CIT - Shamlal Narula**

In popular and also in legal parlance capital is understood as the source of income and income as the fruit of capital. *Misc.***[AIR1964 Pun411,50ITR513]**

Income and profits and gains **Cal.** **Aryasthan Corporation Ltd. - CIT**

The words income and profits and gains should be understood as including losses also so that in one sense profits and gains represent positive income whereas losses represent negative income. In other words, loss is negative profit.Both positive and negative profits are of revenue character. Both must enter into computation, wherever it becomes material, in the same mode of the taxable income of the assessee. *Sec.73.***[172CTR640,253ITR401, 124Taxman516]**

Income as returned by such person **Bom.** **Mansukhlal and Brothers-CIT**

The words income as returned by such person mean income as stated by such person in his return and do not mean income computed or assessed by the Income-tax authorities minus the income aided on the ground of concealment. The maximum penalty does not depend on the amount of concealed income. *Sec.271(1)(c)*. **[AIR1966SC835,57ITR270]**

Income assessed in the foreign country **SC** **CIT-Clive Insurance Co. Ltd.**

The expression income assessed in the [foreign] country in paragraph (iii) of the Explanation to section 49D, in the context in which it was used, meant subjected to tax in the foreign country.

*Sec.5.***[7CTR68,113ITR636]**

Income by way of interest **Mad.** **CIT - S.K. Sundararamier and Sons**

Section 194A of the Income-tax Act, 1961, postulates deduction of the tax at source on the gross amount of interest at the time of credit of the interest amount. The expression, interest can only be gross interest, and it cannot refer to the net interest in the context of crediting of the interest by the person responsible for deducting the tax. Further, a person responsible for paying interest would credit in the account of the payee only the gross interest, and even under the commercial principles or accountancy principles the credit is only of the gross interest and not the net interest.

Sec.*194A.***[240ITR740]**

Income earner **Mad.** **CIT - P.S. Chelladurai**

Any human pursuit which involves the seeking after emoluments or which involves the unrestricted inflow of money into private hands, tends to make the individual concerned not only an income earner but also an I.T. payer and if the individual has a family to maintain and has not altogether renounced it,the individual may be truly regarded as its bread-winner. *Misc.***[145ITR139]**

Income from business **AP.** **Jaldu Manikyala Rao –CIT**

If a transaction is in the nature of trade in general, even though it is not in the same line of trade carried on by the assessee, and even though such transaction is casual in nature,the profit arising there from will be income from business. *Sec.28***[28ITR220]**

2. The principle is that if the owner of a property carries on business with a property owned by him, the income from that property must be assessed as only income from business. **CIT - New India Maritime Agencies P. Ltd., Mad.** *Sec.22,28.* **[207ITR392]**

3. The principle is that if the owner of the property carries on business with a property owned by him,the income from that property must be assessed only as income from business.

[CIT-New India Maritime Agencies (P.) Ltd.Mad. *Sec. 22,28.* **253ITR732, 124Taxman801]**

4. The word business had been defined in section 2(13) of the Income-tax Act, 1961, to include any trade, commerce or manufacture or any adventure or concern in the nature of trade, commerce or manufacture. This definition, being an inclusive one, is indicative of an extensive and expanded meaning rather than a restricted or narrow meaning. In view of this, the activities which may amount to business need not necessarily be by way of trade, commerce or manufacture, or be in the nature of a profession. They may even consist of rendering of services as in the case of selling agents, managing agents, and these services may be of a variegated character. .....when the property has been let out not only as a property but with services which is a complex letting, the income cannot be said to be derived from mere ownership of house property but may be assessable as income from business. If the owner of a property carries on upon the property some activities which result in profits and gains arising, not from the ownership of the property but from the owner's user there of, those profits and gains may be chargeable under section 28 as income from business, apart from the assessment under this section in respect of the income from house property. Thus, if an owner holds a property for a business purpose and receives from his tenants rents which include charges for supplying various services, the owner would be chargeable under this section in respect of the annual value of the property and under section 28 in respect of the profits he makes by rendering the services to the tenants. **[CIT - Sarabhai Pvt. Ltd.,Guj.,**Sec.28. **263ITR197]**

5. Income derived from property would always be termed as Income from house property,but if the property is used as stock-in-trade , then the said property would become or partake of the character of stock,and any income derived from the stock,would be Income from business, and not income from house property.

[CIT-Neha Builders P. Ltd.,Guj.,
Sec.28.**207CTR231,296ITR661, 164Taxman342]**

Income from house property Kar. CIT - K.N.Guruswamy.

From a plain reading of section 22 of the Income-tax Act,1961, it is clear that it is the owner of a house property who becomes liable to be charged under the head Income from house property unless the house property is used by him for the purpose of his own business or profession.The section, which provides for exemption, should be strictly construed. Under section 22 it is only the owner of the property who can claim exemption and that owner must be an assessee and the property concerned must have been used or occupied for the assessee's business. The occupation of the property, in the context, must mean occupation as owner or his own occupation. Sec.22. **[146ITR34,17Taxman87]**

2. Income from house property postulates that what has essentially been let out is house property though some conveniences like fixtures, etc., may have incidentally been let out with it for proper enjoyment of the demised premises. If, however, the letting could not have taken place but for the fixtures, furnitures, etc., in the sense that the tenant or licensee would not have taken the premises but for them and the extent of income attributable to such fixtures or furnitures cannot conveniently be predicated, it would be income chargeable to tax under the residuary head Income from other sources. **CIT - Cawnpore Club Ltd., All.** *Sec.22.*[**34CTR313, 146ITR181]**

3. When an asset is in the nature of land or building capable of being used for any other purpose and when the assessee ceases to use it as a commercial asset either by himself or even through others,the income derived by him by renting out the same would more appropriately fall under the head Income from house property. **[New India Industries Ltd.-CIT.,Guj.,***Sec.22.***203 ITR933]**

4. The question whether a particular letting is business has to be decided in the circumstances of each case.Each case has to be looked at from a businessman's point of view to find out whether the letting was the doing of a business or the exploitation of his property by its owner.It is only a letting out of a property simpliciter without anything more that will fall under the category income from house property. **[CIT-Kongarar Spinners Pvt. Ltd.,Mad.,***Sec.22.***208ITR645]**

5. Income from house property is an artificially defined income and the liability arises from the fact that the assessee is the owner of the property. ..... Since a specific head of charge is provided for income from the ownership of house property, rents or other income from the ownership of house property cannot be brought to tax under any other head. **[CIT-Sarabhai Pvt.Ltd.,Guj.,** *Sec.22.***263 ITR169]**

6. There is a specific head under the Income-tax Act,1961,Income from house property and income received as rent from property ordinarily must be computed under that head unless it is clearly established by the assessee that the said income has been received in the course of business from the property where the business had started. In a case where the business never started or has not started, the assets cannot be considered to be business assets and the transaction would only be exploitation of property by an owner but not exploitation of business assets. **[CIT- Hotel Ratanada International P. Ltd. Raj.,***Sec.22.***293ITR 557,164Taxman292]**

Income from other sources **SC** **M/s.S.G. Mercantile Corporation Ltd.- CIT.**

The residuary head of income can be restored to only if none of the specific heads is applicable to the income in question.It comes into operation only after the proceeding heads are excluded. Sec.56.**[AIR1972SC732,83ITR700]**

2. Under the scheme of the I.T. Act, 1961, any income which does not answer the description of an income under the other heads contemplated by s. 14 is chargeable to tax under the head Income from other sources. This has been made clear by s. 56(1). Sub-section (2) of s. 56 specifies some sources of which income has to be taxed under the head Income from other sources and not otherwise. **CIT - Cawnpore Club Ltd., All.** *Sec.56.* **[34CTR313,146ITR181]**

3.The question as to whether the assessee was exploiting an asset as a commercial asset or as an income yielding property is predominantly a matter of intention. Intention is an inference to be drawn from the relevant facts. Where either by word or conduct, the assessee expresses its intention to go out of business by converting the commercial asset into property, the income that accrues to such an assessee can only be income from other sources and not business income. **[Saswad Mali Sugar Factory Ltd.-CIT,Bom.** *Sec.56.***153 CTR 338,236ITR706,149Taxation523,103Taxman477]**

3. Where surplus funds are parked with the bank and interest is earned thereon, it can only be categorized as income from other sources. Even in a case where the exporter is required to mandatorily keep monies in fixed deposit, in order to avail of credit facility for the export business and interest is earned on fixed deposits for the purpose of availing of credit facilities from the bank, the interest income has to be treated as income from other sources and not business income as it does not have an immediate nexus with the export business. It goes out of the reckoning for purposes of section 80HHC. **[CIT - Mereena Creations.,Del.,** *Sec.56 .r.w.s.80HHC.***330ITR199,189Taxman71]**

Income from property **Bom.** **CIT-Zorostrian Building Society Ltd.**

Section 9 of the Indian Income-tax Act of 1922 states how computation is to be made in respect of the head Income from property Section 9(2) provides that for purpose of this section, the expression annual value shall be deemed to mean the sum for which property might reasonably be expected to let from year to year. The word used is might and not can or is. Reading the two paragraphs of section 9 together, it is clear that income from property is an artificially defined income and the liability arises from the fact that the assessee is the owner of the property. The liability to tax does not depend on the power of the owner to let the property and it also does not depend on the capacity of the owner to receive the bona fide annual value of the property. *Sec.22.*[**102ITR499**]

Income of the owner JK. CIT - Mohd. Amin Tyamboo

The provisions of section 23(2) of the I.T.Act,1961,make it clear that the benefits of the relief in respect of self-occupied property is available only to an individual assessee and no other assessable entity can claim the benefit. The reference to the occupation for the purposes of his own residence shows that the owner must be a natural person, that is, what is known in Income-tax law as an individual. Therefore, when the proviso to section 23(2) speaks of total income of the owner, it is in the sense of an individual that the word owner has been used in it and it has no other significance. It does not limit or restrict the meaning of the expression total income, which has to be understood as defined in section 2(45)....... Therefore, in the proviso to sub-section (2) of section 23, the total income of an owner would include income of his minor child assessable in his hands under section 64. *Sec.23(2).r.w.s.64.* [**14CTR190,125ITR375**]

Income or profits and gains Cal . Eastern Aviation and Industries Ltd. -CIT

The words income or profits and gains should be understood as including losses also, so that in one sense profits and gains represent positive income whereas losses represent negative income. In other words, loss is negative profit. Both positive and negative profits are of revenue character. Both must enter into the computation, wherever it becomes material, in the taxable income of the assessee. *Sec.73.*[**208ITR1023,125Taxation223, 74Taxman641**]

Income profit or gains which have escaped assessment All. Bhawani Prasad Girdhar Lal - ITO

The expression income, profits or gains which have escaped assessment in section 34(1A) of the Income-tax Act covers not only cases where income, profits or gains have totally escaped assessment but also cases where part of the income, profits or gains have been assessed and part only has escaped assessment. *Sec.147.* **[AIR1960All377,40ITR407]**

Income-tax SC CIT - K. Srinivasan

These words in section 2(2)(a)(b) would include surcharge and additional surcharge. *Sec.4,95.***[AIR1972SC491,83ITR346]**

2. Section 2(43) of the Income-tax Act,1961, (as it stood at the relevant time), defined tax as Income-tax and super-tax chargeable under the provisions of that Act.Therefore, what is levied under the charging provisions of that Act, i.e., section 4 thereof alone can be called income-tax. Interest, penalties and fines,which are payable under the other provisions of that Act,cannot be termed as income-tax.They are imposed in addition to Income-tax for the purpose of enforcing the levy of income-tax. **[Soma sumdaram (Pvt.)Ltd.–CIT,Kar.,***Sec.2(43)* **116ITR620]**

3. Income-tax is a charge on an assessee in respect of his total income computed in accordance with the provisions of the Act. **[Indian Rayon Corporation Ltd.-CIT,Bom.***Sec.2(43).***182CTR 247,261ITR98,128Taxman739]**

4. Income-tax is a levy on income. The Act takes into account two points of time at which the liability to tax is attracted, namely, the accrual of the income or its receipt ; but the substance of the matter is the income. If income does not result at all, there cannot be a tax, even though in book-keeping, an entry is made about a hypothetical income...... **[CIT-Orissa State Financial Corporation,Ori,** *Sec.5.***180CTR171,262ITR350,126Taxman157]**

5. ... in the context in which clause (a) of the Explanation to section 115JA occurred,the expression Income-tax was referable to profits reflected in the profit and loss account. The expression Income-tax was used in section 115JA in a general sense and not with reference to the income chargeable under the provisions of the Income-tax Act,1961. No distinction could be drawn between Indian and foreign Income-tax by reference to section 2(43) which defined tax. There was no justification to restrict the scope of the expression Income-tax to Indian Income-tax only. **[Bank of India,AAR,***Sec.2(43).*

295ITR529]

Income-tax authority SC CIT - B.N.Bhatacharjee

The term Income tax authority includes the Income Tax Officer, Assistant Appellate Commissioner and others and not Income Tax Appellate Tribunal. Sec.116.**[AIR1979SC1725,10CTR354,118ITR 461,53Taxation76,1Taxman348]**

2. The power of survey is given to an Income-tax authority under section 133A.Under section 116 of the Act Income-tax authority means a Commissioner, a Joint Commissioner, a Director, a Joint Director, an Assistant Director or Deputy Director or an Assessing Officer and for the purposes of clause (i) of section (1), clause (i) of sub-section (3) and sub-section (5), includes an Inspector ofIncome-tax. **[Reckitt Colman of India Ltd.-ACIT(TDS).Cal.** Sec.116.**172 CTR499,252ITR550,124Taxman496]**

Income-tax free dividend Mad. Barjor Hoshangji Vakil - Mettur Chemical and Ind. Corp.Ltd.

A stipulation contained in the articles or memorandum of association of a company that preference shareholders will be given income-tax-free dividend at a certain rate does not amount to a guarantee on the part of the company that such dividend would be free from Income-tax in the hands of the recipient himself. It only means that the stipulated percentage of dividend will be paid by the company to the shareholders in full without any deduction therefor on account of tax payable by the company. Misc.**[50ITR128]**

Income-tax Officer Cal. Keshab Narayan Banerjee - CIT

The expression Income-tax Officer under section 263 shall include any officer empowered to assess any nature of the proceeding under the Act and it has been defined in clause(25) of section 2 which means a person appointed to be an Income-tax Officer under section 117.The authorities mentioned in sub-section (1) of section 117 include an Income-tax Officer.A reference made to a particular officer is to be treated as ejusdem generis which includes such officers who are empowered to make assessment in respect of the income. Therefore, the application under section 263 cannot be excluded in respect of a person mentioned in section 132 of the Act. *Sec.263.r.w.s.2(25).***[173CTR61,252ITR888,125Taxman299]**

Incur **Ori. Belpahar Refractories Ltd. - CIT**

The word incur means to become liable to, e.g., to incur debt, loss, etc.When an expenditure is said to have been incurred, it may connote actual payment or it could be that the person concerned has merely become liable for payment but has not actually made payment.When a person has made advance payment but has not become liable for the payment,he cannot be said to have incurred any expenditure. <u>*Sec.35(2)(ia).*</u>**[207ITR144]**

India **Bom.** **CIT - Indo Oceanic Shipping Co. Ltd.**

........the circular issued by the Board bearing No. 586, dated November 28,1990, though in the context of deductions in the hands of employees,specifically lays down that, India as defined under section 2(25A) of the Act,does not extend to Indian ships operating beyond the Indian territorial waters. Indian ships operating beyond Indian territorial waters do not come within the term India as defined in section 2(25A) of the Act.Similarly, the word India has also been defined under the General Clauses Act, 1897, under section 3(28).It does not include Indian ships outside territorial waters of the country.Merely because the contract is entered into in India it will not be the conclusive test to decide as to whether an employee was employed in India or outside India. The terms of the contract, the nature of the work, the nature of business and all other relevant facts are required to be considered to decide as to whether the employment was in India or outside India. The provisions of the Merchant Shipping Act are not in pari materia with the provisions of the Income-tax Act.The said Merchant Shipping Act is not even a fiscal legislation. The provisions of the Merchant Shipping Act are essentially invoked in cases of accidents.The provisions are not applicable to the Income-tax Act. The international law refers to the law of the flag in order to decide the nationality of the vessel...... <u>*Sec.2,40A.*</u>**[165CTR404,247ITR 247,114Taxman722]**

Indian concern **Bom.** **CIT - Dorr- Oliver (India) Ltd.**

The word Indian concern means a concern which is Indian in character.A concern, which is not Indian,cannot be termed as Indian concern merely by reason of its location in India. The branches of a foreign concern situated in India cannot be held to be an Indian concern.The expression Indian company has been defined in section 2(26) of the Income-tax Act, 1961, to mean a company formed and registered under the Companies Act, 1956. The above definition clearly goes to show that the Legislature by the use of the expression Indian company really intended to mean companies which are registered in India and not companies which are carrying on business in India.If one takes the analogy of the above definition of Indian company into consideration,the expression Indian concern will also mean a concern Indian in character. *Sec.2(26).*[**116CTR 428,209ITR691,75Taxman8]**

Individual Bom. CIT-Ahmedabad Millowners' Association.

Individual in section 3 means a human being and does not include a company, and the expression other association of individuals does not therefore include an association of companies and the assessee could not be charged to Income-tax as an association of individuals under section 3. *Sec.2(31)4.*[**7ITR 369]**

2. The word individual does not mean only human being and it includes juristic person like a Hindu Idol. **[Sri Sridhar Jiew-ITO,Cal.***Sec.2(31).***AIR 1966 Cal 494,50ITR480]**

3. The word individual in section 3 of the Indian Income-tax Act, 1922, includes within its connotation all artificial juridical persons and this position is made explicit and beyond challenge in the Income-tax Act, 1961. A Hindu deity falls within the meaning of the word individualin section 3 and can be treated as a unit of assessment under that section. An assessment to tax can be made on the deity through the shebaits. The Hindu idol is a juristic entity capable of holding property and of being taxed through its shebaits who are entrusted with the possession and management of its property. **[Jogendranath Naskar-CIT,** *Sec.2(31).***AIR1969SC 1089,74ITR 33]**

4. The view that individual contemplated in section 64(i) of the Income-tax Act,1961,is an individual who has a spouse and, therefore, in section 64(ii) also the individual should be such who has a spouse,is not correct. **[Priti Lata Samanta-CIT,All.,***Sec.64(i).***79ITR18]**

5. The term individual in section 3 of the Wealth-tax Act is not restricted to human beings and includes a corporation constituted under a Central, Provincial or State Act. **[Kerala Financial Corporation-WTO.,Ker.,***Sec.3.WT.Act.***82 ITR477]**

6. The expression individual in section 3 of the Wealth-tax Act, 1957, includes within its ambit Mappilla Marumakkathayam awards and they are well within the purview of the taxing provisions of the enactment.Even after their inclusion in the term individual, section 3 is not violative of article 14 of the Constitution of India. **[WTO- C K. Mammed Kayi.,SC,**<u>*Sec.3.WT. Act.*</u>**21CTR345,129ITR 307,10Taxman489]**

7. Section 3 of the W.T. Act, 1957, creates a charge of wealth-tax in respect of the net wealth of every individual, HUF and company.The word individual includes a body of individuals and hence an association of persons is a taxable entity. Moreover, cl. (b) of sub-s. (1) of s. 4 provides that in computing the net wealth of an individual, there shall be included, as belonging to that individual, whether the assessee is a partner in a firm or a member of an association of persons not being a co-operative housing society, the value of his interest in the firm or association determined in the prescribed manner. This provision clearly lays down that in case an individual is a member of an association of persons, then the value of his interest in the association determined in the prescribed manner shall be included in computing his/her net wealth. Thus, there is a clear provision that the value of the interest of an individual in an association of persons is to be included in his net wealth. **Smt.Prem Lata Agarwal - CWT, All.** <u>*Sec.2(m).WT.*</u> **[29CTR258,142 ITR 586]**

8. An individual referred to in section 64(1)(i) can only be an assessee who is being assessed in his individual capacity and not one who is being assessed in a representative capacity, such as the karta of a Hindu undivided family. **[CIT-Prakashchandra Basantilal,MP.,**<u>*Sec.64(i).*</u>**162ITR536]**

9. It is now well-settled that the word individual does not necessarily and invariably always refer to a single natural person. A group of individuals may as well come in for treatment as an individual under the tax laws if the context so requires. **[CIT-Shri Krishna Bhandar Trust,Cal.,**<u>*Sec.2(31) r.w.s. 164.*</u>**201ITR 989]**

10. The charge of wealth-tax is on the net wealth of three classes of persons, viz., individuals, Hindu undivided family and companies. The three categories of assessee mentioned in section 3 of the Wealth-tax Act, 1957, are clearly mutually exclusive.While the categories of persons subjected to wealth-tax remained the same in section 3 of the Act,the definition of the word company, one of the three categories of persons liable for wealth-tax has been greatly expanded.The context in which the word individual occurs provides the clue for ascertaining the meaning to be assigned to that term in a statute.No single definition can be adopted or applied in all contexts and to all statutes. The fact that the word individual in some circumstances may be wide enough to include an artificial juristic entity like a corporation created by a statute does not lead to the conclusion that the word individual in section 3(1) of the Act includes all juristic persons.If the intention of Parliament was always to regard all juristic entities whose existence in law is traceable to their incorporation under a statute as covered by the word individual, the amendments effected to the definition of the word company would become inexplicable and a futile exercise in redundancy and no such intention can be attributed to Parliament when it brought about those changes in the definition of the word company.Incorporated bodies can be taxed under the Wealth-tax Act, only if they fall within the definition of the word company and not otherwise.The term individual as it is used in the section, is distinct from the artificial incorporated juristic entities and it is only on account of that fact that Parliament considered it necessary to amend the definition of the word company from time to time and expand its coverage with a view to bring incorporated legal entities into the fold of wealth-tax for the purposes of Taxation. **[State Bank of Travancore Employees Union-CWT,Mad.,** *Sec.3.WT.Act***149CTR418,238ITR466]**

Individuals SC CIT - Smt. Sodra Devi

The words any individual and such individual occurring in section 16(3) are restricted in their connotation to mean only the male of the species,and do not include the female of the species (even though by a disjunctive reading of the expression the wifeor a minor childof such individualin S. 16(3) (a) and the expression,by such individual for the benefit of his wife or a minor child or bother in section 16(3)(b) it may be possible in the particular instances of the mothers being connected with the minor children to include the mothers also within the connotation of these words. *Sec.64.***[AIR SC1957832,32ITR615]**

2. The word individuals in entry 86 cannot be stretched so as to include a Hindu Undivided Family. **Jugal Kishore -WTO, All.** *Constitution of India.***[AIR1961All.487,44ITR94]**

3. The word individuals in entry 86 of List I includes undivided families also, as an undivided family is only a collection of individuals; and Parliament has, therefore, power to impose wealth-tax on undivided families under entry 86. **[Sarjerao Appasaheb Shitole-WTO,Mys.,***Constitution of India.***52ITR372]**

4. The word individuals in entry 86 of List I takes within its scope also a Hindu undivided family.Parliament is therefore competent to impose wealth-tax on a Hindu undivided family. **[Vysyaraju Badri Narayanamurthy- CWT,Ori.,**_Sec.3.WT.Act and Constitution of India._**[56ITR298]**

Industrial company Mad. Addl. CIT - Chillies Export House Ltd.

. .....a reading of the definition of industrial company in section 2(6)(c) of the Finance (No. 2) Act,1971,suggests that it is the company which should engage itself in what should be the business of the company and that company must do the processing of goods.
Sec.2.Finance Act1971.**[7CTR230,115ITR73]**

2. A company running a cold storage is not an industrial company. **[DelhiColdStorate(P)Ltd-CIT,SC,**_Sec.2.Finance Act,1973._ **AIR 1991SC2125,191 ITR 656]**

3. A company cannot be held to be an industrial company merely by virtue of its ownership of plant or machinery or factory premises. For that purpose,it must be mainly engaged in the manufacture or processing of goods.Neither the ownership or possession of the manufacturing plant or machinery nor the ownership of the raw materials or the manufactured goods is the determinative factor for that purpose.A company engaged in the manufacture of goods would continue to be an industrial company even if it manufactures or processes goods for a third party for remuneration or consideration,if the income there from is not less than 51 per cent. of its total income. But it would cease to be an industrial company if it suspends or stops the manufacturing activity and hands over the manufacturing apparatus to some third person for use in the manufacture or processing of goods. **[VitaPvt.Ltd.–IT,Bom.,** _Sec.2,(8)FinanceAct1975,_**123CTR256,211 ITR557,131 Taxation 560, Taxman158]**

4. In considering the meaning of the word industrial machinery used in item No. 8 of the Ninth Schedule to the Income-tax Act, 1961, the description of industrial machinery spoken to in item No. 4 of the Fifth Schedule cannot be imported at all. The very fact that the Legislature has chosen to use the expression industrial machinery in one way in one Schedule and in another way in another Schedule, shows that the Legislature itself intended to give different meanings to the same term in the said two Schedules. If really,the Legislature wanted to have the same meaning in both the

Schedules,it would have accordingly worded the said expression in the same way in both the Schedules.Therefore, the meaning of the term industrial machinery used in item No. 8 of the Ninth Schedule could not be the same as the meaning of the said term under item No. 4 of the Fifth Schedule. **[CIT-Indian Textile Paper Tube Co. Ltd.(No.1),Mad.,**<u>*FinanceAct,197.Sec.8(c).*</u>**149CTR117,235ITR70]**

Industrial or commercial profits Kar. AEG Telefunken - CIT

Under article III of the Double Taxation Avoidance Agreement between Germany and India tax shall not be levied on the industrial or commercial profits of an enterprise in that other territory. Sub-clause (3) of article III defines the expression industrial or commercial profits. According to the said definition, it shall not include income in the form of rents, royalties, interests, dividends, management charges, remuneration for labour or personal services or income from the operation of ships or aircraft but shall include rents or royalties in respect of cinematographic films.The remuneration or the amount towards rendering technical services is not specifically excluded from the definition of the term industrial or commercial profits. Once the same is not excluded, it cannot be inferred that there is exclusion. Management charges are quite different from the amounts paid towards technical services. Management charges are paid for managerial supervision or managerial functions done by the concerned person or the company. Management charges would not include charges for rendering technical services unless a particular agreement is shown that management charges include the technical services as managerial duty.In this case there was no such agreement. Service rendered through other technical personnel will not amount to personal service. <u>*DTAA.India-Ger.*</u>**[151CTR222,233ITR129,101Taxman 109]**

Industrial undertaking Ker. P.Alikunju,M.A.Nazeer Cashew Industries – CIT.

Section 54D of the Income-tax Act,1961,which grants an exemption must be construed liberally and the exprssion industrial undertaking occurring in section 54D must be given its popular meaning. An undertaking mentioned in section 54D must be one maintained by a person for take purpose of carrying on his business.Undertaking for the purpose of this section,however,must be an industrial undertaking. The demonstrative adjective industrial qualifying the word undertaking unmistakably and with precision shows that the undertaking must be one which partakes of the character of a business.The word business connotes some real,substantial and systematic or organised course of activity with a set purpose.The word business is a word of wide import and in fiscal statutes must be construed in a broad rather than a restricted sense. The words industrial undertaking, therefore, should be understood to have been used in section 54D in a wide sense, taking in its fold any project or business a person may undertake.

*Sec.54D.***[166ITR804]**

2. The expression industrial undertaking has not been defined in the Income-tax Act,Industry is a term of wide import.Where there is (i)systematic activity;(ii) organised by co-operation between employer and employee; (iii) for the production and/or distribution of goods and services calculated to satisfy human wants and wishes, prima facie there is an industry.Undertaking is in actual effect an activity of man which, in commercial or business parlance, means an activity engaged in with a view to earn profit. **Shankar Construction Co.-CIT.Kar.***Sec.32A.***[189ITR463,56Taxman98]**

3. The expression industrial undertaking for the purpose of the several clauses of section 5(1)of the Wealth-tax Act,1957, is defined by an Explanation as meaning an undertaking engaged in the business of generation or distribution of electricity or any other form of power or in the construction of ships or in the manufacture or processing of goods or in mining.The Explanation contains a definition which is exhaustive.Therefore, unless a particular activity pertains to any of the three broad divisions, it cannot be said to be the activity of an industrial undertaking in terms of the said definition.The Explanation covers a construction company, but such coverage is only for construction undertaking engaged in construction of ships.No other type of construction has a place in it. By necessary implication, all constructions as well as repair of buildings is not comprehended by the expression construction of ships. **[CWT-UrmilaRungta,Cal.,***Sec.5(1)WT.Act.***208ITR552]**

4. The Explanation to section 5(1)(xxxi) of the Wealth-tax Act, 1957, defines an industrial undertaking to mean an undertaking engaged in the business of generation or distribution of electricity or any other form of power or in the construction of ships or in the manufacture or processing of goods or in mining. The manufacturing or processing of goods refers to movable property and there should be a manufacturing or processing of goods as the end product by the industrial undertaking.If an industrial undertaking is engaged in the manufacture or processing of any other item which does not fall in the category of goods, then the exemption cannot be claimed. What constitutes goods has not been defined in the Wealth-tax Act,1957.But if the definition in the Sale of Goods Act,1930, of goods is taken into consideration, which refers only to movable property, a dam cannot be considered as goods. Therefore, a firm engaged in the construction of dams does not constitute an industrial undertaking. **[CWT-Asha Mittal,Raj.,***Sec.5(1)WT.Act.* **209ITR368]**

5. The meaning of the expression industrial undertaking used in section 5(1) (xxxii) of the Wealth-tax Act, 1957, has to be understood as defined in the Explanation to section 5(1)(xxxi) of the Act. According to this definition, the term industrial undertaking for the purpose of the business activity of the assessee means an undertaking engaged in the business of manufacture or processing of goods. The expression engaged in manufacturing postulates the assessee's direct involvement in the manufacture.It may not be necessary that the assessee himself should be personally engaged but it is enough that he employs his own labourers.The expression processing of goods, does not predicate that the assessee should be engaged in all the processes resulting in the end- product. If the assessee has done any processing, i.e., if he is directly involved at any stage of the processing, resulting in the end- manufacture, he is entitled to avail of the exemption. **[CWT-V.O.Ramalingam, Mad.,***Sec.5(1) WT.Act.***[216ITR566]**

6. The Income-tax Act,1961, does not define the expression industrial undertaking. Section 80HH is intended to encourage the setting up of new industrial enterprises and hence will have to be construed liberally in a broad commercial sense keeping its object in view.The basic concepts of industrial undertaking, manufacture, produce occur even in section 32A.If a unit having new machinery amounted to industrial undertaking and the process amounted to manufacture for the purposes of section 32A, it is difficult to see how any different legislative intent could be attributed to identical expressions used in section 80HH. Sub-section (2) of section 80HH refers to industrial undertaking and not to the assessee or his other business. If a new industrial unit is established as a part of an already existing industrial establishment and if the newly established unit is itself an integrated independent unit in which new plant and machinery is put up and that by itself is capable of production of goods independently of the old unit, the said unit could be classified as a newly established industrial undertaking and will qualify for the relief. **[CIT - Chanda Diesels,Bom.,** *Sec.80HH.r.w.s.32A.***216ITR639]**

7. Though there is no definition of the term industrial undertaking in the Income-tax Act, 1961, it is defined under the Wealth-tax Act, 1957. The Explanation to section 5(1)(xxxi) of the Wealth-tax Act defines the term as an undertaking engaged in the business of generation or distribution of electricity or any other form of power or in the construction of ships or in the manufacture or processing of goods or in mining. There is nothing in the language of section 80HH or 80J which would suggest that a different meaning is contemplated for the term industrial undertaking when it relates to a unit engaged in manufacture or production. On the other hand, the conditions to be fulfilled by the industrial undertaking as provided under sub-section (2) of section 80HH would indicate that a unit which is manufacturing or producing articles can be treated as an industrial undertaking without any further qualification. **[CIT-IndianResinsandPolymers.,Ker.,**_Sec.80HH.r.w.s.5.WT.Act._ **[148CTR143,235ITR5]**

8. Under section 5(1)(xxxii) of the Wealth-tax Act,1957, the value of the interest of the assessee in the assets forming part of an industrial undertaking belonging to a firm or an association of persons of which the assessee is a partner or, as the case may be, a member, shall not be included in the net wealth of the assessee. Industrial undertaking had been defined, by way of a definition in the Explanation appended to section 5(1) (xxxi) thereof. The term industrial undertaking, according to the said Explanation, means an undertaking engaged in the business of generation or distribution of electricity or any other form of power or in the construction of ships or in the manufacture or processing of goods or in mining. The first part of the said definition is relatable to generation or distribution of electricity or any other form of power or in the construction of ships. The second part of the definition is relatable to the manufacture or processing of goods or in mining. The expression manufacture or processing of goods, however, is not defined in the Wealth-tax Act. The term manufacture has to be considered with reference to the context of the enactment. Manufacture is a process which results in an alteration or change in the goods which are subjected to such manufacture.A commercially new different article is produced. **[CWT-P.Devasahayam.,Mad.** _Sec.5.WT.Act._**144CTR313,236 ITR 885]**

9...the expression industrial undertaking occurring therein must be given its popular meaning.An undertaking mentioned in section 54D must be one maintained by a person for the purpose of carrying on his business.The demonstrative objective industrial qualifying the word undertaking unmistakably and with precision shows that the undertaking must be one which partakes of the character of a business. The word business connotes some real, substantial and systematic or organised course of activity with a set purpose. Since the word business is of wide import,the words industrial undertaking should be understood to have been used in section 54D in a wide sense, taking in its fold any project or business a person may undertake.

[CIT-Hemsons Industries.,AP., *Sec.54D.* 171CTR527,251ITR693,118Taxman903]

10. The Income-tax Act does not define the expression industrial undertaking. Therefore, reference to its definition in similar enactments or adoption of its ordinary meaning is inevitable. Considering the object of the enactment of sections 80HHA and 80-I, the said expression will have to be construed liberally in a broader commercial sense keeping its object in mind.The concept of Industrial undertaking need not necessarily be confined to manufacture and production of articles and even in the absence of either of them there could be an industrial undertaking.

[Ship Scrap Traders-CIT,Bom.,*Sec.80HHA,8O-I.*168CTR489, 251ITR806,122Taxman29]

11. The expression industrial undertaking in the context of the Income-tax Act and not in the context of the Industrial Disputes Act and if so read, it is clear that the activity should be of production of any article or thing and any activity which primarily concerns production of any article or thing would fall in the category of industrial undertaking. **[Insight Diagnostic & Oncological Research Ins. P.Ltd.-DCIT.Bom.*Sec.32A.*262ITR41,129Taxman 510]**

12. The term industrial undertakinghas not been defined in the Act. Hence, industrial undertakingis to be given the meaning which is understood in common parlance, and which should be interpreted widely.At the same time, the expression is to be construed in the context of section 35D.The activity of construction cannot be treated as manufacturing activity as it does not amount to manufacture or production of an article or a thing. **[Ansal Housing & Construction Ltd. - CIT,Del.,*Sec.35D.*228CTR,262,320ITR420, 185 Taxman 74]**

Information Pun. CIT - Shree Jagan Nath Maheswary

The term information means the act or process of informing, communication or reception, of knowledge. It may be knowledge acquired directly as by observation or study,or derived inferentially, or from communication from others.

<u>*Sec.147.*</u> **[AIR1957Punj226,32ITR418]**

2. The word information in section 34(1)(b) includes information as to the true and correct state of the law and so would cover information as to relevant judicial decisions. **[Maharaj Kumar KamalSingh-CIT,SC,<u>*Sec.147.*</u>[AIR 1959 SC257,35ITR1]**

3. The word information in section 34(1)(b) includes information as to the true and correct state of the law and so would cover information as to relevant judicial decisions. **[Mooljee Sicka & Co.-SecondAddl.ITO,Cal.,<u>*Sec.147.*</u>[AIR 1960Cal.492,40ITR163]**

4. The expression information mean instruction or knowledge derived from an external source concerning facts or particulars, or as to law relating to a matter bearing on the assessment. **[CIT-A.RamanandCo.,SC,<u>*Sec.147.*</u>AIR1961 SC 147,67ITR11]**

5. A judgment of the Income-tax Appellate Tribunal or even of the Appellate Assistant Commissioner in an appeal from an assessment order,taking a different view on the facts of the case, constituted information within the meaning of section 34. On principle there can be no distinction in this respect between information derived from a judgment of the Privy Council rendered on appeal either from an order of the Tribunal or from an answer given in a reference by the High Court and a judgment recorded by the Tribunal or an Appellate Assistant Commissioner against an assessment order in the same assessment proceedings. **[Jawahar Lal Mani Ram-CIT.,All.,<u>*Sec.147.*</u> 48 ITR 837]**

6. If income had escaped assessment owning to the failure of the Income-tax Officer to understand the true implication of a notification, and the Income-tax Officer later on finds that on a correct interpretation of the notification the income was liable to be assessed, he can take proceedings under section 34 for assessment of such income; the word information in section 34 is wide enough to apply to such a case. **[Canara Industrial and Banking Syndicate Ltd.-CIT,Mys.,<u>*Sec.147*</u> 51ITR479]**

7. The in fact he came to know of it subsequently.Information means something that the mind has acquired.In order that an officer should be incapacited from acting under section 34 or section 15 he must have actual knowledge of certain facts,which knowledge he is again using for the purpose of acting under those sections.If actual knowledge was absent-it is immaterial how that actual knowledge was absent-then when the actual knowledge does come about it confers jurisdiction upon the officer to act under those two sections. **[Rajputana Textiles (Agencies) Pvt. Ltd.-ITO EPT., Bom.,*Sec.147.*52ITR1**

8. It at any point of time the Income Tax Officer is equipped with the information that a particular business is not a partnership and that the assessee was the sole proprietor of that, but the entire income from that business was not included in the original assessment,that will be an information within the meaning of Cl.(b) of sub. Section(1) of Section34. **[CIT - Narainji Manji.,Pat., *Sec.147.*AIR1967Pat.409,66ITR322]**

9. Information,in the context in which it occurs in section 147(b) of the Income-tax Act,1961,must mean instruction or knowledge derived from an external source concerning facts or particulars, or as to law relating to a matter bearing on the assessment. Mere change of opinion on the part of the Income-tax Officer cannot constitute information so as to entitle him to initiate proceedings under section 147(b). If the information is as to any fact, it may be received from any person who knows the fact, and it cannot be limited to any particular person, body or authority since such fact may be within the knowledge or possession of anyone and it may be received by the Income-tax Officer from any source. But, in the case of information as to the correct state of the law, the external source from which it may be received must necessarily be of a limited character.Though it may not be possible to define precisely the cases where intimation received by the officer as to the correct state of the law may be regarded as information, opinion as to the state of the law by any and every person cannot constitute information for the purpose of the section. It must be a statement or the expression of the correct state of the law by a person or body or authority competent and authorised the pronounce upon the law so that it is invested with some definiteness and authority. **[Kasturbhai Lalbhai-ITO,Guj.,*Sec.147.*80ITR188]**

10. That the scrutiny note of the Revenue Audit and the letter of the Inspecting Assistant Commissioner constituted information within the meaning of section 147(b) from an external source and the assessments were,therefore,valid. **[CIT - Chand Kanwarji,Del, *Sec.147.*84ITR584]**

11. In order to be information in terms of clause(b) of section 147, the following conditions require to be fulfilled-1.It must be knowledge or instruction concerning facts or particulars or as to law relating to a matter bearing on the assessment; 2.Such knowledge or instruction must come into the possession of the Income-tax Officer after the previous assessment;3.The knowledge or information must be such which leads to the formation of the belief that the income of the assessee had escaped assessment or had been under-assessed;4.The proximate or immediate source of such information and knowledge must be external.5.The fact that such knowledge or information could have been derived during the previous assessment from an investigation of the materials on record but was not in fact derived would not prevent such knowledge or instruction from being information in terms of section 147(b). **[ITO-PanamaPrivateLtd.,Cal.,**<u>*Sec.147.*</u>**97ITR210]**

12. The word information which has been used in section 132(1) of the Income-tax Act, 1961, has been defined in the dictionary as that of which one is apprised or told. The word reason has been defined as a statement of fact employed as an argument to justify or condemn some act. The word conclusion is defined as the judgment arrived at by reasoning; an inference; deduction, etc. In other words,when the information received or the basic facts are harnessed in support of an argument, the resultant effect assumes the shape of a reason and when a number of reasons are considered in relation to each other, the final result is a conclusion. A necessary concomitant of this approach is that the facts constituting information must be relevant to the enquiry. They must be such that a reasonable and prudent man can come to the requisite belief or conclusion there from. If either of the aforementioned elements is missing, the action of the authority shall be regarded as lying outside the ambit and scope of the law and such an action would be liable to be struck down on the basis of what is commonly known as legal malice. **[H.L.Sibal-CIT. P & H.** <u>*Sec.132.*</u>**101ITR 112]**

13. The word information in section 34(1)(b) is of the widest amplitude and comprehends a variety of factors.Nevertheless, the power under section 34(1) (b), however wide it may be, is not plenary because the discretion of the Income-tax Officer is controlled by the words reason to believe. Information may come from external sources or even from the materials already on record or may be derived from the discovery of new and important matter or fresh facts. **[Kalyanji Mavgy & Co.–CIT.SC.**<u>*Sec.147.*</u>**102ITR287]**

14. The word information in section 17(1)(b) of the Wealth-tax Act,1957,is of the widest amplitude and comprehends a variety of factors. Information may come from external sources or even from the materials already on the record. It may consist of oversight or inadvertent mistake committed by the Wealth-tax Officer or he may discover an error apparent on the face of the record from further enquiry or research into facts or law. **CWT -Arundhati Balkrishna Trust. Guj.** <u>*Sec.17(1).WT.Act.*</u>**[108ITR78]**

15. The dividing line between information and change of opinion is rather thin and while their distinction is easy of exposition its practical application is difficult.The expression information which was introduced into the Indian Income-tax Act, 1922, by an amendment of, has not been defined either in the 1922 Act or the 1961 Act.To inform means to impart knowledge and a detail available to the Income-tax Officer in the paper filed before him does not by its mere availability become an item of information. It is transmitted into an item of information in his possession only if and only when its existence is realised and its implications are recognised. **[CIT-Kerala State Industrial Dev.Corp. Ltd.Ker.*Sec.147.*116ITR58]**

16. The opinion of an internal audit party of the Income Tax Department on a point of law cannot be regarded as information.

[Indian and Eastern Newspapers Society,New Delhi-CIT, SC, *Sec.147.*AIR1979SC,119ITR 996,2Taxman 97]

17. The term information signifies instruction or knowledge derived from an external source concerning facts or particulars or as to law, relating to a matter bearing on the assessment.In so far as the term information means instruction or knowledge concerning facts or particulars there is little difficulty, the reason being that, by its inherent nature, a fact has concrete existence.It influences the determination of an issue by the mere circumstance of its relevance. It requires no further authority to make it significant. **[Sterling MachineTools-CIT,All.,*Sec.147.*122ITR926]**

18. In order to be informationin terms of section 8(b) of the Companies (Profits) Surtax Act, 1964, the following conditions must be fulfilled: (a) it must be knowledge or instruction concerning facts or particulars or as to law relating to a matter bearing on the assessment, (b)such knowledge or instruction must come into the possession of the Income-tax Officer after the previous assessment, (c)the knowledge or information must be such which leads to the formation of the belief that the income of the assessee had escaped assessment or had been under-assessed,(d) the proximate or immediate source of such information and knowledge must be external, but the fact that such knowledge or information could have been derived during the previous assessment from an investigation of the materials on record or facts disclosed thereby or from other enquiry but was in fact not derived would not prevent such knowledge or instruction from being informationin terms of the section. Where the Income-tax Officer on his own initiative and on material which was before him at the time of the original assessment changes his opinion and comes to a different conclusion, he would not be acting on informationin terms of the section. **[Stewarts&LloydsofIndiaLtd.-CIT,Cal.,** <u>*Companies(Profits) Surtax Act.*</u>**125 ITR270]**

19. Information means instructive knowledge concerning a matter bearing on the assessment received after the completion of the original assessment. Information may be as to the correct state of facts or of law relating to the taxable income.When section 147(b) of the Income-tax Act,1961,is read as referring to information as to law, what is contemplated is information as to law created by a formal source. **[K.Mansukhram and Sons-CIT,Guj.,** <u>*Sec.147.*</u> **19CTR117,133ITR65]**

20. Information within the meaning of section 147(b) may consist of fact or of law.The information must be fresh or subsequent to the original assessment. The factual information may be derived from external sources or may be gleaned from the material which was already on record at the time of the original assessment.A mere change of opinion on the material already considered by a reappraisal of the same material is not information.The change must be supported by fresh information obtained from the record. Opinion expressed by the department or by the Central Board of Direct Taxes is not law. Law is that which is laid down either by the Legislature or judicial decisions and it is the change in such law which constitutes a fresh information. **[CIT-H.D.Dennis. Bom.**<u>*Sec.147.*</u>**26CTR107,135ITR1,7Taxman231]**

21. The word information as occurring in section 147(b) means not only facts or factual material but includes information as to the true and correct state of the law.The word has been defined to connote instruction or knowledge derived from an external source concerning facts or particulars or as to law relating to a matter bearing on the assessment.But the information as to law, to fall within section 147(b), must be from a formal source, i.e., a competent Legislature or a competent judicial or quasi-judicial authority. Decisions of superior authorities under the Income-tax Act, constitute information relating to law for reopening an assessment under section 147(b).

[ArvindKumar-ITO,MP.,<u>*Sec.147.*</u> **146 ITR437,13Taxman291]**

22. The expression information in the context in which it is used in section 147(b)of the Income-tax Act,1961,means instruction or know -ledgeWhen section147(b) is read as referring to information as to law, what is contemplated is information as to the law created by a formal source.Any statement by a person or body not competent to create or define the law cannot be regarded as law.

[CIT-Union Carbide Corporation,Cal., <u>*Sec.147.*</u>**206ITR402]**

23. The internal audit cannot pronounce the law nor can such pronouncement of law be an information within the meaning of section 147(b) of the Income-tax Act, 1961.The opinion of the Central Board of Direct Taxes interpreting the law does not constitute information within the meaning of section 147(b). **[ITO- Jiyajeerao Cotton Mills Ltd., Cal.,** *Sec.147.***164CTR619, 247ITR 122,116 Taxman16]**

24. Information means the communication or reception of knowledge or intelligence.It includes knowledge obtained from investigation, study or instruction.To inform means to impart knowledge. A detail available in the papers filed before the Income-tax Offer does not by its mere presence or availability become an item of information.It is transmuted into an item of information only if and when its existence is realised and its implications are recognised.Whether a particular fact or material constitutes information in a particular case has to be decided with reference to the facts of that case and there cannot be a definite rule of universal application as to when a particular material will be taken to be an information. Information must be something more than a rumour or gossip or hunch. There must be some material which can be regarded as information, on the basis of which the Assessing Officer can have reason to believe that action under section 147 is called for. Jurisdiction of the court to interfere is very limited, as the court does not act as an appellate authority.No meticulous examination of the information by the court is permissible to decide for itself as to whether action under section 147 is called for. **[Bawa Abhai Singh- DCIT,Del.,***Sec.147.***168CTR521,253ITR83,117Taxman12]**

25. Once a possible view has been taken while allowing the liability on the basis of the provision made and another view is also possible, that does not constitute information, within the meaning of section 147(b) of the Income-tax Act,1961. **[CIT-Sambhar Salt Ltd.,Raj.,** *Sec.147.***183CTR50,262ITR675,131Taxman 241]**

Information technology AAR David Kenneth White - CIT

Information technology is defined as the science or practice of collecting, storing, using and sending out information by means of computer systems and telecommunications.The definitions of the expression do not exclude the field of cellular telephones from the purview of the expression information technology as they also refer to the process of dissemination of information through the medium of computers.The cellular network is a medium for dissemination through a system of computers and is essentially a computer network.The expression information technology,as elaborated in the notification of July 27,also covers the area of cellular networks. <u>Sec.10(5B).</u>**[146CTR384,231ITR464,144Taxation367,98Taxman 143]**

Infrastructure facility **Del.** **Container Corporation of India Ltd. - ACIT**

The term infrastructure facility was defined in section 80-IA(12)(ca) of the Income-tax Act, 1961, to mean a road, highway, bridge, airport, port or rail system or any other public facility of a similar nature as may be notified by the Central Board of Direct Taxes. The Finance (No. 2), 1998, included the words inland water ways and inland ports in the definition of infrastructure facility in sub-section (12), clause (ca), with effect from April 1, 1999. When the entire section was recast and even after several amendments were thereafter made to the section, inland ports continued to enjoy the deduction as infrastructure facility. There is no definition of the words inland ports in any of the dictionaries. But the words inland container depot were introduced in section 2(12) of the Customs Act, 1962, which defines customs port. This was by way of an amendment made by the Finance Act, 1983, with effect from May 13, 1983. Simultaneously, clause (aa) was inserted in section 7(1) of the Act under which the Central Board of Excise and Customs was empowered to issue notification appointing the places which alone shall be considered as inland container depots for the unloading of imported goods and loading of exported goods. The Central Board of Excise and Customs issued a clarification that inland container depots were inland ports. <u>Sec.80IA.</u> **[252CTR67,346ITR140,208 Taxman62]**

Inseparable **MP.** **Addl.CIT-National Newsprint and Paper Mills Ltd.**

The word inseparabledoes not connote that the machinery, plant or furniture should,by its very nature,be inseparable from the building so that the building has also necessarily to be let along with it, or that it should be fixed to the building. The inseparability referred to in the section indicates that the parties should intend that the subject matter of the lease should be enjoyed together and that the letting of the building and of the other assets should be practically one letting. <u>Sec.56.</u>**[7CTR113,114ITR398]**

Install **Cal. CIT - Birla Jute and Industries Ltd.**

The expression install does not mean affixation.The word install has different connotations in different contexts having regard to the nature and character and specified manner of use of a particular plant or machinery for effective and appropriate utilization or operation thereof. The expression used in both sections 32(1) and 32A(1) is in relation to the business, viz., for the purposes of the business.The sections do not make any provision circumscribing the application only in respect of plant required for manufacturing purpose.If it is related to the business of the assessee, namely, used for the purpose of the business, the provisions of sections 32 and 32A would be attracted. *Sec.32(1).*[**180CTR339,260ITR55,133 Taxman337]**

Installed **Mad. CIT - Sri Rama Vilas Service (Private) Ltd.**

..... the word installed in relation to machinery or plant must be considered to mean such installation as that machinery or plant is capable of. Installed means to place an apparatus in position for service or use. *Sec.32.*[**38ITR25]**

2. The expression installed did not necessarily mean fixed in position but was also used in the sense of induct or introduce or placing an apparatus in position for service or use. **[CIT-Indian Turpentine and Rosin Co. Ltd.,All.,***Sec.33.* **75 ITR533]**

3. .... the expression installed must ..... be construed in a wide fashion. LPG cylinder is a plant and if it is brought into use for storing gas, it can be said to be installed in business.

[CIT-Steel Rolling Mills of Hindusthan (P.)Ltd.,Cal.,*Sec.33.* **51 CTR386,164ITR633,26Taxman554]**

4. The expression installed in section 32 of the Income-tax Act,1961,would include placing the apparatus in position for service or use. **[CIT-Instrumentation Ltd.,Raj.,***Sec.32.***201ITR117]**

5. The expression installed in clause (b) of sub-section (2) of section 32A of the Income-tax Act,1961, does not necessarily mean fixed in position.It is used in the sense of induct or introduce or placing an apparatus in position for service or use.**CIT-Bharat Radiators P. Ltd.,Bom.,***Sec.32A(2)(b).***158CTR 519, 239 ITR 608]**

6. An assessee would be entitled to the benefit of investment allowance under section 32A of the Income-tax Act,1961,only on proof that the machinery was not only installed but was functional or was capable of being put to use or was capable of producing the product for which it was acquired.The date from which such machinery becomes capable of producing would be the relevant date for determining the entitlement of the assessee to investment allowance. A distinction has to be drawn between machine and plant. When the assessee claims that the machinery has been installed,he should establish the date of acquisition of that machine; its functional capability to produce the yield for which it was acquired; whether the said machine is complete in itself to produce the yield without being dependent on other machines or accessories and as to when the said machine had become functional in all respects. If the claim of the assessee is that the machine is part of the plant then the assessee would be entitled to such investment allowance not singly in respect of each machinery but on completion of installation of the entire plant and on proof about the functional capability to produce the yield by the plant in its entirety. In case the machine is complete in itself then the date on which the machinery became capable of producing and in case of plant, the date on which the plant becomes capable of producing the yield will be the date relevant for admissibility of the investment allowance as permissible under the provisions of section 32A of the Act. **[Biyar Rubber Pvt. Ltd.-CIT,Kar.,** Sec.32A.**210CTR,191,292ITR251,160 Taxman 377]**

Institution Del. CIT - Charat Ram Foundation

The expressions founder and institution have not been defined in section 13 of the Income-tax Act,1961.The term institution is sometimes used as descriptive of an establishment or place where the business or operation of a society or association is carried on. At other times, it is used to designate an organised body. Sec.13.**[168 CTR261,250ITR64,116Taxman255]**

Instrument Del. Jagdamba Charity Trust–CIT

The word instrument used in section 26 of the Specific Relief Act has a very wide meaning and includes every document by which any right or liability is, or is purported to be created,transferred, limited,extended,extinguished or recorded.There is no reason to exclude a trust deed from its purview. *Sec.11,12.r.w.Specific Relief Act.***[18 CTR 317,128ITR377]**

2. The word instrument has been used in section 184(1) of the Income-tax Act,1961,in the sense of a writing a document or a formal expression in writing of some agreement or obligation, or of some act upon which the rights of the parties are dependent.The word instrument cannot be given too narrow a construction to mean only a regular deed of partnership.A partnership may be evidenced by one document or several documents.In the case of several docum -ents,all such documents together shall constitute the instrument of partnership.The number of documents is not determinative.So long as the terms of the partnership are specified in any document or documents which go to constitute the instrument of partnership, the condition of section 184(1) is fulfilled.It will be a partnership evidenced by an instrument. For the purpose of section 184,it is not necessary that, with every change in the constitution of a firm, a fresh deed of partnership should be executed.The change may be effected by an agreement which may refer to the terms contained in the original deed of partnership and agree to run the partnership in accordance therewith subject to the modification or changes made by such agreement.In such a case,both the agreement and the original deed of partnership would constitute an instrument of partnership within the meaning of section 184(1). **[CIT-Rajmohan Saha.,Gau.,*Sec.184(1).*94CTR278,190ITR236]**

Interest **Ker. Peter John - CIT**

Interest is separate from compensation.Interest,whether it is statutory or contractual, represents the profit the creditor might have made if he had had the use of the money or the loss he suffered because he had not that use. It is something in addition to the capital amount though it arises out of it. *Misc.***[157ITR711]**

2. Clause(28A) of section 2 of the Income-tax Act,1961,defines interest. It is clear from that definition that interest is a word of wide import. It means interest payable in any manner" in respect of moneys borrowed or debt incurred and includes even service fees or other charges in respect of the moneys borrowed. **[British Bank of The Middle East-CIT.Bom.,*Sec.2(28).*149CTR169,233ITR251]**

3. Interest is the return or compensation for the retention by one person of a sum of money belonging to or owed to another.The essence of interest is that it is a payment which becomes due because the creditor has not had his money at the due date.It may be regarded either as representing the profit he might have made if he had had the use of the money, or conversely, the loss he suffered because he had not that use.The general idea is that he is entitled to compensation for the deprivation. It is only interest in the above sense which is deductible under section 36(1)(iii) of the Income-tax

Act,1961.If in the garb of interest, something more is paid over and above interest,that something cannot be allowed as deduction under this section.What is allowable as a deduction under section 36(1)(iii) of the Act is any sum paid by way of interest in the commercial sense.There can be no straitjacket formula.The Income-tax Officer cannot refuse to allow the deduction of interest on the ground that the rate of interest is high or that the assessee could have borrowed money at a lower rate of interest.But if the Income-tax Officer comes to a finding that what is claimed as deduction by way of interest is, in fact, not wholly payment by way of interest but partly interest and partly payment for non-commercial considera-tions, he may allow the deduction of the amount which is interest and disallow the balance which is for extra commercial consideration.

[CIT-HindustanConductorsPvt.Ltd.,Del.,*Sec.36(1)(vii).*240ITR 762,108Taxman 258]

4. The word interest has a basic meaning of advantage or profit and with reference to a loan it means the profit or advantage of the creditor which he gets by giving to another the use of his money. Interest can be described as a consideration paid either for use of money or for forbearance in demanding it after it has fallen due. It is a compensation allowed by law or fixed by parties or permitted by custom or usage, for use of money belonging to another or for delay in paying the money after it has become payable. **[CIT-Saraswati Chemicals and Allied Industries (P.) Ltd., Del.,** *Sec.36(1)(iii).***167CTR150 ITR 235,114Taxman564]**

5. Interest is a consideration paid either for use of money or for forbearance in demanding it, after it has fallen due. It is a compensation allowed by law or fixed by parties or permitted by custom or usage for use of money belonging to another, or for the delay in paying the money after it has become payable.

[CIT - Prem Nath Motors (Pvt.) Ltd. Del. *Sec.201(1A).* **170CTR 424,253ITR705, 120Taxman584]**

6. The definition of interest in section 2(28A) of the Income-tax Act, 1961,after referring to interest payable in any manner in respect of any moneys borrowed or debt incurred, proceeds to include in the term moneys borrowed or debts incurred, deposits, claims and other similar right or obligation and further includes any service fee or other charge in respect of the moneys borrowed or debt incurred, which would include deposit or claim or other similar right or obligation as also in respect of any credit facility which has not been utilised.This statutory definition regards amounts which may not otherwise be regarded as interest as interest for the purpose of the

Act.Even amounts payable in transactions where money has not been borrowed and debt has not been incurred are brought within the scope of the definition as in the case of a service fee paid in respect of a credit facility which has not been utilised. [**Viswapriya Financial Services & Securities Ltd.-CIT,Mad.,** *Sec.2(28A) rws.194A,201(1).***179CTR334,258ITR496,127Taxman385]**

7. ... interest means interest on loans and advances. Secondly, it includes two other items in the definition of the word interest. The two items are commitment charges on unutilised portion of any credit sanctioned for being availed of in India,and discount on promissory notes and bills of exchange drawn or made in India. An extended meaning cannot be given to the term interest. **[CIT-Sahara India Savings and Investment Corp.Ltd.All.** *Interest-tax Act,Sec.4.* **181 CTR 134,261ITR113,129Taxman120]**

8. Interest means interest payable in any manner in respect of any moneys borrowed or debt incurred (including a deposit, claim or other similar right or obligation) and includes any service fee or other charge in respect of the moneys borrowed or debt incurred or in respect of any credit facility which has not been utilised. The meaning of the word interest is thus very wide and would include interest on unpaid purchase price payable in any manner which would include by means of irrevocable letter of credit.

[CIT -Vijay Ship Breaking Corporation.,Guj., *Interest-tax Act.***[185CTR136,264ITR646,134Taxman14]**

9. It is clear from the definition in section 2(28A) of the Income-tax Act,1961, that before any amount paid is construed as interest,it has to be established that the same is payable in respect of any money borrowed or debt incurred.**[CIT Cargill Global Trading Pvt.Ltd.,Del.** *Sec.2(28A).***241CTR443,335ITR94,199Taxman 320]**

Interest / deposit Del. CIT - Sahib Chits (Delhi) (Pvt.) Ltd.

The contribution given by the subscribers/members every month is not a deposit with the chit fund company and the amount of bid disbursed equally among the members is not to be treated as interest payable on money borrowed. Under section 2(28A) of the Income-tax Act,1961,interest is payable in respect of moneys borrowed or debt incurred. All the subscribers/members of the chit contribute moneys each month and a bid takes place among the members.The highest bidder is entitled to take the chit, i.e., the money contributed by all the members, after deducting the bid amount offered by him.It is this bid amount which is distributed among all the subscribers/members equally.This amount is not in respect of any moneys borrowed by the chit fund company or any debt incurred by it and, therefore, would not be covered by the definition of interest as contemplated by the Act. The successful bidder takes the entire amount minus the bid amount and the bid amount is disbursed equally among the members. Therefore, the amount contributed by the members cannot be treated as deposit also. Sec.2(28A),194A. **[328ITR342,185Taxman34]**

Interest chargeable in this act **Cal.** **Jalan & Sons (Private) Ltd. - CIT**

The words interest chargeable under this Act in the proviso mean interest which is not exempt from Taxation and the words payable without the taxable territories refer to interest which is due to be paid outside the taxable territories. *Interest- tax Act.* **[50ITR111]**

Interest of the individual in the property of the family **Guj.** **Kalyanbhai Trikamlal Shah - CWT**

The expression interest of the individual in the property of the family which has been defined for the purpose of section 4 of the Wealth-tax Act,1957, by clause (d) of the Explanation to the section, means the proportion to which the individual would be entitled in the property of the Hindu undivided family if there had been a total partition in the Hindu undivided family as on the valuation date.

Sec.4.WT.Act. **[135ITR750]**

Interested to deny **Nag.** **Governor- General in Council - Mulla Mohommad Bhai**

The words interested to deny in section 42 of the Specific Relief Act mean that the person interested to deny a legal character or a right to any property is a person with a rival claim of some sort and with some interest resembling in its nature that of the person whose legal character or right is denied. They cannot be interpreted to mean any person who might stand to gain financially or otherwise if some person's legal character or right to property were held to be not established. *Specific Relief Act.* **[13ITR10]**

Interlocutory order **Guj.** **Alkesh Subodhchandra Shah-State of Gujarat**

The term interlocutory order in section 397(2) of the Criminal Procedure Code, 1973,has been used in a restricted sense and any order which substantially affects the right of the accused or decides certain rights of the parties, cannot be said to be an interlocutory order so as to bar a revision to the High Court.The orders compelling the appellant to face a trial, without proper application of mind, cannot be held to be an interlocutory matter but one which decides a serious question as to the rights of the appellant to be put on trial.An order rejecting the plea of the accused on a point which, when accepted, will conclude the particular proceeding, will not be an interlocutory order within the meaning of section 397(2). Section 397(3) while eliminating the scope of two successive revision petitions by the same party says that, if a revision application under the section has been made by any person either to the High Court or to the sessions judge, no further application by the same person shall be entertained by either of them. But when once the revision application has been accepted by the sessions court and a finding adverse to the case of the petitioner accused had been recorded, in all fairness and under the law also, it could not be urged that the petitioner accused cannot come before the High Court under a revision.The power to discharge is exercisable under section 245(1) of the Code when the magistrate considers, for reasons to be recorded, that no case against the accused has been made out which, if unrebutted, would warrant his conviction.

*Criminal Procedure Code.***[212ITR255]**

Invalid return Cal. National Insurance Co. Ltd. - CIT

1. Section 139(9) makes a distinction between a defective return and an invalid return. A defective return is not ipso facto to be regarded as an invalid return. It is only when a return contains any of the specified defects and the Income-tax Officer, in his discretion, intimates the defect to the assessee and the assessee fails to rectify the same within the specified period, that the Income-tax Officer may treat the return as an invalid return. The provision in section 139(9) would override the other provisions of the Income-tax Act including section 292B. In a case where any of the specified defects is not removed within the time allowed in which case, the return shall be treated as an invalid return. *Sec.139.***[213ITR862]**

2. The provisions of section 292B of the Income-tax Act,1961,do not authorize the Assessing Officer to ignore a defect of a substantive nature and therefore, the provision categorically records that a return would not be treated as invalid, if the same in substance and effect is in conformity with or according to the intent and purpose of thisAct. **[CIT-Harjinder Kaur, P&H.,**<u>*Sec.292B.*</u>**222CTR254,310 ITR71]**

Invest **Cal. CIT-Birla Charity Trust**

The expression invest in section 13(2)(h) connotes a positive act on the part of the trust whereby the funds of the trust are laid out or committed in any particular property or business or transaction with the object of earning a profit or financial advantage or return. It has to be established that a trust having assets in the form of money or cash or a credit balance in a bank account or in any other form capable of being invested was by a positive act and pursuant to a decision of the trust laid out or committed in a concern of a nature specified, before it can be held that such an investment comes within the mischief of section 13(2)(h).This interpretation is supported by the mode of amendment of section 13 by the Finance Act,1983.Under the amended section, the benefit conferred by section 11 stands withdrawn not only where the funds of the trust are invested or remain invested in a manner other than that prescribed but also where the trust holds any shares in a company other than a Government company or a statutory corporation after a specified date.A positive distinction has been made between a case where the funds of the trust are, or remain, invested in a manner other than prescribed and a case where the trust continues to hold shares in companies other than those excepted. <u>*Sec.13(2)(h).*</u> **[66CTR172,170ITR150,34Taxman504]**

2. The expression invest in section 13(2)(h) of the Income-tax Act, 1961,connotes a positive act on the part of the trust whereby the funds of the trust are laid out or committed in any particular property or business or transaction with the object of earning profit or financial advantage or return.It has to be established that a trust having assets in the form of money or cash or a credit balance in the bank account or in any other form capable of being invested was,by a positive act and pursuant to a decision of the trust, laid out or committed in a concern of a nature specified, before it can be held that such an investment comes within the mischief of section 13(2) (h).Section 13(2)(h) will be attracted only in cases where funds, that is, moneys from actual or available money or cash resources of the trust itself are invested or continue to remain invested for any period during the previous year in any concern in which any person referred to in sub-section (3) of section 13 has substantial interest. **[Sarabhai Foundation-CIT,Guj.**<u>*Sec.13(2)(h).*</u>**209ITR390]**

3. The expression invest connotes a positive act on the part of the trust whereby the funds of the trust are laid out or committed in any particular property or business or transaction with the object of earning a profit or financial advantage or return. What is contemplated is the trust having assets in the form of money or cash or balance in a bank or any other form capable of being invested or by a positive act which pursuant to a decision of the trust is laid out or committed in a concern of a nature specified before it can be held that such an investment comes within the mischief of section 13(2)(h). **[CIT-Sir Shri Ram Foundation, Del.,**<u>*Sec.13(2)(h).*</u>**167CTR349,250ITR55,116Taxman113]**

Investment

The word investments used in item 1 of the Explanation to the Notification has been deliberately used in order to make it clear that not all income, profits and gains from securities but only such income, profits and gains from securities which are held as investment is excluded from the exemption. **[CIT-Bombay State Co-operative Bank Ltd.,Bom.,**<u>*Sec.297(2).*</u>**59ITR31]**

2. The word investment must be understood in the sense in which businessmen understand it: it is a form of income-yielding property. Investment contemplates acquisition of some species of property by laying out money. **[CIT-Aloo Investment Co.Pvt.Ltd.Bom.,** <u>*Sec.109(ii).*</u>**123ITR132]**

3. The expression investment used in r.19(4) of the I.T. Rules,1962, has not been defined and hence the meaning of the word in common parlance should be taken.The expression investment, normally and in the commercial world, is not treated as synonymous with either bank deposits or short-term deposits.Hence,where amounts are deposited in a bank, they would not be investments and the first limb of r.19(4) would not be satisfied,and even if they were investments these amounts would not become ineligible for capital computation,unless it could be shown that income in respect thereof was not taken into consideration in computing the profits of the business.Whether a particular income was part of the income from business would fall to be decided not on the basis of classification of incomes under different heads for purposes of Taxation but on commercial principles. **[Phillips Carbon Block Ltd.-CIT,Cal.,** <u>*Sec.84.r.w.r. 19(4).*</u>**28CTR333,136ITR205]**

4. The expression investment in section 13(2)(h) of the Income-tax Act,1961,denotes a positive act on the part of the trust whereby the funds of the trust are laid out or committed in any particular property or business or transaction with the object of earning a profit.Therefore,where the assessee-trust receives shares of a company by way of donation, it cannot be said that the assessee-trust has dealt with or committed or laid out any part of its existing assets to acquire the said shares. **[J.K.Trust –CWT,Cal.** *Sec.13(2)(h).***205ITR524]**

5. The word investment means to lay out money in business with a view to obtain income or profit. In order to constitute an investment the amount laid down should be capable of resulting in an income or return or profit to the investor and in every case of investment, the intention and positive act on the part of the investor should be to earn such income, return or profit to the investor.The words invest and investment are to be taken in the business sense of laying out of money for interest or profit. **[Anand Charitable Trust-CWT,Del.,***Sec.5(1)(i)WT.Act.***[257ITR275,123Taxman494]**

6. The words funds and investment have different meanings and the investment ought to be made by the trust out of its funds as per the requirement under section 13(2)(h) of the Act and not those investments which have already been made.The value of the investment received by the trust by way of donation cannot be treated as investment within the meaning of section 13(2)(h).**[CIT – Shri Radha Krishna Temple Trust,All.,***Sec.13(2)(h).***277ITR158]**

7. The word investment contained in section 69B deals with investment in bullion, jewellery or other valuable articles. **[CIT-Aar PeeApartmentsP.Ltd.,Del.,***Sec.69B.***55ITR674]**

Investment company Cal. Great Pyramid Insurance Co. Ltd. – CIT

Section 104 of the Income-tax Act,1961, read with section 105(ii) and the definition of investment company in section 109(ii) of the Act (as it stood at that time) applied to a company whose business consists wholly or mainly in dealing in or holding of investments. A company which comes within the scope of those provisions must be one whose primary business must be dealing in or holding of investments. If a company engaged itself in two or more equally or nearly equally important business activities, then it could not be said that the business consisted wholly or mainly in dealing in a particular activity. Even in cases where a company has more than one business activity and one of its activities is more substantial than the others, unless that activity is the primary activity of the company, it could not be said that the company is engaged wholly or mainly in any of its business activities. Section 104 applies only to cases where the primary activity of the company is in dealing in or holding of investments.Where the legislature speaks of the business of holding of investments, it referred to real, substantial or systematic or organised course of activity of investment carried on by the assessee for a set purpose such as the earning of profits.

*Sec.109.r.w.s.104,105.***[102ITR394]**

2. Before a company can be stated to be an investment company under section 109(ii) of the Act, the following have to be found as a fact: (a)that the assessee is carrying on a business, i.e., it is carrying on a systematic or an organised activity with the set purpose, inter alia, of making profits; (b) that the business consists of either dealing in investments, that is dealings in shares, securities, etc.,or holding of investments; and (c) that such business is its primary business. **[Charmugaria Trading Co. Ltd.-CIT,Cal.,***Sec.109(ii).*

110ITR715]

3. In order to treat a company as an investment company under section 23A of the Indian Income-tax Act,1922,the company should have carried out real, substantial or systematic or organised course of activity of investment for a set purpose such as earning of profits. Section 23A applies only in cases when the primary activity of the company is in the dealing in or holding of investments. **[CIT -I.B. Sen and Sons Pvt. Ltd.,Cal.,***Sec.104,109.***121ITR664]**

4. The determination of the question as to whether a company is an investment company within the meaning of the definition in section 109(ii) of the Income-tax Act,1961, prior to its amendment in 1966, would fall in three parts.First, it must be determined as a fact whether any investments as contemplated by the definition were made by the company; second, it has to be determined whether the company could be said to be holding investments, and third, it has to be found whether the business of the company consisted wholly or mainly in the holding of investments. **[CIT-Aloo Investment Co.Pvt.Ltd.,Bom.***Sec.104.***123ITR132]**

5. The expression investment company means a company whose gross total income consists mainly of income which is chargeable under the heads Interest on securities, Income from house property, Capital gains and Income from other sources.

[Eastern Aviation and Industries Ltd.-CIT,Cal.,*Sec.73.r.w.s. 109.***208ITR1023,74Taxman641]**

6. Section 109 of the Income-tax Act, 1961, defines the expression investment company for purposes of sections 104, 105 and 107A. From the definition it is evident that in order to term a company an investment company,its gross total income should consist mainly of income from securities,house property, capital gains,etc.The expression mainly appearing in the definition of investment company in clause(ii)of section 109 means substantially or primary-ly.If the business of the company consists mainly in dealing in goods or merchandise, it cannot be held to be an investment company within the meaning of clause (ii) merely because, for one reason or the other, its income from business happens to fall short of its income from investments, etc.,in a particular previous year. The decisive factor for determining whether a company is an investment company or any other company is, therefore,the true nature of the primary activities of the company. If the activities of the company are such that its total gross income mainly consists of income from securities, etc.,it would be termed an investment company.The word mainly is somewhat akin to wholly and has been used to mean the whole or a substantial portion of the total gross income of the assessee. It cannot be construed to mean not less than fifty-one per cent. **[CIT-Amritlal and Co. Ltd.,Bom.,** *Sec.109.***126CTR142,212ITR540]**

7. Sub-clause (ii) of clause (c) of the Explanation to section 40A(8) of the Act which defines an investment company states that one of the principal business could be acquisition of maketable securities of a like nature, in other words, securities which are in the nature of shares, stock, debentures, etc.,or securities issued by a Government.

[Barkha Investment and Trading Co.-CIT,Guj.*Sec.40A.(8)***200 CTR342,281ITR316,150Taxman523]**

Involved **Ker. CIT - Indian Chamber of Commerce**

In order to take an object of general public utility outside the scope of the definition in section 2(15) of the Income-tax Act,1961,on the ground that it involves the carrying on of an activity for profit, that object must involve the carrying on of any activity for profit. Involve means comprise or imply and the object must, therefore, imply the carrying on of any activity for profit. It is not sufficient, if there is some activity carried on which results in profit. There must be an activity in the form of business because the activity must be for profit and that activity for profit must be involved in the objects of general public utility. Even when an activity is in furtherance of the objects of a trust and even if such activity results in profits, the definition will not be attracted unless the objects involve carrying on of an activity for profit. *Sec.2(15).***[80ITR645]**

Involved in the case SC Santosh Hazari - Purushottam Tiwari

To be a question of law involved in the case there must be first a foundation for it laid in the pleadings and the question should emerge from the sustainable findings of fact arrived at by the court of facts and it must be necessary to decide that question of law for a just and proper decision of the case. Any entirely new point raised for the first time before the High Court is not a question involved in the case unless it goes to the root of the matter. *Constitution of India,C.P.C.***[170CTR160,251ITR84]**

Iron and steel (metal) P&H. CIT - Ludhiana Steel Rolling Mills

In order to determine whether the assessee is entitled to development rebate at the higher rate of 25 per cent. on the ground that it manufactures iron and steel (metal) within the meaning of item No.(1) of the Fifth Schedule to the Income-tax Act,1961,the nature of the article manufactured should be taken into consideration.If iron and steel bars or other raw material has been used for making an article which is known and accepted in common parlance or in the commercial world as a specific article different from iron and steel and that article can no more be treated or understood basically as iron and steel, that article cannot be termed iron and steel (metal). To illustrate, if iron is used for manufacture of shovels or pickaxes,no one would understand, treat or name the shovels or pickaxes as iron and steel. So, the question is whether the finished article can be said to be something basically different from iron and steel. *Sec.33(1).***[180ITR155,45Taxman446]**

Irregularity Del. Remfry & Sons - CIT

Irregularity is an expression of lesser effect as an act which is improper or inefficient by reason, or departure from the prescribed act.It primarily connotes the neglect of order or method and may not be according to regulations.One of the basic distinctions between the two expressions is the resultant effect of its being curable or incurable.Normally, an illegality which goes to the very root of the matter or jurisdiction could hardly fall in the class of those cases, but an irregularity is merely procedural and where it has even been substantially complied with, to bring such a case within the class of cases where irregularity is curable, ould be a fair and just interpretation of the relevant rule. *Misc.* **[195CTR66, 276ITR1,145Taxman22]**

Is **SC** **F.S.Ghandhi - CWT**

The word is cannot be construed to mean has been.The question as to whether the interest in property should be included or excluded from the assets of the assessee under section 2(e)(2)(iii) has to be considered in the light of the nature of the interest on the relevant date, viz., the date on which the interest vests in the assessee *Sec.2(e).WT.Act.* **[84CTR35,184ITR34]**

Is admitted and pending **MP.** **Shree Amarlal Kirana Stores - CIT**

. . .The use of the words is admitted and pending in section 95(i)(c) is important and significant.The deliberate use of the word admitted prior to the words and pending can never be regarded as redundant or otiose ;it being a settled rule of interpretation that every word in the statute has its definite meaning.The words admitted and pending used in sub-clause (c) have their application to all the three types of cases,namely, appeal, reference and writ petition.Indeed, the Legislature was quite aware of the procedure for prosecuting appeals,references and writs provided under the Act and the rules applicable to these three statutory remedies prescribed under the Act and the Constitution.It is for this reason that it specifically used the words admitted and pending after these three types of cases in section 95(i)(c). *(K.V.S.S).* **[180CTR35,259 ITR572,126Taxman512]**

Is included in the profit **SC** **State of Madhya Pradesh-Binod Mills Co. Ltd.**

The words is included in the profits did not mean either included by the managing agent in his return or included by the assessing authority in an order of assessment against the managing agent, but meant liable under the terms of the Ordinance to be charged to tax as part of the profits of the managing agent.*Sec.187.* **[46ITR159]**

Is or was **Bom.** **CIT - B. N. Exports**

The effect of section 10(10D) of the Income-tax Act,1961, is that monies which are received under a life insurance policy are not included in the computation of the total income of a person for a previous year. However, any sum received under a Keyman insurance policy is to be reckoned while computing the total income. For that purpose, a Keyman insurance policy means a life insurance policy taken by a person on the life of another person who is or was in employment as well as on a person on who is or was connected in any manner whatsoever with the business of the subscriber. The words is or was connected in any manner whatsoever with the business of the subscriber are wider than what would be subsumed under a contract of employment. The latter part makes it clear that a Keyman insurance policy for the purposes of clause (10D) is not confined to a situation where there is a contract of employment.

*Sec.10(10D).***[231CTR227,323ITR178,190Taxman325]**

Issue Guj. Madanlal Mathurdas -Chunilal, ITO

The word issue in the first proviso to section 34(1) of the Indian Income-tax Act,1922,as amended in 1956,cannot be equated to serve. *Sec.148.***[44 ITR325]**

Issue and serve Cal. Belland-Banarasi Debi

The two words issue and serve used in section 34 have been used as interchangeable or equivalent expressions and the same meaning has to be given to the word issued as used in section 4.

*Sec.148.***[46ITR28]**

2. The words issue and serve are interchangeable and the word issue has been used in section 148 of the Income-tax Act,1961, in the same sense in which the word serve has been used. Therefore, a notice of reassessment issued against the assessee within the period of limitation but served on the assessee after the period would be without jurisdiction and void.**[Tikka Khushwant Singh -CIT. P&H.***Sec.148.***101ITR106]**

3. The words issue and serve are not synonyms.Their plain dictionary meanings runs directly contrary to any such assumption. The gap between the two may be wide both in point of time and place. A statute may require that the issuance of a general order be conveyed by publication in the locality without individual service. The word issue is to be construed in the context of Section 149 which is an express limitation provision creating a precise bar with regard to reopening of assessments.In sub-section (3) of section 149,the word employed is served in the first line while in the penultimate line the word employed is issued .Thus, in the same short sub-section, the Legislature has used these words as distinct and separate.The hallmark of a limitation provision is that the same must have clear cut and fixed termini at both ends. Section 149 fixed the terminus a quo from the end of the relevant assessment year, i.e., on the 31st March of the said year. On the other hand, the terminus ad quem under clauses (a) and (b) is fixed at 4 years, 8 years and 16 years, from the fixed date of 31st of March of the relevant assessment year. Clearly enough, if the terminus a quo is fixed as the relevant assessment year, namely, 31st March of the said year, the other terminus must equally be fixed with regard to the fixed date of the issuance of the notice, which is precise and predictable.The plain scheme of sections 148 and 151 is that the satisfaction and the sanction of the Commissioner or the Board on the reason recorded by the Income-tax Officer is necessary before the notice under section 148 is sent out. If the word issued used in both these sub-sections is read as served, it will lead to the strange phenomenon that even after the Income-tax Officer has recorded his reason and issued the notice, the sanction therefor may be recorded before its service on the assessee. **[CIT-SheoKumariDebi,Pat.**<u>*Sec.149.r.w.s.144.*</u>**157 ITR13,24Taxman 77]**

Issued **Cal. M.M.Ispahani Ltd. - CIT**

The word issued used in section 297(2)(d)(i) of the Income-tax Act, 1961, should be interpreted in its wider meaning as including the service of a notice under section 34. Therefore, the issue of a notice under section 34 of the Indian Income-tax Act, 1922, can be said to initiate a proceeding which could be said to be pending when the Act of 1961 came into force so as to bring it within the scope of section 297(2)(d)(i) of the later Act. <u>*Sec.297(2).*</u>**[75ITR479]**

Issued and served **AP. CIT - Kailasa Devi and Rukmini Bai**

Under section 148 read with section 149 of the Income-tax Act,1961, a notice under section 148 for reassessment should not only be issued but also served within the period prescribed under section 149 of the Act. The words issued and served were consistently interpreted as interchangeable under section 34 of the Indian Income-tax Act, 1922. Section 148 of the Income-tax Act,1961, is in similar terms and there is no difference between section 34 of the Indian Income-tax Act,1922 and the 1961 Act except that the old section 34 had been split into various sections. *Sec.148.* **[105ITR479]**

2. The expressions issued and served are used as interchangeable terms and in the legislative practice of our country they are sometimes used to convey the same idea.**[Banarsi Debi -ITO. SC.** *Sec.148.***53ITR100]**

J

Jeep **Mad.** **Crompton Engg. Co. (Madras) Ltd. - CIT**

Merely because a jeep has four-wheel propulsion and is used through jungles and hilly terrains,it would not make it different from a motor car particularly when the motor car and the jeep have certain common features such as (a) both are mounted on wheels upon which they run over the surface of the land; (b) both are guided and controlled by a person riding upon or in them; (c) both are designed and intended to carry one or more persons; and (d) both are propelled by power not from any source external to themselves but which is for the time being stored and generated within themselves and both are self moving vehicles. Sec.32.r.w.s. 43(1).**[92CTR207,193ITR483,106Taxation86]**

Jewellery **Guj.** **CWT - Jayantilal Amratlal**

The dictionary meaning of the term jewellery is clear enough to include ornaments. Therefore, it does not mean that ornaments made of gold, silver or platinum and studded with precious stones can be considered to be jewellery only from April 1,1972. It cannot, therefore,be contended that ornaments would be entitled to exemption from wealth-tax under section 5(1)(viii) for assessment years subsequent to the year 1962-63 on the ground that such ornaments are included in the definition of jewellery only from April1,1972. Sec.5(1).WT.Act.**[102ITR105]**

2. The ordinary meaning of the word jewellery is not so wide as to cover all ornaments. The ordinary meaning of the word as known now will only embrace precious and semi-precious stones and gold and silver ornaments which contain precious or semi-precious stones.It is in this sense that the word jewellery as used in clause (viii) of section 5(1) of the Wealth-tax Act,1957, has to be understood before 1st April,1972. The inclusive definition contained in the Expln.to clause(viii) of section 5(1),which became effective from 1st April, 1972; was not added merely as a matter of abundant caution. It was clearly intended to give a wider meaning to the word jewellery w.e.f. April1,1972.The very fact that the words but not including jewellery were retrospectively added in section 5(1)(viii) w.e.f.April 1,1963, by the same Finance Act,gives out the clear intention of Parliament that the wider meaning of the word jewellery as contained in the Expln.was not to be applied for any assessment year prior to 1st April,1972. **[CWT-Smt.Tarabai Kanakmal,***Sec.5(1).WT.Act.* **31CTR297,140 ITR374]**

3. Jewellery is defined in the Oxford English Dictionary as jeweller's work; gems or ornaments made or sold by jewellers; especially precious stones in mountings.The New Webster's English Dictionary defines jewellery as jewels; articles made of gold, silver, precious stones or similar materials for personal adornment. As the Oxford English Dictionary suggests, the word jewellery may be derived from either the word jewel or the word jeweller. As derived from the word jewel,jewellery suggests an ornament that contains a precious stone or stones.As derived from the word jeweller, jewellery suggests an ornament made by a jeweller.Whereas the former derivation excludes ornaments of gold, silver and the like not containing precious stones from the ambit of jewellery, the latter does not, for, ornaments of gold, silver and the like, though devoid of precious stones, are still the handiwork of a jeweller.The word jewellery, then, is capable of two meanings, one embracing only ornaments made of precious metals which contain precious stones and the other embracing ornaments made of precious metals regardless of whether or not they contain precious stones.

[CWT- Godavaribai R. Podar, Bom.,*Sec.5(1).WT.Act.***63CTR152, 169ITR 245,32Taxman416]**

4. Jewellery will not include gold ornaments not studded with precious or semi-precious stones for any assessment year prior to 1-4-1972. **[CWT-MeghajiGirdharilal,MP.***Sec.5(1).WT.Act.***177 ITR297]**

5. In the absence of any statutory definition, jewellery had to be given its natural meaning, and understood in the same way as in ordinary parlance.Therefore, if without reference to Explanation 1, it could be held that jewellery includes not only ornaments studded with gems, precious stones, etc., but also includes ornaments made of gold, silver, platinum, or any other precious metal, the subsequent insertion of Explanation 1 must be understood to be by way of abundant caution. The fact that the amendment to section 5 (1)(viii) was effected retrospectively, while Explanation 1 was added prospectively, cannot be of much significance in interpreting the word jewellery in section 5(1)(viii). The dictionary meaning of jewellery is wide enough to include articles of gold and other precious metals used for personal adornment. The fact that jewels or precious stones are embedded in such articles will not make any difference. The distinction sought to be made between ornament and jewellery is artificial. There is no reason to impute an intention to the Legislature that by excluding the jewellery from the exemption clause,it purported to exclude only ornaments studded with precious or semi-precious stones and not ornaments made of gold, silver or other precious metal. The word jewellery has been used in a generic sense as to include both the above categories. In the context of the Wealth-tax Act, it stands to reason that the Legislature did not intend to include within the purview of exemption valuable articles such as jewellery while granting exemption for the value of furniture,household utensils, wearing apparel intended for personal or household use of the assessee. **[CWT-SureshMohanThakur,Pat.,**Sec.5(1).WT.Act.**211ITR811]**

6. The term jewellery is not confined in ordinary parlance to only those ornaments which have precious stones embedded in them. It covers all articles of value used for adornment. Explanation 1 to section 5(1)(viii) of the Wealth-tax Act,1957, has been introduced by the Finance (No. 2) Act,1971,partly to clarify this position.The Explanation provides an extensive definition of jewellery.It includes ornaments made of gold, silver, platinum or any other precious metal whether or not containing any precious or semi-precious stones. It covers such items which may or may not be sewn into any wearing apparel.It also includes precious and semi-precious stones whether or not set in any furniture, utensils or other article, or worked or sewn into any wearing apparel. The Explanation may have extended the meaning of jewellery to cover, for example, precious stones by themselves or precious stones set in furniture or utensils.But in so far as it includes ornaments made of gold, silver, platinum or any other precious metal or alloy, it is merely clarificatory in nature. ..Jewellery, even before the coming into force of Explanation1,would include gold ornaments.

[CWT-Smt. Binapani Chakravarty,SC,Sec.5(1).WT.Act.**125CTR 119,214ITR 721]**

Jewellery held for personal use **AP.** **CIT - Trustees of HEH The Nizam's Wedding Gifts Trusts**

Under section 2(14) of the Income-tax Act,1961, prior to its amendment with effect from April 1,1972, jewellery held for personal use by the assessee was excluded from the definition of capital asset. A close scrutiny of the context in which the expression jewellery held for personal use occurred showed that it should be an article that should be associated with the person of the possessor. It was, therefore, clear that an intimate connection between the effects and the person of the assessee must be shown to exist to render them personal effects. According to this expression, any jewellery held by the assessee did not come under personal effects. It was only such jewellery that was normally, ordinarily and commonly used that came under personal effects but not the jewellery calculated to give a pride of possession on special ceremonial occasions. *Sec.2(14),45.* **[40CTR88,154ITR573,18Taxman99]**

Journal **Bom. Ethnor Ltd. - CIT**

Printing and distribution of journals ordinarily involves publication of article/articles on any subject written by one or many. A journalis ordinarily published periodically and would contain information in general on any topic or any research work done in a particular field. *Sec.37(3).***[181CTR550,260ITR401,126Taxman408]**

Jurisdiction **Ker. CIT - S.M.Syed Mohamed**

Jurisdiction is the power,right or authority to take cognizance and decide any matter according to law.An authority can take cognizance of conferment of jurisdiction or power only when it is brought to its notice or when it is consciously made known to it. The cognizance of conferment of jurisdiction is a pre requisite for the exercise of the power. *Sec.271,274.***[216ITR331]**

2. ....jurisdiction has reference to the power of the court or tribunal over the subject matter, over the res or property in contest, and to its authority to render the judgment or decree it assumes to make. It is in this sense that publication of the notice in the Official Gazette confers jurisdiction on the competent authority.......

[CIT-Pearl Mech. Engg. & Foundry Works P.Ltd.,SC,*Misc.*189 CTR289,267ITR1,136Taxman586]

Jurisdictional fact **SC Arunkumar-Union of India**

A jurisdictional fact is a fact which must exist before a court, Tribunal or an authority assumes jurisdiction over a particular matter.A jurisdictional fact is one on the existence or non- existence of which depends the jurisdiction of a court,a Tribunal or an authority.It is the fact upon which an administrative agency's power to act depends. If the jurisdictional fact does not exist, the court, authority or officer cannot act. If a court or authority wrongly assumes the existence of such fact, the order can be questioned by a writ of certiorari. The underlying principle is that by erroneously assuming the existence of such jurisdictional fact, no authority can confer upon itself jurisdiction which it otherwise does not possess. The existence of the jurisdictional fact is thus the sine qua non or condition precedent for the exercise of power by a court of limited jurisdiction.......If the jurisdictional fact exists,the authority can proceed with the case and take an appropriate decision in accordance with law.Once the authority has jurisdiction in the matter on existence of the jurisdictional fact, it can decide the fact in issue or adjudicatory fact.A wrong decision on a fact in issue or on an adjudicatory fact would not make the decision of the authority without jurisdiction or vulnerable provided the essential or fundamental fact as to existence of jurisdiction is present. *Misc.*

[205CTR193,286ITR89,155Taxman659]

Justified **Guj.** **Patel Engineering Co. Ltd.**

The word justified is a word of wide import. Something could be said to be justified if it is proved or shown to be fair or right or according to justice or backed by sufficient reasons.

Sec.215.r.w.r. 40. **[44CTR158,151ITR542]**

K

Karta of the Hindu undivided family **SC** **CIT - Shri Om Prakash**

When the karta of a Hindu undivided family is a partner in a partnership firm, he has a dual capacity ; qua the partnership, he functions in his personal capacity and qua third parties, in his representative capacity. Under the Income-tax Act,1961,when he is assessed in respect of the income derived by him from the partnership firm as a partner, it is in his representative capacity as karta of the Hindu undivided family and not as an individual as such.That is because his capacity vis-a-vis his spouse/minor children who are members of the Hindu undivided family is that of karta and not as individual though vis- a-vis other partners of the partnership firm he functions in his personal capacity. This being the position, the income of a karta's spouse/minor child cannot be included in the computation of his total income for that is the income of the Hindu undivided family and, not his individual income. Section 64 of the Income-tax Act,1961, will be attracted only when an assessees' own income is being assessed and not that of a Hindu undivided family. If a karta is brought within the ambit of individual in section 64(1),the share income of the spouse of the karta and his minor children will,in effect, be included in the income of the Hindu undivided family which is not what is contemplated by section 64(1)(i) and (ii) which is impermissible.

<u>*Sec.64(1).*</u>**[155CTR206,238ITR1044,105Taxman619]**

Kept ready for use **Ker** **CIT - Air Travel Enterprises India Ltd.**

A perusal of sections 2(7) and 2(35), 66, 72,73 and 74 of the Motor Vehicles Act,1988, shows that in order to say that the vehicle was kept ready for use for carrying persons for hire, the vehicle must be registered as a contract carriage and it is only on obtaining the

contract carriage permit that it can be said that the vehicle was kept ready for use related to the business of the assessee, namely, used for the purpose of the business, the provisions of sections 32 and 32A would be attracted. *Sec.32.r.w.Motor Vehicles Act.*
[188CTR33,265ITR537,136Taxman194]

Know-how Guj. CIT - Elecon Engineering Co. Ltd.

Know-how is a peculiar kind of asset.It is the accumulated fund of knowledge acquired by years of observation, research, experimentation and experience.The whole of it is not in an intangible form even while it is in the process of being acquired and very often it takes a physical form as it grows in the shape of formulae, drawings,patterns,blue prints, specifications and so on. The material form it takes not only facilitates preservation, collation and ready reference but also makes it perceptible and visible and easily capable of being transmitted to other section.
Sec.32.r.w.s.43(3). **[96ITR672]**

Knowledge and experience AAR P. No. 25 of

The words knowledge and experience have not been defined in the Act. A person having specialised knowledge and experience is known as an expert in common parlance who can be considered as a technician...provided his specialised knowledge and experience pertains to the prescribed spheres.Special knowledge may be obtained either by education or by special experience.Possession of a certificate issued by a recognised body is not necessary for knowledge and experience. *Sec.10(5B).r.w.s.245Q.*
[153CTR216,237I827TR,151Taxation566,104Taxman372]

L

Land SC State of West Bengal -Kesoram Industries Ltd.

Land cannot be assigned a narrow meaning so as to confine it to the surface of the earth. It includes all strata above and below.

Constitution of India. **[187CTR219,266ITR721]**

Land and building/ annual value method Bom. Jehangir Mahomedali Chagla - CED

There are principally two methods of valuation of buildings: (1) Land and building method, and (2) annual value or rental method. In the first method, which is commonly known as contractors' method, the market value of the land is separately ascertained and to that is added the value of the building determined by ascertaining the prime cost of the building and deducting there from depreciation. By far the more prevalent, especially in urban areas, is the annual value method. The annual value is arrived at in three ways, namely: (1) actual rent fetched by land or building where it is actually let, (2) where it is not let, rent based on hypothetical tenancy, and (3) where either of these two modes are not available, by valuation based on capital value from which annual value has to be found by applying a certain percentage which may not be the same for lands and buildings. *Estate Duty*[155ITR637] **Literary work Kar. CIT - Samsung Electronics Co. Ltd.**

Land appurtenant thereto AP. CIT - Zaibunnisa Begum

The expression land appurtenant thereto occurring in section 54 has not been defined. It must, therefore, be understood in its popular and non-technical sense. It is not possible to accept the contention that clause (b) of the Explanation to section 5(1)(ivc) of the Wealth-tax Act,1957, defining land appurtenant for the purpose of that clause should be considered equally applicable for the purpose of understanding that expression occurring in section 54 of the Income-tax Act. The Explanation in the Wealth-tax Act is only for the purpose of section 5(1)(ivc) because it is specifically stated so. The meaning assigned to that expression in the Urban Ceiling and Regulation Act is also not relevant.The tax authorities will have to determine the extent of land appurtenant to a building transferred, taking into consideration a variety of circumstances that may be relevant for the purpose. It is not possible to lay down infallible tests to be applied as the tests would vary depending upon the facts and circumstances of each case. For instance: (1) If the building together with the land is treated as an indivisible unit and enjoyed as such by the persons occupying the building, it is an indication that the entire extent of land is appurtenant to the building; (2) If the building has extensive lands appurtenant thereto and even if the building and the land have been treated as one single unit and enjoyed as such by the occupiers, an enquiry could be made to find out whether any part of the land contiguous to the building can be put to independent user without causing any determent to the enjoyment of the building as such. Such an enquiry should be conducted not based on any artificial considerations but from the point of view of the persons occupying the building. The number of persons or different branches of families residing in the building, the requirements of the persons occupying the building, consistent with their social standing, etc., are relevant for the purpose.If any surplus is arrived at on such enquiry, then the extent of such surplus land may not qualify to be treated as land appurtenant to the building; (3) if there is any evidence to indicate that any portion of the land contiguous to the building was applied to user other than the enjoyment of the building, then that provides a safe indication regarding the extent of land applied for such user.For instance, the land used by the occupiers for commercial or agricultural purposes although forming part of the land adjacent to the building, does not qualify to be treated as land appurtenant to the building; (4) if the owner or occupier is deriving any income from the land which is not liable to be assessed as income from house property under section 22 of the Income-tax Act, then the extent of such land does not qualify to be treated as land appurtenant to the building; and (5) any material pointing to the attempted user of the building for purposes other than the effective and proper enjoyment of the house would also afford a safe guide to determine the extent of surplus land not qualifying to be treated as land appurtenant to the building. *Sec.5(1).WT.Act.* **[151ITR320,20Taxman120]**

2. A perusal of section 54 of the Income-tax Act,1961, makes it clear that the exemption under the provision is available only where a building or land appurtenant thereto is sold and within two years immediately preceding the date on which the transfer took place, was being used by the assessee or a parent of his mainly for the purposes of his own or the parent's own residence and the assessee has within a period of one year before or after that date purchased, or has within a period of two years after the date constructed, a house property for the purpose of his own residence.Land appurtenant to the building implies that the ownership of the building and the land appurtenant should be of the same person. If the building is owned by one person and the land is owned by another person then it will be the case of land adjoining the building and by no stretch of imagination can it be called land appurtenant to the said building. **[P.K. Lahiri-CIT,All.,** *Sec.54.***275ITR17,146Taxman349]**

Land appurtenant to **Mad. CIT - Smt. M.Kalpagam**

The question whether certain land is appurtenant to a house is one of fact.The following five tests can be applied to understand the meaning of the words land appurtenant more precisely : (1) If the building together with the land is treated as an indivisible unit and enjoyed as such by the persons occupying the building, it is an indication that the entire extent of land is appurtenant to the building.(2) If the building has extensive lands appurtenant thereto and even if the building and the land have been treated as one single unit and enjoyed as such by the occupiers, an enquiry could be made to find out whether any part of the land contiguous to the building can be put to independent user without causing any detriment to the enjoyment of the building as such. Such an enquiry should be conducted not based on any artificial considerations but from the point of view of the persons occupying the building.The number of persons or different branches of families residing in the building, the requirements of persons occupying the building, consistent with their social standing, etc., are relevant for the purpose.If any surplus is arrived at on such enquiry, then the extent of such surplus land may not qualify to be treated as land appurtenant to the building; (3) If there is any evidence to indicate that any portion of the land contiguous to the building was applied to user other than the enjoyment of the building, then that provides a safe indication regarding the extent of land applied for such user. For instance, the land used by the occupiers for commercial or agricultural purpose although forming part of the land adjacent to the building, does not qualify to be treated as land appurtenant to the building; (4) If the owner or occupier is deriving any income from the lands which is not liable to be assessed as income from house property under section 22 of the Income-tax Act,1961, then the extent of such land does not qualify to be treated as land appurtenant to the building; and (5) Any material pointing to the attempted user of the land for purposes other than the effective and proper enjoyment of the house would also afford a safe guide to determine the extent of surplus land not qualifying to be treated as land appurtenant to the building. <u>*Sec.54.*</u>

[143CTR336,227ITR733,93Taxman283]

Landowner/trader Ker Attukal Shopping Complex P. Ltd. - CIT

Ownership of property and leasing it out may be done as a part of business, or it may be done as a land owner.Whether it is the one or the other must necessarily depend upon the object with which the act is done.When a company is formed with the specific object of acquiring properties not with the view to leasing them as property but to selling them or turning them to account even by way of leasing them out as an integral part of its business, it cannot be said to treat them as landowner but as trader. <u>*Sec.22,28.*</u>

[178CTR469,259ITR567,125Taxman881]

Landscape Raj. CIT - Udaipur Distillary Co. Ltd. (No.1)

The word landscape had multiple meanings. However, the term was not used for making the land.Therefore, the expenses described as landscaping expenses would ordinarily refer to expenses incurred for laying ground to give the effect of natural scenery.It was ordinarily capital in nature resulting in improvement of the land value and also was not connected in any sense with research and development activity relating to the business of the assessee. It was an expense not in the regular course of cultivation activity but related to making permanent improvement of uneven land to level it for future cultivation for the assessee's purposes. <u>*Sec.37.*</u>

[186CTR1,268ITR305,134Taxman398]

Lease/hire Guj. Bhagwati Appliance - ITO

There is a basic difference between lease and hire.This difference is borne out by the basic difference in the meaning of the expression property and the expression possession.A transaction of hire is essentially a contract of bailment of a vehicle.In the case of a hire, only a licence is given to the hirer to use the vehicle for a temporary period the vehicle so hired.In the case of lease, the hirer has an option to buy the equipment which is one of the main distinguishing features between the words hire and lease. *Sec.32.*

[337ITR286,199Taxman131]

Legal representative Ker. Rajeevi R. Hegde - ITO (Agrl.)

The term legal representative means one who stands in the place of and represents the interests of another, i.e., one who is entitled to take by descent or distribution. It denotes one on whom the status of a representative is fastened by reason of the death of his ancestor.If on the death of a person, some of his rights fell on or accrue to another as successor, that another is a legal representative. If there is an executor or administrator for the estate of the deceased, he is the legal representative, and, in other cases, the heirs or legatees on whom his proprietary rights devolve are the legal representatives.

Kerala Agricultural Income-tax Act.
[20CTR34,127ITR855]

Letting Bom. CIT - Bhandara Zilla Sahakari Kharedi Vikri Sangh Ltd.

Normally,co-operative societies which have storage facilities receive goods from their members, agriculturists or others for storage and charge certain fees depending upon the quality of the goods delivered for storage. In such cases the godowns are not let out to the others but possession of the godowns is retained by the societies and the societies merely take charge of the goods brought by the member, agriculturists or others and stores the same. Thus, most of the activity of the society consists in taking custody of the goods of others and storing the same in their godowns or warehouses.It must be taken that the Legislature when it introduced the above provisions was quite aware of the usual activities of co-operative societies which maintained the godowns and warehouses for storage purposes. It cannot be taken to be the intention of the Legislature to except only the rents received from letting out of the godowns and not the amounts received by the co-operative society for permitting the user of the godowns for storage by its members, agriculturists or others. In the setting in which the word letting occurs, it should be understood as having a wide and comprehensive sense so as to include even the mere user of the godowns either by the society or others. *Sec.80P(2).*[212ITR124]

Levy Bom. CIT - Zoroastrian Building Society Ltd.

Levy as used in the third proviso to section 9(2) of the Indian Income-tax Act, 1922, means impose and not collect. *Sec.23.*

[27ITR218]

2. The three expressions levy,assessment and collection are of the widest significance and embrace in their broad sweep all such proceedings for raising money by the exercise of the power of Taxation. **[Lakshman Shenoy -ITO, SC, *Misc.*34ITR275]**

3. The language of the proviso to section 23(1) of the Income-tax Act, 1961, shows that there must have been a levy by the local authority in respect of the property,which is in the occupation of a tenant and to the extent to which such a levy is borne by the owner, he would be eligible for the deduction envisaged by the proviso.The word levied is a significant one in the construction of this provision. Its import cannot be ignored.Section 104 of the Madras City Municipal Act,1919,also contemplates a levy .The property tax is not automatic so as to require the property owner to pay the tax voluntarily.Hence, the liability to tax arises at the time of actual levy and the deduction under the proviso to section 23(1) is justified only in the year in which the levy is made. **[CIT - L. Kuppuswamy Chettiar, Mad., *Sec.23.*132ITR416]**

4. The term levy is an expression of wide import: it includes both imposition of a tax as well as its quantification and assessment...... The word levied is a wide and generic expression. One can say with as much appropriateness that the Income-tax Act levies a tax on income as that the Income-tax Officer levies the tax in accordance with the provisions of that Act. It is an expression of wide import and takes in all the stages of charge, quantification and recovery of duty, though in certain contexts, it may have a restricted meaning. In the context of section 3(1), the word levied, admittedly, means charged as well as assessed.The words levy and collection in section 3(3) cannot be construed differently from the words levied and collected used in section 3(1).Section 3(3), therefore, also covers the entire gamut of section 3(1) and cannot be construed as becoming operative at a somewhat later stage. Its operation cannot be excluded in determining the scope of the charge...

[Ujagar Prints-Union of India,SC, *Additional Duties of Excise (Goods of Special Importance)Act.* **179ITR317]**

Liability Del. Addl.CIT-Minerals & Metals Trading Corp. of India

In view of the second sub- para of rule 7(1) of Part III of Schedule VI to the Companies Act,1956, the expression liability should be understood to mean all liabilities including disputed or contingent liabilities. *Companies Act.* **[25CTR228,134ITR78]**

Liability in praesenti Cal. CIT - Remington Rand India Ltd.

When the liability for payment of gratuity is ascertained by actuarial valuation in which all contingencies are taken into consideration,such liability is in praesenti and the amount so set apart is a permissible business expenditure in the year concerned for an assessee following the mercantile system of accounting.

Sec.37. **[49CTR242,159ITR922,21Taxation220,78Taxman69]**

Liable Guj. ACIT-Narmada Chematur Petrochemicals Ltd.

The word liable has been defined to mean obliged in law or equity; subject; answerable; responsible. Thus, liability is the state of being liable; having responsibility or obligation, e.g., to make payment of debts. *Sec.43B,145.* **[233CTR265,327ITR369,194Taxman103]**

Liable to tax AAR Abdul Razak A. Meman

That liable to tax connoted that a person was subject to one of the taxes mentioned in article 2 of the Double Taxation Avoidance Agreement in a Contracting State and it was immaterial whether the person actually paid the tax or not. Article 4(1) postulated the existence of tax liability in praesenti by reason of domicile, residence, place of management,place of incorporation or other criteria of a similar nature on the date of making that claim under the law of the State of which the person is claiming to be a resident. Where,however, the tax liability of a person in the concerned State was to arise in the future, the person would become resident as and when the tax net of the State was so spread as to cover such person.

DTAA-India-UAE. **[195CTR534,276ITR306,146Taxman115]**

Likely to amount to Cal. Manik Chand Nahata - ITO

The expression likely to amount to a lakh of rupees or more in section 34 of the Indian Income-tax Act,1922, means that the Income-tax Officer must form some kind of belief or even a suspicion that the amount of escaped income for the year may amount to rupees one lakh or more in the aggregate before the notice under section 34 is issued. Where there is no material in the records of the case to establish that the amount of escaped income amounted to rupees one lakh or more, the Income-tax Officer would have no right to reopen the proceedings under the old Act beyond the expiry of eight years from the relevant assessment year.

Sec.147. **[78ITR204]**

Lineal descendant Raj. CIT - Dhannalal Devilal

A son or a grandson is a lineal descendant of his mother or grandmother respectively......The words lineal descendant have to be interpreted in their natural meaning without importing any notion as to whether a Hindu female can form a line of succession. A son will be a lineal descendant of the mother as well as of his grandmother irrespective of whether the mother or the grandmother can form a line of succession in Hindu law.Wives or widows of male members of an undivided Hindu family and unmarried daughters of male members are members of the family, though they may not be coparceners. *Misc.* **[29ITR165]**

Literary work Kar. CIT - Samsung Electronics Co. Ltd.

A literary work is entitled to copyright. In India, section 2(o) of the Copyright Act, 1957, defines literary work as including computer programmes, tables and compilations including computer database. Therefore, computer software has been recognised as copyright work in India also. The Legislature has treated literary work like books and other articles separately from computer software within the meaning of the copyright under the Copyright Act. *DTAA-USA.* **[245CTR481,345ITR494,203Taxman477]**

Loan and Deposit Mad. A.M. Shamsudeen -Union of India

There is a distinction between a loan and a deposit. In the case of a loan, it is the duty of the debtor to seek the creditor and repay the money to him or to repay the money. *Sec.269SS.* **[164CTR466, 244ITR266,103Taxman286]**

Local authority **All. U. P. Jal Nigam - CIT**

The Supreme Court in the case of Union of India- R.C.Jain, AIR 1981 SC 951 has laid down the following five ingredients, which are required to be fulfilled cumulatively before an authority can be said to be a local authority, in the light of the definition of local authority as given under section 3(31) of the General Clauses Act, 1897.(1) The authority must have separate legal existence as a corporate body. It must be a legally independent entity. (2) The body must function in a defined area and ordinarily, wholly or partly,directly or indirectly be elected by the inhabitants of the area. (3) The body must enjoy a certain degree of autonomy, with freedom to decide for itself questions of policy affecting the area administered by it. (4) The body must be entrusted by statute with such Governmental functions and duties as are usually entrusted to municipal bodies, such as those connected with providing amenities to the inhabitants of the locality, like health and education services, water and sewerage, town planning and development, roads, markets, transportation, social welfare services, etc. Broadly such body may be entrusted with the performance of civic duties and functions, which would otherwise be Governmental duties and functions. (5) The body must have the power to raise funds for the furtherance of their activities and the fulfilment of their projects by levying taxes, rates, charges, or fees. Essentially, control or management of the funds must vest in such body.

Sec.10(20). **[338ITR248]**

2.The expression local authority appearing in the third proviso to section 9(2) of the Indian Income-tax Act,1922, must be interpreted in the context in which it appears. The fact that it is not defined in the Income-tax Act can only mean that there can be no rigid restraint in interpreting the expression. The term property used in the provision includes property situate outside the taxable territory and as such the local authority referred to in the same provision must mean the local authority which can exercise the power of levying tax in respect of that property.

[CIT - R.Venugopala Reddiar,Mad., *Sec.23.* **58ITR439]**

3.The expression local authority has not been defined in the Income-tax Act and so the definition given in section 3(31) of the General Clauses Act will hold good for construing the expression. The definition given in section 3(31) of the General Clauses Act is as follows:Local authority shall mean a municipal committee, district board, body of port commissioners or other authority legally entitled to or entrusted by the Government with the control or management of a municipal or local fund. A municipal committee, district board or a body of port commissioners come squarely within the definition of a local authority. The position as regards other authorities is not so clear. Such other authorities must first be shown to be authorities. The term authority has not been defined by the General Clauses Act. The dictionary meaning of the word authority is a person or body having a legal right to command and be obeyed. **[Calcutta State Transport Corporation-CIT,Cal.,** *Sec.23* **108ITR922]**

4.The expression local authority has not been defined in the Income-tax Act. The meaning given to it in the General Clauses Act, therefore, applies. In order that an authority may be termed local authority , it is essential that the authority must have separate legal existence as a corporate body.They must not be mere governmental agencies but must be legally independent entities. Next,they must function in a defined area and must ordinarily, wholly or partly, directly or indirectly be elected by the inhabitants of the areas. Next, they must enjoy a certain degree of autonomy, with freedom to decide for themselves questions of policy affecting the area administered by them. The autonomy may not be complete and the degree of dependence may vary considerably, but there must be an appreciable measure of autonomy. Next, they must be entrusted by statute with such governmental functions and duties as are usually entrusted to municipal bodies. Broadly they may be entrusted with the performance of civic duties and functions which would otherwise be Governmental duties and functions. Finally, they must have power to raise funds for the furtherance of their activities by levying taxes, rates, charges or fees. This may be in addition to moneys provided by the Government or obtained by borrowing or otherwise.What is essential is that control or management of the fund must vest in the authority. **[Krishi Utpanna Bazar Samiti - ITO, Bom.,** *Sec.10(20).* **158ITR742]**

5. Section 10(20) of the Income-tax Act,1961,applies to a local authority.Even if a body is an authority within the meaning of article 12 of the Constitution, it would not be enough to attract the exemption in section 10(20) of the Act. It must be a local authority.

The expression local authority is not defined in the Income-tax Act. Its definition is,however, contained in the General Clauses Act, 1897, in clause (31) of section 3. An authority, in order to be a local authority,must be of like nature and character as a municipal committee, district board or body of port commissioners, possessing, therefore, many, if not all, of the distinctive attributes and characteristics of a municipal committee, district board or body of port commissioners, but, possessing one essential feature, viz., that it is legally entitled to or entrusted by the Government with, the control and management of a municipal or local fund. The authority must have separate legal existence as a corporate body, it must not be a mere governmental agency but must be a legally independent entity, it must function in a defined area and must ordinarily, wholly or partly, directly or indirectly, be elected by the inhabitants of the area. It must also enjoy a certain degree of autonomy either complete or partial, must be entrusted by statute with such governmental functions and duties as are usually entrusted to municipal bodies such as those connected with providing amenities to the inhabitants of the locality like health and education, water and sewerage, town planning and development, roads, markets, transportation, social welfare services, etc. Such a body must have the power to raise funds for furtherance of its activities and fulfilment of its objects by levying taxes, rates, charges or fees. It must have the power to affect persons and their rights even where they do not choose to deal with it, the power of compulsion. **[Calcutta State Transport Corporation -CIT,SC,***Sec.10(20).***132 CTR283,219ITR515,85Taxman402]**

6. Section 10(20) of the Income-tax Act,1961,was amended with effect from April 1,2003 , and an Explanation was added to it. The most striking feature of the Explanation is that it provides an exhaustive meaning of the expression local authority. The word means used in the Explanation leaves no scope for addition of any other entity as a local authority to those listed in the Explanation. In other words, even if an entity constitutes a local authority for purposes of the General Clauses Act, 1897, or for purposes of any other enactment for that matter, it would not be so construed for purposes of section 10(20) of the Act unless it answers the description of one of those entities enumerated in the Explanation. Reliance upon the provisions of the General Clauses Act, 1897, would, in fact, be permissible only in cases where the enactment in which the expression occurs does not itself provide for a definition or meaning to be given to the same. While interpreting a taxing statute, one has simply to look to what is clearly stated therein.

There is, in fiscal statutes, no room for any intendment nor is there any equity about the levy sanctioned under it.

[Agricultural Produce Market Committee - CIT, Del., <u>*Sec.10(20).*</u> 205CTR1,294ITR549, 156Taxman286]

7. The expression local authority has not been defined in the Income-tax Act.The definition of local authority fell for consideration of the Supreme Court in Union of India -R. C. Jain AIR 1981 SC 951.It held that a proper and careful scrutiny of section 3(31) of the General Clauses Act, 1897, suggests that an authority, in order to be a local authority, must be of like nature and character as a Municipal Committee, District Board or Body of Port Commissioners, possessing, therefore, many, if not all, of the distinctive attributes and characteristics of the Municipal Committee, District Board or Body of Port Commissioners, but, possessing one essential feature, namely, that it is legally entitled to or entrusted by the Government with, the control and management of a municipal or local fund. The authorities must have separate legal existence as corporate bodies. They must not be mere Governmental agencies but must be legally independent entities. They must function in a defined area and must ordinarily, wholly or partly, directly or indirectly, be elected by the inhabitants of the area. They must enjoy a certain degree of autonomy, with freedom to decide for themselves questions of policy affecting the area administered by them. The autonomy may not be complete and the degree of dependence may vary considerably but, an appreciable measure of autonomy there must be. They must be entrusted by statute with such Governmental functions and duties as are usually entrusted to municipal bodies, such as those connected with providing amenities to the inhabitants of the locality, like health and education services, water and sewerage, town planning and development, roads, markets, transportation, social welfare services etc.Broadly they may be entrusted with the performance of civic duties and functions which would otherwise be Governmental duties and functions. They must have the power to raise funds for the furtherance of their activities and the fulfilment of their projects by levying taxes, rates, charges, or fees.This may be in addition to moneys provided by the Government or obtained by borrowing or otherwise.What is essential is that control or management of the fund must vest in the authority. **[Sone Command Area Development Agency -UOI, Pat.,** <u>*Sec.10(20).*</u>**333ITR102]**

Lorry **Mad. CIT - Popular Borewell Service**

The rig and compressor mounted on a lorry and used by the assessees for drilling bore-wells, consist of three distinct items, viz., rig, compressor and lorry, and the rig and compressor do not form an integral part of the motor lorry. Further, the rig and compressor are not necessary for operating a motor lorry. Again, the rig and compressor are mounted on a lorry for the purpose of conveniently transporting them from place to place for sinking bore- wells. Merely because the rig and compressor are mounted on a lorry to facilitate their easy and convenient transport from one place to another,it cannot be said that the rig and compressor either constitute integral parts of a lorry by themselves or can be appropriately called or known as a lorry, as understood in common parlance. Therefore, the rig and compressor used for drilling bore-wells, though mounted on a lorry, cannot be held to fall under motor lorry occurring in entry No. III(ii)D(9) of Part I of Appendix I to the Income-tax Rules. *Sec.32.*[**94CTR240,194ITR12]**

2.The origin of the word lorry is uncertain. Lorry means (i) a large strong motor vehicle for transporting goods, etc., (ii) a long flat low wagon,or (iii) a truck used on railways or tramways, as per the Concise Oxford Dictionary. Lorry or truck would mean not only any motor vehicle designed to carry freight or goods but also to perform special services like fire fighting. Motor vehicles like fire trucks, fork lift trucks and crane trucks which are designed for special services fall within the category of motor trucks (also called motor lorries). **[GujcoCarriers-CIT,Guj.,***IT Rules.R 5,Appex I,***174 CTR324,256ITR50, 122 Taxman 206]**

Loss Cal. David Mitchell - CIT

Loss must mean loss arising out of the exigencies of some transaction by reason of the circumstances attending it and not a deprivation of property caused by a voluntary and spontaneous act of the person who is deprived of or rather deprives himself of the property. *Misc.***[AIR1956Cal.516,30ITR701]**

Loss brought forward Del. CIT - Sumi Motherson Innovative Engineering Ltd.

The term loss brought forward means the loss on the last date of the immediately preceding year,which is to be brought forward to the financial year in question. What happens during the course of the year is not relevant. *Sec.115JB.***[336ITR321,195Taxman353]**

Lottery **P&H. CIT - Sanjiv Kumar**

The element of chance is one of the important relevant factors for considering whether a particular scheme falls within the definition of the word lottery.A lottery and a wagering contract are two distinct things.A scheme may amount to a lottery though none of the competitors is a loser. A scheme would be a lottery even if the prize money came out of the interest earned from the subscribers' contribution. *Sec.80TT.* **[123ITR187]**

2. The term lottery has been defined in the Corpus Juris Secundum which says that pooling the proceeds derived from chances or tickets taken or purchased and then allotting such proceeds or a part of them or their equivalent by chance to one or more such takers or purchasers are indicia of a lottery. Hence, it is necessary that the winner must be not only a contributor to the prize amount, but must also be a participant in the lottery. All the ingredients which are set out in the definition in Corpus Juris Secundum must be present to identify the winner and the winnings of the lottery. **[Visveswaraiah Lucky Centre-CIT, Kar.,** *Sec.80TT.r.w.s.2 (24).* **93CTR261,189ITR698,56Taxman80]**

3. Lottery is a chance for a prize against a price and, therefore, the element of purchase of a lottery ticket must be present and, secondly,the purchaser of a lottery ticket has a right to participate in the draw. Undoubtedly, it is a sale of goods and lastly it is an income liable to tax. **Commercial Corporation of India Ltd. -ITO,Bom.,***Sec.194B.r.w.s.2(24).***201ITR348]**

4. Three things must concur to establish a thing as a lottery: a prize or prizes; the award or distribution of the prize or prizes by chance; and the payment either directly or indirectly by the participants of a consideration, for the right or privilege of participating.

[Director of State Lotteries - ACIT, Gau., *Sec.28,194B* **156 CTR 464,238ITR1,150Taxation506,108Taxman88]**

5. The word lottery embraces the elements of procuring through lot of chance, by the investment of a sum of money or something of value,some greater amount of money or value. The expressions winnings from lotteries and winnings from horse races have been used in different contexts at different places in the statute.

[G.N.Pant-CIT,Del.,*Sec.80TT.***161CTR326,248ITR718,116 Taxman30]**

6. Before a scheme can be regarded as a lottery, there must be the element of distribution of prizes which should be by chance or lot and such distribution should be among those who had paid a price for participating in the scheme. Mere gratuitous distribution without any price having been paid by the participants for acquiring the chance and receiving a prize that is ultimately distributed, would not amount to a lottery. That appears to be the reason why the Explanation was added under section 2(24)(ix) of the Income-tax Act, 1961, with effect from April 1,2002, to bring within the purview of section 194B, winnings from prizes awarded to any person by a draw of lots under any scheme or arrangement, whether or not the persons taking part in that arrangement, had paid a price for acquiring the chance of winning the prize. The Explanation lays emphasis on the winnings awarded by a draw of lots or chance. **[CIT - Deputy Director of Small Savings,Mad.** *Sec.194B.r.w.s.2(24).***266ITR27, 136 Taxman 261]**

7. The essential elements that go to constitute a lottery are : (1) a prize or some advantage in the nature of a prize; (2) distribution thereof by chance, and (3) consideration paid or promised for purchasing the chance. The chance of a person getting the prize cannot be treated as part of the bargain unless independent consideration is there with respect to the prize awarded. **[Canaan Kuries and Loans(P.) Ltd. - ITO,Ker.,***Sec.194B.***194 CTR413, 272ITR534, 141Taxman615]**

Lull in business **Bom.** **Hindustan Chemical Works Ltd. - CIT**

There is a marked distinction between lull in business and going out of business . A temporary discontinuance of business may, in certain circumstances, give rise to an inference that a business is going through a lean period of transition and it could be revived if proper circumstances arise. But where an assessee decides to dispose of its property, the plant and machinery are dismantled and taken away from the factory premises, there is not even the slightest chance of the assessee re-starting production, and there is no finance available, and even the licences ceased to be effective, it cannot be held that there was a lull in business which was of a temporary nature. *Sec.32,22.***[124ITR561]**

Lump sum **Tata Yodogawa Ltd. - CIT**

The words lump sum as used before the word consideration in section 35AB only exclude periodical or turnover based payments like royalty etc., and any one-time payment for the know-how would fall within the expression lump sum if it is fixed and specified in the agreement, although it may be payable in installments. <u>*Sec.35AB.*</u>**[335ITR53,196Taxman353]**

M

Machinery Mys. Mangalore Ganesh Beedi Works - CIT

As the word machinery is not defined in the Income-tax Act, the word must be understood in its ordinary sense, and the word machinery when used in ordinary language prima facie means some mechanical contrivances which by themselves or in combination with one or more other mechanical contrivances by the combined movement and interdependent operation of their respective parts generate power or evoke, modify, apply or direct natural forces with the object in each case, of effecting so definite and specific a result.

*Sec.32.***[52ITR615]**

2. Definition of machinery adopted by the Privy Council in Corporation of Calcutta v. Chairman, Cossipore and Chitpore Municipality [1932] I.L.R.49 Cal. 190, namely, machinery means some mechanical contrivances which by themselves or in combination with one or more contrivances by the joint movement and interdependent operation of their respective part generate power or evoke,modify, apply or direct natural forces with the object in each case of effecting so definite and specific a result,although not given in a tax case, should prevail under the Income-tax Act also,as the word machinery was an ordinary and not a technical word. **[CIT - Mir Mohammad Ali, SC,***Sec.32.***53 ITR165]**

3. In order to be termed machinery, it is essential that there must be some mechanical contrivances which,by themselves or in combination with one or more other mechanical contrivances,by the combined movement and interdependent operation of their respective parts, generate power or evoke, modify, apply or direct natural forces. **[India Leaf Spring Mfg.Co. P. Ltd. - CIT,AP.,** *Sec.32(1) Sch. IX,***175ITR639,40Taxman123]**

4. Machinery is that which produces goods, articles or things, or that which assists the manufacturing process. **[CIT-Technico Enterprise Pvt. Ltd.,Cal.,***Sec.32(1)(ii).***119CTR 25,206ITR36, 73 Taxman204]**

5. The word machinery when used in ordinary language prima facie means some mechanical contrivances which, by themselves or in combination with one or more other mechanical contrivances, by the combined movement and interdependent operation of their respective parts generate power, or evoke, modify, apply or direct natural forces with the object in each case of effecting definite and specific result. The word machinery must mean more than a collection of ordinary tools. **[CIT-Indian Textile Paper Tube Co. Ltd.,Mad.,**<u>*Sec.32(1)(vi); Sch.V*</u>**138CTR342,234ITR47]**

Machinery and plant used in any business **Bom.** **CIT- Indian Card Clothing Co. Pvt. Ltd.**

The words machinery or plant used in any business occurring in section 15C means machinery or plant used in a business in India and not machinery or plant used in a business anywhere else in the world.The words in India do not occur in the relevant provision but the provision refers to a building the transfer of which is contemplated and such transfer contemplated is of a thing existing in India. Since the word building has been used together with the words machinery and plant,the machinery and plant the transfer of which is contemplated, must be of machinery or plant which has been used in India.Secondly, the object of the provision appears only to be that the benefit of section 15C should not be made available twice. <u>*Sec.80J.*</u>**[6CTR202,110ITR103]**

Made **Del.** **Tea Consultancy and Plantation Services (India) Pvt. Ltd. - Union of India**

The expression made should be given its normal meaning and there is no scope for giving it a restricted meaning or substituting it by the expression received. If the intent of the Legislature was that the application would be entertained by the Board as valid and in time only if it was received before the 1st day of October of the relevant assessment year then it was expected to use the word received or at least submitted in place of the word made. This only indicates that an act which is within the control of the assessee should be completed and he should not be able to exercise any control or tamper with the required documents after the cut-off date. Once the application is sent by the assessee by registered post through the post office he obviously loses control over the applications and cannot in any way interject or interfere with the delivery of the documents to the addressee.This aspect of the case can also be examined from another point of view. The applications to the Board are to be submitted by different persons/assessees from all over the country. It would be impracticable to expect every assessee to travel from various parts of the country and deliver the documents by hand to the office of the Board. The practice of the Board in the past has been to receive such applications by post. Practice which has prevailed over a long period and has been accepted by all concerned without violation of any specific provision would be a good practice. A good practice in law can be a ground for providing protection to the person who has acted as per the practice. *Sec.80MM.* **[194CTR481,278ITR356,144Taxman604]**

Made for the first time All. Bela Singh Daulat Singh - CIT

Made for the first time means made before the grant of registration; every application made before the grant of registration is an application made for the first time under the Act and every application made subsequently is for renewal. *Sec.184.***[62ITR250]**

Mainly Mad. CIT - Kamala Ranganathan

The word mainly in section 54 of the Income-tax Act,1961, means in the main, principally or chiefly. *Sec.54.***[82CTR251,186ITR536]**

2. A perusal of section 54 of the Income-tax Act, 1961, shows that the word mainly is not to be confused with wholly and has to be understood as principally. A careful consideration of the language of section 54 of the Act shows that it cannot be restricted only to new construction but must be extended to constructions which are in the nature of remodelling. **[CIT-A.R.Mathavan Pillai,Ker.,** *Sec.54***132CTR374,219ITR696,88Taxman292]**

Maintains a dwelling place Mad. S.M.Zackariah Sahib - CIT

The expression maintains a dwelling place in section 4A(a)(ii) of the Indian Income-tax Act, 1922, connotes the idea that the assessee owns or has taken on rent or on a mortgage with possession a dwelling house which he can legally and as of right occupy if he is so minded. The expression has maintained for him a dwelling place would cover a case where the assessee has a right to occupy or live in a dwelling place though the expenses of maintaining the dwelling place are not met by him in whole or in part. *Sec.6.***[22ITR 359]**

2. The expression maintains a dwelling place in section 4A(a)(ii) of the Income-tax Act, 1922, connotes a system, or course of conduct on the part of the assessee, by which he either keeps up a dwelling place at his expense,or has a dwelling place kept up for him by others. Neither the de jure title, nor the de facto possession of the dwelling would by itself constitute a setting apart and keeping ready of a dwelling place to be used as a residence or home. To maintain a dwelling place is different from having a dwelling place. The concept of a residence or home ready in hand is implicit in the section. What is referred to as dwelling is not a physical structure, or a brick and mortar construction but the organisation of a household. The word is in the second part of section 4A(a)(ii) does not necessarily refer to the place of dwelling mentioned in the first part, nor does it merely contemplate a bare physical presence of the individual. The real meaning of the section is that the individual must have maintained or had maintained for him a residence in the taxable territories for 182 days or more and must have resided in the territories at any time in the year of account, however short the duration of the residence might have been. A common thread of animus of residence in the taxable territories holds both the parts of the section together. **[J.M.Abdul Aziz-CIT,Mad.,** *Sec.6.* **48ITR620]**

Maintenance **Mad. CIT - South India Viscose Ltd.**

Maintenance is not defined in section 37(4)(i) of the Income-tax Act, 1961, which provides for disallowance of expenditure incurred by the assessee after February 28, 1970, on the maintenance of any residential accommodation in the nature of a guest house.The term maintenance normally refers to keeping a thing in the condition that it was and keeping it in a condition that enables the thing to be used in the way in which it was intended to be used. The rent paid for a guest house, in the absence of any other indication in section 37(4)(i), cannot be regarded as part of maintenance, especially when rent is specifically dealt with in another section, section 30, which expressly provides for deduction of such rental. *Sec.30,37(4).*

[259ITR107,134Taxman727]

Maintenance/Gift **SC CGT - B.S. Apparao**

.........by reason of section 20 of the Hindu Adoptions and Maintenance Act, 1956, a Hindu was bound during his lifetime to maintain his daughter. His obligation extended to providing maintenance which was defined under section 3(b) of that Act to include provision for food, clothing, residence, education and medical attention and treatment, and in the case of an unmarried daughter, the duty also extended to reasonable expenses incidental to her marriage. The transfers by the assessee to meet his obligation under the law were not gifts within the meaning of the Gift-tax Act,1958. *Sec.3(b).G.T.Act.*[**165CTR11,248ITR103,116 Taxman202]**

Making an assessment Bom. Maneklal Chunilal & Sons Ltd. - CIT

The expression making an assessment in section 26(2) of the Indian Income-tax Act, 1922, before its amendment by the Income-tax (Amendment) Act, 1939, does not merely mean passing an order of assessment but it means initiating proceedings necessary for making an assessment and it covers the whole period during which assessment is being made. *Sec.143.*[**24ITR375]**

Making up an accounts Guj. Bhailal Tribhovandas and Co. - CIT

Making up of accounts means ascertaining the profits or loss of a particular period. *Sec.3.*[**68ITR136]**

Making up of an account Mad. CIT - John Peter

In the absence of any definition of the words making up of an account in the Income-tax Act, 1961, it is open to find out the meaning of the words used with reference to the dictionary, keeping in view the purposes of the Income-tax Act. *Sec.3(1)(a).* **[162CTR283,243ITR561,118Taxman]**

Mala fide Del. Bansal Exports Private Ltd.-Union of India

The action of the Income-tax Department at times may be erroneous but all erroneous actions cannot be said to be mala fide. While passing orders, an error can be committed by the officers of the Department but such an error cannot always be said to be in bad faith. *Sec.293.r.w.s.132(3).* **[130CTR550,217ITR83,133Taxation9,84Taxman373]**

Manager All. CIT- Official Liquidator of the Agra Spg. and Weaving Mills Co. Ltd.

The word manager in section 2(12) of the Act is wide enough to include liquidators of a company. <u>*Sec.2.*</u>**[2ITR79]**

Manufacture Bom. CIT -Tata Locomotive and Engineering Co. Ltd.

The word manufacture has a wider and also a narrower connotation. In the wider sense it simply means to make,or fabricate or bring into existence an article or a product either by physical labour or by power, and the word manufacturer in ordinary parlance would mean a person who makes, fabricates or brings into existence a product or an article by physical labour or power. The other shade of meaning, which is the narrower meaning,implies transforming raw materials into a commercial commodity or a finished product which has an entity by itself,but this does not necessarily mean that the materials with which the commodity is so manufactured must lose their identity.Thus, both the words manufacture and produce apply to the bringing into existence of something which is different from its components. <u>*Sec.80J.*</u> **[68ITR325]**

2.The broad distinction between manufacture and production is that manufacture involves bringing into existence of a new product, a product which is of a different chemical composition or whose integral structure is different from the raw materials.Production, as distinguished from manufacture, is nothing except bringing into existence a product after processing the raw materials in a manner which may not change the inherent quality or chemical composition of the raw material. **[Singh Engineering Works Pvt.Ltd.-CIT, All.,**<u>*Sec.80-I.*</u>**7CTR201,119ITR891]**

3. In defining an industrial company the Legislature has advisedly used different expressions, viz., manufacture, processing or mining. Therefore, the Legislature is not treating the manufacture of goods as the same as processing of goods.Manufacture implies a transformation or an alteration of goods.**[CIT-Radha Nagar Cold Storage (P.) Ltd.,Cal.,**<u>*Misc.*</u>**18CTR166,126ITR66]**

4. The prevalent and generally accepted test to ascertain that there is manufacture is whether the change or the series of changes brought about by the processes take the commodity to the point where commercially it can no longer be regarded as the original commodity but is, instead, recognised as a distinct and new article that has emerged as a result of the processes.The principles are clear. But difficulties arise in their application in individual cases. There might be border line cases where either conclusion with equal justification may be reached. Insistence on any sharp or intrinsic distinction between processing and manufacture results in an oversimplification of both and tends to blur their interdependence.

[Ujagar Prints-Union of India, SC, *Constitution of India.* **179ITR 317]**

5.The word manufacture includes any process incidental or ancillary to the completion of a manufactured product. Hence, any process by which an object becomes new commercial goods, having a distinctive name, character or use would be manufacture. Manufacture means bringing into existence new goods. Manufacture implies a change and more than a mere change; there must be transformation. **[Narne Tulaman Manufacturers Pvt. Ltd.-Collector of Central Excise., SC,***Central Excises and Salt Act.* **183ITR577]**

6. Manufacture, in its ordinary connotation, meant production of an article which was commercially different from the basic component by which the item was manufactured, that the activity of the assessee amounted to manufacturing or processing of goods and that the assessee was an industrial undertaking entitled to the benefit of deduction under section 80 .....the production of a cinematograph film amounted to manufacture of an article or goods within the meaning of section 104(4)(a) (as it then stood) and the activity of the assessee was an 'industrial undertaking within the meaning of section 80J. Further, even from a common sense point of view, film production had to be considered as a manufacturing activity and the undertaking had to be considered as an industrial undertaking. **[CIT-D.K.Kondke,Bom.,***Sec.104.r.w.s.80J.* **192ITR128, 57 Taxman13]**

7. The Legislature has used the word manufacture and also the word processing to indicate two different types of activities. There may be some overlapping in certain cases. It may be that certain activities may fall within the meaning of the word processing as well as manufacture. But, generally, the two expressions must not be used as synonymous so as to make the word processing otiose and meaning less. **[CIT-Sky Room Pvt. Ltd., Cal.,** *Misc.* **195ITR763,62 Taxman 236]**

8. The prevalent and generally accepted test to ascertain whether there is manufacture is to find whether the change or the series of changes brought about by the application of processes take the commodity to the point where, commercially, it can no longer be regarded as the original commodity but is, instead, recognised as a distinct and new article that has emerged as a result of the processes. The principles are clear. But difficulties arise in their application in individual cases. There might be borderline cases where either conclusion with equal justification may be reached. **[CIT- Kanam Latex Industries (P.) Ltd., Ker.,** *Sec.80HH,80J.* **203ITR542]**

9. The words manufacture and production apply to a case which brings into existence something different from its components.The Legislature has used different expressions, manufacture, processing or mining and, therefore, it is apparent that the Legislature is not treating manufacture of goods as the same as processing of goods. Manufacture implies a transformation or alteration of goods. Manufacture means production of an article for use from raw, semi-raw or prepared materials by giving these materials new form, quantities or properties or combination, whether by hand, labour or machinery. **[Appeejay Pvt.Ltd.–CIT,Cal.,** *Sec.80J(4)(iii).***206ITR367]**

10. Manufacture involves the bringing into existence of a new product which may have a different physical or chemical composition and is understood differently in common and commercial parlance. **[CIT-Best Chem& Limestone Industries Pvt. Ltd., Raj.,** *Sec.32A.***210ITR883]**

11. The word used in the section is manufacture. Commonly, manufacture is the end result of one or more processes through which the original commodity is made to pass. The nature and extent of processing may vary from one case to another, and indeed there may be several stages of processing and perhaps a kind of processing at each stage.With each process suffered, the original commodity experiences a change. But it is only when the change, or a series of changes, take the commodity to the point where commercially it can no longer be regarded as the original commodity but instead is recognised as a new and distinct article that a manufacture can be said to take place. **[CIT - Oceanic Products Exporting Co., Ker.,** *Misc.***131CTR305,219ITR293, 85Taxman554]**

12. If a new substance is brought into existence or if a new or different article having a distinctive name, character or use results from particular process or processes, such process or processes would amount to manufacture.**[Chillies Export House Ltd.-CIT, Mad.,** *Misc.***140CTR9,220ITR411]**

13. The prevalent and generally accepted test to ascertain whether there is manufacture is to find out whether a change or the series of changes brought about by the application of processes to the commodity,are such where commercially, it can no longer be regarded as the original commodity, but is, instead recognised as a distinct and new article that has emerged as a result of the processes. **[CIT-Kanam Latex Industries P. Ltd.,Ker.,** *Misc.***132CTR 178, 221ITR1, 86Taxman466]**

14. The expression manufacture involves the concept of changes effected to a basic raw material resulting in the emergence of, or transformation into, a new commercial commodity. Whether an article is converted into a different article depends on several criteria and one of the essential tests is whether in a commercial sense, the original article has ceased to exist and a new article has taken its place. It is, however, not necessary that the original article or material should have lost its identity completely: all that is important is whether, what has emerged as a result of the operations is a different commercial commodity,having its own name, identity, character or end use.This determination is essentially one of fact and has to be arrived at on a consideration of all relevant factors such as the quality and nature of the original article, the extent and magnitude of the operations carried out on, or in relation to, it and the commercial identity, character and use of the article produced. **[Arthur E. Newell - CIT,AAR,** *Sec.80J,80HH.***137CTR47,223ITR776,90Taxman26]**

15. Manufacture implies a change but every change is not manufacture, although every change in the article is the result of treatment, labour and manipulation. To bring about change qualifying as manufacture something more is necessary and that something is transformation, i.e., a new and different article, having a distinct name, character or use, must emerge. Where the commodity retains a continuing substantial identity through the processing stage it cannot be said that it has been manufactured.

[CIT- Lucky Mineral Pvt. Ltd.,Raj.,*Sec.80HH.***134CTR541,226 ITR245,87Taxman215]**

16. Manufacture is a process which results in an alteration or change in the goods which are subjected to such manufacture. A commercially new different article is produced.

[CWT-P. Devasahayam, Mad., *Sec.5(1)(xxxi).WT.Act.* **144CTR313, 236ITR885]**

17. Commonly, manufacture is the end result of one or more processes through which the original commodity is made to pass.

The nature and extent of processing may vary from one case to another, and indeed there may be several stages of processing and perhaps a different kind of processing at each stage. But it is only when the change or a series of changes takes the commodity to the point where commercially it can no longer be regarded as the original commodity but instead is recognised as a new and distinct article that a manufacture can be said to take place. Manufacturing thus involves the consumption of one or more articles in order to produce a different article. Consumption is necessary in the process of manufacture and there can be no manufacture without consumption. **[CIT-Sacs Eagles Chicory,Mad.,** *Sec.80 HH,80-I 80J.***164CTR45,241ITR319,107Taxman463]**

18. The word manufacture has various shades of meaning but unless defined under the Act, it is to be interpreted in the context of the object and the language used in the sections. In the context of the provisions which deal with grant of investment allowance or deduction under section 80J it is apparent that it is used to mean production of a new article or bringing into existence some new commodity by an industrial undertaking. It would not be applicable in cases where only processing activity is carried out. Further, such production activity must be by an industrial undertaking and not by the assessee having mainly trading activity.

[Hotel and Allied Traders Pvt. Ltd.- CIT,SC,*Sec.-80J.***162CTR 310,245ITR538,112Taxman46]**

19. The word manufacture is to be understood in a wide sense. Manufacture would imply a change and a transformation. A new and a different article must emerge having a distinct and different character and use. **[CIT - Darshak Ltd., Kar.,** *Sec.80- I***165CTR17,247ITR489,118Taxman863]**

20. That extracting granite from quarry and cutting it to various sizes and polishing was manufacture or production of any article or thing and the assessee's business activity was an industrial undertaking for the purpose of granting reliefs under sections 32A and 80-I. **[Indian Poultry - CIT, SC,** *Sec. 32A,80- I.***166CTR503,250ITR664,116Taxman193]**

21. The word manufacture has not been defined in the Income-tax Act. In the absence of a definition, the word manufacture has to be given a meaning as is understood in common parlance. It is to be understood as meaning the production of articles for use from raw or prepared materials by giving such materials new forms,qualities or combinations whether by hand labour or machines. If the change made in the article results in a new and different article then it would amount to manufacturing activity. **[Aspinwall and Co. Ltd.-CIT,SC,**<u>*Sec.32A.*</u>**170CTR68,251ITR323,118Taxman771]**

22. The expression manufacture has in ordinary acceptation a wide connotation.It means the making of articles, or material comercially different from the basic components,by physical labour or mechanical process. However, the word manufacture appears in the company of the word production which has a wider connotation than the word manufacture. The word production or produce when used in juxtaposition with the word manufacture takes in, bringing into existence new goods by a process which may or may not amount to manufacture.The associated words are indicative of the mind of the Legislature. Where a word is doubtful or ambiguous in nature the meaning has to be ascertained by considering the company in which it is found and the meaning of the word associated with it. **[Ship Scrap Traders - CIT, Bom.,** <u>*Sec.80HHA,80I.*</u> **251ITR806,122Taxman29]**

23. In the definition of the word manufacture in a taxing statute, it should be understood in its commercial sense.Manufacture is complete, as soon as by the application of one or more processes, the raw material undergoes some change and a new substance or article is brought into existence. **[CIT - Vinay Kumar Sigtia,Ori.,**<u>*Sec.32A.*</u>**262ITR686,134Taxman528]**

24. The word manufacture can be used both as a verb and as a noun.When it is used as a verb it refers to the process or a centering action having a starting point and an end point. The expression manufacture when used as a noun, has a fixed point, namely, when the process culminates into the end-point, i.e., the process ends up in the final result. The expression to produce article is used in the very same section. This clearly shows that the Legislature has in mind the production of articles.There is no doubt, that the use of the word or between two expressions, viz., begins to manufacture and begins to produce articles shows that they are interchangeable and denote the same meaning. The Legislature has not used the expression to include process of manufacturing, as is evident on comparing the expression used in section 80P(2)(f). The intention is clear, namely, the expression begins to manufacture is used as a noun and that is when the undertaking achieves the end product out of the process of manufacture. **[Kanodia and Sons - CIT,All.,**<u>*Sec.80J.*</u>**201CTR442,281ITR255]**

25. The word manufacturehas not been defined in the Income-tax Act. In the absence of a definition, the word manufacturehas to be given a meaning as is understood in common parlance. It is to be understood as meaning the production of articles for use from raw or prepared materials by giving such materials new forms, qualities or combinations whether by hand labour or machines. If the change made in the article results in a new and different article then it would amount to manufacturing activity. **[CIT-Premier Tobacco Packers P. Ltd., Mad.,** *Sec80HH.***203CTR201,284ITR222]**

26. The expression manufacture involves the concept of changes effected to a basic raw material resulting in the emergence of, or transformation into, a new commercial commodity. But it is not necessary that the original article or material should have lost its identity completely. All that is required is to find out whether as a result of the operation in question, a totally different commodity had been produced having its own name, identity, character or end use. **[CIT - Jamal Photo Industries (I) P. Ltd., Mad.,** *Sec.80I.* **285ITR209]**

27. The word production or produce, when used in juxtaposition with the word manufacture, takes in bringing into existence new goods by a process which may or may not amount to manufacture. It also takes in all by- products, intermediate products and residual products which emerge in the course of manufacture of goods. **[CIT- Tara Agencies, SC,** *Sec80HH.***210CTR454,292ITR444,161 Taxman337]**

28. Whether a particular activity amounts to manufacture or not would always be a question dependent upon the facts and evidence on record and would be based on the facts of each case. The only issue which would call for an inquiry is as to whether correct tests have been applied after finding the facts on record. Broadly, the requirement is that the raw material must be, in the first instance, subjected to a process of such a nature that it cannot be termed to be the same as the end-product after the raw material undergoes the process of manufacture. In other words, the goods purchased as raw material should go in as inputs in the process of manufacture and the result must be manufacture of other goods. The article produced must be regarded by the trade as a new and distinct article having an identity of its own,an independent market after the commodity is subjected to the process of manufacture.The nature and extent of the process would vary from case to case, and in a given case, there may be only one stage of processing, while in another case,there may be several stages of processing,and perhaps, a different kind of process at every stage. **[CIT - Alfa Lamination, Guj,***Sec.80IB.***225CTR212,329ITR348]**

29. In each case the question whether there is any process which amounts to manufacturehas to be decided by examining the process undertaken by the assessee...If one examines this thermo mechanical process in detail, it would be clear that texturising and twisting of yarn constitutes manufacturein the context of conversion of POY into texturised yarn.The thermo mechanical process also brings about a structural change in the yarn itself, which is one of the important tests to be seen while judging whether a process is manufacture or not.Therefore, twisting and texturising of partially oriented yarn amounts to manufacturein terms of section 80-IA of the Income-tax Act,1961. **[Emptee Poly - Yarn P. Ltd.,SC,*Sec.80IA.*229CTR1,320ITR665]**

30. The court had to move along with the times when there was technological advancement as in computer science.In each case when an issue of this nature arose for determination the Department had to study the actual process undertaken by the assessee.If an operation/process rendered a commodity fit for use for which it would otherwise not be fit, the operation/process fell within the meaning of the word manufacture. Applying this test, the assessee had undertaken an operation which rendered the blank CD fit for use for a purpose for which it would otherwise not be fit. The blank CD was an input.By the duplicating process undertaken by the assessee, the recordable media which is unfit for any specific use got converted into the programme which is embedded in the master media and thus the blank CD got converted into recorded CD by such an intricate process. The duplicating process changed the basic character of a blank CD dedicating it to a specific use.Without such processing, blank CDs would be unfit for their intended purpose.Therefore, processing of blank CDs dedicating them to a specific use constituted manufacturein terms of section 80- IA(12)(b) read with section 33B of the Act.The marketed copies of the CD were goods and the process in which they became goods was covered by manufacture or processing of goods. **[Oracle Software India Ltd.,SC, *Sec. 80IA.*228CTR,443,320ITR546,187Taxman275]**

31. The Supreme Court in India Cine Agencies v. CIT 308 ITR 98 has laid down the test to determine what activity would amount to manufacture. The Supreme Court stated that the test to determine whether a particular activity amounts to manufactureor not is : do new and different goods emerge having distinctive name, use and character.The moment there is transformation into a new commodity commercially known as a distinct and separate commodity having its own character, use and name, whether it be the result of one process or several processes manufacture takes place. In CIT v. Sesa Goa Ltd. 271 ITR 331 the Supreme Court observed that the word production or produce when used in just a position with the word manufacture takes in bringing into existence new goods by a process, which may or may not amount to manufacture. It also takes in all the by-products, intermediate products and residual products,which emerge in the course of manufacture of goods. **[Chiranjjeevi Wind Energy Ltd., Mad.,**_Sec.80IB._**333ITR192]**

32. In the EXIM policy manufacture was given a very wide definition to take in even processing involving conversion of something to another with distinct name, character and use. Since the purpose of exemption under section 10A was to give effect to the EXIM-Policy of the Government, the definition of manufacture contained in the EXIM Policy was applicable for the purpose of the provision.Manufacture as defined under the EXIM Policy had a wide and liberal meaning covering tea blending as well and so much so, blending and packing of tea qualifies for exemption under section 10A. Besides this, the assessee-industry in the special economic zone engaged in the same process of blending and packing of tea was specifically brought under the exemption clause through incorporation of section 2(r) of the Act, in the provisions of section 10AA of the Act. Therefore, the later amendment was only clarificatory and the definition of manufacture contained in section 2(r) of the Act, incorporated in section 10AA of the Act with effect from February 10, 2006, which was essentially the same as the definition contained in the EXIM Policy, applies to section 10A also.

[GirnarIndustries-CIT,Ker.,_Sec.10A.r.w.s.10AA._**230CTR401,338 ITR277,187Taxman136]**

Manufacture and produce **Mad.** **CIT - S.S.M. Finishing Centre**

Under the Income-taxAct,1961,there is no definition of the expressions manufacture and produce. It, therefore, follows that those expressions have to be understood as words of ordinary import so as to mean to bring into being or existence a product falling under item No. 32 of the Fifth Schedule to the Income-tax Act,1961.

Sec.33(1)(b).**[85CTR189,186ITR597]**

2. The expressions manufacture and produce have not been defined in the Income-tax Act. The dictionary meaning of manufacture is transform or fashion new materials into a changed form for use. In common parlance, manufacture means production of articles from raw or prepared materials by giving these materials new forms, qualities, properties or combinations, whether by hand labour or by machinery. In other words, it means making of articles or materials commercially different from the basic components by physical labour or mechanical process. In its ordinary connotation, manufacture signifies emergence of new and different goods as understood in relevant commercial circles.So far as the meaning of the word produce is concerned,though the word produce has a wider connotation than the word manufacture, when used in juxtaposition with the word manufacture, it takes in bringing into existence new goods by a process which may not amount to manufacture.

[CIT-Abdul Ahad Najar,J&K.,<u>*Sec.80J.*</u>
16CTR273,248ITR744, 114Taxman655]

3. A plain reading of the Explanation to section 5(1)(xxxii) of the Wealth-tax Act, 1957, shows that in order that the assessee's share in the value of assets forming part of an industrial undertaking belonging to a firm can be exempt from inclusion in his taxable wealth, the pre-requirement is that the firm must be engaged in the business of (i) generation of electricity or any other form of power,or distribution of electricity or any other form of power, or (2) construction of ships, or (3) manufacture of goods,(4) or processing of goods or (5) in mining. Though the definition is of an industrial undertaking, no definition has been given of the words manufacture or processing. According to the ordinary dictionary meaning, the term manufacture means a process which results in an alteration or change in the goods which are subjected to the process of manufacturing leading to the production of a commercially new article. The word process means anything done requiring continuous and regular action or succession of actions leading to the accomplishment of some result but one of the requirements is that the activity should involve some operation on some material for conversion into some other stuff. What is necessary in order to characterise an operation as processing is that the commodity must, as a result of the operation,experience some change. The words engaged in the manufacture in the said Explanation postulate the assessee's direct involvement in the manufacture. However, it may not be necessary that the assessee should be personally engaged in the manufacture. It is sufficient if he employs his own labourers.In cases where the assessee gets the goods manufactured by an outside agency, he cannot be said to manufacture the goods, merely because the assessee pays for the manufacture.

[CWT-Mohinibai Kanaiyalal, Guj., Sec. 5 (1)(xxxii).WT.Act.**157 CTR298,240 ITR636,105Taxman69]**

Manufacture or processing of goods **Ker.** **Hotel and Allied Trades Pvt. Ltd. - CIT**

The assessee, running a hotel business, is not eligible to be assessed to tax at the concessional rate of tax as an industrial company as the activity carried on in preparing articles of food from raw materials in a hotel would not constitute manufacture or processing of goods. Sec.32(1)(v).**[149CTR224,238ITR226,102Taxman470]**

Manufacture or produce **Kar.** **V.M.Salgaocar Bros. (P.) Ltd. - CIT**

The principles to be borne in mind when considering a claim for special deduction under section 80J of the Income-tax Act,1961, are: (a) The terms manufacture or produce appearing in section 80J of the Income-tax Act,1961, not having been defined under the Act, shall have to be given their meaning as understood in the ordinary parlance; (b) The term produce is wider in its amplitude than the term manufacture, for while the latter can be used only in regard to inanimate objects, and articles, the former can be used even in regard to animate things and objects; (c) All manufacturing or production activities involve some process or the other but all processes need not necessarily amount to production or manufacturing of an article. It is only when a change or series of changes take the commodity subjected to such processes to a point where it can no longer be regarded as the original commodity but is instead recognised as a new and a distinct article, that such a process may be said to have resulted in a manufacture or production of a new article. In other words, where there is no difference in the identity of the original commodity and the processed article, it is not possible to say, that a commodity has been manufactured; (d) Whether an article has been manufactured or produced will depend upon whether the same is commercially different from the commodity out of which the same has been produced. This would vary from case to case.....The test is not whether what is produced as a result of the process carried out in the plant becomes more saleable from an otherwise less saleable article.Simply because a process carried out on a particular article adds to its value or improves its marketability on account of processes like shining, polishing, removal of impurities, etc., this may not by itself be sufficient to hold that the products so finished are commercially different from the ones on which such a process has been carried out. Value addition therefore, does not by itself constitute production or manufacture and cannot be made the sole test for determining whether an article has been produced or manufactured.

Sec.80J.[**130CTR1,217ITR849,85Taxman106,130Taxation 675]**

Manufacture or production Mad. CIT - Veena Textiles Pvt. Ltd.

Mere purchase of textiles already manufactured by another and dyeing or printing or otherwise processing it resulting in the retention of its identity as a cloth material would not be comprehended within the expression manufacture or production of textiles.

Sec.33.[**155ITR794]**

2. The original block did not remain marble block ;it became a slab or tile. Blocks were converted into polished slabs and tiles resulting in the emergence of a new and distinct commodity.Such an activity was something beyond manufacture and brought a new product into existence. The stepwise activity constituted manufacture or production in terms of section 80-IA.

[ArihantTilesandMarbles Pvt. Ltd.,SC, *Sec.80IA(2)(iii).r.w.s.2(29BA).***320ITR79,186Taxman 439]**

3. In order to determine whether any particular business activity amounts to manufacture or production for the purpose of tax incentives under the Act, each case is required to be examined in the light of facts and circumstances of that very case. The most important aspect of this exercise should be the analysis of the process involved in the activity and an enquiry into the nature of transformation that the product has undergone to find out whether it is distinct in identity from the raw commodity involved in its manufacture.

[CIT-HLS IndiaLtd.,Del.,*Sec.80IA/IB.***242CTR1,335ITR292]**

4. The activities undertaken by the assessee, namely, computer data processing services and sale of computer stationery amounted to manufacture or production of any article or thing and the assessee was an industrial undertaking within the meaning of section 80-I of the Income-tax Act, 1961. **CIT-Business Information Processing Services, Raj.** *Sec.80I.* **[345ITR548]**

Manufacture or production of articles. Pat. CIT-Natraj Processing Industries

A new industrial undertaking can claim special deduction under section 80J of the Income-tax Act, 1961, if it is engaged in the manufacture or production of articles. The broad test for determining whether a process is a manufacturing process is whether it brings about a complete transformation of the old components so as to produce a commercially different article or commodity.

<u>*Sec.80J.*</u>**[203ITR 833]**

Manufacturing Bom. CIT-Pressure Piling Co.(India)P.Ltd.

The essence of a manufacturing process is the conversion of raw material into entirely a new commodity or a new thing. The place or the site where the manufacturing process or the process of production of something is carried out is not relevant for determining whether the product produced or manufactured is an article. The mere fact that an article is manufactured or brought into being at the site itself would not be material for determining whether the thing produced or manufactured is an article. It is not necessary that an article should be manufactured in a factory alone. It is also not necessary that all articles must necessarily have the quality or the possibility of being sold and purchased across the counter or that they must necessarily be transportable in order to be classified as an article. <u>*Sec.84.*</u>**[126ITR333,1Taxman406]**

2. Whether a particular activity is manufacturing activity or not is dependent upon several factors, and no strait- jacket formula or general principle can be applied. Though manufacture implies a change, every change is not a manufacture. There must be a transformation of some kind and a new and different article must emerge having distinctive features. **[CIT-S.L.Agarwala and Co.,Ori.,<u>*Sec.32A*</u>197ITR239]**

3.The essence of manufacturing is that what is made shall be a different thing from that out of which it is made despite the original material not losing its identity completely. **[Ramit Kumar Sharma, AAR, <u>*Sec.80IC(2).*</u>221CTR621,309ITR344,177Taxman 240]**

Manufacturing process All. CIT - Sultan and Sons Rice Mill

The expression manufacturing process should be interpreted in its ordinary sense and should not be confined or restricted to the actual manufacturing alone. The processes which are intimately connected with actual manufacturing process will also be within the expression. The words employed ten or more workers in the manufacturing process normally would cover the entire process carried on by the industrial undertaking of converting the raw material into finished goods. The work of ten or more persons employed in the manufacturing process should be integrally connected with the manufacturing process. The work should be reasonably connected with and be part of the manufacturing. The various processes starting from purchase of the raw material till the sale of finished goods form an integral part of the manufacturing process and the workers and labourers employed in these processes are workers employed in the manufacturing process. *Sec.80HH, 80J.*[**193CTR444,272ITR181,145Taxman506]**

Market value Mad. Asher Textiles Ltd. - CIT

The market value means market value at the commencement of the year when the opening stock has to be valued and at the close of the year when the closing stock has to be valued and not any intermediate valuation. *Sec.145.*[**22ITR125]**

2.The expression market value has not been defined in the Act. Market value means the price that a willing purchaser would pay to a willing seller for a property having due regard to its existing condition, with all its existing advantages and its potential possibilities when laid out in its most advantageous manner. In order to exercise the power of acquisition of immovable property under section 269C of the Income-tax Act,1961,burden lies on the revenue to prove and establish that the apparent value adopted in the sale deed fell short of the fair market value by more than fifteen per cent. **[Joseph Vallooran - CIT, Ori.,** *Sec.269C.***108ITR544]**

Marketing Kar. Addl.CIT-Ryots Agrl. Produce Co-op. Marketing Soc. Ltd.

The expression marketing appearing in clause (c) of section 81(i) (now section 80P) of the Income-tax Act, 1961, is of wide import and generally means the performance of all business activities involved in the flow of goods and services from the point of initial agricultural production until they are in the hands of the ultimate consumer. In order to make agricultural produce fit for marketing, it may have to be transported or processed, but all the activities involved are understood as amounting to a single activity, namely, marketing, and not independent activities such as transporting, processing, selling, etc. The marketing functions may involve exchange functions such as buying and selling, physical functions such as storage, transportation, processing and other commercial functions such as standardisation, financing, market, intelligence, etc.

Sec.80P. **[115ITR709]**

2. The term marketing cannot be restricted to the buying and selling activity. It includes all activities connected with the process of taking over the agricultural produce of its members and handing over the marketable commodities to the purchasers and all the intermediate processes connected with the marketing of the agricultural produce of the members.

[CIT-Haryana State Co-op.Supply & Marketing Federation Ltd.,P&H., *Sec.80P(2)(a)(iii).* **79CTR94, 182 ITR 53]**

3. For the purposes of section 80P(2)(a)(iii) of the Income-tax Act, 1961, so long as agricultural produce handled by the assessee belonged to its members it is entitled to exemption in respect of the profits derived from the marketing of the same. Whether the members came by the produce because of their own agricultural activities or whether they acquired it by purchasing it from cultivators is of no consequence for the purpose of determining whether the assessee is entitled to the exemption. The only condition required for qualifying the assessee's income for exemption is that the assessee's business must be that of marketing, the marketing must be of agricultural produce and that agricultural produce must have belonged to the members of the assessee-society before they came up for marketing by it,whether on its own account or on account of the members themselves. Section 80P does not in effect limit the scope of the exemption to agricultural produce raised by members alone but includes agricultural produce raised by others but belonging to co-operative societies. The contrast in the said provision is with reference to the marketing of agricultural produce of the members of the society against that purchased from non-members.....The expression marketing is an expression of wide import. It involves exchange functions such as buying and selling, physical functions such as storage, transportation, processing and other commercial activities such as standardisation, financing, marketing intelligence, etc. Such activities can be carried on by an apex society rather than a primary society....Section 80P of the Income-tax Act,1961, is introduced with a view to encouraging and promoting the growth of the co-operative sector in the economic life of the country and in pursuance of the declared policy of the Government. The correct way of reading the different heads of exemption enumerated in the section would be to treat each as a separate and distinct head of exemption.Whenever a question arises as to whether any particular category of an income of a co-operative society is exempt from tax what has to be seen is whether the income fell within any of the several heads of exemption. If it fell within any one head of exemption, it would be free from tax notwithstanding that the conditions of another head of exemption are not satisfied and such income is not free from tax under that head of exemption.

[Kerala State Co-operative Marketing Federation Ltd.– CIT, SC, *Sec.80P(2)(a)(iii).* **147CTR29,231ITR814,98Taxman413]**

4. Marketing means the transfer of goods from the point of agricultural production to the hands of the consumer. So long as the end-product which reaches the consumer is an agricultural produce, the assessee will be entitled to special deduction. Marketing is a comprehensive term.It does not mean merely buying and selling.It includes processing which may be necessary for making the agricultural produce marketable.

[Karnal Co-operative Sugar Mills Ltd. - CIT, P&H., *Sec.80P.* **170 CTR590,253ITR 659,119Taxman795]**

5. The expression marketing appearing in section 81(i)(c) (now section 80P) of the Income-tax Act,1961, is of wide import and generally means the performance of all business activities involved in the flow of goods and services from the point of initial agricultural production until they are in the hands of the ultimate consumer. In order to make agricultural produce fit for marketing, it may have to be transported or processed but all the activities involved are understood as amounting to a single activity, namely, marketing,and not independent activities such as transporting, processing,selling etc.The marketing functions may involve exchange functions such as buying and selling, physical functions such as storage, transportation, processing and other commercial functions such as standardisation, financing, market, intelligence, etc. **[CIT- Ryots Agrl. Produce Co-op. Marketing Soc. Ltd., Kar.,** *Sec.80P.* **323ITR666]**

Marketing of commodities **Kar. Addl.CIT - Karnataka State Warehousing Corp.**

The expression marketing of commodities in section 10(29) of the Income-tax Act,1961, must not be construed in a narrow sense; it includes every activity of purchase, selling and distribution as also warehousing. *Sec.10(29).***[125ITR136]**

2.The term marketing of commodities as used in section 10(29) excludes an activity done by the assessee for its own benefit, because it refers to income from letting of godowns or warehouses for purposes, inter alia, of facilitating the marketing of commodities. Obviously, if the owner of the warehouse does the activity for facilitating for his own benefit, he will not get any income from letting of the warehouse. In the last clause, the term marketing clearly excludes the idea of an activity done for one's own self.The word marketing has been used in the wider sense to include the various activities which generally go to form the trade of marketing.

[U.P. State Warehousing Corporation - ITO, All., *Sec.10(29).***94ITR129]**

Matching **Bom** **Taparia Tools Ltd. - JCIT**

In order to determine the net income of an accounting year under the mercantile system of accounting, the revenue and other incomes are matched with the cost of resources consumed. The matching is to be done on accrual basis.Under this matching concept,revenue and income earned during an accounting period irrespective of actual cash in-flow is to be compared with expenses incurred during the same period irrespective of actual out-flow of cash.

*Sec.2(28A),36(1)(iii),43(2).***[260ITR102]**

Matching Principle **SC** **J. K. Ind. Ltd. - Union of India**

Taxes on income are treated as expense which is recognized in the same period as revenue is recognized as an expense. This is called the matching principle. Such matching results in what is called the timing difference. The tax effect of timing differences is included as tax expense in the statement of the profit and loss and as deferred tax asset or as deferred tax liability in the balance- sheet. In short, deferred tax should be recognized for timing differences. This is the basic mandate of Accounting Standard 22.This mandate is based on an important principle of accounting,namely, that every transaction has a tax effect. Accounting Standard 22 is right in stipulating that the tax effect of such timing differences should be included in the tax expense in the statement of profit and loss and deferred tax liability deferred tax asset in the balance- sheet.

Companies Act.**[213CTR301,297ITR176,165Taxman323]**

Material facts SC Associated Stone Industries (Kotah) Ltd. - CIT

The duty of the assessee is only to disclose fully and truly all material facts necessary for his assessment for the relevant year. The expression material facts.... refers only to primary facts and the duty of the assessee is to disclose such primary facts. There is no duty cast on the assessee to indicate or draw the attention of the Income-tax Officer to what factual or legal, or other inferences can be drawn from the primary facts disclosed.

*Sec.147.***[138CTR260,224ITR560,137Taxation172,90Taxman553**

Maximum marginal rate Ker. CIT - C.V.Divakaran Family Trust

Maximum marginal rate is defined in Explanation 2 thereto as the rate of tax applicable in relation to the highest slab of income provided for associations of persons in the relevant Finance Act.The definition is not capable of any doubt, and the only meaning that it admits of is that the rate on the maximum slab of income for associations of persons is to be treated as the maximum marginal rate of tax for the purposes of section 164.

*Sec.164.***[173CTR399,254ITR222,122Taxman405]**

May Cal. Imperial Chemical Industries Ltd.-CIT

The expression may must normally be construed as an enabling provision unless the expression is coupled with certain duty to the donee of the power when it becomes obligatory for him to exercise the enabling power.In those circumstances the expression may is construed to mean must. Parliament has deliberately used the expression may only in some sub-clauses of section 297(2) of the Income-tax Act, 1961.Due significance and weight must be given to the choice of language by Parliament. In clause (a) of section 297(2), though a power has been given to the Income-tax Officer to proceed under the provisions of the Indian Income-tax Act, 1922, there is no duty, as such, cast upon him in the sense that he must proceed only under the old Act and not under the new Act if the situation in a particular case so warrants. There is no compelling obligation on the revenue authorities to proceed only under the old Act in case a return is filed under the old Act. Hence, the expression may in section 297(2)(a) must be construed only as an enabling provision. It entitles the Income-tax Officer to resort to the Indian Income-tax Act, 1922, but where he chooses to proceed under the Income-tax Act, 1961, his action is not illegal. Where a return of income for the assessment year 1961-62 is filed before April, and the assessment is made under the Act of 1961 the rectification of the assessment under section 154 of the Act of 1961 is legal. It need not be treated as an action under section 35 of the Act of 1922 to confer validity. An appeal from such an order of rectification is competent.

Sec.297(2).**[116ITR516]**

May become due **Mad.** **Buddha Pictures - ITO (Fourth)**

The essential criterion for notice under section 46(5A) is that on the date of service of notice the person should be under an existing obligation to pay amounts to the assessee. A notice issued under section 46(5A) upon an alleged garnishee cannot prevent him from entering into contract with the assessee thereafter and from paying him any money under that contract. The words may become due in section 46(5A) mean may become payable and not may become entitled.

Sec.226.**[52ITR321]**

May pass such orders thereon as it thinks fit **Guj.** **Natwarlal & Co. - CIT**

The words may pass such orders thereon as it thinks fit in section 33(4) of the Indian Income-tax Act, 1922, refer to the order that the Appellate Tribunal may pass in the appeal and do not give power to the Tribunal to give a direction to the Income-tax Officer to take action in respect of a completed assessment of any earlier assessment year.

Sec.254.**[50ITR783]**

May presume **SC** **P. R. Metrani - CIT**

May presume leaves it to the discretion of the court to make the presumption according to the circumstances of the case.

Misc.**[287ITR209]**

Mean **Kar.** **Patil Vijaykumar - Union of India**

There are two forms of interpretation clauses. In one, when the word defined is declared to mean something, the definition is explanatory and prima facie restrictive.In the other,where the word defined is declared to include something,the definition is extensive. A definition or interpretation clause which extends the meaning of a term does not take away its ordinary meaning.The words salary and profits in lieu of salary have, therefore,to be understood as comprehensive of not only such things as they signify according to their nature and import but also those things which the interpretation clause declares that they shall include. *Sec.17(3)(ii).* **[151ITR48]**

Member **Bom.** **Jalgaon District Central Co- op. Bank Ltd.- UOI**

The expression member is not defined in the Income-tax Act. A co-operative society has to be established under the provisions of law made by the State Legislature.The definition of the expression member is given under section 2(19) of the Maharashtra Co-operative Societies Act. As per the definition, member means a person joining an application for registration of a co-operative society, which is subsequently registered or a person duly admitted to membership of a society after registration and includes a nominal, associate or sympathiser member. The rights and privileges of a duly registered member under the Maharashtra Co-operative Societies Act,1960, and the rights and privileges of a co-operative society under section 194A(3)(v) of the Act are two distinct things. *Sec.2(19) Maharashtra Co-operative Societies Act.* **[184 CTR 343,265ITR423, 134Taxman1]**

Merchandise **SC** **CIT - B. Suresh**

Telecasting rights fell in the category of articles of trade and commerce and hence within the category of merchandise and the transfer of the said rights by way of lease fell within the meaning sale and would attract section 80HHC. *Sec.80HHC* **[222CTR,513,313ITR149,178Taxman457]**

Metal **Mad.** **Addl. CIT - Trichy Steel Rolling Mills Ltd.**

The addition of the word metal after the words iron and steel in item I of the Fifth Schedule to the Income-tax Act,1961, is for the purpose of emphasising that the change should not be such as to alter the character of the iron and steel used from iron and steel to any other specified finished article. If there is no such change, iron and steel will continue as iron and steel (metal). Sec.33(1)(b)(B)(1)(a),80-I,Sch.-**[7CTR316,118ITR39]**

Method employed Lah. Ganga Ram Balmokand - CIT

There is no warrant for restricting the interpretation to be put on the words method employed as used in the second alternative dealt with in the proviso to section 13. In ordinary parlance these words convey the idea of 'the manner in which accounts are kept' and they are used in the same sense in the said proviso. If the accounts tendered by the assessee are found to be incorrect or incomplete, if they are cooked or fictitious, it will not be wrong to say that the method of accounting is such that it is not possible to deduce the total income of the assessee there from. Sec.145.**[5ITR464]**

Minerals SC. Stonecraft Enterprises - CIT

The word minerals in sub-section (2)(b) of section 80HHC must be read in the context of mineral oil and ores with which it is associated.These three words taken together are intended to encompass all that may be extracted from the earth.All minerals extracted from the earth, granite included, must, therefore, be held to be covered by the provisions of sub-section (2)(b) of section 80HHC,and the exporter thereof is, therefore,disentitled to the benefit of that section. Sec.80HHC(2)(b).
[153CTR86,237ITR131,103Taxman490]

Mines and Mining AAR John A. Sayre - CIT

The Income-tax Act has not defined mines or mining.The other tax Acts like the Wealth-tax Act, the Gift-tax Act, etc., have also not provided any definition of mines.But if the provisions of the Income-tax Act are examined, it will be seen that special provision has been made for petroleum and gas which have not been treated as part of mines or mining. Mineral has not been used in a broad sense so as to include mineral oil and mining does not include exploration or extraction of oil. It cannot be laid down as a general proposition that mining includes prospecting for and extraction of mineral oil.

It cannot also be said that legislative practice in this country is to use mineral in this sense. In the Constitution itself in the Seventh Schedule mines and minerals have not been used as inclusive of oil fields and mineral oil. Schedule VII, List I, entry 53 deals with regulation and development of oil-fields and mineral oil resources, petroleum and petroleum products, and other liquids and substances declared by Parliament by law to be dangerously inflammable, whereas entry 54, List I, deals with regulation of mines and mineral development.

Sec.10(5B),293A.[**151CTR651,236ITR652,103Taxman78**]

Mistake **Ker.** **M.Kumaran-First Addl.ITO**

A mistake is an omission made not by design, but by mischance, and a mistake apparent is a mistake that is manifest, plain or obvious, a mistake that can be realised without a debate or dissertation. Section 35 will not enable a general revision or review of the order passed or a reconsideration of the factual conclusions reached in a case. *Sec.263.*[**33ITR290**]

2. Mistake is an ordinary word, but in Taxation law, it has a special signification.It is not an arithmetical error alone that comes within its purview.It comprehends errors which, after a judicious probe into the record from which it is supposed to emanate, are discerned. It does not cover any mistake which may be discovered by a complicated process of investigation, argument or proof. The line of demarcation is neither firm nor fixed. Sometimes, an evident error which does not require any extraneous matter to show its incorrectness is confused with an erroneous view of law on a debatable point or a wrong exposition of the law or a wrong application of the law or a failure to apply the appropriate law. In the former case, it is an error apparent from the record and in the latter case, it is not so apparent.Discovery of an error on the basis of assessment due to initial mistake, say, in determining the written down value which happened on account of misapplication of the law can provide a ground for rectification, but where it is debatable whether there was any mistake or misapplication of the law, rectification may not be permissible. It is different from the provisions under Order XLVII,rule 1, of the Code of Civil Procedure, in the sense that rectification of any mistake in the case of the Revenue is when the mistake is apparent from the record and in the case of a review, as in the said provision of the Code of Civil Procedure, it is in the case of an error apparent on the face of the record. **[CIT-E.I.D.ParryLimited,*Sec.154* 216ITR489, 131Taxman495]**

Mistake apparent from the record SC Second Addl.ITO - Atmala Nagaraj

...a mistake which becomes apparent only from the record of the firm is not a mistake apparent from the record so far as the assessment of the partner is concerned. *Sec.263.***[46ITR609]**

2. Though section 35(1) of the Act empowered the Income-tax authorities to rectify mistakes apparent from the record within four years from the date of the assessment order sought to be rectified, a mistake which becomes apparent only from the record of the firm was not a mistake apparent from the record so far as the assessment of the partner was concerned.**[Ram Bhagat-CIT, Pun.,*Sec.263.* 61ITR146]**

3. Section 154 of the Income-tax Act,1961, is not meant for preferring a claim which the assessee had omitted to prefer in the assessment proceedings. The section can operate only on the facts which are already on the record and cannot be resorted to to introduce new facts. Therefore, where an assessee had failed to claim the rebate allowed under the Act, in its return or even at the appellate stage, it cannot be held that there was any mistake, much less a mistake apparent from the record, which would enable the assessee to resort to section 154 of the Act. **[Anchor Pressings (P.) Ltd.-CIT,All.,*Sec.154.*100ITR347]**

4. The jurisdiction of the Income-tax Officer to make an order of rectification under section 154 of the Income-tax Act, 1961, depends upon the existence of a mistake apparent on the face of the record. A mistake is apparent on the face of the record when it is glaring, obvious or self-evident. The language of section 154 makes it clear that a mistake capable of being rectified under this provision is not confined to clerical or arithmetical mistakes only. The provision would not, however, cover a case where a mistake has to be discovered by a complicated process of investigation, argument or proof. **[VedParkashMadanLal-CIT,*Sec.154.*102ITR 213]**

5. It is a well-settled position of law that in order to be a mistake apparent from the record of the case, it must be an error apparent, obvious and glaring. Mere complexity of the problem or that some genuine argument is necessary to discover the error may not be sufficient to oust the jurisdiction of the taxing authorities to rectify such a mistake.None the less, it should be one which could be discerned with some precision after a judicial probe into the assessment records without long drawn process of reasoning and on which no two reasonable contrary opinions are conceivable. **[PadmavatiJaykrishnaCWT,Guj.,**<u>*W.T.Act.5.r.w.s.35*</u>**105ITR115]**

6. A mistake apparent from the record must for the purpose of section 154 of the Income-tax Act,1961, be an obvious and patent mistake and not something which could be established by a long drawn process of reasoning on points on which there might conceivably be two opinions. **[Lilavatiben Harjivandas Kotecha. J.V. Shah ITO,Guj.,**<u>*Sec.154.*</u>**11CTR97,122ITR863]**

7. As the assessee,who had been previously assessed, was taken to be one who had not been previously assessed and penal interest charged on that basis for not filing the estimate of advance tax, it was a mistake apparent from the record liable to be rectified. **[CIT-N.D.Georgopoules,Mad.,**<u>*Sec.154.r.w.s.139(2)*</u>**125 ITR630]**

8. A mistake apparent from the record must be an obvious and patent mistake and not something which can be established by a long drawn process of reasoning on points on which there may be conceivably two opinions. If there are two views and the mistake is required to be substantiated by lengthy arguments and by placing reliance on decisions, it will not be an error or mistake that can be said to be either apparent on the face of the record or apparent from the record. **[StadmedP.Ltd.-CIT,Cal.,**<u>*Sec.254(2).*</u> **140 ITR 361]**

9. The application of a wrong provision of the Act or the erroneous application of the same to the facts of the case, which do not call for such application, will amount to a mistake apparent from the record for the purpose of section 154 of the Income-tax Act,1961.**[CIT-Sundaram Textiles Ltd., Mad.,** <u>*Sec.154.*</u> **149 ITR525]**

10. Section 254(2) of the Income-taxAct,1961, enables the concerned authorities to rectify any mistake apparent from the record. An oversight of a fact cannot constitute an apparent mistake rectifiable under this section. Similarly, failure of the Tribunal to consider an argument advanced by either party for arriving at a conclusion, is not an error apparent on the record, although it may be an error of judgment. **[Perfetti Van Melle India P. Ltd.-CIT,Del.,** <u>*Sec.254 (2).*</u> **212 CTR 137,296ITR595,164Taxman493]**

11. Sub-section (1) of section 154 of the Income-tax Act,1961,clearly states the circumstances under which rectification can be made. A mistake must be apparent from the records, meaning thereby no external help either on fact or in law is required to detect such mistake. The mistake must be so obvious that it can easily be corrected, to wit an arithmetical mistake, a wrong quotation of section,etc. **[Hindustan Lever Ltd.-JCIT,Cal.,** <u>*Sec.154.*</u>**204 CTR 125,284 ITR 42]**

Mistake apparent on the face of the record **P&H. Maya Ram Jia Lal - CIT**

Where, by misreading a section, a wrong view is taken and a wrong calculation is made, it would certainly come within the purview of section 154 of the Income-tax Act,1961,as a mistake apparent on the face of the record.<u>*Sec.154.*</u>**[43CTR138,152ITR608,18Taxman512]**

2. The jurisdiction of the Income-tax Officer to make an order of rectification depends upon the existence of a mistake apparent on the face of the record. Such a mistake must be an obvious and patent mistake and not something which can be established by a long drawn process of reasoning on points on which there may be conceivably two opinions.A decision on a debatable point of law cannot be a mistake apparent from the record. If the determination of the mistake depends on the interpretation of the provisions of the Act, it cannot be a mistake apparent on the face of the record. **[CIT-Calcutta Steel Co. Ltd., Cal.,** <u>*Sec.154.*</u>**72CTR185,174ITR 521,41Taxman117]**

3. Section 154 of the Income-tax Act, 1961,only confers power of rectification if there is a mistake apparent on the face of the record. Glaring, obvious, patent and apparent mistakes are those for which no investigation into facts or determination of law or discussion of debatable points are involved, and to establish which long drawn argument would not be necessary and in respect of which no two opinions are possible. **ITSC-Netai Chandra Rarhi and Co., Cal.** <u>*Sec.154.*</u> **[193CTR430,271ITR514,142Taxman446**

4. A patent,manifest and self-evident error which does not require elaborate discussion of evidence or arguments to establish it, can be said to be an error apparent on the face of the record and can be corrected while exercising certiorari jurisdiction. An error cannot be said to be apparent on the face of the record if one has to travel beyond the record to see whether the judgment is correct or not. An error apparent on the record means an error which strikes one on mere looking and does not need a long drawn out process of reasoning on points on which there may be conceivably two opinions. Such error should not require any extraneous matter to show its incorrectness. To put it differently, it should be so manifest and clear that no court would permit it to remain on record.If the view accepted by the court in the original judgment is one of possible views, the case cannot be said to be covered by an error apparent on the face of the record. **[Saurashtra Kutch Stock Exchange Ltd, SC,** *Sec.254(2).***219 CTR,90,305ITR227,173 Taxman322]**

Mistake apparent on the record **Bom.** **Nandlal Mangaram Pamnani - G. Lakshminarasimhan**

A mistake apparent on the record must be an obvious and patent mistake and not something which can be established by a long drawn process of reasoning on points on which there may be conceivably two opinions. A decision on a debatable point of law is not a mistake apparent from the record. *Sec.154.***[82ITR1]**

2. A mistake apparent on the record must be an obvious and patent mistake and not something which can be established by a long drawn process of reasoning on points on which there may conceivably be two opinions. **[T S Balaram -Volkart Brothers, Bombay,SC,***Sec.154.***AIR1971SC2204,82ITR50]**

3. A mistake apparent on the record must be an obvious and patent mistake and not something which can be established by a long-drawn process of reasoning on points on which there may be conceivably two opinions. A decision on a debatable point of law is not a mistake apparent from the record. **[Pieco Electronics and Electricals Ltd.-CIT,Cal.,** *Sec.154.***205ITR469]**

4. A mistake apparent on the record must be an obvious and patent mistake and not something which can be established by a long drawn process of reasoning on points on which there may be conceivably two opinions. **Netai Chandra Rarhi and Co.-ITSC, Cal.** *Sec.154.* **[186CTR706,263ITR186,133Taxman640]**

Money **SC** **CIT - Kasturi and Sons Ltd.**

......the word money used in section 41(2) of the Act has to be interpreted only as actual money or cash and not as any other thing or benefit which could be evaluated in terms of money. *Sec.41(2).* **[153CTR1,237ITR24,103Taxman342]**

Money lent at interest and brought into India in cash or kind **Kar.** **Meturit A.G. - CIT**

The expression money lent at interest and brought into India in cash or kind would not mean only currency in various forms. Cash and kind are always understood to mean one as ready money and the other as not merely other types of currency but as goods or commodities as distinguished from money. Sec.9,58. **[82CTR345,184ITR 257]**

Month All. CIT - Laxmi Rattan Cotton Mills Co. Ltd.

The word month occurring in section 271(1)(a) must be taken to mean a period of thirty days. Section 271(1)(a) was enacted for the purpose of imposing a penalty on an assessee who has not filed his returns within the prescribed time and its object was to serve as a deterrent for such lapses.Penalty is imposable for every month during which the default continues. If the meaning ascribed to this word in the General Clauses Act, i.e., if the English calendar month is adopted, it may in some cases lead to a defaulting assessee escaping penalty altogether.For example, if the time given to an assessee to file his returns is up to the 31st of January of a particular year and he files it on the 27th of February, he would not be liable to pay any penalty. Such a result is not contemplated by the sub-section.The sub-section in clear and unambiguous terms makes every assessee liable for penalty during the period of default. In the circumstances, month should be taken to mean a period of thirty days and not an English calendar month. *Sec.271(1)(a).* **[97ITR285]**

2. The word month occurring in section 271(1)(a) of the Income-tax Act, 1961, has to be reckoned according to the British calendar according to section 3(35) of the General Clauses Act, 1897. Accordingly, when an assessee who was granted time for filing his return for the assessment year 1961-62, till January 15, 1962, actually filed his return on February 15,1962, it had filed its return on the last day of the default, and hence the default had not lasted for a month for the purpose of levying penalty under section 271 (1) (a) of the Income-tax Act, 1961,and therefore, no penalty can be imposed on such assessee under section 271(1)(a). **[CIT-Kadri Mills (Coimbatore)Ltd.,Mad.,*Sec.271(1)(a).*106 ITR 846]**

3. The expression month has not been defined in the Income-tax Act,1961, and it has been used in different contexts in different sections of the Act. Therefore, the word month occurring in section 271(1)(a) has to be reckoned according to the English calendar month, according to section 3(35) of the General Clauses Act,1897. **[CIT- Brijlal Lohia and Mahabir Prasad Khemka,Cal.,*Sec.271 (1)(a).*124ITR485]**

4. The term month occurring in the Income-tax Act,1961, and the Income-tax Rules,1962, has not been defined and, therefore, the definition of that term, if any, found in the General Clauses Act, 1897,has to be applied in ascertaining the meaning of that term occurring in the Act and the Rules. Section 3(35) of the General Clauses Act,1897,defines the term month as a month reckoned according to the British Calendar. The context of the Act and the Rules do not provide for not applying the term month occurring in the General Clauses Act. On the definition of the term month occurring in section 3(35) of the General Clauses Act, the British calendar for the month of August,1976 ,would also include August 31,1976. **[B.V.Aswathaiah and Brothers-ITO,Kar.,**<u>*Sec.139(8).*</u>**155ITR422]**

5.The word month occurring in section 271(1)(a) of the Income-tax Act,1961, has to be reckoned according to the British calendar according to section 3(35) of the General Clauses Act,1897. **[CIT-S.L.M. Maneklal Industries Ltd.,Guj.,** <u>*Sec.271(1)(a).*</u> **274 ITR485]**

More and less **Bom** **Gilbs Computer Ltd.**

The expressions more and less in section 253(6) of the Act have to be given their natural meaning. Negative income cannot be more. It will always be less. In that event the language of sub-section (6)(a) would be attracted. If the total income can be considered even to be a loss then the absence of it will not be covered by either clause (a), (b) or (c) of sub-section (6). It will be clause (d) of sub-section (6) which will apply. <u>*Sec.253(6).*</u> **[226CTR,19,317ITR159]**

More reliable data **Bom.** **CIT -Great Eastern Life Insurance Co. Ltd.**

The scope of the expression more reliable data occurring in rule 8 of the Schedule is not confined to the data required for arriving at the higher of the two computations under rules 2(a) and 2(b) of the Schedule.The separate actuarial valuation statement for the Indian business of such a company must be regarded as containing more reliable data for the purpose of rule 8. <u>*Sec.44.*</u>**[13ITR141]**

2. In assessing the profits and gains of the Indian branch of an insurance company not resident in British India recourse must be had in the first instance to rule 8 of the Schedule to the Indian

Income-tax Act, 1922, which is the only rule dealing specifically with the subject. In the absence of more reliable data the profits and gains must be computed on the proportionate basis laid down in the rule. In that case no other rule is relevant except for the purpose of computing the total world income which has to be apportioned.If there is more reliable date, that is data more reliable for computing the profits and gains of the Indian business of the assessee than those on which the proportionate rule is based, then rule 8 passes out of the picture and by virtue of section 10(7) of the Act,the computation of profits and gains must be made under the other rules in the Schedule which are appropriate. **[CIT- Great Eastern Life Assurance Co. Ltd., PC,*Sec.44.*[17ITR173]**

Mutatis - mutandis Pun Khushi Ram Raghunath Sahai - CIT

The phrase Mutatis Mutandis has its usual meaning, that is, that only such verbal changes are to be made in the rules mentioned in ... make the principles embodied in these rules applicable to applications under. *Misc.*[**AIR1953Pun300,20ITR170]**

Mutual association Ori Cuttack Club Pvt. Ltd. - CIT

If contributors to an association as well as the participants in the surplus arising out of the business carried out by the association happen to be the same, such an association is known as a mutual association. *Misc.* **[196ITR407]**

Mutual concern Bom Ismailia Grain Merchants Association - CIT

A concern will be a mutual concern where the income is derived from the contribution of all the numbers and the income or part of it is spent for the benefit of all the members; in other words, there must be an identity between the contributors of the participators in that fund. *Misc.*[**AIR1958Bom32,31ITR433]**

2. Where the assessee is found to be a mutual concern, the income which it receives from its members is not liable to tax. This is founded on the principle that no one can make a profit by transacting with oneself. The primary condition of mutuality between the assessee and its members is that the assessee which collects money

from its members, must apply the same for their benefit not as shareholders having an interest in its profits but as persons themselves who have put up the fund by contributing to it. There must be a thread of agency for acting for the contributors for achieving the objectives. The identity of individuals as contributors and participants is not essential but what is essential is the identity of character of contributors and participants. **[CIT - Adarsh Co-op. Housing Society Ltd., Guj.,** *Misc.***[213ITR677]**

Mutuality Guj Sports Club of Gujarat Ltd. – CIT

One of the essentials of mutuality is that the contributors to the common fund are entitled to participate in the surplus, thereby creating an identity between the participants and the contributors. Once such identity is established, the surplus income would not be exigible to tax on the principle that no man can make a profit out of himself. The principle of mutuality is not destroyed by the presence of transactions which are non- mutual in character. The principle of mutuality can, in such cases,be confined to transactions with members.The two activities can, in appropriate cases, be separated and the profits derived from non- members can be brought to tax. *Sec28(iii),44A.***[67CTR233,171ITR504,37Taxman38]**

2. The cardinal principle to apply the test of mutuality is that all the contributors to the common fund must be entitled to participate in the surplus and that all the participators in the surplus must be contributors to the common fund. In other words, there must be complete identity between the contributors and the participators. **CIT - Bombay Oilseeds and Oil Exchange Ltd., Bom** *Sec. 28(iii).* **[113 CTR 404, 202 ITR 198,71 Taxman 351,116 Taxation232]**

3. Where a number of persons combine together and contribute to a common fund for the financing of some venture or object and in this respect have no dealings or relations with any outside body, then any surplus returned to those persons cannot be regarded in any sense as profit. There must be complete identity between the contributors and the participators. If these requirements are fulfilled, it is immaterial what particular form the association takes. Trading between persons associating together in this way does not give rise to profits which are chargeable to tax. Where the trade or activity is mutual, the fact that, as regards certain activities, certain members only of the association take advantage of the facilities which it offers, does not affect the mutuality of the enterprise. The law recognises the principle of mutuality excluding the levy of Income-tax from the income of such business to which the above principle is applicable. A perusal of section 2(24) of the Income-tax Act,1961, shows that the Act recognises the principle of mutuality and has excluded all businesses involving such principle from the purview of the Act, except those mentioned in clause (vii) of that section. The three conditions, the existence of which establishes the doctrine of mutuality are (1) the identity of the contributors to the fund and the recipients from the fund, (2) the treatment of the company, though incorporated as a mere entity for the convenience of the members, in other words, as an instrument obedient to their mandate, and (3)the impossibility that contributors should derive profits from contributions made by themselves to a fund which could only be expended or returned to themselves. **Chelmsford Club - CIT, SC** *Sec.2(24).* **[159CTR235, 243ITR89,109Taxman]**

4. In order to fulfil the requirement of mutuality, a mutual association has to establish, as an essential requirement, the identity between participators and contributors to the fund. However, the fact that an association satisfies the norm of mutuality in respect of the receipts of contributions from its members does not necessarily lead to the conclusion that every activity of the association satisfies the test of mutuality. An association may engage in activities which can be described as mutual and in other activities which are not mutual. The two activities can in appropriate cases be separated and the profits derived from transactions which do not fulfil the requirements of mutuality can be brought to tax. *Misc*.**[328ITR362,192Taxman238]**

Mutuality principle **Mad.** **CIT - South Indian Film Chamber of Commerce**

The underlying idea behind section 28(iii) of the I.T. Act, 1961, is that there must be a business from which income is derived and in the course of such business specific services must be rendered to its members. The concept behind this provision is to cut at the mutuality principle being relied upon in support of a claim for exemption when the assessee actually derives income or makes profits as a result of rendering specific services to its members in a commercial way. Section 28(iii) would not apply to tax such income from business so long as the business is held under a trust as otherwise the benefit sought to be conferred under section 11 would be destroyed and the exemption provision would be stultified. *Sec.28(iii).* **[129ITR22]**

N

Necessarily **Bom.** **Vodafone International Holdings B.V- UO I**

The expression necessarily means inevitably or as a matter of a compelling inference. <u>*Misc.*</u>**[329ITR126]**

2. Parliament has used the word necessarily in Explanation 1 to section 147 of the Income-tax Act,1961.The expression necessarily means inevitably or as a matter of a compelling inference. **[3i Infotech Ltd.- ACIT.,Bom.,** <u>*Sec.147*</u> **235 CTR 240,329ITR257, 192Taxman 137]**

Net value of the assets of the business as a whole **Guj.** **CWT - Raipur Manufacturing Co. Ltd.**

When the words net value of the assets of the business as a whole are used together, they can only convey that what is required to be ascertained is the value of the assets of the business considered as a whole after taking into account the liabilities of the business. <u>*Sec.2(m.)WT.Act.*</u>**[52ITR482]**

Net wealth **Bom.** **Seth Ramnath K. Daga - CWT**

Interest in property which is available to the assessee for a period not exceeding six years from the valuation date is not an asset within the meaning of section 2(e) of the Wealth-tax Act,1957, prior to its amendment w.e.f. April 1,1965, and the value thereof cannot be included in the net wealth of the assessee. <u>*Sec.2(e) WT.Act.*</u> **[17CTR11,127ITR371]**

Net wealth tax **SC** **CWT - P.C.Oswal**

.. a net-wealth tax is imposed upon individuals, groups of individuals like Hindu undivided families and companies. The tax is not upon the assets as such but is upon individuals and companies with reference to the capital value of the assets held by them and falls wholly within entry 86 of List I of Schedule VII to the Constitution of India.

Constitution of India. **[110CTR221,200ITR614,67Taxman3]**

New **Ker.** **CIT - Cochin Company**

The word new when used with reference to a machine means new-made,or brought into existence for the first time, and stands in antithesis to the word used. *Sec.32.* **[49ITR310]**

New machinery **SC** **Cochin Company – CIT**

The word new may have different meanings, but when, used with reference to a machine, its meaning is new made or brought into existence for the first time:which is one of the meanings given in the Shorter Oxford Dictionary. *Sec.32 .* **[67ITR199]**

No appeal shall lie **SC** **CIT - Filmistan Ltd.**

The meaning of the words no appeal shall lie in the proviso is not that no memorandum of appeal can be presented. All that it means is that the appeal will not be held to be properly filed until the tax has been paid. *Sec.246.* **[AIR1961SC1134,42ITR163]**

No income liable to tax **Bom.** **CIT - Allied Silk Mills**

The expression no income liable to tax contemplated by section 271 (3)(b) of the Income-tax Act,1961, is the income which the assessee believes to be his income and not the income as finally assessed by the Income-tax Officer. It is possible that an assessee may not consider a particular item to be his income and the Income-tax Officer may hold otherwise. In such a case if what the assessee considers to be his income is less than the maximum not chargeable to tax, he is not required to file a voluntary return even if his income as finally assessed is more than such maximum and he will not be liable to penalty for failure to file his return. It is essential that the belief of the assessee must be bona fide.

Sec.271(3)(b). **[26CTR132,140ITR428]**

No profits or gains chargeable for that year **SC** **CIT - Virmani Industries Pvt. Ltd.**

The words no profits or gains chargeable for that year in section 32(2) of the Income-tax Act, 1961, are not confined to profits and gains derived from business. They refer to the totality of the profits or gains computed under the various heads and chargeable to tax. It is, therefore, clear that effect must be given to depreciation allowance first against the profits and gains of the particular business whose income is being computed under section 28 and if the profits of that business are not sufficient to absorb the depreciation allowance, the allowance to the extent to which it is not absorbed would be set off against the profits of any other business and if a part of the depreciation allowance still remains unabsorbed, it would be liable to be set off against the profits and gains chargeable under any other head and it is only if some part of the depreciation allowance still remains that it can be carried forward to the next assessment year. But where any part of the depreciation allowance remains unabsorbed after being set off against the total income chargeable to tax, it can be carried forward to the following year and set off against the year's income and so on for succeeding years.The method adopted by the statute for achieving this result is that the carried forward depreciation allowance is deemed to be part of and stands on exactly the same footing as the current depreciation for the assessment year. Sec.32(2).**[129CTR189,216ITR607,83Taxman343]**

Non recovery receipt **Bom.** **Mehboob Productions Pvt. Ltd. - CIT**

The words non-recurring receipt in section 4(3)(vii) of the Indian Income-tax Act, 1922, do not mean that the receipt is a single one or which has, in fact, not been repeated, but only that there is no claim or right in the recipient to expect its recurrence. Merely because the mode of payment is one that would ensure to the assessee receipt of the amount in driblets, it would not necessarily characterise the receipt as a recurring receipt. *Sec.10(3).***[106ITR758]**

Non recurring nature **All.** **Rani Amrit Kunwar - CIT**

The words non- recurring nature in section 4(3)(vii) means not that the payments have as a matter of fact not recurred but that they are not bound to recur. *Sec.10(3).***[14ITR561]**

Non-resident Guj. Chimanbhai K. Patel - CWT

The term non-resident has not been defined in the Wealth-tax Act. However, if the term is construed in its popular sense or as having the same meaning as is ascribed to it in the Income-tax Act, it would clearly not include within its ambit a person who is not ordinarily resident. A person not ordinarily resident is not entitled to the rebate under rule 3 of Part II of Schedule I to Wealth-tax Act.

*Sec.5,6.WT.Act.***[49CTR104,156ITR373]**

2. .....the term non-resident as defined in section 2(30) of the Act has to be read in the contextual setting of section 10(4A) of the Act and if read in that manner the term non-resident will have to include within its fold the meaning as prescribed under the provisions of the Foreign Exchange Regulation Act and the rules there under. Thus, section 10(4A) of the Act will apply to a person who is not resident in India, that is to say, a non-resident will be a person residing outside India. **[Rambhai L. Patel – CIT,Guj.,** *Sec.2(30).r.w.s.10(4A).***171CTR 16, 252ITR 846,129Taxman866]**

Not a business loss Cal Raghubir Prasad Gupta - CIT

An amount paid to the customs authorities by way of fine in lieu of confiscation of goods imported without valid licence cannot be said to be expenditure wholly and exclusively laid out for the purpose of the business of an assessee .... An infraction of law is not a normal incidence of business and a penalty paid for an infraction of the law in not a business loss in the commercial sense.

*Sec.28(i),37(1).***[120ITR789]**

Not being a charge created by the assessee voluntarily Bom. CIT-Tarachand Kalyanji

The phrase in section 24(1)(iv) of the Income-tax Act,1961, as amended in not being a charge created by the assessee voluntarily has reference to the charge which comes into being by operation of law or by virtue of an order of the court or by an act of parties other than the assessee,such as when the assessee gets a property already subject to a charge. The question whether the charge was voluntary or involuntary will have to be decided with reference to the acts relating to the creation of such charge. If the charge is created voluntarily, it remains so, whether it is created before the amendment or after the amendment. If the benefit of the deduction is taken away as a result of the amendment, then from the date of the amendment such a deduction cannot be allowed.

Sec.24(1)(iv).[204ITR43]

Not involving the carrying on of any activity.... **SC** **Addl.CIT -Surat Art Silk Cloth Mfg. Asso.**

The true meaning of the last ten words in section 2(15), viz., not involving the carrying on of any activity for profit, is that when the purpose of a trust or institution is the advancement of an object of general public utility, it is that object of general public utility and not its accomplishment or carrying out which must not involve the carrying on of any activity for profit. So long as the purpose does not involve the carrying on of any activity for profit, the requirement of the definition would be met and it is immaterial how the monies for achieving or implementing such purpose are found, whether by carrying on an activity for profit or not. *Sec.2(15).* **[13CTR378,121ITR1,2Taxman501]**

2. It is well established that the expression not involving the carrying on of any activity for profit qualifies or governs the fourth category of charitable purposes mentioned in the definition clause, viz., the advancement of any other object of general public utility and further that it is the object of general public utility and not its advancement or accomplishment which must not involve the carrying on of any activity for profit.

[Addl. CIT -Etawah District Exhibition & Cattle Fair Association, All. *Sec.2(15).***19CTR337,131ITR461]**

3. The true meaning of the words not involving the carrying on of any activity for profit used in section 2(15) of the Income-tax Act, 1961,is that when the purpose of a trust or institution is the advancement of an object of general public utility,it is that object of general public utility and not its accomplishment or carrying out which must not involve the carrying on of any activity for profit.It has to be seen whether the predominant object of the activity involved in carrying out the object of general public utility is to subserve the charitable purpose or to earn profit.Where profit making is the predominant object of the activity,the purpose, though an object of general public utility,would cease to be a charitable purpose.But where the predominant object of the activity is to carry out the charitable purpose and not to earn profit, it would not lose its character of a charitable purpose merely because some profit arises from the activity

[CIT-MP Anaj Tilhan Vyapari Mahasangh. MP., Sec.2(15).**68 CTR3,171TR677, 88Taxation62]**

4. In section 2(15) of the Income-tax Act,1961, the words not involving the carrying on of any activity for profit qualify or govern only the last head of charitable purpose and not the earlier three ones. If the purpose of a trust or institution be relief of the poor, education or medical relief, the requirement of the definition of charitable purpose would be fully satisfied, even if an activity for profit is carried on in the course of the actual carrying out of the primary purpose of the trust or institution.

[CIT- Secretary Regional Comm., National Sports Club, Gau, Sec.2(15).**180ITR648]**

Not ordinarily resident Mad. Manibhai S. Patel - ITO

If an assessee resides in the taxable territories for more than two years, then he is not ordinarily resident in the taxable territories, or in other words, if an assessee does not reside in the taxable territories for more than two years he is entitled to the exemption which not ordinarily resident given to him.

Sec.6. **[AIR1953 Mad.746, 23ITR27]**

2.A person is not ordinarily resident in any year if he has not been resident in nine out of the ten years preceding that year; he need not establish that he was not resident in nine out of the ten years.In order to be ordinarily resident in any year an individual has to be resident in nine out of the ten years preceding that year.

[P.B.I.Bava-CIT,T&C.,Sec.6.**27ITR463]**

Not resident in the territories Cal. Anglo-Indian Jute Mills Co. Ltd. - S.K. Dutt

The word resident in the expression not resident in the territories in section 18(3B) must be construed in accordance with the definition of the word resident in section 4A(c) of the Act and cannot be read as meaning residence in the physical or any other sense.

Sec.6.**[30ITR525]**

Not wholly used for the purpose of the business profession, etc. **Pat.** **Dalmia Cement Ltd.**

The words not wholly used for the purposes of the business, profession, etc. do not mean not used through the year or during the whole of the year in question. They mean that the building, machinery etc. have not been used exclusively for the purpose of the profession or vocation, that is to say, they have been used d for other purposes also. *Sec.32.* **[AIR1946Pat39,13ITR415]**

Notification **Mad.** **Asia Tobacco Co. Ltd. - Union of India**

The intendment of a notification in the Official Gazette is that in the case of either grant or withdrawal of exemption, the public, must come to know of the same. Notify, even according to ordinary dictionary meaning, would be to take note of, observe, to make known, publish, proclaim, to announce, to give notice to, to inform. It would be a mockery to state that it would suffice the purpose of notification if the notification is merely printed in the Official Gazette, without making the same available for circulation to the public or putting it on sale to the public. Neither the date of the notification nor the date of printing, nor the date of the Gazette counts for notification within the meaning of the rule, but only the date when the public gets notified in the sense, the concerned Gazette is made available to the public. The date of release of the publication is the decisive date to make the notification effective. Printing the Official Gazette and stacking them without releasing to the public would not amount to notification at all. *Misc.* **[45CTR306,155 ITR568]**

Notwithstanding anything contained in **Mad.** **Kalpana Lamps and Components Ltd. - DCIT**

When a clause begins with notwithstanding anything contained in this Act or in some particular provision/provisions in the Act, it is with a view to give the enacting part of the section, in case of conflict, an overriding effect over the Act or provision mentioned in the non obstante clause. It conveys that in spite of the provisions or the Act mentioned in the non obstante clause, the enactment following such expression shall have full operation. It is used to override the mentioned law/provision in specified circumstances.

Sec.43B. **[175CTR549,255ITR491,125Taxman1045]**

2. Section 44D starts with an overriding expression notwithstanding anything to the contrary contained in sections 28 to 44C...This means that section 44D has application in respect of royalties and technical fees in the course of a business and that its special provisions take precedence over sections 28 to 44C and override these provisions.That means section 44BB is also superseded in respect of computation of income by way of royalties or fees for technical services received from an Indian concern. The proviso to section 44BB excluding the application of that section to cases covered by section 44D is consistent with and complementary to this.This double safeguard provided by the statute shows that section 44D includes within its purview royalties and technical service fees arising in the course of business.

[No.P/6of,AAR,<u>*Sec.44D.*</u>**148CTR481,234ITR371,146Taxation222 100Taxman206]**

Notwithstanding anything to the contrary in sec. 30 to 39	**Bom.**	**CIT - Yorkshire Insurance Co.**

Section 40 commences with the words Notwithstanding anything to the contrary in sections 30 to 39. This takes in section 32 which deals with depreciation. Section 32 provides that in respect of depreciation of, inter alia, buildings owned by an assessee and used for the purposes of its business, a deduction on account of depreciation would be allowed in the manner provided therein. The deduction on account of depreciation is, therefore, an allowance. The words of section 40(a)(v) indicate that what shall not be deducted in computing income chargeable under the head of profits and gains of business and profession in the case of any assessee shall be, inter alia, any allowance in respect of assets of the assessee used by an employee.There is nothing in those words which can lead to the conclusion that an allowance in the form of a deduction for depreciation is not within the ambit of section 40(a)(v).

<u>*Sec.40.*</u> **[162ITR565]**

O

Object of general public utility **Bom. CIT - Grain Merchants' Association of Bombay**

The expression object of general public utility in section 4(3) of the Indian Income-tax Act means an object of public utility which is available to the general public as distinct from any section of the public. *Sec.2(15).* **[6ITR427]**

2.The Satsangis, being followers of one religion, are a cross section of the public, and a charitable trust for the benefit of Satsangis as such must be deemed to be a trust for an object of general public utility. **[CIT-RadhaswamiSatsangSabha,All.,** *Sec.2(15).***25ITR472]**

3.The expression object of general public utility was not restricted to objects beneficial to the whole of mankind. An object beneficial to a section of the public was an object of general public utility. To serve as a charitable purpose, it was not necessary that the object should be to benefit the whole of mankind or even all persons living in a particular country or province.

[CIT- Andhra Chamber of Commerce,SC,*Sec.2(15)***[55ITR722]**

4.The expression object of general public utility in section 2(15) would include only those objects which promote the welfare of the general public and not the personal and individual interests of some persons. **[CIT- Bombay Suburban Electric Supply Co. Pvt. Ltd., Bom.,** *Sec.2(15)* **106ITR752]**

Obtained **Guj. CIT - Rashmi Trading**

The words obtained, whether in cash or in any other manner whatsoever, any amount in respect of such loss or expenditure (incurred in any previous year) in section 41 clearly refer to the actual receiving of that amount. The cash may be actually received or it may be adjusted by way of an adjustment entry or a credit note or in any other form when the cash or equivalent of the cash can be said to have been received by the assessee. But it must be the obtaining of the actual cash which is contemplated by the legislature when it used the words has obtained, whether in cash or in any other manner whatsoever, any amount in respect of such loss or expenditure. <u>*Sec.41.*</u>**[103ITR312,42Taxation99]**

2.....the term obtained used in sub-section (1) of section 41 of the Act could not be given the meaning capable of being obtained. While considering the provisions of section 41(1), the system of accounting followed by the assessee was of no relevance or consequence.

[Travancore Chemical & Manufacturing Co. Ltd.-CIT, Ker., <u>*Sec.41(1).*</u>[150CTR256,237 ITR 821, 101 Taxman639

Obtained, whether in cash or in any other manner **Cal.** **Goodricke Group Ltd. - CIT(No. 2)**

The words obtained, whether in cash or in any other manner, whatsoever, any amount in respect of such loss or expenditure incurred in any previous year in section 41(1)(a) of the Income-tax Act,1961, refer to the actual receiving of cash of that amount. The amount may be actually received or it may be adjusted by way of any adjustment entry or a credit note or in any other form when the cash or the equivalent of the cash can be said to have been received by the assessee. But it must be the obtaining of the actual amount which is contemplated by the Legislature when it used the words has obtained, whether in cash or in any other manner whatsoever, any amount in respect of such loss or expenditure in the past. The question whether the liability is actually barred by limitation is not a matter which can be decided by considering the assessee's case alone but has to be decided only if the creditor is before the concerned authority. In the absence of the creditor, it is not possible for the authority to come to a conclusion that the debt is barred and has become unenforceable. There may be circumstances which may enable the creditor to come with a proceeding for enforcement of the debt even after expiry of the normal period of limitation as provided in the Limitation Act. <u>*Sec.41(1).*</u>**[338ITR116,199Taxman402]**

Occupation **CET - Manorama Sarabhai**

A person running a school must be regarded as carrying on occupation even when it is carried on with the motive of doing social service or social good without taking any remuneration thereof.

<u>*Expenditure Tax Act.*</u>**[AIR1966Guj166,59ITR262]**

2. The word occupation is a word of large and indefinite import and its meaning is not susceptible of any precise or definite formulation. No universal test can be laid down for determining when an activity amounts to an occupation and when it does not. But there are certain features which are definitely indicative of what is occupation. Activity in a specific line of endeavour which engages or occupies time and attention of a person and which is carried on with a certain amount of continuity or regularity in the sense that it is not momentary-not an isolated or semi-occasional and temporary adventure in that line of endeavour would certainly constitute occupation.The absence of profit- making motive will not take an activity out of the category of occupation. **[CET Ambalal Sarabhai, Guj.,***ExpenditureTaxAct.***AIR1970Guj141,3ITR 78]**

3. Since an occupation is that with which a person occupies himself, there is no reason why politics cannot be an occupation provided it is taken up as a career.Though a profit motive is absent in politics, profit motive is not an essential requisite of occupation,for a person, well endowed with the goods of the world,may prefer to pursue a profession or occupation without receiving any remuneration for his service. **[P.V.G.Raju-CET,AP.,***Expenditure Tax Act.***79ITR430]**

4. Occupation is of wider import than vocation or profession. Occupation is that with which a person occupies himself either temporarily or permanently or for a considerable period with continuity of activity.The profit motive or earning of income is not an essential ingredient to constitute the activity termed as business, profession, vocation or occupation. **[P.V.G. Raju, Raja of Vizianagaram - CET.,AP.,***ExpenditureTax Act.***86 ITR 267]**

5. The word occupation generally means the trade or calling by which a person seeks to get his livelihood. The word vocational so means the same thing. **[Father Epharam,Ker.,** *Sec.10(3).r.w.s.2(24).***75CTR142,176ITR78]**

Occupy **Del.** **CIT - Modi Industries Ltd.**

The term occupy appearing in section 22 of the Income-tax Act, 1961, refers to occupation directly by the assessee or through an employee or an agent but such occupation by the employee, etc., must be subservient to and necessary for the performance of the duties in connection with the business of the assessee.

*Sec.22.***[128CTR315,210ITR1,73Taxman691]**

2. The term occupy refers to occupation directly by the assessee himself or through an employee or agent but such occupation by the employees, etc., within the meaning of the exception in the said section, must be subservient to and necessary for the performance of the duties in connection with the business of the assessee. **[CIT-T.V. Sundaram Iyengar and Sons Ltd..Mad.,**<u>Sec.22.</u>**193 CTR 50,271ITR79,145Taxman380]**

Of its members **SC** **Kerala State Co-op.Marketing Federation Ltd. - CIT**

For the purposes of section 80P(2)(a)(iii) of the Income-tax Act,1961, so long as agricultural produce handled by the assessee belonged to its members it is entitled to exemption in respect of the profits derived from the marketing of the same. Whether the members came by the produce because of their own agricultural activities or whether they acquired it by purchasing it from cultivators is of no consequence for the purpose of determining whether the assessee is entitled to the exemption. The only condition required for qualifying the assessee's income for exemption is that the assessee's business must be that of marketing, the marketing must be of agricultural produce and that agricultural produce must have belonged to the members of the assessee-society before they came up for marketing by it, whether on its own account or on account of the members themselves.Section 80P does not in effect limit the scope of the exemption to agricultural produce raised by members alone but includes agricultural produce raised by others but belonging to co-operative societies.The contrast in the said provision is with reference to the marketing of agricultural produce of the members of the society against that purchased from non-members.

<u>*Sec.80P(2)(a)(iii).*</u>**[147CTR29,231ITR 814]**

Office **Bom.** **Maharaj Shri Govindlalji Ranchhodlalji - CIT**

Office means a position which requires the person holding it to perform certain duties and discharge certain obligations. An office may not depend on any law, or any contract or any mandate from the State or any authority. <u>*Sec.4,56.*</u>**[34ITR92Bom]**

Office appliance **Guj.** **Mehsana District Co-op. Milk Producers Union Ltd. - CIT**

Unless an appliance is capable of being primarily used in an office, it cannot be termed office appliance. <u>*Sec.32.*</u>**[171CTR350,256 ITR 322,121Taxman689]**

Office appliances **Guj** **CIT - Tarun Commercial Mills Ltd.**

The word appliance is qualified by the word office and, therefore, some meaning must be given to the word office and unless an appliance is capable of being primarily used in the office,it cannot be termed as office appliance. It must be, therefore, an appliance which is generally used in office as an aid or facility for the proper functioning of the office. It is difficult to lay down any formula for determining what are office appliances, but the initial test for determining the nature of the article is what is known as the test of common or popular parlance as understood by a person dealing with those articles. The second test would be the principal and primary use for which the goods are required and for which the same are capable of being used. The third test is what is known as the commercial test in seeing how the articles or goods are known in the world of trade and commerce. On application of any of these tests it is difficult to agree that the electrical fans and air-conditioners would be office appliances. By no stretch of imagination in trade and commerce or in popular parlance can they be said to be office appliances or equipments. Merely because these appliances are fixed in office premises they do not become, by that fact, office appliances.

*Sec.32,33.***[38CTR148,151ITR75,72Taxation196, 16Taxman18]**

Office premises **Guj.** **CIT - Statronics and Enterprises P. Ltd.**

The words office premises have not been defined in the Income-tax Act. The word office would share its character with the activities carried on in the said premises. Industrial premises would not come within the expression office premises if the said premises are used for data processing. *Sec.32(1)(iia).***[207CTR 96,288ITR455]**

Omission **Bom.** **Pannalal Nandlal Bhandari - CIT**

Omission is a colourless word which merely refers to the not doing of something, and if the assessee in fact does not make a return, it is an omission on his part, whether the law casts any obligation upon him to make a return or not.

*Sec.139,147.***[AIR1956Bom557, 30ITR57]**

2. A person cannot be said to have omitted or failed to disclose something when of such thing, he had no knowledge. **[P.R.Mukherjee-CIT,Cal.,**_Sec.147_**AIR 1956Cal.197,30ITR 535]**

On behalf of **Mys. G.T.Rajamannar - CIT**

The words on behalf of in section 41(1) and the proviso must be construed as equivalent to for the benefit of. _Sec.147._**[51ITR339]**

On or before a particular day **Mad. N.S. Balasubramaniam -State of Madras**

The use of the phrase on or before a particular day in a lease deed indicates that the lessee has the option of paying the rent before that date but the lessor can demand and enforce payment only on that date but not before. The rent will, therefore, accrue to the lessor only on that date and not earlier.

Madras AgriculturalIncome-tax Act. **[90ITR 377]**

On the occasion of the marriage **Mad. CGT - Dr.Neelambal Ramaswamy**

...the expression on the occasion of the marriage in section 5(l)(vii) could not be given any restricted meaning and if the gift was associated with the event of the marriage or if the reason for the gift or the immediate cause thereof was the marriage, it would be covered by the said expression. The relationship between the gift and the marriage was as the relevant factor and not the time of making the gift.Therefore,the gifts made by the assessee were entitled to exemption under section 5(l)(vii) of the Gift-tax Act.

Sec.5(1)(vii)G.T.Act.**[52CTR411,164ITR369]**

2.Though the word occasion has reference to the time of the marriage, it conveys the idea of association with the event,for example, the marriage event.The expression on the occasion of marriage is not equivalent in meaning to the phrase on the same date as.A daughter-in-law is not a dependent at the time the marriage is solemnised. She is certainly not a dependent at that juncture of time.The relationship starts after the marriage is solemnised.Before solemnisation of the marriage,there is no relationship stricto sensu.Even though the word relative is to be understood in a broader sense that cannot be said to include a prospective relative.Therefore, a gift to the daughter-in- law even subsequent to the marriage cannot be said to be a gift to a relative dependent for support and maintenance on the occasion of marriage. At the time of marriage, there is no question of a prospective daughter-in-law being dependent for support and maintenance. This thin line of distinction is to be made, otherwise the purpose of enactment of section 5(1)(vii) would be rendered illusory. Hence, a gift made to a daughter-in-law immediately after marriage would not be entitled to exemption under section 5(1)(vii).

[CGT-K.B.B.Subudhi, Ori, Sec.5(1) (vii) G.T.Act.**201ITR741]**

On which a penalty has been imposed Sec.271(1) **Del.** **V.S.Malhotra - Union of India**

The words in section 288(4), on which a penalty has been imposed under this Act other than a penalty imposed on him under clauses (i) and (ii) of sub-section (1) of section 271 contemplate a penalty imposed for an actual infringement of the Income-tax Act,1961, and not a deemed infringement of that provision brought about by the fiction created in section 297(2)(g). *Sec.271.r.w.s.288(4).***[88ITR110]**

One house used for residential purposes by **All.** **Shiv Narain Chaudhari - CWT**

Exemption can be claimed under section 5(1)(iv) of the Wealth-tax Act,1957, in respect of only one house used for residential purposes by the assessee.Two separate buildings situate in two different localities cannot be regarded as one house even if the assessee-Hindu undivided family were using both these buildings exclusively for residential purposes.Several self- contained dwelling units which are contiguous and situate in the same compound and within common boundaries and having unity of structure could be regarded as one house. *Sec.5(1)(iv)WT.Act.***[107ITR104]**

Operation **Mad.** **Anglo-French Textile Co. Ltd.-CIT**

It cannot be contended that the word operation in section 42(3) connotes a cross-section of the entire business including the purchase, manufacture and sale of goods resulting in the realisation of profits.Systematic and regular purchases of raw materials through an established agency in British India would come within the import of the term operation. *Sec.9.***[18ITR888]**

2. It is not every business activity of a manufacture that comes within the expression operation to which the provisions of section 42(3) are attracted. These provisions have no application unless according to the known and accepted business notions and usages the particular activity is regarded as a well-defined business operation.Activities which are not well defined or are of a casual or isolated character would not ordinarily fall within the ambit of this rule. **[Anglo-French Textile Co. Ltd.-CIT,SC,***Sec.9.***23ITR101]**

Operations carried out in India **AP.** **Addl. CIT - Skoda Export**

The Explanation to section 9(1)(i) of the Income-tax Act,1961, makes it clear that in the case of a business of which some operations are carried out in India and some outside India, the income of the business deemed to accrue or arise in India shall be only such part of the income as is reasonably attributable to the operations carried out in India. The Explanation contemplates a business comprising several operations some of which are carried out outside India. In such a situation, only the operations carried out within India shall be taken into account and a reasonable portion thereof shall be treated as income accruing or arising in India. The operations which are carried out outside India shall not be taken into account for this purpose.

Sec.9(1). **[172ITR358,37Taxman306]**

Opinion **Del.** **VLS Finance Ltd. - CIT**

Opinion means something more than mere retailing of gossip or hearsay; it means judgment or belief, that is, a belief or a conviction resulting from what one thinks on a particular question. It means : judgment or belief based on grounds short of proof. If a man is to form an opinion and his opinion is to govern, he must form it himself on such reasons and grounds as seem good to him.

*Sec.281B.***[163CTR343,246ITR707,112Taxman295]**

Opposed to public policy **MP.** **Addl. CIT- Badrinarayan Shrinarayan Akodiya**

Any donations to political parties or Government sponsored funds with the object of gaining some advantage in future would not be a deductible expense; such an action would be opposed to public policy. <u>*Sec.37(1).*</u>**[101ITR817]**

Or **P&H CIT - Punj. State Co-op. Bank Ltd.**

Clause (a)(i)of sub-section (2) of section 80P talks of a co-operative society engaged in carrying on the business of banking or providing credit facilities to its members. The carrying of on the business of banking by a co-operative society or providing credit facilities to its members are two different types of activities which are covered under this sub-clause. The word or used in this sub-clause cannot be read as and. If the literal reading of the whole of a section or sub-section or a clause is quite clear and there is no ambiguity, then the plain meaning to the section should be given effect and the word or should not be read as and. <u>*Sec.80(P)(2)(a).*</u>**[300 ITR24]**

Or otherwise **Ker. Narayanan and Co. - CIT**

The words or otherwise should be given the widest possible meaning and import. In its general sense, the expression transfer of property connotes the passing of rights in the property from one person to another. In the case of sale, the entire bundle of rights of the transferor passes on to the transferee. There are transactions wherein there may be a reduction of the exclusive interest in the totality of rights of the original owner in favour of others.

<u>*Sec.185.*</u>**[134CTR105,223ITR209,88Taxman99]**

2. The words or otherwise in section 10(10AA) of the Income-tax Act, 1961, are of very wide amplitude. These words shall draw the restricted meaning qua the immediately preceding word superannuation. ...The phrase or otherwise will cover only such an eventuality when there is severance of relationship of employer and employee and contract of employment. The words or otherwise will not cover such cases where there is no severance of relationship of employer and employee and the assessee continues to be under the employment of the same employer. **[CIT-AshokKumarDixit,All.,** <u>*Sec.10(10AA),17(1)(va).*</u>**195CTR375, 273ITR126, 144Taxman504]**

Or the extinguishment of any rights **Cal. Shaw Wallace and Co. Ltd. - CIT**

Under section 45 of the I.T. Act, 1961,a capital gain,or in the negative sense,a capital loss,can only occur on the transfer of a capital asset.A transfer always involves more than one party. There cannot be a transfer by a person to himself. Therefore, the expression or the extinguishment of any rights therein in section 2 (47) of the Act must mean the extinguishment of rights as a result of some operation involving more than one person. A consideration must be involved in a transfer before there can be any capital gain or capital loss. Sec.2(47)r.w.s.45.**[119ITR399]**

Order Bom. Petlad Bulakhidas Mills Co. Ltd. - Raj Singh

The expression order in section 33A(2) of the Indian Income-tax Act means an order of which the party affected has actual or constructive notice. The right to make an application for revision is given to an assessee against an order, and that right can only be effectively exercised if the party affected had knowledge, either actual or constructive, of that order. If the assessee had neither actual nor constructive knowledge,it cannot be said that there is an order within the meaning of section 33A(2) against which the assessee could possibly make an application for revision. Limitation should not be computed under section 33A(2) from a date earlier than that on which the assessee actually knew of the order or had an opportunity of knowing of the order. Sec.263.**[37ITR264]**

2.The word order in the expression from the date of the order sought to be amended in section 154 of the Income-tax Act,1961, is not qualified in any way. It does not necessarily mean the original order; it could be any order including an amended or rectified order.

[Salem Co-operative Spg. Mills Ltd.-CIT, Mad., Sec.154.**148 CTR112,230ITR139,93Taxman534]**

3.The expression order for the purposes of section 264 has a wide connotation. Any order passed by an authority subordinate to the Commissioner, other than an order to which section 263 applies, is subject to the revisional jurisdiction under section 264. A determination on an application under section 197 requires an order to be passed by the Assessing Officer after application of mind to the circumstances which are germane under section 197 and the Rules framed under sub-section (2A). Such an order would be subject to the Commissioner's revisional jurisdiction under section 264.

[Larsen and Toubro Ltd.-ACIT(TDS),Bom.,Sec.264.**232 CTR295,326ITR514,190Taxman373]**

Order of assessment Mad. CIT-T.V.Sundaram Iyengar and Sons Ltd.

The expression order of assessment under section 244(1A) of the Act should be given a wider meaning to envelop an order passed under section 104 of the Act as well. In so far as the assessee who paid the money in pursuance of the order under section 104 of the Act is concerned,whether the liability was imposed by virtue of the provisions of the Income-tax Act or by virtue of an order passed under section 104 of the Act,he is deprived of the use of the money as long as the order remains in force and when the said order is set aside or modified, the assessee is entitled to the refund of the amount.Though liability under section 104 is imposed by an order of the Income-tax Officer, the order has a statutory backing and further there is a statutory compulsion to pay the tax by virtue of the order and hence, there is absolutely no reason to restrict the meaning of the expression, order of assessment found in section 244(1A) of the Act only to an order of assessment made under section 143 or 144 of the Act. That apart section 104 of the Act uses the expression assessee is liable to Income-tax and the marginal note of the section also indicates that it is an Income-tax on undistributed income of certain companies. When the Act itself postulates that the payment made under section 104 of the Act is income- tax,the denial of interest under section 244(1A) of the Act on the ground that the order under section 104 of the Act is not an order of assessment is not justifiable in law.

Sec.244(1A). **[147CTR15,236ITR524]**

Order passed by the ITO **SC** **T.N.Civil Supplies Corporation Ltd.- CIT**

There is no scope for limiting the phrase order passed by the Income-tax Officer in section 263 to exclude orders passed by the Income-tax Officer on the directions of a superior authority either under section 144A or under section 144B.

Sec.263. **[180CTR307,260ITR82,129Taxman69]**

Order prejudicial **Cal.** **Smt. Phool Lata Somani - CIT**

The expression order prejudicial means the prejudicial effect of an order passed by the revising officer on the merits.

Sec.263. **[276ITR216]**

Ordinarily **Cal.** **Charki Mica Mining Co. Ltd. – CIT**

Section 30(1) gives a statutory right of appeal to the assessee relating to the amount assessed under section 23 of the Act. Section 30(2) provides that such appeal shall ordinarily be presented within 30 days of the receipt of the notice of demand. The word ordinarily only means that appeals should be filed within 30 days from the receipt of the notice of demand relating to the assessment and that, if under special circumstances, it cannot be filed within the prescribed time, the Appellate Assistant Commissioner can condone the delay in filing the appeal. It cannot be held to mean that in some cases the time will not begin to run until the receipt of the assessment order by the assessee. In all cases the period of limitation starts to run from the date of receipt of the notice of demand. Sec.246.**[111ITR193]**

Ordinarily resident Mad. V.E. Periannan - CWT

Section 5(1)(xxxiii) of the Wealth-tax Act,1957, is a provision meant to encourage persons of Indian origin or citizens of India who have lived abroad for a long time and acquired assets there and who decide ultimately to settle down in India permanently. Such persons have been granted exemption under the Wealth-tax Act in respect of the assets brought by them to India and reinvested in India. Section 5(1)(xxxiii) uses the words ordinarily resident in a foreign country with reference to the persons who are eligible to claim the benefit under the section and the further qualification required to be met by them is couched in language which leaves no doubt about the intention of the Legislature. The second qualification required for such persons is that they should return to India with the intention of permanently residing therein. Those words employed in the section clearly indicate that it was not a provision made to benefit persons who ordinarily reside in India who choose to go abroad for a short time and return to their original permanent home. Such persons are not those contemplated by the Legislature when this provision was incorporated. Though the word ordinarily is not defined under the section or elsewhere in the Act, the true scope of that word does not pose any major problem of interpretation as that word has to be understood in the light of the other words used in the section. Ordinarily resident in a foreign country must be read along with the other words which require an intention to permanently reside in India after return. Ordinarily in this context refers to residence of long duration outside the country. The duration being long enough for the person to regard himself as being ordinarily resident in the country outside India and not to regard India as his permanent place of residence.A person who normally resides in India and for whom India is a permanent residence cannot claim the benefit of the section merely by travelling abroad and residing abroad for a period of one year and thereafter returning to his own country.

Sec.5(1)(xxxiii).WT.Act. **[240ITR723]**

Ordinarily resident in India **Guj.** **Pradip J. Mehta - CIT**

Section 6(6) of the Income-tax Act,1961, does not define ordinarily resident in India, but describes not ordinarily resident in India. It resorts to the concept of resident in India for which the criteria are laid down in section 6(1) of the Act. Ordinarily resident for the purposes of Income-tax connotes residence in a place with some degree of continuity and apart from accidental or temporary absences. When an individual has been a resident in India for nine out of ten preceding years, then in order to escape tax on his foreign income, he must not have been in India for seven hundred and thirty days or more in the aggregate during the preceding seven years. The test is one of presence and not absence from India and the length of presence will determine when an individual is not ordinarily resident in India. In order that an individual is not an ordinarily resident, he should satisfy one of the two conditions laid down in section 6(6)(a) of the Act, the first condition is that he should not be resident in India in all the nine out of ten years preceding the accounting year and the second condition is that he should not have during the seven years preceding that year, been in India for a total period of seven hundred and thirty or more days.

Sec.6. **[175CTR394,256ITR647,123Taxman11118]**

Original cost **Pat.** **Motiram Roshan Lal Coal Co.**

The expression original cost appearing in sub-clause (vi) of clause 2 of section 10 of the Income-tax Act means the original cost paid by the assessee and not that paid by the predecessor in business of the assessee. *Sec.32.***[1ITR329]**

Original cost price **Mad.** **Asher Textiles Ltd. - CIT**

When valuing the closing stock of a trader according to the market value or cost price, whichever is lower at the option of the trader, the cost price should be taken as meaning original cost price and not a notional cost price. *Sec.145.*[**22ITR125]**

Other assets **Cal.** **Calcutta Insurance Ltd. - CIT**

Organisation expenses are not other assets within the meaning of rule 3(b) and setting off a portion of the profits against organisation expenses cannot be regarded as writing off the amount to meet loss on the realisation of those assets.Other assets contemplated by rule 3(b) are assets of the nature of investments which may appreciate or depreciate about which it is appropriate to speak of realisation and realisation of which may result in gain or loss. *Sec.32.* **[21ITR404]**

Other association of individual **Cal.** **Keshardeo Chamria**

The words other association of individuals in section 3 of the Indian Income-tax Act must be construed according to the ejusdem generis rule with reference to the word firm preceding it and they do not cover the members of an undivided Hindu Mitakshara family after a preliminary decree for partition has been made. The members of such family are in the same position as members of a Dayabhaga family and can be individually assessed in respect of their shares.

*Sec.4.***[5ITR246]**

2. The words other association of persons in section 3 of the Indian Income-tax Act,1922, have to be construed in their plain ordinary meaning and not ejusdem generis with the word firm immediately preceding or the other words going before that word. An association which produces income, profits or gains is assessable to tax by force of section 3. It is unnecessary in order to constitute an association that there should be any mutual rights or obligations among the members enforceable in a court of law. So long as the object and purpose of an association is to carry on business which is not illegal (the case of an association to commit crime being distinguishable as an instance of an association formed for a criminal or unlawful object) its income is not immune from Taxation because of the employment of unlawful means for achieving the income. The only test to be applied is whether the income falls within the purview of the charging section. **[K.P.G.B.U.G.M.S.S.A Mohamad Abdul Kareem & Co.-CIT,Mad.,***Sec.4* **16ITR412]**

Other educational institution **AP.** **CIT-Gurukul Ghatkeswar Trust**

An educational society, or a trust, or other similar body running an educational institution, solely for educational purposes, and not for the purpose of profit, can be regarded as other educational institution coming within section 10(22) of the Income-tax Act,1961. *Sec.10(22).*[**332ITR611]**

Other legal proceeding **SC** **S.V.Kondaskar,Official Liquidator - ITO**

The expression other legal proceeding in sub-section (1) and the expression legal proceeding in sub-section (2) of section 446 of the Companies Act, 1956, convey the same sense and the proceedings contemplated by both the sub-sections must be such as can appropriately be dealt with by the winding up court. The liquidation court cannot perform the functions of the Income-tax Officer while assessing the amount of tax payable by the assessee, even if the assessee be a company which is being wound up by the court. There is no principle on which the liquidation court should be vested with the power to stop assessment proceedings for determining the amount of tax payable by the company which is being wound up. The liquidation court would have full power to scrutinise the claim of the revenue after Income-tax has been determined and its payment demanded from the liquidator. It would be open to the liquidation court then to decide how far under the law the amount of Income-tax determined by the department should be accepted as a lawful liability on the funds of the company in liquidation. At that stage the winding up court can fully safeguard the interests of the company and its creditors. *Companies Act.*[**83ITR685]**

Other person **Del.** **S. R. Batliboi and Co.**

The words other person employed in the section must only be construed as referring to the other person having dealings or transactions with the party who is being searched or whose material is being seized.... *Sec.132* **[224CTR,369,315ITR137,181Taxman9]**

Other place of their work **Cal.** **CIT - Chemcrown (India) Ltd.**

The expression other place of their work used in Explanation 2 to section 37(2A) of the Income-tax Act,1961, would include any place where any of the employees of the assessee is asked to perform his work in connection with the employer's business. It has to be read in its context and given a simple meaning. It is not limited to expenditure on food and beverages provided to the employees of the assessee including any expenses either in the regular course or at the normal place of work or by way of terms of employment or otherwise. *Sec.37(2A).***[182CTR133,262ITR177,133Taxman579]**

Other sources Ori. Ramachandra Mardaraj Deo - CIT

Under section 2(6AA) (c) of the Income-tax Act, income from other sources can be earned income only if it is attributable to the direct personal effort of the assessee. The income must have been brought into being immediately by the personal exertion of the assessee and not by the exertion of an intermediate agency. *Sec.56.***[27ITR667]**

Otherwise transferred Mad. CIT - P.K.Ramaswamy Raja

On a reading of the provisions of section 32A(5), 34(3)(b), 155(4A) and 155(5) of the Income-tax Act,1961, it is clear that the intention of the Legislature is to withdraw the benefit of development rebate given to the assessee in respect of certain machinery if it is sold or otherwise transferred by the assessee within the stipulated period. The expression otherwise transferred is a very wide expression and takes within its sweep transfer of the assets from the assessee to another person by any means or mode whatsoever. *Sec.32A(5)(a), 34(3).r.w.s.155(4)(A).***[141CTR250,223ITR324,95Taxman173]**

2. The words otherwise transferred occurring in section 32A(5)(a) should bear an appropriate meaning in the context of the main provision, i.e., section 32A of the Act. Section 32A(5)(a) is closely linked to section 32A(1) of the Act. Keeping in view the purpose for which the relief by way of investment allowance is afforded under section 32A(1) of the Act, in cases where the machinery or plant is not wholly used by the assessee for the purpose of business carried on by him, for the specified period and such user is given over to another, it can be stated that the machinery or plant is otherwise transferred by the assessee to another person. **[CIT- Nipa Twisting Works, Guj.,***Sec.32A(5)(a).***[183CTR465, 263ITR697, 130 Taxman649]**

Out house **Cal.** **Consolidated Tea and Land Co. (India) Ltd. - CWT**

If there is a dwelling house or a main building and there is a house near that building or adjacent to that building and that will qualify for the expression out- house as contemplated by the section. It may also be that the main building may not be the dwelling house of the person who is entitled to exemption of the out-house. *Sec.2(e)(ii), 5(1)(ix).WT.Act.* **[76ITR584]**

Outstanding **Cal.** **CWT - Banarashi Prasad Kedia**

In order to be outstanding the amount must be such which the assessee was obliged to pay prior to the relevant valuation date and not an amount which the assessee had the right to pay subsequent to the valuation date. An amount which according to the installment scheme the assessee has an option or right to pay on a date subsequent to the relevant valuation date cannot in that context be said to be outstanding on the relevant valuation date. Though the expression outstanding according to the dictionary meaning is unsettled or unpaid, the expression outstanding, in the context of section 2(m)(iii)(b) of the Wealth-tax Act, would require careful handling and should be construed in relation to the expression amount payable in consequence of an order by the Income-tax Officer. *Sec.2(m)(iii).WT.Act.***[77ITR159]**

2. The expression outstanding in section 2(m)(iii)(a) and (b) has to be construed in the background of the phrase amount of tax . . . payable in consequence of an order and in that context it must mean remaining unpaid after the obligation to pay is incurred.

[CWT- J. K.Cotton Manufacturers Ltd.,SC, *Sec.2(m)(iii).WT.Act.***39CTR158,146ITR552, 16Taxman18]**

Own, ownership **SC** **Mysore Minerals Ltd. - CIT**

Section 32 of the Act allows certain deductions, one of them being depreciation of buildings, etc., owned by the assessee and used for the purposes of the business or profession. The terms own, ownership and owned are generic and relative terms. They have a wide and also a narrow connotation. The meaning would depend on the context in which the terms are used. CIT v. Podar Cement Pvt. Ltd. 226 ITR 625 (SC), is a case under the Income-tax Act and has to be taken as a trend-setter in the concept of ownership. Assistance from the law laid down therein can be taken for finding out the meaning of the term owned as occurring in section 32(1) of the Act. The term owned as occurring in section 32(1) of the Income-tax Act must be assigned a wider meaning. Anyone in possession of property in his own title exercising such dominion over the property as would enable others being excluded there from and having the right to use and occupy the property and/or to enjoy its usufruct in his own right would be the owner of the building though a formal deed of title may not have been executed and registered as contemplated by the Transfer of Property Act, the Registration Act, etc. Building owned by the assessee, the expression as occurring in section 32(1) of the Income-tax Act, means the person who having acquired possession over the building in his own right uses the same for the purposes of the business or profession though a legal title has not been conveyed to him consistently with the requirements of laws such as the Transfer of Property Act and the Registration Act, etc. Generally speaking depreciation is an allowance for the diminution in the value due to wear and tear of a capital asset employed by an assessee in his business. The very concept of depreciation suggests that the tax benefit on account of depreciation legitimately belongs to one who has invested in the capital asset and is utilising the capital asset and thereby losing gradually the investment caused by wear and tear, and would need to replace the same by having lost its value fully over a period of time. It is well-settled that there cannot be two owners of the property simultaneously and in the same sense of the term. The intention of the Legislature in enacting section 32 of the Act would be best fulfilled by allowing deduction in respect of depreciation to the person in whom for the time-being vests the dominion over the building and who is entitled to use it in his own right and is using the same for the purposes of his business or profession. Assigning any different meaning would not subserve the legislative intent.

*Sec.32(1).***[156CTR1,239ITR775,106Taxman166]**

Own, ownership and owned	**J&K.**	**CIT-Jammu and Kashmir Tourism Dev.Corp.**

The terms own, ownership and owned have meanings, both wide and narrow, and the term owned as occurring in section 32(1) must be assigned a wider meaning. Anyone in possession of property in his own title exercising such dominion over the property as would enable others being excluded there from and having the right to use and occupy the property and/or to enjoy its usufruct in his own right would be the owner of the buildings though a formal deed of title may not have been executed and registered as contemplated by the Transfer of Property Act, the Registration Act, etc. The expression building owned by the assessee as occurring in section 32(1) means the person who having acquired possession over the building in his own right uses the same for the purposes of the business or profession though legal title has not been conveyed to him consistently with the requirements of laws such as the Transfer of Property Act and the Registration Act, etc., but nevertheless is entitled to hold the property to the exclusion of all others. The very concept of depreciation suggests that the tax benefit on account of depreciation legitimately belongs to one who has invested in the capital asset, is utilising the capital asset and thereby losing gradually investment caused by wear and tear and would need to replace the same by having lost the value fully over a period of time.

*Sec.32(1).***[166CTR554,248ITR94,114Taxman734]**

Owned **AP.** **Mysore Minerals Ltd. - CIT**

The term owned as occurring in section 32(1) of the Income-tax Act, 1961, must be assigned a wider meaning. Any-one in possession of property in his own title exercising such dominion over the property as would enable others being excluded there-from and having the right to use and occupy the property and or to enjoy its usufruct in his own right would be the owner of the buildings though a formal deed of title may not have been executed and registered as contemplated by the Transfer of Property Act, 1882, the Registration Act, 1908,etc.

*Sec.32(1).***[156CTR1,239ITR775,152Taxation488, 106Taxman166]**

2. The term owned occurring in section 32(1) must be assigned a wider meaning. Anyone in possession of a property in his own title exercising such dominion over the property as would enable others being excluded there from and having the right to use and occupy the property and/or to enjoy its usufruct in his own right would be the owner though he may not have formal documents recognized as documents of title under the provisions of the respective law governing the subject. A lease does not extinguish the right of ownership of the lessor; nor does the lessee acquire any right of ownership thereon. **[S.B.I.Home Finance Ltd.-CIT,Cal.,** *Sec.32(1).***199CTR124, 280ITR6, 147Taxman103]**

3. The terms own, ownership and owned were considered by the Supreme Court as generic and relative in the decision in Mysore Minerals Ltd.v.CIT [239ITR775].The terms have a wide and narrow connotation. The meaning would depend on the context in which the terms are used. In the case of CIT v. Podar Cement P. Ltd. [226 ITR 625 SC], which has to be taken as a trend-setter in the concept of ownership, the Supreme Court held that the term owned as occurring in section 32(1) of the Income-tax Act,1961, must be assigned a wider meaning. Anyone in possession of prop-erty in his own title exercising such dominion over the property, as would enable others being excluded there from and having the right to use and occupy the property and/or to enjoy its usufruct in his own right would be the owner of the building though a formal deed of title may not have been executed and registered as contemplated by the Transfer of Property Act, the Registration Act, etc.

[UniversalRadiatorsLtd.-CIT,Mad.,*Sec.32(1).***281ITR261]**

Owned by the assessee A.P. CIT - Orient Longman (P.) Ltd.

A plain reading of section 32 of the Income-tax Act,1961, shows that to claim depreciation in respect of buildings, machinery, plant or furniture, they must be owned by the assessee and they must be used by the assessee for the purpose of his business or profession. For purposes of claiming benefit under section 32 of the Income-tax Act, the requirement of ownership by the assessee will be deemed to be fulfilled if the assessee has the dominion and control over the property in his own right. Where the assessee has paid full consideration and has been put in possession of the property and has been in exclusive possession and enjoyment of the property as absolute owner there of, the mere fact that no title deed has been executed in his favour, will not deprive him of the right to claim depreciation under section 32.

*Sec.32.***[145CTR142,227ITR68, 98Taxman189]**

2. That the expression owned by the assessee in section 32 of the Act had not been used in the sense of the property, complete title in which vested in the assessee. The assessee would be considered to be an owner under section 32 if he was in a position to exercise the rights of an owner not on behalf of the person in whom the title vested, but in his own right.

[CIT - Navdurga Transport Co., All.,
Sec.32 **149CTR219,235ITR158]**

3. Though the term owned by the assessee is used in section 32(1) there is nothing to show that the assessee should have remained the owner of the asset in question for the entire previous year in question. **[CIT - Fluid Controls Mfg. Co.,Guj.,** S*ec.32(1).***196CTR1,280ITR86,147Taxman139]**

Owner of a building Mad. CIT- Madras Cricket Club

In order that a person may be assessed as the owner of a building under section 9, it is not necessary that he should also be the owner of the land on which the building stands. *Sec.22.***[2ITR209]**

Owner Cal. Official Assignee for Bengal (Estate of Jnanendra Nath Pramanik)

In view of the provisions of section 17 of the Presidency Towns Insolvency Act, the Official Assignee was the owner of the property and he could rightly be assessed in respect of the income from such property under section 9 of the Indian Income-tax Act.

*Sec.22.r.w.Presidency Towns Insolvency Act.***[5ITR233]**

2. Before a person can be taxed as an owner under section 9 of the Indian Income-tax Act,1922, it must be decided that he is in fact the owner of the property in question and this decision rests with the Income-tax Officer subject to the rights of appeal under sections 30 and 31. **[Keshardeo Chamaria,Cal.,** *Sec.22.***5ITR246]**

3. Though the word owner in section 9 of the Indian Income-tax Act, might very well cover the legal owner, the equitable owner or even the owner of a limited estate carved out of a larger estate, it is not correct to say that in that section the expression the 'owner of the property' means the owner of the annual value of the property. In the case of a Hindu impartible estate, the owner of the property' for the purposes of section 9 is the Hindu undivided family and not its encumbent for the time being.

[CIT-Diwan Bahadur Diwan KrishanKishore,Lah.,*Sec.22.***7ITR427]**

4. Owner in section 9 of the Act means the owner of the property itself and not the owner of the annual value. A receiver appointed by court is merely managing the property for the time being under orders of the court and the property in his hands is in custodia legis for the person who can make a title to it.The receiver cannot therefore be assessed as owner of the property under section 9.His position is different from that of a trustee or official assignee.

[Raja P.C. Lall Choudhary-CIT,Pat.,<u>*Sec.22.*</u>**16ITR123]**

5. A person who is a full owner who is entitled to the whole income but who is under certain restrictions with regard to the disposition of the properties will be a owner within section 9 of Income Tax Act.

[Nawab Bahadur of Murshidabad-CIT, Cal., <u>*Sec.22.*</u>**AIR1956 Cal.242,28ITR510]**

6. A person may be assessed as the owner of a property under section 9 of the Act even though he has no right to alienate the property or his full right of ownership is in some other way subject to contractual or other limitations.**[Sri Ganesh Properties Ltd.-CIT,Cal.,**<u>*Sec.22.*</u>**44ITR606]**

7. For the purpose of Sec. 22, the owner must be that person who can exercise the rights of the owner, not on behalf of the owner but in his own right.

[RR Jodhamal Ruthiala-CIT,SC, <u>*Sec.22.*</u>**[82 ITR570]**

8. Builders build and sell flats to individual buyers. Upon payment of full consideration, buyer of a flat has a right to possess it to the exclusion of others.A buyer has exclusive right of possession, enjoyment and disposal of the flat. He has right to control, handle and dispose. This right may not be perfect in the sense that there has been no change of ownership from the builder to the buyer by a registered deed of conveyance, but his right as owner is in no way encumbered. The flat is not charged with any real right towards a third person. Accordingly, the word owner appearing in section 22 of the Income-tax Act, which is comprehensive and generic, must be construed in the setting of socio- economic development in the concept of ownership......For the purposes of section 22 of the Income-tax Act, 1961, the owner must be that person who could exercise the rights of the owner, not on behalf of the owner, but in his own right. Though equitable considerations are irrelevant in interpreting tax laws, those laws, like all other laws, have to be interpreted, reasonably and in consonance with justice. The meaning to be given to the word owner must not be such as to make that provision capable of being made an instrument of oppression. It must be in consonance with the principles underlying the Act.

[Madgul Udyog – CIT,Cal., <u>*Sec.22.*</u>**184ITR484]**

9. The flat which has been allotted to the assessee by the promoter of the multi- storeyed building against payment and of which the assessee had the full user and possession and from which the assessee cannot be evicted by the promoter must be held to be belonging to the assessee.

[CWT-Satyabama Ganeriwalla, Cal., *Sec.5(1)(iv).WT.Act.* **196ITR401]**

10. The expression owner used in section 32 of the Act has been considered by taking into account all its phases and aspects. The owner need not necessarily be a lawful owner entitled to pass on the title of the property to another.

[CIT-General Marketing andManufacturing Co. Ltd.,Cal., *Sec.32.***132CTR50, 222ITR574, 86 Taxman488]**

11. ...... under the common law owner means a person who has got valid title legally conveyed to him after complying with the requirements of law such as the Transfer of Property Act, the Registration Act, etc., in the context of section 22 of the Income-tax Act, 1961, having regard to the ground realities and further having regard to the object of the Income-tax Act, namely, to tax the income, owner is a person who is entitled to receive income from the property in his own right. The requirement of registration of the sale deed in the context of section 22 is not warranted.

CIT - Podar Cement Pvt. Ltd. SC *Sec.22.* **[AIR1997SC 2523141CTR67,226ITR 625, 140Taxation261,92Taxman 541]**

12. In order to get the benefit of depreciation allowance in respect of a building or property, the assessee must be the owner of the property. Without a valid registered document, the right, title and interest in immovable property does not pass to the transferee. Therefore, even where the assessee has paid the full consideration and is in exclusive possession, in the absence of a registered deed of conveyance in the assessee's favour, the assessee is not entitled to get the benefit of depreciation under section 32 of the Income-tax Act,1961.

[CIT - A.B.C.India Ltd.,Gau., *Sec.32.***143CTR155, 226ITR733,94Taxman7]**

13. The liability to pay Income-tax as owner is on the person who receives or is entitled to receive the income from the property in his own right. The requirement of registration of the sale deed in the context of section 22 of the Income-tax Act,1961, is not warranted. Although under the common law owner means a person who has got a valid title legally conveyed to him after complying with the requirements of law such as the Transfer of Property Act, the Registration Act, etc., in the context of section 22 of the Income-tax Act having regard to the ground realities and further having regard to the object of the Income-tax Act,namely, to tax the income, owner is a person who is entitled to receive income from the property in his own right.

[A.P.Small Scale Industries Dev.Corp.-CIT,AP, *Sec.22.* **233ITR453]**

14. Section 22 of the Income-tax Act,1961, which uses the word owner deals with assessment of income from house property while section 32(1) deals with allowing depreciation on buildings, etc., owned by an assessee. In both the provisions, the words owner or owned by are used in the context of enjoyment of the property by the owner. So long as a person continues to enjoy the property as an owner, he can make out the case for assessment of the income from the property under section 22 and claim depreciation thereon as per the provisions of section 32(1). Thus, the context in which the words owner or owned by are used, relates to deriving benefit by way of income or by personal use of the property. The real test is the right to enjoy the property as owner.This interpretation advances the object of the provision. Secondly,this interpretation also upholds the principle that in taxing statutes,when there is need for interpret-tation, it should be in a manner favouring the assessee. ......that though under the common law owner means a person who has got valid title legally conveyed to him after complying with the requirements of law such as the Transfer of Property Act, the Registration Act, etc., in the context of section 22 of the Income-tax Act having regard to the ground realities and having regard to the object of the Income-tax Act, namely, to tax the income, owner is a person who is entitled to receive income from the property in his own right. ....the requirement of registration of sale deed in the context of section 22 is not warranted. What is to be seen is who is in a position to enjoy the property or reap benefits from it as owner. An assessee seeks to have the income from his property assessed under section 22 only because he derives the income in his own right. The right to sell the property does not come in the picture. It is not germane to the issue. If for all practical purposes the assessee has a right to enjoy a property as owner he will be deemed to be the owner of the property even if a formal document conferring the title to the property in his favour is yet to be executed. The provisions of the Act have reference to the enjoyment of the property by the assessee as owner and nothing else........ the term owned as occurring in section 32(1) of the Act must be assigned a wider meaning. Any one in possession of property in his own title exercising such dominion over the property as would enable others being excluded there from and having the right to use and occupy the property and/or to enjoy its usufruct in his own right would be the owner of the buildings though a formal deed of title may not have been executed and registered as contemplated by the Transfer of Property Act, the Registration Act, etc. Building owned by the assessee, the expression as occurring in section 32(1) of the Act, means the person who having acquired possession over the building in his own right uses the same for the purposes of the business or profession though a legal title has not been conveyed to him consistently with the requirements of laws such as the Transfer of Property Act and the Registration Act but nevertheless is entitled to hold the property to the exclusion of all others.

[Gowersons Publishers (Pvt.) Ltd.- CIT, Del.,<u>*Sec.22,32.*</u>**240TR 191]**

15. The meaning of the term owner in section 32 of the Income-tax Act,1961, cannot be any different from what it is for the purpose of section 22. The object of the Act is to tax the income. The manner of computation of income is laid down in the Act.Under section 32 depreciation is one of the items to be considered in computing income. The allowance permitted under that provision is available to the person whose income is sought to be taxed. A person cannot be held to be an owner for the purpose of taxing his income under section 22 even when he is not the complete owner but be denied the benefit of depreciation under section 32 by restricting the meaning of the term owner in section 32 to persons who are complete legal owners.

[Tamilnadu Dairy Dev. Corp. Ltd.-CIT, Mad.,<u>*Sec.32.*</u>**156CTR409,240ITR191,107Taxman298]**

16. The word owner has not been defined in the Income-tax Act. Under the general law a person derives right, title and interest in respect of an immovable property only when a registered instrument is executed in his favour. However, registration of the deed alone does not make a person the holder of title if the same had been done in violation of any statute or if certain formalities as required under the statute have not been complied with. ...Owner is a person who is entitled to receive income from the property in his own right. Although the aforementioned decision was rendered in the context of section 22 of the Act having regard to the underlying principles enumerated therein, the same principles would also apply in the matter of interpretation of section 32.

[Ledo Tea Co. Ltd.-CIT,Cal.,<u>*Sec.32.r.w.s.22.*</u>**241ITR605]**

17. Owner means a person who has got valid title legally conveyed to him after complying with the requirements of law such as the

Transfer of Property Act, Registration Act, etc. But in the context of section 22 of the Income-tax Act, 1961, having regard to the object of the Act,namely, to tax the income, owner is a person who is entitled to receive income from the property in his own right.

[CIT-Ajit Kumar Roy,Cal.,<u>*Sec.22*</u>**170CTR187,252ITR468,121 Taxman240]**

18. The meaning of the word owner in the context of section 22 has been held to be a person who is entitled to receive income in his own right.3

[Balraj-CIT,Del.,<u>*Sec.22.*</u>**173CTR452,254ITR22,123 Taxman290]**

19. The concept of depreciation suggests that the tax benefit on account of depreciation legitimately belongs to one who has invested in the capital asset, is utilising the capital asset and is gradually losing investment thereby caused by wear and tear and would need to replace the same having lost its value fully over a period of time. There cannot be two owners of a property simultaneously and in the same sense of the term. Where the possession of a property is acquired, with a right to exercise such necessary control over the property acquired which it is capable of, it is the intention to exclude others which evinces an element of ownership.

[National Industrial Corporation Ltd.-CIT,Del., <u>*Sec.32.*</u>**177CTR194,258ITR575,124Taxman413]**

20. In CIT v.Podar Cement Pvt. Ltd. [226 ITR 625], the Supreme Court held that though under the common law owner means a person who has got valid title legally conveyed to him after complying with the requirements of law such as the Transfer of Property Act, the Registration Act, etc., in the context of the Income-tax Act, 1961, having regard to the ground realities and further having regard to the object of this Act, namely, to tax the income, owner is a person who is entitled to receive income from the property in his own right.

[CIT-Fazilka Dabwali TPT.Co. Pvt. Ltd.,P&H,<u>*Sec.22*</u>**270ITR398,144Taxman 376]**

21. The word owner has different meanings in different contexts and when a transporter of goods sells the goods and receives money for that, not on behalf of the real owner, it becomes the owner for the purpose of tax.

[D.N.Singh-CIT,Pat.,<u>*Sec.69A.*</u>**324ITR304]**

22. Owner is a person who is entitled to receive income from the property in his own right. In order to claim the benefit of section 32 of the Act, it is not necessary that the assessee should be a complete owner.The expression owner used in section 32 of the Act has been considered by taking into account all its facets and aspects.The owner need not necessarily be a lawful owner entitled to pass on the title of the property to another.

[CIT-Smt.A. Sivakami, Mad., *Sec.32.*322 ITR64]

23. The definition of owner of house property as contained in section 27 of the Income-tax Act,1961, was amended by inserting clause (iiia) by the Finance Act of so as to make the position clear and explicit. The amended provision reads as under:(iiia) A person who is allowed to take or retain possession of any building or part thereof in part performance of a contract of the nature referred to in section 53A of the Transfer of Property Act,1882, shall be deemed to be the owner of that building or part thereof. **[CIT-Nandanam Constructions,AP., *Sec.27.*137CTR501,222ITR737]**

24. An owner of property is the person who can exercise the rights of the owner not on behalf of the owner but in his own right. **[CIT-Sahney Steel and Press Works (P.) Ltd.,AP,*Sec.32.*63CTR275, 168ITR811, 32Taxman96]**

25. Anyone in possession of property in his own title exercising such dominion over the property, as would enable others being excluded there from and having the right to use and occupy the property and/or to enjoy its usufruct in his own right would be the owner of the building though a formal deed of title may not have been executed and registered as contemplated by the Transfer of Property Act, the Registration Act,etc. **[Universal Radiators Ltd.-CIT,Mad.,*Sec.32.*281ITR261]**

26. Before a person can be assessed under section 22 of the Income-tax Act,1961, it is not necessary that he must be the owner of the property by virtue of a sale deed in his favour. What is being taxed under section 22 is the income from house property or the annual value of the property of which the assessee is the owner.The focus of the section is on the receipt of income from house property.If, in a given case, it is found as a fact that the assessee is in occupation of the building as owner to all intents and purposes, except the sale deed in his favour, then he is liable to tax under section 22. **[Kala Rani-CIT,P&H,*Sec.22.*130ITR321]**

27. It is not necessary for the purpose of assessment under section 22 of the Income-tax Act, 1961, that an assessee should be the absolute owner of the properties and in given cases even a life estate holder can be equated to the owner of the property so long as he or she was in a position to enjoy the property or the income there from. However, even a life estate holder must have some vestige of rights in the property before he or she can be treated as the owner of the property. **[Mrs. M.P.Gananambal-CIT,Mad.,** *Sec.22.***136 ITR 103]**

Ownership **Cal. Madgul Udyog - CIT**

The term ownership is not merely a word of technical legal meaning but should be interpreted in its broadest possible meaning. It consists of a bundle of rights. What rights would constitute ownership in a given case would depend on the context in which it is used. Broadly, ownership rests in one who has dominion over the property which is the subject of ownership. If the ownership itself is the criterion of assessment under section 22 of the Income-tax Act,1961, one must find out whether an assessee has income from such property or not. Mere interest of a person in the property will not be enough. *Sec.22.***[184ITR484]**

P

Paid **SC** **J.Dalmia - CIT**

That the expression paid in section 16(2) did not contemplate actual receipt of the dividend by the member. In general, dividend could be said to be paid within the meaning of section 16(2) when the company discharged its liability and made the amount of dividend unconditionally available to the member entitled thereto.

Indian Companies Act. **[53ITR 83]**

2. A dividend is not paid within the meaning of section 16(2) of the Income-tax Act, 1922, as soon as it is declared, nor does it remain unpaid until it is actually received in cash by the shareholder. It is paid when the company discharges its liability and makes the amount of the dividend unconditionally available to the shareholder entitled to it. **[R.R.Khandelwal -CIT,Bom.,** *Sec10.***58ITR14]**

3. ......Notional payments are thus excluded from the ambit of section 285.Paid according to the dictionary (Chambers 20th Century Dictionary, New Edition, reprint of page 909) means paid in full; satisfied.The notional debit and credit as per the mercantile system of accounting would, thus, not be payment in terms of section 285. **[R.M.Donde, ITO-Mukundrai Kuberdas Katakia, Bom.,***Sec.285 .r.w.s.194A.***76CTR67,176ITR381]**

4. Word paid has been defined in section 43(2) to mean actually paid or incurred according to the method of accounting on the basis of which the profits and gains of business are computed. **[CIT-Gurunathan,Mad.,***Sec.43(2),30,37.***211 ITR174]**

5. The term paid appearing in section 36(1)(iv) of the Act means not only actually paid but also incurred according to the method of accounting upon the basis of which profits and gains are computed under the head Profits and gains of business or profession. The assessee was following the mercantile system of accounting. Paid occurring in section 36(1)(iv) has to be given the same meaning as has been assigned to it by the Legislature in section 43(2) of the Act unless the context otherwise requires.

[CIT-Tata Hydro Electric Supply Co. Ltd., Bom.,<u>*Sec.36(1)(iv).*</u> **132CTR137, 219ITR178,85Taxman444]**

6. The word paid in section 36(1)(iii) should be construed to mean paid in accordance with the method of accounting followed by the assessee.

[Taparia Tools Ltd.-JCIT,Bom. <u>*Sec.36(1)(iii).*</u> **260ITR102]**

Paint **Bom. CIT - Hardcastle and Waud Manu. Co. Ltd.**

The definition of the word paint makes it clear that paints are necessarily in liquid form. The opinion of experts in the field is also that powder coating for appliances,electrical equipment and other products to protect them from corrosion and improve their looks are essentially plastics. Ordinary paints are applied with brush, while powder coatings are applied by electrostatic spray or by fluidized techniques.

<u>*Sec.32A-Sch XI.*</u>**[142CTR168,233ITR732,97Taxman532]**

Parent **Kar. South Kanara Central Co-op.Wholesale Stores - CIT**

Section 54 of the Income-tax Act,1961, which deals with profit on sale of property used for residence,by its very terms, speaks of the residence of an individual or his parent. The expression parent can only be understood in relation to a human being and not in the case of an artificial person or society. <u>*Sec.45.r.w.s.54.*</u>**[114ITR298]**

Part of the profits of the partnership **Cal. Lovelock and Lewes - CIT**

A firm as such is not a person except for the limited purpose of Taxation. The firm is the compendious name for the partners who constitute it. Therefore, when the firm pays to its partners it pays to itself. Likewise, when the firm extends to the partners a benefit, the benefit avails to the firm itself. That is also the underlying rationale of section 40(b) of the Income-tax Act,1961, which gives the mandate that where the firm pays to the partners any salary, commission, remuneration or interest, the said payments should be treated as part of the profits of the partnership firm and should be added to its profits where such payments are debited to the profit and loss account. Sec.40(b). **[120CTR477,208 ITR95,122Taxation402,76Taxman501]**

Particular source All. Shiv Narain Sarraf - CIT

The proviso to section 2(11)(i)(a) refers to the case where an assessee was assessed for an earlier assessment year on income from a particular source. The words particular source in the proviso refer to a known or disclosed source; an undisclosed source can never be said to be a particular source. Therefore, the latter clause of the proviso is not applicable when an assessee was assessed for an earlier assessment year on income from an undisclosed source.

Sec.3. **[62ITR711]**

Particulars Chg. CIT - Vijay Kumar Jain

The meaning of the word particulars used in section 271(1)(c) would embrace the details of the claim made. Where no information given in the return is found to be incorrect or inaccurate, the assessee cannot be held guilty of furnishing inaccurate particulars.

Sec.271(1)(c). **[232CTR255,325ITR378]**

2. The meaning of the word particulars used in section 271(1)(c) would embrace the details of the claim made.Where no information given in the return is found to be incorrect or inaccurate, the assessee cannot be held guilty of furnishing inaccurate particulars. By no stretch of imagination can making an incorrect claim tantamount to furnishing inaccurate particulars. There can be no dispute that everything would depend upon the return filed by the assessee,because that is the only document where the assessee can furnish the particulars of his income. When such particulars are found to be inaccurate, the liability would arise. To attract penalty, the details supplied in the return must not be accurate, not exact or correct, not according to the truth or erroneous. **[Reliance Petro-products Pvt. Ltd., SC,***Sec.271(1)(c).***230CTR320,322ITR158]**

Partition AP. Satish Chandra Modi (HUF) - CIT

A transaction can be recognised as a partition under section 171 only if, where the property admits of a physical division, a physical division of the property has taken place; and where the property does not admit of a physical division, then such division, as the property admits of, should take place to satisfy the test of a partition under section 171; and if a transaction does not satisfy the above additional conditions it cannot be treated as a partition under the Act even though under the Hindu law there has been a partition total or partial.This means that the first basic requirement that has to be satisfied before the additional conditions are satisfied, is that the transaction should be a valid partition under the Hindu law applicable to the Hindu undivided family. Sub-sections (2) and (3) of section 171 of the Act state how such a finding of total or partial partition is to be given. A claim of partition should be made by the petitioner at the time of making the assessment in respect of the relevant assessment year.

Sec.171.[**128CTR155,216ITR717, 82Taxman208**]

Pass and pronounce Del. CIT - Sudhir Choudhrie

The fine distinction between the expressions pass and pronounce should not be stretched to the extent that it offends the basic rule of law. The expression pass in common parlance or its usual sense means to deliver. The term pronounce means to proclaim, to utter formally, to utter rhetorically, to declare, to utter articulate. Pronouncement in relation to a judgment requires the authority to apply its mind and arrive at a conclusion whether there is any cause to modify or remit the award. Further, the phrase pronounce judgment would itself indicate judicial determination by reasoned order, for arriving at a conclusion that decree in terms of the award be passed.

Sec.254.[**196CTR528,278ITR490,147Taxman306**]

Passes SC CED - Aloke Mitra

The word passes in section 5 may be taken as meaning changing hands on death regardless of its destination.Whenever property changes hands on death, the State is entitled to step in and take a toll of the property as it passed without regard to its destination or to the degree of relationship, if any,that may have subsisted between the deceased and the person or persons succeeding.

Estate Duty Act.[**126 ITR599,4Taxman19**]

Passing of an order All. Dilip Kumar Agarwal - CIT

In order to find out the meaning of the phrase passing of an order, we have to keep in mind that the nature of the order means a direction or a command. An order which is not communicated to the person who is obliged to comply with or execute the order, is no order in the eyes of law and becomes an order only when it is communicated to such person. The word pass, inter alia, means to put in circulation, to transfer from one person to another. Therefore, the use of the word passing makes the intention clear that the order has to be communicated to the declarant and the order would be deemed to have been passed only when it is communicated to the declarant... the mere preparing of the draft and signing it in the authority's own office did not amount to passing of an order. The order would legally stand passed when it was passed over to the declarant for compliance for payment of the amounts determined payable by the designated authority...

KVSS **[166CTR334,247ITR765,117Taxman335]**

Payable **Lah.** **Vir Bhan Bansi Lal - CIT**

Where notice under section 28 of the Indian Income-tax Act for the imposition of penalty has been issued before the assessment order is made,a penalty under the section may be imposed on a date subsequent to the date of the assessment order and even after the Income-tax assessed has been paid by the assessee. The word payable in the latter portion of the section means to which he has been assessed, whether the amount has been paid or not.

*Sec.271(1).***[6ITR616]**

2. The word payable not merely relates to fixation of the liability but also the liability to pay that amount.**[Inayat Hussain-Union of India,Bom.,***Sec.222* **122ITR227]**

3. The word payable is somewhat indefinite in import and its meaning must be gathered from the context in which it occurs; payable generally means that which should be paid. In the context of section 273A the word payable would mean that the assessee is liable to pay a particular sum as penalty. Even if the liability to pay penalty is discharged, it would not mean that he is not entitled to get relief under section 273A as otherwise no assessee would pay the penalty till the proceedings under section 273A are over.

[Garden Silk Weaving Factory-CIT, Guj.,*Sec.273A.***213ITR10]**

Payable and chargeable **Del.** **CIT - Vasavi Pratap Chand**

The words payable and chargeable occurring in section 2(m)(ii) of the Wealth-tax Act,1957, before and after its amendment in, are not interchangeable and they carry different meanings. Wealth-tax shall not be payable in respect of certain assets and such assets shall not be included in the net wealth of the assessee. Even in the Income-tax Act, the words chargeable and payable have been used in different provisions thereof.The assets and property may be chargeable to tax but tax may not be payable having regard to the exemption provided for in the Schedule appended to the Act. The exemption clause is to be strictly construed and cannot be interpreted in such manner so as to take away a benefit to which the assessee may otherwise be entitled. *Sec.2(m)(ii).WT.Act.*
[172CTR190,255ITR517,122Taxman792]

Payment Mad. CIT - Alagusundaram Chettiar

The word payment in section 2(6A)(e) of the Act means the act of paying and does not either expressly or by necessary implication restricted to payment of a sum of money towards a pre-existing liability or by way of discharge of an existing obligation or by way of payment to a person by way of hire or wages to which the payee was already entitled. *Sec.2(22).***[6CTR410,109ITR508]**

Payment made on periodical or ultimate settlement of the contract for purchase and.. Bom. CIT - Kamani Tubes Ltd.

From a reading of section 73 of the Contract Act, 1872, and particularly illustration (c) to the section, it is abundantly clear that failure to accept the goods in terms of the contract on the stipulated date amounts to a breach of the contract and in such a case, the other party is entitled to receive compensation for loss or damage caused by such breach. In the case of a contract for the sale of goods,the measure of damages upon a breach by the buyer is the difference between the contract price and the market price at the date of the breach. Such payment cannot be termed as payment made on periodical or ultimate settlement of the contract for purchase and sale of the commodity otherwise than by actual delivery or transfer of the commodity contemplated by section 43(5) of the Act. That would happen only in cases falling under section 63 of the Indian Contract Act, 1872.

Sec.28,43(5)73. **[207ITR298,121Taxation285,75Taxman55]**

Penal Mad. Cement Distributors Pvt. Ltd. - IAC

The word penalise somewhat ambiguous in its scope and content. A penalty may be the subject-matter of a breach of statutory duty or it may be the subject-matter of a complaint, depending on the context in which it is used in a given statute.All penalties do not flow from an offence as is commonly understood, but all offences lead to a penalty. Sec.271,277.**[87ITR163]**

2. The dictionary meaning of the word penal is imposing, constituting a punishment, a penalty; to put under a disadvantage.

Dr. S.Reddappa-Union of India, Kar. Sec234. **[149CTR521, 232ITR62]**

Penalty SC Shiv Dutt Rai Fateh Chand - Union of India

....the word penalty used in article 20(1) of the Constitution could not be construed as including a penalty levied under the sales tax laws by the departmental authorities for violation of statutory provisions.A penalty imposed by the sales tax authorities was only a civil liability though penal in character. Constitution of India **[148ITR664]**

Penalty for an infraction of the law MP. Simplex Structural Works - CIT

If the amount of penalty is such an expenditure which the assessee would have been required to incur, even if he had not broken the law, such an expenditure cannot, in the true sense, be termed as penalty for an infraction of the law. Sec.37(1).**[140ITR782]**

Pending Guj. CIT - Kashiram Textile Mills (P.) Ltd.

A proceeding is in pendency from the date of its institution and remains pending until its conclusion before the concerned authority. The entire period is to be treated as pendency of the proceedings concerned. That is the ordinary meaning of the word pending. Proceedings for assessment which commence from the date of filing or issuing the notice to file a return remain pending until a final effective order is passed by the assessing authority on the basis of which tax can be recovered or the aggrieved party can exercise his right of appeal or challenge it in whichever forum permissible. Until that stage is reached, it cannot be said that the assessment proceeding before the Income-tax Officer has come to an end. Sec.144A.**[170CTR1,252ITR162]**

2. . . . the expression pending in section 95(i)(c) in relation to revisions, means factually pending and merely because the revision petitions were not commenced within the period of limitation did not detract from the fact that the proceedings were pending... **[Shatrushailya Digvijaysingh Jadeja-CIT,Guj.,KVSS.177CTR 508,259ITR149,132Taxman644]**

3. The word pending occurring in the second proviso to section 153A of the Act is qualified by the words on the date of initiation of the search, and makes it abundantly clear that only such assessment or reassessment proceedings are liable to abate. The pendency of an appeal in the Tribunal against the order of assessment against which an appeal has been decided by the Commissioner (Appeals) is not a continuation of the proceedings of assessment. An appeal under the Income-tax Act lies to the Appellate Tribunal on a question of law. Even if it is pending on the date of search the reassessment proceedings will not abate. The abatement of any proceedings has serious effect inasmuch as it takes away all the consequences that arise thereafter. The material found in the search may be a ground for notice and assessment under section 153A of the Act but that would not efface or terminate all the consequences, which arise out of the regular assessment or reassessment resulting in a demand or proceedings for penalty. **CIT - Smt. Shaila Agarwal, All.** Sec.153A. **[246CTR266,346ITR130, 204 Taxman276]**

Pension **Cal. CIT - Dipali Goswami**

Pension generally is paid in consideration of past services. Pension is paid to an employee following his retirement from service due to age or disability or to the surviving dependants of an employee entitled to such pension. It may also represent a portion of employee's retirement income accumulated in the pension fund to which the employee had to contribute.Pension is one of the very important terms and conditions of employment which is earned by an employee by rendering requisite period of service and its receipt is one of the incidents of employment. In a sense, the benefits by way of pension and gratuity are in the nature of deferred wages which are paid at the time of retirement or thereafter. The fact that the amount is paid to the widow of an employee does not alter the nature of the payment and the widow is merely the recipient as a nominee of her husband. The right to pension either to the employee or to his widow necessarily flows from past employment of the employee. It flows as part of the consideration of the contract of employment and not de hors or independent of the contract of employment. *Sec.17,15.* **[156ITR36]**

Per annum **Cal.** **CIT - Oyster Packagers (P.) Ltd.**

The relief under section 80J of the Income-tax Act,1961,to a new industrial undertaking is available only for the first five years and that is the reason why the words per annum have been used. Per annum means for each year or annually. The exemption or concession will be available for each year to the extent of 6 per cent. of the capital employed. There is no warrant for the proposition that the expression per annum postulates the relief being granted only for the periods during which the assets were actually in use during the relevant previous year. *Sec.80J.* **[152ITR471]**

Performing specific services **Cal.** **Calcutta Stock Exchange Asso. Ltd. - CIT**

Performing specific services is an expression,for stronger and more definite than rendering service and it connotes the actual doing of definite acts in the nature of services.One may confer a benefit on another by some act or omission or even by merely maintaining a particular altitude and in such circumstances it may well be said that he has rendered a service to the latter.But in order that a person may be correctly described as performing specific services for another, he should execute certain definite tasks in the interest and for the benefit of the latter under an arrangement of a direct character. *Misc.* **[29ITR687]**

Period within which the ITO has to complete one **Mad.** **R.M.P.R.Viswanathan Chettiar - CIT**

The time limit of four years which sub-section (2) of section 34 provides is the period within which the Income-tax Officer has to complete one stage of the proceedings, that is, the assessment of the income and the determination of the tax payable. That stage can be completed by the Income-tax Officer himself within that period and it is not necessary that the terms of the order of assessment should be communicated to the assessee within that period. *Sec.147.* **[25ITR79]**

Periodically / Ultimately Raj. CIT - Aditya Mills Ltd.

Section 43(5) of the Income-tax Act,1961,lays down that a speculative transaction means a transaction in which a contract for the purchase or sale of any commodity is periodically settled otherwise than by the actual delivery or transfer of the commodity. The said section is not restricted to a contract where the settlement is only in respect of the entire contract. The word periodically or ultimately makes it clear that the provisions of section 43(5) are applicable where a part of the contract or the entire contract has been settled otherwise than by actual delivery of the goods.

*Sec.43(5).***[209ITR933]**

Permanent establishment AP. CIT - Visakhapatnam Port Trust

The expression Permanent establishment used in the Double Taxation Avoidance Agreements postulates the existence of a substantial element of an enduring or permanent nature of a foreign enterprise in another country which can be attributed to a fixed place of business in that country. It should be of such a nature that it would amount to a virtual projection of the foreign enterprise of one country into the soil of another country.

*Misc.***[144ITR146,15Taxman72]**

2. ..the term permanent establishment means a fixed place of business through which the business of an enterprise is wholly or partly carried on. **[Dun and Bradstreet Espana S.A.,AAR,** *Sec.44D,115A,195.***193CTR9,272ITR99,142Taxman284]**

Permissible deduction SC Indian Aluminium Co. Ltd. - CIT

..... to be a permissible deduction, there must be a direct and intimate connection between the expenditure and the business, i.e., between the expenditure and the character of the assessee as a trader, and not as owner of assets, even if they are assets of the business, needs to be qualified by stating that if the expenditure is laid out by the assessee as owner- cum- trader, and the expenditure is really incidental to the carrying on of his business, it must be treated to have been laid out by him as a trader and as incidental to his business. Sec.37.**[1CTR51,84ITR735]**

Perquisite Guj. CIT - S.G.Pgnatale

A perquisite is something which arises by reason of a personal advantage. A mere reimbursement of a necessary disbursement would not amount to a perquisite. *Sec.17(2)(ii).* **[124ITR391]**

2. The definition of perquisite in sub-cl.(ii) of section 17(2) of the I.T. Act, 1961, extends the meaning of that term by including therein the value of any concession in the matter of rent respecting any accommodation provided to the assessee by his employer.If the employer gives no concession to the employee in the matter of rent, there can be no question of any perquisite. A case of concession in the matter of rent would arise when the rent normally payable for the accommodation is higher than the rent paid.

[Officers'Association,Bhilai Steel Plant-Union of India, MP, *Sec.17(2)(ii).***139ITR937,6Taxman281]**

3. The definition of perquisites in section 17(2) begins with the words perquisite includes.The definition is, therefore, comprehensive.Cash allowances are included in the ordinary meaning of perquisites and would fall under section 17(2). City compensatory allowance, bad climate allowance,shift allowance and incentive bonus are perquisites within the meaning of section 17(2) and, therefore, they constitute taxable receipts. **[M.Krishna Murthy-CIT,AP,**Sec.*17(2)(ii).***75CTR36,152ITR163]**

4. The word perquisite covers the value of rent free accommodation,the value of any benefit or amenity granted or provided free of cost or at concessional rate in certain stipulated cases. Parliament inserted clause (vi) in section 17(2) by the Finance Act No. 14 of 2001 with effect from April 1,2002 ,by stating that even the value of any fringe benefit as may be prescribed by the Central Board of Direct Taxes be included within perquisite as defined under section 17(2). The words as may be prescribed not only refer to value, but also to itemisation/ identification of a particular benefit or amenity. Hence, the entire policy of Parliament is to cover all types of benefits/amenities, including substantial and fringe benefits within the word perquisite. The policy is discernible from the scheme of sections 14, 15, 16 and 17 of the Income-tax Act. Further, the Legislature has left it to the Central Board of Direct Taxes as the Central Board of Direct Taxes is an expert rule making authority. The concept of perquisite, so as to include benefit, can be understood, but not defined. Hence the Legislature has left it to the Central Board of Direct Taxes to define the benefit or amenity and to assign it a value so that the latter could form part of the total income. **[National Federation of Insurance Field Workers of IndiaUnion of India,Uttr.,**<u>*Sec.17(2)(ii).*</u> **187CTR180, 265ITR84,135Taxman307]**

5. A perquisite is a privilege, gain or profit incidental to an employment in addition to regular salary or wages. It is a benefit or an advantage received by the holder of an office over and above his salary. The benefit received by an employee is incidental to employment in excess of or in addition to the salary. In computing a perquisite in the matter of residential accommodation, the fundamental question of applicability of section 17(2) of the Act still remains. It cannot be gainsaid that section 17(2) would apply only if there is a perquisite.Indisputably, the definition of perquisite is inclusive in nature and takes within its sweep several matters enumerated in clauses (i) to (vii). Section 17(2)(ii) declares that the value of any concession in the matter of rent respecting any accommodation provided to the employee by his employer would be a perquisite. Nevertheless, it must be a concession in the matter of rent respecting any accommodation provided by the employer to his employee. It is, therefore, clear that before section 17(2)(ii) can be invoked or pressed into service and before calculation of concession as per rule 3, even after its amendment in 2001, is made, the authority exercising power must come to a positive conclusion that it is a concession...... A benefit or facility which furthers the commercial interest of the employer would not per se become a perquisite. Such facility of accommodation furthers the commercial interest of the employer by having a satisfactory work force which but for such accommodation, would not have been available. **[Arunkumar - Union of India,SC,**<u>*Sec.17(2)(ii).*</u>**205CTR193,286 ITR89,155 Taxman659]**

Perquisites **Ker. Aspinwall and Co. Ltd. - CIT**

Section 17(2) of the Income-tax Act,1961, provides what is known as an inclusive definition of the term perquisites as distinguished from the provisions of clause (b) of Explanation 2 to section 40A(5) of the Income-tax Act,1961,which is self-explanatory. It would appear that five categories of situations which are understood to be perquisites show that the term perquisites has certain essential characteristics and features. Rent-free accommodation and concession in the matter of rent in regard to accommodation provided to the employee by the assessee would be a perquisite. <u>*Sec.17(2).r.w.s.40A*</u> **[132CTR365,220ITR611]**

2. The term perquisites has been defined in an inclusive and expansive manner in the sub-clauses of section 17(1). Any sum paid by the employer in respect of any obligation which, but for such payment, would have been payable by the assessee (employee) is treated and regarded as a perquisite. A distinction has been drawn between the term perquisite and wages. Perquisite means a benefit and an advantage received in addition to or over and above what is paid as wages and salary. It is something extra, in addition to, and other than wages/pay. **CIT - Telsuo Mitera, Del.** _Sec.17(2)._ **[345ITR256]**

Person **All. Mohd. Ishaq - ITO**

The word person in section 16(3)(b) Income tax Act is wide enough to cover all juristic entities which may be capable of having property vested in them, of receiving property and of transferring property. God almighty is a person. _Sec.64._**[AIR1951All512, 19ITR70]**

2. A firm is a person, a firm and an individual or group of individuals may form an association of persons within the meaning of section 3 of the Income Tax Act. **[M. M Ipoh – CIT, SC,** _Sec.2(31)._ **AIRSC317,67ITR106]**

3. The expression person in sections 276, 276A and 277 of the Income-tax Act,1961, is not used in the sense in which it is defined in section 2(31) of the Act. **[Kapurchand Shrimal -TRO, SC,** _Sec.276,277.r.w.s.2(31)._**72ITR623]**

4. The definition of person in section 3(42) of the General Clauses Act, 1897, applies to the word person occurring in section 4 of the Indian Income-tax Act, 1922, and an unincorporated association or a body of persons, like a firm, is a person for the purposes of income -tax. **[Bhai Sunder Dass and Sons-CIT, Del.,** _Sec.2(31)._**85ITR28]**

5. The description of personin section 2(31) of the Income-tax Act, 1961 makes no distinction between a firm or its partners, association of persons or body of individuals, whether incorporated or not, or the members of the association or the body of individuals in their individual capacity. The charging section of the Act,that is, section4,views each category of the taxable entity alike as a distinct and different unit. In the case of a firm it would mean a firm or a partner. Similarly,in the case of an association of persons or body of individuals, it would mean the association of persons or the body of individuals or the members. **[Rodamal Lalchand -CIT,P&H, *Sec.2(31)*109ITR7]**

6. In the context in which the expression person occurs in section 277 of the Income-tax Act,1961,it is evident that it must be given its ordinary dictionary meaning which would include any individual who fails to carry out the duty imposed on him by the specific provisions of the Act.It is not used in reference only to an assessee as defined in the Act. **[M.R.Pratap -V.M. Muthukrishnan, ITO,Mad.,*Sec.277*110ITR655]**

7. The word person which has been defined in section 2(xviii) of the Gift-tax Act,1958, includes a Hindu undivided family or a company or association or a body of individuals or persons, whether incorporated or not. This definition of the word person is not an exhaustive one and the categories specified therein do not exhaust the category of other persons,who answer the description of a person. **[CGT-S.B.SugarMills,All.,*Sec.2.G.T.Act* 120ITR126]**

8. .... the definition of the expression person occurring in section 2(31) of the Income-tax Act, 1961, is a very crucial definition because it is with reference to the categories of entities specified in section 2(31) that the liability to tax under the Act is determined. If a person is not capable of being considered as a person within the meaning of section 2(31),then no liability attaches. If the State or the Government cannot be regarded as a person for the purpose of section 2(31) and,consequently, is immune from Taxation, whether on the grounds of sovereignty or otherwise,it is natural to extend the same logic to understand the expression person wherever it occurs in the Act.There is no reason to give a different expression to the word person which occurs in the Explanation to section 32(1)(vi).The Explanation to section 32(1)(vi) does not use the expression person in isolation. It uses that expression in the company of the other qualification, namely, person resident in India. Therefore, what we have to understand for the purpose of the Explanation is the consolidated expression person resident in India. A peremptory look at the expression person resident in India itself would indicate that the person referred to in the Explanation must be one who is capable of residing in India or, in the alternative, is a person resident in India as provided in section 6 of the Act. The Government is not a person capable of having residence either on its own or in terms of section 6. **[CIT-Dredging Corporation of India,AP.,*Sec.32 (1)(vi).* 174ITR682,39Taxman301]**

9. Section 3 of the Wealth-tax Act,1957, imposes a charge for every year commencing on and after the 1st day of April,1957, of a tax in respect of net wealth on the corresponding valuation date of every individual,Hindu undivided family and company.The difference between the definition of a person contained in section 2(31) of the Income-tax Act,1961, and the persons who are taxable under section 3 of the Wealth-tax Act has to be noted. An association of persons or a body of individuals, whether incorporated or not, comes within the definition of person for the purpose of the Income-tax Act,whereas only three categories of persons are contemplated under section 3 of the Wealth-tax Act. **[CWT - Mulam Club, Ker.,**<u>*Sec.3.WT.Act.*</u> **191ITR370]**

10. Section 22(1) of the Gift-tax Act,1958, deals with a class of persons who can prefer appeals under that section. Section 22(1) in the opening part gives such right subject to the provisions of sub-section (1A) to any person. In contrast, the other sections in the Chapter which deal with appeals and revisions confer the right only on assessees.The marked difference in the expression used in the provisions indicates the legislative intention to confer the right of appeal under section 22 to a larger class than assessees. The term any person used in section 22(1) of the Act is therefore required to be construed as comprising persons other than assessees if they object to orders of the nature referred to in sub-section (1) of section 22 which orders affect them adversely.... the term person in section 22(1) would include the donee against whom recovery proceedings were taken under section 29 of the Act. **[CGT - A.C. Mahesh, Mad.,**<u>*Sec.22.G.T.Act.*</u>**172CTR647, 252ITR440,122Taxman269]**

11. When a person fails to pay to the credit of the Central Government,the tax deducted at source by him, then he is.. punishable with rigorous imprisonment for a term which shall not be less than three months but which may extend to seven years and with fine. The reference to person in section 276B is with regard to the requirement of such person to pay to the credit of the Central Government the tax deducted at source collected by him. Section 192... A reference to the provisions of section 204(i) and (iii) would make it clear that in case the payments are to be made by a company, then the expression person responsible for paying shall include the principal officer thereof. The managing director of a company,by the very nature of his office,would be a principal officer of the company.

[CIT Sushil Suri, Del., <u>*Sec.276B.r.w.s.204(i)(iii).*</u> **206CTR662,303 ITR86,160Taxman31]**

Person concerned	**Mad.**	**Gulab and Co.-Supdt.of Central Excise (Pre.), Trichy**

The words person concerned in section132(5)can refer only to the person against whom a warrant is issued and not the person who was in actual possession.That does not mean that the person who was in physical possession and from whom the actual seizure was effected need not be given any opportunity of being heard. He is a person who is entitled to be heard both under the provisions of the Act and under the principles of natural justice. *Sec.132(5).* **[98ITR581]**

Person responsible for paying **PH.** **Baldeep Singh - Union of India**

Section 194A(1) of the Income-tax Act,1961,refers to person responsible for paying any income by way of interest.Section 204 defines the expression person responsible for paying. According to this section, the person responsible for paying,in the case of payments of income chargeable under the head Salaries other than payments by the Central Government and the Government of a State, is the employer himself,in the case of interest on securities, the borrower,and regarding any other sum chargeable under the Act, the payer himself. *Sec.194A(1).***[199ITR628]**

Person who is substantially interested in the company **Mad.** **CIT T.P.S.H. Sokkalal**

The expression person who is substantially interested in the company is defined in section 2(32) of the Income-tax Act, 1961, to mean a person who is the beneficial owner of shares, not being shares entitled to a fixed rate of dividend whether with or without a right to participate in profits, carrying not less than twenty per cent. of the voting power.The definition is relevant in considering the applicability of section 2(22)(e) as the said provision would apply only in the case of advance or loan made by a company in which the public are not substantially interested,to a person having a substantial interest in the company. Section 153 of the Companies Act, 1956, provides that no notice of any trust, express, implied or constructive,shall be entered on the register of members or of debenture-holders.There is no provision of law which would deprive the shareholder from exercising the right of voting when the shares are standing in the name of the guardian for the benefit of minors and the right to vote is an essential and important characteristic of a share.The expression voting power found in section 2(32) is significant and it signifies the right to vote which is inherent in the shares. Sec.2(31).**[236ITR981]**

Personal effect SC H.H. Majaraja Rana Hemant Singhji – CIT

An intimate connection between the effects and the person of the assessee must be shown to exist to render them personal effects within the meaning of that expression used in clause (ii) of the exceptions in section 2(4A) of the Indian Income-tax Act, 1922. The legislature intended only those articles to be included within the expression personal effects which were intimately and commonly used by the assessee. Sec.2(14).**[103ITR61]**

2. The Legislature intended only those articles to be included within the expression personal effects which were intimately and commonly used by the assessee.**[CIT-Benarashilal Kataruka,Cal.,** Sec.2(14),45.**185ITR493]**

Personal effects Raj. Maharaja Rana Hemant Singhji - CIT

In order to constitute an article part of personal effects, it is necessary that the article must be associated with the person of the possessor and must more or less have intimate relation with the possossor. Merely because gold sovereigns, silver coins and bullion are placed before Goddess Lakshmi at the time of Puja, they do not became articles of personal use of the assessee. Sec.2(14). **[AIR1976SC662,77ITR1007]**

Personal expenses SC State of Madras - G.J. Coelho

Every expense to discharge a personal obligation did not become a personal expense within section 5(e).Personal expenses included only expenses on the person of the assessee or to satisfy his personal needs such as clothes, food, etc., or for purposes not related to the business for which the deduction was claimed. *Madras Plantations Agricultural Income-tax Act.***[53ITR186]**

Personally Mad. CIT - Rao Bahadur Ravula Subba Rao

The word personally in rule 6 of the Income-tax Rules as framed under section 59 of the Indian Income-tax Act,1922, would exclude a duly authorised agent of a partner from signing an application on behalf of the partner under section 26A of the Indian Income-tax Act. Sec.187.**[14ITR232]**

Persons **Cal.** **Aminchand Pyarelal - CGT**

A firm is not included in the definition of persons under section 2 (xviii) of the Gift-tax Act,1958. Hence, a firm is not assessable as an entity under the Gift-tax Act and a notice issued to a firm would not be valid. *Sec.2(xviii).*[**185ITR264**]

Persons interested **Guj.** **CIT-Premanand Industrial Co-op. Service Soc. Ltd.**

The definition of persons interested in section 269A(g) of the Income-tax Act,1961, is an inclusive definition. In section 3(b) of the Land Acquisition Act, 1894, the expression person interested includes all persons claiming an interest in the compensation to be made on account of the acquisition of land. The first part of section 3(b) of the Land Acquisition Act is in pari materia with the definition in section 269A(g) of the I.T. Act, and since both these definitions occur in the context of acquisition of property whatever interpretation has been given to the words persons interested in the context of the Land Acquisition Act will also help in considering the meaning to be attributed to the words persons interested in the context of Chapter XXA of the I.T. Act.*Sec.269.* [**16CTR6,124ITR722**]

Perverse decision **Cal.** **Kejriwal Enterprises - CIT**

When an authority draws a conclusion which cannot be drawn by any reasonable person or authority on the disclosed state of facts, then a perverse decision is entered and a perverse decision is wrong in law. *260A.*[**181CTR 305,260 ITR 341,133Taxman749**]

Perversity **Del.** **CIT - D. S. Promoters and Developers P. Ltd.**

Perversity has been defined as indicative of an action, opinion or conclusion which could not reasonably be arrived at. Even an incorrect conclusion would be perverse or mala fide only if it is patently deliberate. *Sec.260A.*[**330ITR291,183Taxman153**]

Philanthropy **Ker.** **CIT - Pulikkal Medical Foundation P. Ltd.**

The word philanthropy is not defined in the Income-tax Act.The meaning of the words philanthropic purposes includes activities promoting goodwill to mankind or activities beneficial to humanity at large as opposed to activities solely for the benefit of a few individuals. .... In order to achieve the main philanthropic objects,the hospital may do some profit earning business provided such profit is appropriated towards the expansion and development of the hospital or to start another institution with the same philanthropic objectives.The real test is to find out what the dominant or primary purpose of the institution is. If the primary purpose is philanthropic,the inclusion of some objects for earning profit for the implementation of the primary object would not alter the character of that primary object. In other words, this will not be a ground for holding that the hospital is not existing solely for philanthropic purposes. All cumulative factors will have to be taken into consideration in order to decide whether the institution exists for philanthropic purposes and not for purposes of profit. Neither the fortuitous factor of having large surplus in any particular year, nor the fact of diverting some income to objects which are not philanthropic in themselves would be decisive of the matter.

<u>Sec.10(22A).</u>**[210ITR299]**

Plant Bom. Jayasingrao Piraji Rao Ghatge - CIT

The primary meaning of the word plant is machinery, apparatus, fixtures, etc., employed in carrying on a business or trade or a mechanical or other industrial business.The primary meaning of the word,therefore, has connection with mechanical or industrial business or manufacture of finished goods from raw materials. Even in the extended meaning of the word, it would only cover an asset representing capital investment in a manufacturing trade or business because, according to the said extended meaning, it must be something which represents capital invested in the means of carrying on business exclusive of its raw materials or the manufactured product.Neither the primary meaning nor the extended meaning of the word plant given in the dictionary will cover the container of the stock-in-trade of a business as a plant within the meaning of section 10(2)(vi) of the Indian Income-tax Act.

<u>Sec.32.</u>**[AIR1963Bom66,46ITR1160]**

2. The word plant in section 10(2)(vib) of the Income-tax Act, 1922, should be given the same popular meaning as machinery. If the plant in combination with other appliances in the business effectuates and perpetuates the trade or commerce in question, then such induction or introduction of such a plant should be deemed to be such that they are placed in a position for service or use in the business. **[Sundaram Motors Private Ltd.-CIT,Mad.,**<u>Sec.32.</u>**71 ITR578]**

3. The definition of plant in section 10(5) is very wide and poles, cables,conductors and switch-boards for distribution of electricity can be treated as plant within the meaning of section 10(2)(vib) of the Act. **[CIT - Indian Turpentine and Rosin Co.Ltd.,All.,** *Sec.32.***75ITR533]**

4. Sanitary and pipeline fittings fell within the definition of plant in section 10(5). **[CIT-Taj Mahal Hotel,SC,** *Sec.32.***AIR1972SC168, 82ITR44]**

5. The word plant in its ordinary meaning is a word of wide import and it must be broadly construed having regard to the fact that articles like books and surgical instruments are expressly included in the definition of plant in section 43(3) of the Act. It includes any article or object, fixed or movable, live or dead, used by a businessman for carrying on his business. It is not necessarily confined to an apparatus which is used for mechanical operations or processes or is employed in mechanical or industrial business.It would not, however, cover the stock-in-trade,that is,goods bought or made for sale by a businessman. It would also not include an article which is merely a part of the premises in which the business is carried on. An article to qualify as plant must furthermore have some degree of durability and that which is quickly consumed or worn out in the course of a few operations or within a short time cannot properly be called plant. But an article would not be any the less plant because it is small in size or cheap in value or a large quantity thereof is consumed while being employed in carrying on business.In the ultimate analysis the enquiry which must be made is as to what operation the apparatus performs in the assessee's business.The relevant test to be applied is: does it fulfil the function of plant in the assessee's trading activity? Is it the tool of the taxpayer's trade ? If it is, then it is plant,no matter that it is not very long lasting or does not contain working parts such as a machine does and plays a merely passive role in the accomplishment of the trading purpose. **[CIT - Elecon Engineering Co. Ltd., Guj.,***Sec.43(3).***96ITR672]**

6. In common parlance the word plant includes within its ambit buildings and equipment used for manufacturing purposes. The definition of plant in section 43(3)is inclusive and does not exclude things normally included in it. **[CIT - Kanodia Cold Storage, All.,***Sec.43(3).***100ITR155]**

7. The definition of plant in section 43(3) of the Income-tax Act, 1961,is only inclusive and not exhaustive. Adopting the dictionary meaning of plant, a safe deposit vault in a bank is apparatus or fixture employed in carrying on a trade or business which is not its stock-in-trade and would fall within the meaning of a plant. Therefore,such safe deposit vault will be entitled to development rebate as provided in section 35 of the Income-tax Act,1961.**[CIT-Union Bank of India Ltd.,Bom.,*Sec.43(3)*102ITR270]**

8. The definition of plant in section 43(3) of the Income-tax Act, 1961,being an inclusive definition, the term should be given its ordinary meaning as understood by the common man. The various decisions also go to show that the expression plant has to be construed in the context of particular kind of trade or manufacture carried on by the assessee and if in the context it could be taken as plant as understood in the popular sense it would be eligible for the allowance of depreciation and development rebate.The dictionary meaning of the word plant comprehends buildings employed in carrying on trade or other industrial business and hence the special reinforced concrete foundation for the purpose of locating or installing the rotary kiln in the assessee's factory would come within the scope of the expression plant and would be entitled to depreciation and development rebate.**[Addl.CIT-Madras Cements Ltd.,Mad.,*Sec.43(3)*110 ITR 281]**

9. In view of the Supreme Court decision in CIT v.Taj Mahal Hotel (82 ITR 44)where the term plant has been construed in a wide manner the assessee was entitled to development rebate in respect of capital items like coal tubs, cast iron pipes, winding and guiding ropes, etc., subject to the assessee's compliance with the requirements of law in respect of appropriate reserves. **[CIT-Swadeshi Mining and Manu.Co. Ltd.,Cal., *Sec.33A.*116ITR259]**

10. The definition of plant in section 43(3) of the Income-tax Act, 1961,is of wide amplitude so as to take in even a well, provided that the well was dug for the purpose of carrying on the business of the assessee. **[CIT-Warner Hindustan Ltd.,AP., *Sec.43(3).*7CTR228,117ITR15]**

11. Plant has been held to include whatever apparatus or instrument used by a businessman for carrying on his business- not his stock-in-trade which he buys or makes for sale or the place where the business is carried on; but all goods and chattels, fixed or movable, live or dead, which he keeps for permanent employment in his business. **[CIT-Bank of India Ltd.,Bom., *Sec.33. r.w.s. 43(3).* 8CTR230,118ITR809]**

12. Section 43(3) of the Act states that plant includes ships, vehicles,books, scientific apparatus and surgical equipment used for the purpose of the business.The definition is not an exhaustive one. As time passes it is not only tangible assets that depreciate but also intangible assets like technical knowledge that become obsolete as progress is made in scientific research.Moreover, when technical know-how necessary for the business is acquired by incurring expenditure, there is no justification in denying appropriate deduction in respect of its cost while computing taxable profits if it can be brought under the heading Plant.Hence, the word plant, when it is construed liberally, includes within its meaning such designs and blue-prints also. **[Nippon Electronics (P.) Ltd.–CIT, Kar.,***Sec.43(3).***116ITR231]**

13. The word plant, in its ordinary meaning, is a word of wide import and, in the context of section 32 of the I.T.Act, 1961, it must be broadly construed.It includes any article or object,fixed or movable,live or dead,used by a businessman for carrying on his business. It is not necessarily confined to an apparatus which is used for mechanical operations or processes or is employed in mechanical or industrial business. It would not, however, cover the stock-in-trade,that is,goods bought or made for sale by a businessman.It would also not include an article which is merely a part of the premises in which the business is carried on as distinguished from a part of the plant with which the business is carried on. An article to qualify as plant must furthermore have some degree of durability and that which is quickly consumed or worn out in the course of a few operations or within a short time cannot properly be called plant. But an article would not be any the less plant because it is small in size or cheap in value or a large quantity thereof is consumed while being employed in carrying on the business.In the ultimate analysis the inquiry which must be made is as to what operation the apparatus performs in the assessee's business. The relevant test to be applied is: Does it fulfil the function of plant in the assessee's trading activity? It is the tool of the taxpayer's trade? If it is, then it is plant no matter that it is not very long- lasting or does not contain working parts such as a machine does and plays a merely passive role in the accomplishment of the trading purpose. Therefore, there is no reason to exclude from the wide meaning of the term plant objects such as drawings, patterns, designs, etc, which, are embodiments of know-how and serve the purpose of teaching at long range. Having regard to the legislative intent to give a wide meaning to the word plant in section 32, the material record of know-how is clearly included therein. **[CIT-Emco Electro Pvt. Ltd., Bom.,** *Sec.32.***118 ITR 864,53Taxation165]**

14. In order to find out if a building or structure or part thereof constitutes plant, the functional test must be applied. It must be seen whether the subject- matter involved, that is, the building or structure or part there of, constitutes an apparatus or a tool of the taxpayer or whether it is merely a space where the taxpayer carries on his business.If the building or structure or part thereof is something by means of which the business activities are carried on, it would amount to a plant but where the structure plays no part in the carrying on of those activities but merely constitutes a place within which they are carried on, it cannot be regarded as a plant. **[CIT-Kanodia Warehousing Corporation, All.,**<u>*Sec.33r.w.s.43(3).*</u> **121ITR996]**

15. The word plant in section 33 of the Income-tax Act,1961,must be given a wide meaning. The fact that even books have been included in the definition of plant in section 43(3) shows that the meaning intended to be given to plant is wide.It is well-settled that neither the word plant nor the word machinery is confined to a self-contained unit-plant includes part of a plant, e.g.,the engine in a vehicle;machinery includes part of a machinery and building includes part of a building. The word includes is often used in an interpretation clause in order to enlarge the meaning of the words or phrases occurring in the body of the statute.When it is so used, these words and phrases must be construed as comprehending not only such things as they signify, according to their nature and import,but also those things, which the interpretation clause declares, that they shall include. **[CIT-Tata Hydro Electric Power Supply Co. Ltd., Bom.,**<u>*Sec.33.*</u>**122ITR288]**

16. The word plant in its ordinary meaning is a word of wide import and it must be broadly construed having regard to the fact that articles like books and surgical instruments are expressly included in the definition of plant in section 43(3) of the Act. It includes any article or object, fixed or movable, live or dead,used by a businessman for carrying on his business. It is not necessarily confined to an apparatus which is used for mechanical operations or processes or is employed in mechanical or industrial business.It would not,however, cover the stock-in-trade,that is, goods bought or made for sale by a businessman. It would also not include an article which is merely a part of the premises in which the business is carried on.An article to qualify as plant must, furthermore,have some degree of durability and that which is quickly consumed or worn out in the course of a few operations or within a short time cannot property be called plant. But an article would not be any the less plant because it is small in size or cheap in value or a large quantity thereof is consumed while being employed in carrying on business. In the ultimate analysis, the inquiry which must be made is as to what operation the apparatus performs in the assessee's business.The relevant test to be applied is: does it fulfil the function of plant in the assessee's trading activity ?Is it the tool of the taxpayer's trade ?If it is, then it is plant,no matter that it is not very long-lasting or does not contain working parts such as a machine does and plays a merely passive role in the accomplishment of the trading purpose.

[CIT - National Air Products Ltd., Del.,<u>*Sec.43(3).*</u>**18CTR300,126ITR196]**

17. The word plant includes within its ambit buildings and equipment used for manufacturing purposes.The definition of plant in section 43(3) of the Income-tax Act,1961, is inclusive and does not exclude things normally included in it.Where a building with insulated walls is used as a freezing chamber,though it is not machinery or part there of, it is part of the air-conditioning plant of the cold storage and will be entitled to depreciation at 15 per cent. on its written down value.Therefore, the thermocole insulation in a cold storage is plant and is also entitled to depreciation at 15 per cent.

[CIT-Yamuna Cold Storage, P&H., <u>*Sec.43(3)*</u>**129ITR728]**

18. The word plant in its ordinary meaning is a word of wide import and it must be broadly construed having regard to the fact that articles like books and surgical instruments are expressly included in the definition of plant in section 43(3) of the Act. It includes any article or object, fixed or movable, live or dead, used by a businessman for carrying on his business.It is not necessarily confined to an apparatus which is used for mechanical operations or processes or is employed in mechanical or industrial business. It would not,however, cover stock-in-trade, that is,goods bought or made for sale by a businessman. It would also not include an article which is merely a part of the premises in which the business is carried on. An article to qualify as plant must furthermore have some degree of durability and that which is quickly consumed or worn out in the course of a few operations or within a short time cannot properly be called plant. But an article would not be any the less plant because it is small in size or cheap in value or a large quantity thereof is consumed while being employed in carrying on business. In the ultimate analysis, the inquiry which must be made is as to what operation the apparatus performs in the assessee's business. The relevant test to be applied is, does it fulfil the function of plant in the trading activity? Is it the tool of the taxpayer's trade? If it is, then it is plant, no matter that it is not very long lasting or does not contain working parts such as a machine does and plays merely a passive role in the accomplishment of the trading purpose. **[CIT-Mcgaw Ravindra Laboratories(India)Ltd.,Guj.,**Sec.43(3). **132ITR401]**

19. The definition of plant in section 43(3) of the Income-tax Act, 1961, includes ships, vehicles, books, scientific apparatus and surgical equipment used for the purposes of the business or profession. Books would fall within the definition of plant provided they satisfied the qualification that they were tools of the business or profession.In the larger sense,sheets of paper written or unwritten brought together in a form would be a book, but in the definition of plant books are referred to in a more restricted sense, restricted by the indication of the functional use of the books. In the sense in which it is used in the Income-tax Act it must be one used for the purpose of the business or profession. In the case of a lawyer books relating to law and allied subjects would fall within the definition. In the case of a medical practitioner books relating to medicines and allied subjects would similarly fall within the definition of books. Therefore, the definition must satisfy a dual test, viz., that it should be a book in form as well as functionally. **[Catalysts and Chemicals India (West Asia) Ltd.-CIT, Ker.,**Sec.43(3). **137ITR110**

20. Roads constructed by the assessee in the premises of the factory would not constitute plant as defined in section 43(3)of the Income-tax Act,1961.They must be treated as building for purposes of section 32. **[CIT-Sandvik Asia Ltd., Bom.,** Sec.43(3).r.w.s.32. **33CTR128,144ITR585,15Taxman286]**

21. The word plant includes apparatus or instruments used by a businessman in carrying on his business. The intention of the Legislature was to give a wide meaning to the term plant as it included articles like books and surgical instruments within the term. In determining whether an article is a plant, the enquiry must be as to what operation it performs in the assessee's business and whether it fulfills the function of a plant. Air conditioners or electric fans are instruments which would advance the performance of business of the assessee. They are entitled to be included within the term plant. The part of the structural premises where the business is carried on is excluded from the purview of the plant. Office premises are no doubt part of the premises where the business is carried on. But all the fixtures and fittings in the office premises cannot be said to be part of the premises in which the business is carried on. Fixtures like air- conditioners or fans cannot be said to be part of the premises in which the business is carried on. They cannot also be considered to be office appliances. The word appliance is a qualified by the word office and, therefore, some meaning must be given to the word office and unless an appliance is capable of being primarily used in the office, it cannot be termed as office appliance.It must be,therefore,an appliance which is generally used in office as an aid or facility for the proper functioning of the office. It is difficult to lay down any formula for determining what are office appliances, but the initial test for determining the nature of the article is what is known as the test of common or popular parlance as understood by a person dealing with those articles. The second test would be the principal and primary use for which the goods are required and for which the same are capable of being used. The third test is what is known as the commercial test in seeing how the articles or goods are known in the world of trade and commence. On application of any of these tests it is difficult to agree that the electrical fans and air-conditioners would be office appliances. By no stretch of imagination in trade and commerce or in popular parlance can they be said to be office appliances or equipments.Merely because these appliances are fixed in office premises they do not become, by that fact, office appliances. They are capable of being adopted for the purposes for which they are meant,namely, for maintaining a particular bearable climatic temperature in laboratories,workshops, surgical and nursing homes and even in private residential buildings.

[CIT-Tarun Commercial Mills Ltd.,Guj., *Sec.32,33.***38CTR148, 151ITR75,16 Taxman18]**

22. The term plant is a term of wide import. The definition in section 43(3) itself is an inclusive definition and must be given a liberal construction and not a literal construction. One of the tests to be applied to determine whether a particular item is plant or not, is to ascertain whether it is an item by which business is carried on or only an item used in carrying on the business. That determination depends upon the type of business the assessee is carrying on.The question whether roads laid by an assessee constitute plant has to be determined on the facts and circumstances of each case. It depends upon the particular situation of the roads,the use to which they were put and in particular whether they were an integral part of the factory in which the business was carried on.

[CIT-Coromandel Fertilisers Ltd., AP.,*Sec.43(3).* **156ITR283,18Taxman411]**

23. ...plant was not necessarily confined to an apparatus which was used for mechanical operations or process or was employed in mechanical or industrial business. But in order to qualify as plant , the particular article had to have some degree of durability. The test to be applied was: Did the article fulfil the function of a plant in the assessee's trading activity ? Was it a tool of his trade with which he carried on his business ?If the answer was in the affirmative, it would be a plant.(ii)that the drawings,designs,charts,plans,processing data and other literature comprised in the documentation service as specified in clause 3 constituted a book and fell within the definition of plant in section 43(3)of the Income-tax Act, 1961. The purpose of rendering such documentation service by supplying these documents to the appellant was to enable it to undertake its trading activity of manufacturing theodolites and microscopes and these documents had a vital function to perform in the manufacture of these instruments; in fact, it was with the aid of these complete and up-to date set of documents that the appellant was able to commence its manufacturing activity and these documents really formed the basis of the business of manufacturing the instruments in question.That by themselves these documents did not perform any mechanical operations or processes did not militate against their being a plant since they were in a sense the basis tools of the assessee's trade having a fairly enduring utility, though owing to technological advances they might or would in course of time become obsolete. The capital asset acquired by the appellant, viz., the technical know-how in the shape of drawings, designs, charts, plants, processing data and other literature, fell within the definition of plant and was, therefore,a depreciable asset.

[Scientific Engineering House (Pvt.) Ltd. - CIT,SC, *Sec.43(3).* **49CTR386,157ITR86,23Taxman66]**

24. It was the intention of the Legislature to give the expression plant a wide meaning and that is why articles like books and surgical instruments were expressly included in the definition of plant.

[CIT-Steel Rolling Mills of Hindusthan(P.)Ltd.,Cal., *Sec.33.***51CTR386,164ITR633,26Taxman554]**

25. The definition of plant in section 43 of the Income-tax Act,1961, is an inclusive definition and the intention of the Legislature to give it a wide meaning is evident from the fact that articles like books and surgical instruments have been expressly included in the definition of plant .This inclusive definition of plant must be understood to mean,in its ordinary sense, as including all apparatus used by a businessman for carrying on his business but not as stock-in-trade.

[CIT - Jai Drinks (P.) Ltd.,Raj., *Sec.43(3).***68 CTR 41,173ITR100,36Taxman330]**

26. The term plant has not been defined in the Act, though section 43(3) defines that term for the purposes enumerated in that section and other allied sections as including ships,vehicles,books, scientific apparatus and surgical equipment used for the purposes of the business or profession. An inclusive definition enlarges the ordinary meaning and brings into its fold what may not normally fall within its ordinary meaning. Storage tanks and bore-wells are not included in the definition of the term plant occurring in section 43(3) of the Act. Whether bore-wells and storage tanks are plant or not depends on their relation to the nature of the business carried on by an assessee.There cannot be any hard and fast rule in such matters and each case has to be decided on the facts and circumstances of that case only but with due regard to the above principle. When this test is properly applied to each case, it may turn out that in the case of one assessee, they may be plant but in the case of another they may not be plant. **[CIT-Motor Industries Co. Ltd., Kar.,** *Sec.43(3).***173ITR374]**

27. The definition of the expression plant, in section 43(3) of the Income-tax Act, 1961, is a wide definition as in section 10(5) of the Indian Income-tax Act of 1922. A tube-well is an apparatus with equipment necessary for drawing water from subterranean sources and where water is necessary not only for production but also for industrial labour and where such tube-well is required by an assessee in its capacity as an employer as also in its capacity as a businessman, the tube-well falls within the meaning of the term plant under section 43(3). **[CIT-Hindusthan Motors Ltd.,Cal.,** *Sec.43(3).***170ITR431]**

28. ...the books, designs and formulae constituted plant and the assessee was entitled to depreciation and development rebate in respect of them. **[Porritts and Spencer (Asia) Ltd. - CIT, P&H.,** *Sec.32.***180ITR211,47Taxman 45]**

29. The word plant is to be given a very wide meaning. In its ordinary sense, it includes whatever apparatus is used by a businessman for carrying on his business but it does not include his stock-in-trade which the tradesman keeps for permanent employment in his business. (2) But the building or the setting in which the business is carried on cannot be plant. (3) The thing need not be part of the machine used in the manufacturing process but could be merely an apparatus used in carrying on the business but having a degree of durability. (4) Merely because the asset has a passive function in the carrying on the business, it cannot be said that it is not plant. It may have a passive or an active role.(5) The subject must have a function in the trader's operation and if it has, it is prima facie a plant unless there was good reason to exclude it from that category. It must be a tool in the trade of the businessman.(6) Gross materiality or tangibility is not necessary and, in fact, intangible things like ideas and designs contained in a book could be plant. They fall under the category of intellectual storehouse. (7) In considering whether a structure is plant or premises, one must look at the finished product and not at the bits and pieces as they arrive from the factory. The fact that a building or part of a building holds the plant in position does not convert the building into plant. A piecemeal approach is not permissible and the entire matter must be considered as a single unit unless,of course, the component parts can be treated as separate units having different purposes.(8) The functional test is a decisive test. **[CIT-Sri Krishna Bottlers Pvt. Ltd.,AP.,**<u>*Sec.32(1)(ii).*</u>**71CTR88,175ITR154,90Taxation208,40 Taxman15]**

30. In order to find out whether a building or structure or a part thereof constitutes plant, the functional test must be applied. If it is found that the building or structure constitutes an apparatus or a tool of the taxpayer by means of which the business activities are carried on, it would amount to plant but where the structure plays no part in the carrying on of those activities but merely constitutes a place within which they are carried on, the building cannot be regarded as a plant. **[S.K.Tulsi and Sons - CIT,All.,** <u>*Sec.32,33.*</u>**90CTR99, 187ITR685]**

31. In its ordinary sense plant includes whatever apparatus is used by a businessman for carrying on his business; not his stock-in-trade which he buys or makes for sale, but all goods and chattels, fixed or movable, live or dead,which he keeps for permanent employment in his business.In order to decide whether a particular object is an apparatus,an enquiry has to be made as to what operation it performs.The functional test is,therefore,essential at any rate as a preliminary.In other words, the test would be: Does the article fulfil the function of plant in the assessee's trading activity ?Is it a tool of his trade with which he carries on his business ? If the answer is in the affirmative, it will be plant. It is not necessary that, to constitute plant, the asset should be directly engaged in the manufacture of articles. What is properly to be regarded as plant can only be answered in the context of the particular industry concerned.**[Tribeni Tissues Ltd. -CIT,Cal.,** <u>*Sec.32A.r.w.s.43.*</u>**190ITR487]**

32. The definition in section 43 of the Income-tax Act,1961, does not, in any way, restrict the purport and scope of the expression plant.On the other hand, it widens it. The principles applicable are as follows: (1) The definition of plant in section 43(3) should be given a wide meaning as it is an inclusive definition. (2) All buildings are not plant despite the dictionary meaning which includes buildings; but a building or structure is not per se to be excluded from the ambit of the expression plant. (3) If the concrete construction or building is used as the premises or setting in which the business is carried on in contradistinction to the fulfilling of the function of a plant, the building or construction or part thereof is not considered a plant. The true test is in whether it is the means of carrying on the business or the location for so doing. (4) In order for a building or concrete structure to qualify for inclusion in the term plant, it must be established that it is impossible for the equipment to function without the particular type of structure. (5) The particular apparatus or item must be used for carrying on the assessee's business and must not be his stock-in-trade. The matter has to be considered in the context of the particular business of the assessee, e.g., books are a lawyer's plant but a bookseller's stock-in-trade. **[CIT-Mazagaon Dock Ltd.,Bom.,*Sec.43(3).*191ITR460]**

33. The definition of plant in section 43(3) of the Income-tax Act, 1961,is an inclusive definition as the very definition indicates that articles like books are also included in the definition of plant. Under the provisions of section 32A,the plant should be owned by the assessee and should be wholly used for the purpose of business and the purpose of the business is again a phrase of wide import. The concept of plant cannot be limited to the actual installation of machinery which produces goods by itself. Anything which is used for the purpose of business including any installation which facilitates the production or increases the efficiency of the business will be plant. **[CIT-Electronics Research Industries Pvt. Ltd., Kar., *Sec.43(3).* 97CTR 51,192 ITR 20,59 Taxman 46]**

34. There is no definition of plant in the Income-tax Act,1961; but in its ordinary sense, it includes whatever apparatus is used by a businessman for carrying on his business-not his stock-in-trade which he buys or makes for sale ; but all goods and chattels,fixed or movable,live or dead, which he keeps for permanent employment in his business. **[CIT-Oil India Ltd.,Cal.,*Sec.33.*198ITR701]**

35. There is no definition of plant in the Income-tax Act, 1961; but, in its ordinary sense, it includes whatever apparatus is used by a businessman for carrying on his business,not his stock-in-trade which he buys or makes for sale, but all goods and chattels, fixed or movable, live or dead, which he keeps for permanent employment in his business.The definition of plant is an inclusive one. The very fact that even books have been included shows that the meaning intended to be given to plant is wide. .... the screening wall and the ceiling of the auditorium having been constructed with requisite installations so as to have a proper control of the sound effect and for the efficient screening of the films had to be treated as part of plant but no other part of the building could be included in the said term. The furniture, fittings and fixtures consisting of wooden walls including false ceiling and wooden panelling of the walls and the chairs would come within the purview of plant .... However,the case of the chairs outside the auditorium would be different and they would not come within the definition of plant. **[Santosh Enterprises-CIT,Kar.,**<u>*Sec.32,33.*</u>**200ITR353]**

36. The Act has defined plant in an inclusive manner. Conceptually speaking, anything that is a tool of trade is plant. But one thing is clear, viz., that the term tool of trade signifies close and direct connection between the tool and the assessee's trade, that is to say, the plant must be employed directly in the manufacture or production of goods, etc. If a particular item can properly be called a tool of trade, then it is plant. On the contrary, a particular item, even though called plant in popular parlance, will not be plant if it is just a part of the setting in which the trade is conducted. For example, an air-conditioner installed in the office premises of an assessee cannot be called plant .... If, on the contrary,an air-conditioner is indispensable for the manufacture of, say, medicines, then the air-conditioner becomes the tool of the assessee's trade and can properly be regarded as plant. **[CIT - Technico Enterprise Pvt. Ltd., Cal.,** <u>*Sec.32(1)(ii),32A(1).*</u>**119CTR25,206ITR36]**

37. For any article or thing, component or object to be termed as plant ...the article or component, as the case may be, must be used by the assessee as a self-contained unit and not as a part or attachment of a bigger unit. This, however, does not mean that the article or object would cease to be plant for the purpose of depreciation as a part of the bigger unit. All that it would mean is that while it may qualify for depreciation as a part or extension of the bigger plant of which it becomes a part, it would not be entitled to be termed as a plant in itself to qualify for the allowance in its own right. **[Pathange Poultry Farm-CIT,Kar.,** <u>*Sec.32.r.w.s.43.*</u> **210ITR668]**

38. The term plant in section 43(3) of the Income-tax Act,1961, should be given a wide meaning, as it is an inclusive definition. A building or structure is not to be excluded, per se, from the ambit of the expression plant. However, if a concrete construction or building is used as premises or a setting in which the business is carried on in contradiction to its fulfilling the function of a plant, the building or construction will not be considered a plant.And if the equipment cannot function without such a structure, the structure would form a part of the plant. **[CIT - Electro Metallurgical Works (P.) Ltd.,Bom.,** *Sec.43(3).* **207ITR 494]**

39. Plant will mean and include apparatus used by a businessman for carrying on his business. It is not confined to apparatus used for mechanical operations or processes or industrial business. Hence, fans installed in the office premises of the assessee constituted plant which was eligible for depreciation. **[CIT-Parke Davis (India) Ltd., Bom.,** *Sec.32,43.* **128CTR,214ITR587]**

40. The expression plant has been defined in clause (3) of section 43 of the Income-tax Act,1961, to include ships, vehicles, books, scientific apparatus and surgical equipment used for the purposes of the business or profession. It is an inclusive definition. Its intention is to enlarge the meaning of the expression plant occurring in the Act to include not only such items as are commonly known as plant but also those which are enumerated therein. To decide whether a particular item is plant or not, one of the tests applied is the common parlance or trade or commercial parlance test. Another test that is often applied for that purpose is the functional test.An item would not qualify to be plant even if it satisfied the functional test, if on the application of the premises test it is found to be used as or part of the premises or place upon which the business was conducted. **[Siemens India Ltd.-CIT,Bom.,** *Sec.43(3).* **217ITR622,129Taxation 201]**

41. The principles applicable with regard to plant are :

(1) The definition of plant in section 43(3) of the Income-tax Act, 1961, should be given a wide meaning, as it is an inclusive definition.

(2) All buildings are not plant despite the dictionary meaning which includes building; but a building or structure is not per se to be excluded from the ambit of the expression plant.

(3) If the concrete construction or building is used as the premises or setting, in which the business is carried on, in contradistinction to the fulfilling of the function of a plant, the building or construction or part thereof is not considered a plant. The true test is whether it is the means of carrying on the business or the location for so doing.

(4) In order for a building or concrete structure, to qualify for inclusion in the term plant, it must be established, that it is

impossible for the equipment to function without the particular type of structure.

(5)The particular apparatus or item must be used for carrying on the assessee's business and must not be his stock-in-trade .The matter has to be considered in the context of the particular business of the assessee, e.g., books are a lawyer's plant but a book seller's stock-in-trade. **[CIT - Lawly Enterprises (P.) Ltd., Pat.,** *Sec.43(3).***225ITR154,88Taxman320]**

42. The word plant in its ordinary meaning is a word of wide import and it must be broadly construed having regard to the fact that articles like books and surgical equipment are expressly included in the definition of plant in section 43(3) of the Act. It includes any article or object, fixed or movable, live or dead, used by a businessman for carrying on his business.It is not necessarily confined to an apparatus which is used for mechanical operation or processing or is employed in mechanical or industrial business. It, however, does not cover the stock-in-trade or an article which is merely a part of the premises in which business is carried on. To reach a correct conclusion whether a given item is plant or not, the inquiry which must be made is as to the operation the apparatus performs in the assessee's business.The relevant test to be applied is : does it fulfil the functions of plant in the assessee's trading activities ? is it the tool of the taxpayer's trade ? If it is, then it is plant. No matter that it is not very long- lasting or does not contain working parts such as a machine does and plays merely a passive role in the accomplishment of the trading purpose. So, the main test is whether a given item is such that without it business cannot be carried on. **[Harijan Evam Nirbal Varg Avas Nigam Ltd. - CIT,All.,***Sec. 43(3).***131CTR178,229ITR776,85Taxman456]**

43. ....the expression plant has not been extensively defined and section 43(3) of the Income-tax Act,1961, provides only an inclusive definition, which has the effect of adding items mentioned thereunder to the meaning of the word plant as understood in its ordinary sense. The word plant as defined in the Oxford English Dictionary means fixtures, implements, machinery and apparatus used in carrying on any industrial process. The word apparatus would mean the equipment needed for a particular purpose or function, and the word equipment would mean the necessary articles, etc., for a purpose.The word plant in its ordinary sense would,therefore,mean the equipment needed for a particular purpose or function. Such equipment can be any article, which may be necessary for a purpose. In the context of business, therefore, a plant would mean any equipment or article necessary for the purpose of that business.The articles or equipment which it may become doubtful to read into the plain meaning of the word plant are added to that meaning so that no doubt may arise in that regard. **[CIT - Saurashtra Bottlings Pvt. Ltd.,Guj., *Sec.43(3).*147 CTR 115,232ITR270]**

44. Plant is a word of wide import. Plant would include an article or object, fixed or movable, used by a businessman for carrying on his business and it is not necessarily confined to an apparatus which is used for mechanical operations or processes or is employed in mechanical or industrial business. The test would be : does the article fulfill the function of a plant in the assessee's trading activity ? Is it a tool of his trade with which he carries on his business ? If the answer is in the affirmative patterns and dies, and electrical installations would fall within the definition of plant. The expression installed in clause (b) of sub-section (2) of section 32A of the Income-tax Act, 1961, does not necessarily mean fixed in position. It is used in the sense of induct or introduce or placing an apparatus in position for service or use. **[CIT-Bharat Radiators P. Ltd.,Bom.,*Sec.32A.*158CTR519,239ITR608]**

45. Plant as such has not been defined in the Act. Section 43(3) of the Act merely contains an inclusive definition which says that plant shall include ships, vehicles,books,scientific apparatus and surgical equipment used for the purposes of the business or profession.Obviously it is an inclusive definition. Its intention is to enlarge the meaning of the expression plant occurring in the Act to include not only such items as are commonly known as plant but also those which are enumerated therein.To decide whether a particular item is plant or not, one of the tests applied is the common parlance or trade or commercial parlance test.Another test that is often applied for that purpose is the functional test.In common parlance electrical fittings,etc.,are never regarded as plant. They are ordinarily described and referred to as fittings. Even applying the dictionary meaning,these items cannot be regarded as plant. **[CIT - Hoechst Dyes and Chemicals P. Ltd., Bom., *Sec.43(3).*157CTR58,240ITR1,106Taxman684]**

46. Plant is defined in section 43(3) of the Act. It includes all motor vehicles including motor car. **[CIT- Travancore Cements Ltd.,Ker.,*Sec.43(3).*157 CTR 395,240ITR816,108Taxman313]**

47. From a reading of section 43(3) of the Act, unless the context otherwise requires, the expression plant includes vehicles also. **[Hyderabad Deccan Cigarettes Factory -CIT,AP.,*Sec.43(3).* 152CTR23,236ITR615, 152Taxation 693]**

48. The word plant is given an inclusive meaning under section 43 (3) which nowhere includes buildings. The Rules prescribing the rates of depreciation specifically provide for grant of depreciation on buildings, furniture and fittings, machinery and plant and ships. Machinery and plant include cinematograph films and other films and other items .... In the context of the legislative scheme under section 32,even though the word plant may include building or structure in certain set of circumstances as per the dictionary meaning, to say that a building used for running the business of hotel or a cinema would be plant under the Act would be inconsistent with the provisions of section 32 and the legislative intent.

[CIT- Anand Theatres, SC,*Sec.32.*AIR2000SC2356,160CTR492, 244ITR192,110Taxman338]

51. Section 43(3) of the Income-tax Act,1961 defines plant by way of an inclusive definition thereby intending to enlarge the meaning of the expression.In order to qualify as plant the article must have some degree of durability and if the article fulfils the function of a plant in the assessee's trading activity or is a tool of his trade with which he carries on business it would be plant.

[JCIT -Anatronics General Co. (P.) Ltd.,Del.,*Sec.43(3).*154CTR 210,247ITR25, 113 Taxman511]

49. Section 43(3) of the Income-tax Act,1961,defines plant in very wide terms.Plant would include any article or object fixed or movable, live or dead,used by a businessman for carrying on his business and it is not necessarily confined to an apparatus which is used for mechanical operations or processes or is employed in mechanical or industrial business.In order to qualify as plant,the article must have some degree of durability. **[CIT-Hindustan Insecticides Ltd.,Del.,*Sec.43(3).*171CTR370,253ITR520,120Taxman 605]**

50. The word plant is defined in section 43(3) of the Act. It is not an exhaustive definition. It is an inclusive definition.It includes vehicles among other things used for the purpose of the business or profession. **[CIT-Birla Jute and Industries Ltd.,Cal., *Sec.43(3).* 180CTR339,260ITR55,133Taxman337]**

51. When a building is an apparatus or a tool of taxpayer as against merely a space from where the taxpayer carries on business it can be treated as plant. If a building is an integral part for carrying on the business of manufacture it would be a plant whereas if the structure plays no part in carrying on any of the activities relating to manufacture, it would merely constitute a place where the business was carried on and it could not be recognised as plant. When the assessee has with a definite purpose, considering the nature of business carried on by the assessee, constructed a building with specific required design keeping in view specified technical requirement,without which the assessee's business could not be carried out,the building in question would qualify to be treated as plant. **[Nowrangroy Metals Pvt. Ltd.-JCIT (Asst), Gau.,**<u>*Sec.32.r.w.s.43.*</u>**183CTR269,262ITR231,129Taxman657]**

52. Cinema theatres are premises and are not plant for the purposes of depreciation and extra shift allowance under sections 32 and 43(3) of the Income-tax Act,1961.The cinema building is not a tool or an apparatus for carrying on business activity. **[CIT-A.B.A.Sons,SC,**<u>*Sec.32.r.w.s.43(3).*</u>**186 CTR77,264ITR469]**

53. The word plant has been given an inclusive meaning in section 43 of the Income-tax Act,1961.In order to find out whether a building is a plant or not the following tests would be decisive : (i) Functional test is a decisive test; (ii) An item which falls within the category of building cannot be considered to be a plant. Buildings with particular specification for atmospheric control like moisture or temperature are not plants; (iii) In order to find out as to whether a particular item is a plant or not, the meaning which is available in the popular sense, i.e., the meaning people conversant with the subject matter would attribute to it, has to be taken ; (iv) The term plant would include any article or object, fixed or movable, live or dead, used by a businessman for carrying on his business and it is not necessarily confined to any apparatus which is used for mechanical operations or process or is employed in mechanical or industrial business.The article must have some degree of durability;(v)The building in which the business is carried on cannot be considered to be a plant; (vi) The item should be used as a tool of the trade with which the business is carried on. For that purpose the operations it performs have to be examined.**[CIT-Navodaya,Ker.,**<u>*Sec.43.*</u>**186CTR357,271ITR173,135Taxman258]**

54. Where a particular structure is merely helpful for carrying on the activities of the assessee it may not be a plant but if it is an integral part of the plant and machinery or a portion of that building is an integral part of the plant and machinery that should be considered as plant. **[CIT-R.G. Ispat Ltd., Raj.,** <u>*Sec. 32,32A.*</u>**186CTR262,272ITR383,135Taxman456]**

55. Unless, on the facts, it is established, that the nursing home is a tool or an apparatus, for carrying on business activity, for which allowance of depreciation is claimed,it may not be possible to extend the benefit to the entire building and, if at all, the benefit can be extended to the portion of the building or the building,as the case may be, which could be treated as a plant. **[CIT -Kamala Selvaraj,Mad.,*Sec.32.*273ITR154]**

56. The word plant is given an inclusive meaning in section 43(3) of the Income-tax Act,1961,which nowhere includes buildings.There is a well-established distinction between the premises in which the business is carried on and the apparatus with which the business is carried on. The latter category would fall within the ambit of the phrase plant. The premises cannot be termed as plant. The building in which the business is carried on might be well suited to the business or it could have been built for the business but it would not be a plant. The suitability is the reason why the business is carried on there but it does not make it a thing with which the business is carried on. If a building is merely a setting or place to accommodate some apparatus, it cannot be termed as plant but if it plays an important role in carrying on the business then it will fall within the definition of the term plant. It would be a plant if it is a tool of the trade with which one carried on his business. **[CIT - Shivalik Poultries,P&H.,*Sec.43(3).*274ITR529]**

57. The word plant must be given a very wide meaning and it includes whatever apparatus is used by a person for carrying on his business. **[Associated Bearing Co. Ltd.-CIT,Bom., *Sec.32(1)(ii)r.w.s.43(3).*286 ITR 341]**

58. .....no depreciation was allowable in respect of gratuity liability even if it was regarded as capital expenditure.Gratuity liability was neither building, machinery, plant or furniture nor an intangible asset of the kind mentioned in section 32(1)(ii). Even the deeming meaning given to plant in section 43(3) did not include gratuity liability. **[CIT-HooglyMillsCo.Ltd.,SC, *Sec.32(1) (ii).*203CTR 356,287ITR333]**

62. In determining whether an item constitutes plant the courts have applied the durability and/or functional tests.If a thing itself is durable(in the sense it can be used and re-used as non-inter-dependent, interconnected a non-consumable thing) and has functional utility in the trade or business of the assessee to advance his business interest, such thing would be a plant. If it is durable, but cannot effectively stand alone without functional integration with other similar or dissimilar components or units, it would not qualify as a plant. **[CIT - Vijaya Enterprises,AP., *Sec.31(1)(ii).*332ITR235]**

59. The definition of plant in section 43 of the Income-tax Act, 1961, is inclusive. Everything which is not building, furniture or fittings used for the purpose of business is a plant. If it is a building, it cannot be a plant. But, in certain circumstances, a plant itself can be a building.If a building itself is the place with which the business is carried on, it is a plant. But, if the building is a place from where the business is carried on it is only a building notwithstanding special arrangements.**[CIT-Padmavathi Hatcheries P.Ltd.,AP.,** *Sec.43.***241 CTR171,335ITR325]**

Plant and machinery Mys. M. Ganesh Beedi Works - CIT

An electric stove is both a plant and a machinery.Diesel engine fitted to a lorry is a machinery.*Sec.32.***[AIR1964Mys235, 52ITR615]**

2.The jeep and motor cycle could by no stretch of imagination be treated as plant and machinery.**[CIT-General Fibre Dealers (P.)Ltd.,Cal.,***Sec.32A.***164 CTR392,248ITR622]**

Plant installed Ker. CIT - Calicut Wyanad Motor Services P. Ltd.

The term plant installed applies to motor vehicles, which are inducted or introduced into the business. *Sec.33A.***[38ITR479]**

2. The expression plant installed does not necessarily mean something fixed in position, but it should be construed widely in the sense of inducted or introduced.**[CIT-Baker Mercer India P. Ltd., Bom.,** *Sec.32A.***196ITR667]**

Point of law Gau. CIT - Basanta Kumar Agarwalla

A point of law cannot be equated with the expression question of law . The question must be a disputed or disputable question of law. The object of a reference is to get a decision from the High Court on a problematic or debatable question and not on an obvious and simple point of law, although somehow the determination is somewhere linked up with a provision of law. The meaning of the term question,in the context, means a subject or point of investigation,examination or debate,a problem, as a delicate or doubtful question. The Tribunal need not refer every point of law . The Tribunal is obliged to refer only a question of law which calls for investigation, examination, debate or when it is a dubious problem.However, if a point of law decided by the Tribunal is positive,certain, definite and sure, there is no obligation on the part of the Tribunal to refer the matter, as the point cannot be termed as a question of law. When a decision is apparently correct and there is no scope for any debate or dispute or difference, it does not fall within the expression a question of law. Sec.256,260. **[25CTR117,140ITR418]**

Possession AAR Jasbir Singh Sarkaria

Possession contemplated by clause (v) of section 2(47) need not necessarily be sole and exclusive possession. So long as the transferee is, by virtue of the possession given, enabled to exercise general control over the property so as to make use of it for the intended purpose, the mere fact that the owner has also the right to enter the property to oversee the development work or to ensure performance of the terms of the agreement does not introduce incompatibility. The concurrent purpose of the owner who can exercise possessory rights to a limited extent and for a limited purpose and that of the buyer developer who has a general control and custody of the land can very well be reconciled. Clause (v) will have its full play even in such a situation. There is no warrant to postpone the operation of clause (v) and the resultant accrual of capital gains to a point of time when the concurrent possession will become exclusive possession of the developer transferee after he pays full consideration. Possession given to the developer need not ripen into exclusive possession on payment of the installments in entirety for the purpose of determining the date of transfer. It is enough if the transferee has, by virtue of that transaction, a right to enter upon and exercise acts of possession effectively pursuant to the covenants in the contract. That amounts to legal possession. *Sec.2(47)(v).***[212CTR,107,294ITR196, 164Taxman108]**

Power Mad. CIT - V.Saraswathi.

An industrial undertaking engaged in the business of generation or distribution of electricity or any other form of power is entitled to exemption under section 5(1)(xxxii) of the Wealth-tax Act,1957. According to the Oxford Dictionary, power means vigour of energy, mechanical energy as opposed to hand labour, the capacity for exerting mechanical force. It is no doubt true that power would include any form of energy as commonly understood. But, power would not include the source of power. Power can be generated through various sources. Coal, kerosene, oil and gas constitute sources of power and they would not by themselves become power or energy. If it is power, it can be directly applied to get energy and if it is fuel, it should be oxidised for getting energy. Therefore, gas which is fuel cannot be considered as power. *Sec.5(1)(xxxii).WT.Act.* **[142CTR175,229ITR82]**

2. Power means authority to give one to another to act for him or to do some specified acts. The expression power is not always synonymous with jurisdiction. Section 120 uses the expression jurisdiction and section 127 uses the expression power having different meaning and content. While exercising the powers under section 127 the Chief Commissioner is not exercising any jurisdiction but exercising power of transferring a case from one officer to another. Section 158BG also confers power on the Assistant Commissioner or Deputy Commissioner, as the case may be, to make block assessment. Power under section 127 has to be exercised in public interest and in the best interest so as to facilitate effective and co-ordinated investigation. **[K.V. Kader Haji-CIT,Ker.,***Sec.127***.189CTR313, 268ITR465,140Taxman527]**

Practicable **Guj.** **Hasanand Pinjomal - CIT**

Clause (j) of rule 6DD sets out four circumstances in which the rigour of the rule contained in sub-section (3) of section 40A has to be relaxed. We are concerned in this case with one of these conditions, namely, where payment in the manner provided in section 40A(3) is not practicable. The meaning of the word practicable in the ordinary parlance must prevail in the context of rule 6DD(j). If the object of the enactment, namely, to relax the rigour of sub-section (3) of section 40A in genuine and bona fide cases to avoid hardship and harassment, is borne in mind, the adoption of the ordinary meaning which is quite wide would be justified, because it would advance the cause rather than defeat it. Accordingly, the word practicable in rule 6DD(j)(2) must be held to signify, as stated in the dictionary and in Corpus Juris Secundum, that which is feasible, that is to say, capable of being put into practice, done, or accomplished with the available means and resources. In determining the practicability for the purposes of rule 6DD(j)(2), regard will have to be had to the facts and circumstances of each case. *Sec.40A(3).***[6CTR 486,112ITR134]**

Preferring an appeal **SC** **CIT - B.N. Bhattachargee**

Purposefully interpreted, preferring an appeal means more than formally filing it but effectively pursuing it. *Sec.245.* **[10CTR354,118ITR461,Taxman348]**

Prejudice to the interests of the Revenue **Mad.** **CGT-O.E.Arumugha Mudaliar**

The expression prejudice to the interests of the Revenue have not been defined, but it must mean that the orders of assessment challenged are such as are not in accordance with law, in consequence whereof the lawful revenue to the State has not been realised or cannot be realised. It can mean nothing else...... The phrase prejudicial to the interests of the Revenue has to be read in conjunction with an erroneous order passed by the AO.Every loss of revenue as a consequence of an order of the AO. can not be treated as prejudicial to the interest of the Revenue. Sec.24.G.T.Act. **[237ITR789,154Taxation467]**

2.The expression prejudice to the interests of the Revenue have not been defined, but it must mean that the orders of assessment challenged are such as are not in accordance with law, in consequence whereof the lawful revenue to the State has not been realised or cannot be realised. It can mean nothing else. **[CWT-N.T. Rama Rao, AP.,**Sec.2(e),(2)(ii),5(1)(vi),25.W.T.Act.**182CTR25, 261ITR611,129Taxman430]**

Prejudicial **Lah.** **Nanhe Mal Janki Nath - CIT**

The word prejudicial in section 33 need not have the same meaning as it has in section 66(2). In the former section it is obviously used in the narrower sense of prejudice occasioned to the assessee by the order of the Commissioner himself. Sec.246.**[8ITR437]**

2. An order made by the Commissioner under section 33 can only be said to be prejudicial to the assessee when he is, as a result of it, in a different and worse position than that in which he was placed by the order under review.**[CIT -Tribune Trust, PC,**Sec.246.**16ITR214]**

Prejudicial to the assessee **Ker.** **K.C.Luckose - ITO**

The term prejudicial to the assessee in section 33 of the Indian Income-tax Act, 1922, corresponding to section 264(1) of the 1961 Act, was explained by the Privy Council in the case of CIT v. Tribune Trust 16 ITR 214.It appears that an order made by the Commissioner under section 33 can only be said to be prejudicial to the assessee when he is, as a result of it, in a different and worse position than that in which he was placed by the order under review. As the incidence of the tax on the assessee after the order passed on revision did not certainly leave him in a worse position than what he was in prior to the revision, it cannot be said that the order of the Commissioner was prejudicial to the assessee.

Sec.264(1).[**105ITR418**]

Prejudicial to the interest of the revenue **All.** **Addl. CIT-Saraya Distillery**

....section 142 of the Act, read with section 215, indicates that at the time of regular assessment, interest has to be charged from persons who have paid advance tax on their own estimate in an amount which falls short of the assessed tax by more than 75 per cent. there of, and the officer is not required to pass a formal order to that effect. The officer has only to calculate the interest and make a demand for it. If the officer does not calculate the interest and make a demand for it, the revenue would be deprived of the interest and that would be prejudicial to the interest of the revenue. The words prejudicial to the interest of the revenue must mean that the orders of assessment challenged are such as are not in accordance with law, in consequence whereof the lawful revenue due to the State has not been fixed or cannot be realised.

Sec.215.r.w.s.263.[**7CTR382,115ITR34**]

2. The words prejudicial to the interests of the Revenue are of wide import and they should not be limited to a case where the order passed by the Income-tax Officer can be considered to be one prejudicial to the Revenue administration as such. The question whether an order of the Income-tax Officer is prejudicial to the interests of the Revenue would depend upon the facts of each case and there can be no universal formula applicable to finding out any such prejudicial error. **[Malabar Industrial Co. Ltd.-CIT, Ker.,** *Sec.263.***198ITR611,62Taxman25]**

3. What is meant by the words prejudicial to the interests of the Revenue has not been defined.However, giving the ordinary meaning to the words used in the statute,they must mean that the orders under consideration are such as are not in accordance with law and, in consequence whereof, the lawful revenue due to the State has not been realised or cannot be realised. The well settled principle in considering the question as to whether an order is prejudicial to the interests of the Revenue or not is to address oneself to the question whether the legitimate revenue due to the exchequer has been realised or not or can be realised or not if the orders under consideration are allowed to stand. For arriving at this conclusion, it becomes necessary and relevant to consider whether the income in respect of which tax is to be realised has been subjected to tax or not or if it is subjected to tax, whether it has been subjected to tax at the rate at which it could yield the maximum revenue in accordance with law or not. If the income in question has been taxed and legitimate revenue due in respect of that income had been realised, though as a result of an erroneous order having been made in that respect, the Commissioner cannot exercise the powers for revising the order under section 263 merely on the basis that the order under consideration is erroneous. If the material in that regard is available on the record of the assessee concerned, the Commissioner cannot exercise his powers by ignoring that material which links the income concerned with the tax realisation made thereon. The two questions are inter- linked and the authority exercising the powers under section 263 is under an obligation to consider the entire material about the existence of income and the tax which is realisable in accordance with law and further what tax has in fact been realised under the assessment orders. **[CIT - Minalben S. Parikh,Guj.,** Sec.263.**215 ITR81]**

4. The phrase prejudicial to the interests of the Revenue is not an expression of art and is not defined in the Act. Understood in its ordinary meaning it is of wide import and is not confined to loss of tax. The scheme of the Act is to levy and collect tax in accordance with the provisions of the Act and this task is entrusted to the Revenue. If due to an erroneous order of the ITO..... The phrase prejudicial to the interests of the Revenue has to be read in conjunction with an erroneous order passed by the AO. Every loss of revenue as a consequence of an order of the AO. cannot be treated as prejudicial to the interests of the Revenue, for example, when an ITO. adopted one of the courses permissible in law and it has resulted in loss of revenue, or where two views are possible and the ITO. has taken one view with which the CIT does not agree, it cannot be treated as an erroneous order prejudicial to the interests of the Revenue unless the view taken by the ITO. is unsustainable in law. **[Malabar Industrial Co. Ltd.-CIT, SC,** Sec.263.**159CTR1,243ITR83,109Taxman66]**

5. The phrase prejudicial to the interests of the Revenue has to be read in conjunction with an erroneous order passed by the Assessing Officer and every loss of revenue as a consequence of an order of the Assessing Officer cannot be treated as prejudicial to the interests of the Revenue.When an Assessing Officer adopts one of the courses permissible in law and it has resulted in loss of revenue, or where two views are possible and the Income-tax Officer has taken one view with which the Commissioner does not agree, it cannot be treated as an erroneous order prejudicial to the interests of the Revenue unless the view taken by the Income-tax Officer is unsustainable in law. **[CIT-Arvind Jewellers,Guj.,** *Sec.263.***177CTR546,259ITR502,124Taxman615]**

6. The provisions of section 263 of the Income-tax Act,1961, cannot be invoked to correct each and every type of mistake or error committed by the Assessing Officer. It is only when an order is erroneous that the section will apply and an incorrect assumption of facts or an incorrect application of law will satisfy the requirement of the order being erroneous. If due to the erroneous order of the Income-tax Officer, the Revenue is losing tax lawfully it will be prejudicial to the interests of the Revenue. The phrase prejudicial to the interests of the Revenue has to be read in conjunction with an erroneous order passed by the Assessing Officer.Every loss of revenue as a consequence of an order of the Assessing Officer cannot be treated as prejudicial to the interest of the Revenue. **[Simplex Concrete Piles (India) P. Ltd. - CIT,Cal.,** *Sec.263.* **261 ITR611]**

7. The phrase prejudicial to the interests of the Revenue in Sec. 263 of the Act, has to be read in conjunction with the expression erroneous order passed by the AO. Every loss of revenue as a consequence of an order of the AO. cannot be treated as prejudicial to the interests of the Revenue. For example,when the AO. adopts one of two courses permissible in law and it has resulted in loss of revenue, or where two views are possible and the AO. has taken one view with which the CIT does not agree, it cannot be treated as an erroneous order prejudicial to the Revenue, unless the view taken by the AO. is unsustainable in law. **[CIT-Max India Ltd.,SC,** *Sec.263.***295ITR282,166Taxman188]**

Premium Ker. Baby Marine (Eastern) Exports - ACIT

Bruton's Legal Theasaurus defines premium as a noun as under :

Amount over par, bonus, bounty, charge, beyond normal, charge to excess, excessive charge, extra, incentive, increased value,over-charge, prize.

Collins Cobuild English Language Dictionary more graphically defines premium to mean as under:

A premium is a sum of money that is added to something, for example to someone's earnings or to the price of goods, in order to encourage people to work hard or to produce the goods.Likewise, we may advert to other dictionaries as well. As for instance in Black's Law Dictionary premium is defined as :A reward for an act done. Brown - Board of Police Commissioner's of City of Los Angeles (58 Cal. App. 2d 473, 136 p.2d 617, 619). See also Bonus.A bounty or bonus; a consideration given to invite a loan or a bargain, as the consideration paid to the assignor by the assignee of a lease, or to the transferer by the transferee of shares of stock, etc. So stock is said to be 'at a premium when its market price exceeds its nominal or face value. The excess of issue (or market) price over par value.

*Sec.80HHC.***[184CTR151,262ITR88,131Taxman546]**

Prescribed **MP. Suresh Deole - CCIT**

A perusal of section 34AB of the Wealth-tax Act,1957,makes it clear that power rests with the authorities to prescribe different qualifications for valuers of different classes of assets, and, therefore, a person who possesses the qualifications as prescribed can be registered as a valuer.Now, the word prescribed should always be construed to mean prescribed by rules by the rule-making authority.When the Legislature uses the word prescribed in an enactment,it is always a normal expression for conferring power on the executive to make rules. *Sec.34AB.WT.Act.* **[91CTR178,188ITR741,61Taxman240]**

Presumption **SC P. R. Metrani - CIT**

A presumption is an inference of fact drawn from other known or proved facts. It is a rule of law under which courts are authorized to draw a particular inference from a particular fact. It is of three types,(i) may presume, (ii) shall presume and (iii) conclusive proof.

Misc.**[203CTR290,287ITR209,155Taxman186]**

Previous year **SC Anglo - French Textiles Co. Ltd.-CIT**

The expression previous year substantially means an accounting year comprised of a full period of twelve months and usually corresponding to a financial year preceding the financial year of assessment. It also means an accounting year comprised of a full period of twelve months adopted by the assessee for maintaining his accounts but different from the financial year and preceding a financial year. For purposes of the charging sections of the Act unless otherwise provided for it is co-related to a year of assessment immediately following it,but it is not necessarily wedded to an assessment year in all cases and it cannot be said that the expression previous year has no meaning unless it is used in relation to a financial year. In a certain context it may well mean a completed accounting year immediately preceding the happening of a contingency. *Sec.3.*[**23ITR82]**

2. A perusal of section 3 of the Income-tax Act,1961, shows that, ordinarily,previous year means the financial year immediately preceding the assessment year.But,where the accounts of the assessee have been made up to a date within the said financial year, then, at the option of the assessee,the period of twelve months ending on such date shall be the accounting year.

[CIT - J.B. Lohada,AP.,*Sec.3.***174ITR318,39Taxman290]**

3. The expression previous year should be so construed as to be applicable uniformly to all cases. It cannot be said that a previous year should be taken as the financial year in which the application is made provided the stay of the applicant up to the date of the application or the estimated stay of the applicant in India in that financial year exceeds 182 days and that it should be the previous year preceding that financial year in case it is not possible to determine the duration of the stay of the applicant in India in the financial year in which the application is made. It appears more practical and reasonable, for purposes of determining the residential status of an applicant under section 245Q, to look at the position in the earlier previous year, i.e., the financial year immediately preceding the financial year in which the application is made. This is a period with reference to which the residential status of the applicant in every case can be determined without any ambiguity whatsoever.

[Monte Harris -CIT, AAR, *Sec.245Q.***135 CTR 472, 222ITR562,135Taxation121,89Taxman92]**

4. The term previous year is now defined as : previous year means twelve months ending on the 31st day of March of next preceding the year for which the assessment is to be made.

[Parry Agro Industries Ltd.-State of Tamilnadu., Mad., *Sec.3.* **176CTR314,255ITR194]**

5. The reference to previous year in sub-section (2), in the context, can only refer to every previous year in which the assessee seeks to carry forward the loss and other adjustable sums.

[CIT- Fab Exports (P.)Ltd.,Mad.,*Sec.3.***176CTR129,258ITR56, 131 Taxman790]**

Previously used for any purpose **Raj.** **Kanhiyalal Rameshwar Das - CIT**

The words previously used for any purpose are wide enough and cover all old and used machinery, no matter who used such machinery previously and from which source it was acquired. There is no justification for reading into sub-section 4(ii), the words by the assessee himself after the last word purpose occurring therein. Had sub-section 4(ii) been intended to mean that machinery previously used by a person other than the assessee,if acquired by the assessee for his new business would qualify for tax relief, the Legislature would not have felt the necessity of inserting Explanation 1, for this Explanation has obviously been inserted by way of an exception to sub-section 4(ii) to provide tax relief to an assessee who acquires old machinery from another person who had used it outside India.

Sec.80J(4).[156ITR463,22Taxman455]

Prima facie **Kar.** **Motor Industries Co. Ltd.-JCIT (Asst)**

Prima facie consideration has to be prima facie which means at the first sight or the first impression. That impression may be varied or modified or even destroyed at the stage of final hearing. The final hearing approach has not to be the approach at the first impression namely, prima facie impression, in terms of section 143(1)(a) of the Income-tax Act, 1961. *Sec.143(1)(a).*[292ITR70]

Prima facie case **Bom.** **Inayat Hussain -Union of India**

A prima facie case does not mean a case which will succeed, but a case which is not such as is apparently barred by any provisions of law and in respect of which something could be said in favour of the plaintiff. *Sec.281.*[122ITR227]

Prima facie case **P&H.** **ITO-Emerson Paul Plastic Company**

A prima facie case is not made out where the evidence is totally unworthy of credit or the same is patently absurd or inherently improbable. It is not possible to define the expression prima facie case because it will vary from case to case. The standard of test, proof and judgment which are to be applied finally before finding the accused guilty or otherwise is not exactly to be applied. At this stage, even a very strong suspicion founded upon materials before the Magistrate which lead him to form a presumptive opinion as to the existence of the factual ingredients constituting the offence alleged may justify the framing of a charge. In determining whether a prima facie case had been made out, the evidence of the witnesses is entitled to a reasonable degree of credit. *Sec.276C.* **[191ITR560]**

Prima facie inadmissible **Ker. CIT-Kerala Solvent Extractions Ltd.**

The question whether an item is prima facie inadmissible or not has to be found out from the return and statement of accounts furnished by the assessee. If the deduction claimed is not allowable even according to the statement of the assessee, then it is a prima facie inadmissible deduction which can be added back in the proceedings under section 143(1)(a) of the Act. *Sec.143(1)(a).*
[217CTR311,306ITR54,173Taxman155]

Principal officer **Del. ITO - Delhi Iron Works (P) Ltd.**

A director is not included within the ambit of sub-clause (a) of section 2(35) of the Income-tax Act,1961. In case the Income-tax Officer seeks to prosecute the director along with the company for an offence punishable under section 276B of the Act, he has to issue a notice under sub-clause (b) of section 2(35) of the Act expressing his intention to treat the director as principal officer of the company. However, it would not be necessary to issue a separate notice or communication to all the directors that they are to be treated as principal officers. It would be sufficient compliance if in the show-cause notice issued to the company it was mentioned that the directors were to be considered as principal officers of the company under the Act. *Sec.2(35)r.w.s.276B.*
[331ITR5,195Taxman372]

Principle of mutuality **Pat. CIT - Ranchi Club Ltd.**

No person can trade with himself and make an assessable profit. If, instead of one person, more than one combine themselves into a distinct and separate legal entity for the purposes of rendering services to themselves or for the supply of refreshments, beverages, entertainment, etc.,by overcharging themselves, the resulting surplus is not assessable to tax if the surplus is to be refunded to the members.The contributors to the common fund and the participators in the surplus must be an identical body. That does not mean that each member should contribute to the common fund or that each member should participate in the surplus or get back from the surplus precisely what he has paid. What is required is that the members as a class should contribute to the common fund and participators as a class must be able to participate in the surplus. It is immaterial whether the surplus is paid back to the members in cash or is put to reserve for development and for providing better amenities to the members. When the body of individuals is incorporated into a company or formed into a registered society, what is essential is that it should not have dealings with an outside body which result in surplus. The participation of the members in the surplus must be in their character as contributors to the common fund or as consumers and not as shareholders getting dividends on their share amount or as debenture holders earning interest. In all cases of incorporation as a company or as a registered society, the proper mode of regarding the company or the registered society is that it is a convenient instrument or medium for enabling the members to conduct a social club, the objects of which are immune from every taint of commerciality. The property of the incorporated company or a registered society, for all practical purposes in this behalf, is considered as property of the members.A members' club formed for social intercourse and for either recreation or for cultural activities cannot be considered to trade for profits so as to make its surplus taxable in law when it over-charges its members for the supply of refreshments, beverages or amenities to its members. Such supplies are not sales as there is no element of transfer of property in them. But this principle cannot have any application in respect of the surplus received from dealings with non-members. It is not difficult to conceive of cases where one and the same concern may indulge in activities which are partly mutual and partly non-mutual.The principle of mutuality can, in such cases, be confined to transactions with members.

Misc. **[196ITR137]**

Procedure for the assessment SC Kalawati Devi Harlalka - CIT

The word assessment can bear a very comprehensive meaning; it can comprehend the whole procedure for ascertaining and imposing liability upon the taxpayer and there is nothing in the context of Sec. 297 of the 1961 Act,which compels the court to give the expression procedure for the assessment a narrower meaning.

Sec.297. **[66ITR680]**

Proceeding for assessment SC S.S.Gadgil - Lal and Co.

A proceeding for assessment is not a suit for adjudication of a civil dispute... The Income-tax authorities who have power to assess and recover tax are not acting as judges deciding a litigation between the citizen and the State: they are administrative authorities whose proceedings are regulated by statute,but whose function is to estimate the income of the taxpayer and to assess him to tax on the basis of that estimate. Tax legislation necessitates the setting up of machinery to ascertain the taxable income and to assess tax on the income, but that does not impress the proceeding with the character of an action between the citizen and the State. *Sec.147.***[53ITR231]**

Proceedings **Cal.** **Keshab Narayan Banerjee - CIT**

Since proceeding has not been qualified in section 263, it cannot exclude any proceeding and cannot be confined only to mean a proceeding for assessment by the Income-tax Officer. *Sec.263.***[173 CTR61,252ITR888,125Taxman299]**

Proceedings are dropped **Raj.** **Smt.Badan Kanwar Medical Trust-CWT**

The assessing authority may on the basis of documents submitted or even on verbal submissions come to the conclusion that the proceedings are dropped. ...the recording of the words that proceedings are dropped amount to an order. .... in a case where a plea raised by an assessee is accepted and proceedings are dropped it is not necessary that the order should be communicated and served on the assessee. *Sec.25.WT.Act.***[214ITR130]**

Proceedings for the assessment **Cal.** **Kalawati Devi Harlalka - CIT**

The expression proceedings for the assessment in section 297(2)(a) has a wide connotation and embraces within its scope the various proceedings relating to assessment as envisaged in Chapter IV of the Act of 1922 including proceedings by way of appeal, reference and revision in a case where the return of income has been filed before the commencement of the Act of 1961. Clause (c) of section 297(2) does not restrict the scope of clause (a) to proceeding for original assessment. *Sec.297(2)(a).***[62ITR544]**

Proceedings resulting in revision **Gau. CIT - Assam Roofing Ltd**

The expression proceedings resulting in revision appears in sub-section (2) of section 244A of the Income-tax Act,1961, notwithstanding the fact that under section 240 refund following an appellate order is automatic and does not depend on a claim being lodged to the said effect or a refund proceeding being initiated. The use of the expression in the above context is both significant and conspicuous.To read the same only to situations covered by sections 238 and 239 of the Act which contemplate a claim for refund to be made in the situations specified therein would be against the scheme under the Act.The expression proceeding referred to in sub-section (2), more reasonably, would mean any proceeding as a result of which refund has become due. Viewed from this perspective the expression proceeding in sub-section (2) may take within its ambit an appeal proceeding consequential to which refund may have become due. Sec.244A(2).**[330ITR87]**

Process **Guj. CIT - Lakhtar Cotton Press Co. (Pvt.) Ltd.**

The activity contemplated by the word process in general,requiring only continuous and regular action or succession of actions leading to the accomplishment of some result but it is not one of the requisites that the activity should involve some operation on some material in order to its conversion into some other stuff. Therefore, what is necessary in order to characterise an operation as processing is that the commodity must, as a result of the operation, experience some change. *Misc.***[142ITR503,66Taxation21]**

2. Every process does not amount to manufacture of articles or goods. In order to constitute manufacture, as a result of the processing of a commodity, a commercially different and distinct commodity should emerge.The test to be applied is whether the commodity subjected to the process can no longer be regarded as the original commodity but is recognised in the trade as a new and distinct commodity. **[CIT-Ravi Ratna Exporters,Bom., Sec.80J.212ITR588]**

3. There is no statutory definition of term process in the Income-tax Act. The plain dictionary meaning of the term is to subject to some special process or treatment, to subject (especially raw material) to a process of manufacture,development or preparation for the market,etc.,to convert into marketable form as livestock by slaughtering, grain by milling,cotton by spinning, milk by pasteurising, fruits and vegetables by sorting and repacking. **[CIT - Ashwinkumar Gordhanbhai and Bros. Pvt. Ltd.,Guj.,** *Misc.***212ITR614]**

Processing **Mad. Chillies Export House Ltd. -CIT**

If a new substance is brought into existence or if a new or different article having a distinctive name, character or use results from particular process or processes, such process or processes would amount to manufacture. Processing is a continuous and regular action or succession of actions leading to the accomplishment of some results. *Misc.***140CTR9,220ITR411,137Taxation370]**

2. The word processing used in the definition of industrial company in section 2(7)(c) of Finance Act,1978, has not been defined in the Income Tax Act,1961, and it must, therefore, be interpreted according to its own natural meaning. Websters' Dictionary gives the following meaning of the word process: to subject to some special process or treatment, to subject (especially raw material) to a process of manufacture, development or preparation for the market, etc., to convert into marketable form as livestock by slaughtering, grain by milling, cotton by spinning, milk by pasteurizing, fruits and vegetables by sorting and repacking. Where, therefore, any commodity is subjected to a process or treatment with a view to its development or preparation for the market, as, for example , by sorting and repacking fruits and vegetables, it would amount to processing of the commodity. The nature and extent of processing may vary from case to case; in one case the processing may be slight and in another it may be extensive; but with each process suffered, the commodity would experience a change. Wherever a commodity undergoes a change as a result of some operation performed on it or in regard to it, such operation would amount to processing of the commodity. The nature and extent of the change is not material. The question is not whether there is manual application of energy; or there is application of mechanical force. Whatever be the means employed for the purpose of carrying out the operation, it is the effect of the operation on the commodity that is material for the purpose of determining whether the operation constitutes processing.

[G.A. Renderian Ltd.-CIT,Cal. *Misc.***32CTR318,145ITR 387,12Taxman160]**

3. The word processing has in one sense a wider meaning than the term manufacture.At some point, processing and manufacturing may merge, but where the commodity retains its substantial identity through the processing stage, it will be said to have been processed and not manufactured. In a manufacture, a commodity undergoes a change as a result of some operation. It should be so transformed as to lose its original character. A manufacture may involve several processes but the ultimate product that emerges can no longer be regarded as the original commodity, but is a new and distinct commodity. Whenever a commodity undergoes a change as a result of some operation performed on it such operation would amount to processing of the commodity. It is immaterial whether there is manual application of energy or mechanical force is employed. Whatever be the means employed for the purpose of carrying out the operation, it is the effect of the operation on the commodity that is material for the purpose of determining whether the operation is processing or manufacture.

[Koshy's P.Ltd.-CIT,Kar.,*Misc.*154ITR53]

4. The word processing has not been defined and must, therefore, be interpreted according to its plain natural meaning. According to the dictionary meaning of the term, where any commodity is subjected to a process or treatment with a view to its development or preparation for the market, as for example, by sorting and repacking fruits and vegetables, it would amount to processing of the commodity.The nature and extent of processing may vary from case to case; in one case, the processing may be slight and in another, it may be extensive; but with each process suffered, the commodity would undergo a change. **[CIT-Datacons(P.) Ltd., Kar.,*Misc.*47CTR162,155ITR66, 21Taxman 341]**

5. What is necessary in order to characterise an operation as processing is that the commodity must, as a result of the operation, experience some change.Keeping of goods in a cold storage plant does not bring about any change, whatsoever, in the goods stored therein. On the contrary, they are kept intact, in the same nature and form in which they are originally stored.Hence,running of a cold storage plant does not involve the processing of goods stored therein.

[Delhi Cold Storage(P.) Ltd.-CIT,Del., *Misc.*45CTR24,156ITR97,19Taxman 412]

6. Wherever a commodity undergoes a change as a result of some operation performed on it or in regard to it, such operation would amount to processing.**[CIT-Kutch Oil and Allied Industries Pvt. Ltd.,Guj.,** *Misc*

7. What is necessary in order to characterise an operation as processing is that the commodity must, as a result of the operation, experience some change. However, the nature and extent of processing may vary from case to case; in one case, the processing may be slight and in another it may be exclusive; but, with each process suffered, the commodity would experience a change. Wherever a commodity undergoes a change as a result of some operation performed on it or in regard to it, such operation would amount to processing of the commodity. The nature and extent of the change is not material. **[CIT-Ahmed A. Fazalbhoy (P.) Ltd.,Bom.,*Misc.*189ITR663]**

8. In common parlance, processing is understood as an action which brings forth some change or alteration of the goods or material subjected to the act of processing. **[Delhi Cold Storage P. Ltd. - CIT,SC,*Misc.*191ITR656]**

9. Wherever a commodity undergoes a change as a result of some operation performed on it or in regard to it, such operation would amount to processing of the commodity. The nature and extent of the change is not material. Whatever be the means employed for the purpose of carrying out the operation, it is the effect of the operation on the commodity that is material for the purpose of determining whether the operation constitutes processing.**[CIT-S.P.Jaiswal Estates (P.)Ltd.,Cal.,*Sec.32A*196ITR179]**

10. Processing is a continuous and regular action or succession of actions leading to the accomplishment of some results.**[Chillies Export House Ltd.-CIT,Mad.,*Misc.*140CTR9,220ITR411]**

11. The word processing has been used in different contexts in different Acts.The said word is of wide amplitude. It has various shades of meaning and it has been interpreted in different ways under various enactments. The word processing has not been defined under the Act. The Supreme Court has laid down in Chowgule -Union of India 47 STC 124 that wherever a commodity undergoes a change as a result of some operation performed on it or in regard to it, such operation would amount to processing of the commodity. The nature and extent of the change is not material. **[North Koel Kendu Leaves and Mahulam Leaves -UOI, Pat., *Sec.206C.*228ITR630,100Taxman172]**

12. Tendu leaves are natural forest produce.The general features of trade in tendu leaves consist of the purchase of tendu forest units from the State Governments and thereafter by processing the tendu leaves they are marketed. The business of processing of tendu leaves comprises obtaining the rights of collection and purchase of raw tendu leaves from the concerned forest department.There are other processes undertaken like collection of tendu leaves, pruning, prevention of diseases, plucking of green leaves, arrangements of pudas at collection centres, drying the same by solar energy, classification/ gradation of bidi leaves,putting them in the bags and thereafter their transportation.These activities do not amount to processing of tendu leaves within the meaning of section 206C of the Income-tax Act,1961. **[Natwarlal -Union of India,MP.,** Sec.206C.**151CTR644,233ITR490,102Taxman49]**

13. The integrated activities of processing, preservation and packaging of the fruits were embraced within the scope of sub-section (11A) of section 80- IB. The extraction of juice and oil from the fruits or further converting the homogenised juice into fruit powder and adding the substance meant for preservation would legitimately fall within the sweep of the expression processing. The fact that the fruit assumed a different form or that a series of operations were involved in preparing the mixed juices did not take it out of the bounds of processing.**[Delna Rustom Boyce,AAR,** Sec.80-IB(11A).**318 ITR 455,185Taxman180]**

14. In Chowgule and Co. v. Union of India [47 STC 124 SC] the Supreme Court held that when different brands of tea were mixed for the purpose of producing a tea mixture of a different kind and quality according to a formula evolved by the assessee, there was plainly and indubitably processing of the different brands of tea, because these brands of tea experienced, as a result of mixing, qualitative change, in that the tea mixture which came into existence was of different quality and flavour than the different brands of tea which went into the mixture. **[Stewart Holl (India) Ltd.-CIT,Cal.,** Misc.**338ITR194].**

Processing, manufacture and production **Bom.** **CIT-Sterling Foods (Goa)**

The three expressions processing, manufacture and production used in various taxing statutes, are not inter-changeable expressions. Though often used in juxtaposition, they convey different concepts and refer to different activities. Processing is a much wider concept. The nature and extent of processing may vary from case to case. Every process does not tantamount to manufacture. It is only when the process results in the emergence of a new and different article having a distinctive name, character or use, that manufacture can be said to have taken place. Similarly, production is wider than manufacture. As a result, every production need not amount to manufacture though every manufacture can be characterised as production. On a careful reading of section 80HH of the Income-tax Act, 1961, in the light of the scheme thereof and other provisions of the Act, it is clear that the Legislature intended to extend the benefit of deduction under section 80HH only to the industrial undertakings which manufacture or produce articles. This section was not intended to be applied to industrial undertakings which are engaged in processing of goods not amounting to manufacture or production of articles. *Sec.80HHC.***[213ITR851,79Taxman381]**

Processing / production **Cal.** **S.B. Cold Storage and Industries P.Ltd. - CIT**

The expressions processing and production are not identical. The Legislature in different statutes has used the said expressions in different contexts and has made a distinction between the two. Section 32A of the Income-tax Act,1961, lays down that there should be manufacture or production of an article or thing. An assessee which stores potatoes in its cold storage plant carries out an operation of processing within the meaning of the said expression as understood in legal parlance. The object of putting the goods in cold storage is mainly to preserve their original condition and not to produce any thing new. By such preservation no new article is brought into existence. It also cannot be said that by reason of such processing, something which was not marketable to start with becomes marketable. *Sec.32A.***[59CTR235,166ITR646, 30Taxman418]**

Processing of goods **Del.** **Addl. CIT - Kalsi Tyre P. Ltd**

The expression processing of goods used in section 2(6)(d) of the Finance Act,1968,refers to a wide category of activities.The expression means the subjection of the goods to some special process or treatment. It may be for the purpose of manufacture, for the purpose of development,for preparation for the market or for conversion into marketable form such as livestock by slaughtering, grain by milling, cotton by spinning, milk by pasturizing and fruits and vegetables by sorting and repacking. *Misc.***[131ITR636]**

2. In the processing of goods the original article need not lose its identity altogether but some changes are brought into it. When a cloth is dyed, printed and cut into pieces, the original identity of the cloth is not lost but a change is brought about as a result of going through the different processes and the end product is different from the feed-in material. **[CWT -Radhey Mohan Narain, Cal., <u>*Sec.5(1)(xxi)(xxii)WT.Act.*</u>135ITR372]**

3. In respect of processing of goods, it is not necessary that all the processes resulting in the end product must be carried out by the assessee himself. If an assessee has done some process which ultimately has brought about the end product, such an assessee will be an industrial company entitled to the concessional rate of tax. **[CIT-Oricon P. Ltd.,Bom.,<u>*Misc.*</u>38CTR212,151ITR296,16 Taxman100]**

Produce AP. CIT - Sri Venkateswara Hatcheries (P.) Ltd.

The word produce as used in ordinary parlance and its dictionary meaning shows that in the case of the word, no distinction is made between animate and inanimate objects.Even if a living object is brought forth by human effort,it can be said to have been produced. <u>*Sec.80J.*</u>**[71CTR80,174ITR231,39Taxman279]**

2. The word produce means bringing into existence new goods by a process which may or may not amount to manufacture. It also takes in all by-products, intermediate products and residual products which emerge in the course of manufacture of goods. The next word to be considered is articles.The word is not defined in the Act or the Rules.It must, therefore, be understood in its normal connotation-the sense in which it is understood in the commercial world. It is equally well to keep in mind the context since a word takes its colour from the context.The word articles is preceded by the words manufactures or produces. **[CIT-VijayaRetreaders,Ker. <u>*Sec.80-I.*</u>170CTR307,253ITR53, 119 Taxman395]**

Production Cal. CIT - Shaw Wallace and Co. Ltd.

The word production has not been defined in the Act. The expression production of an article means, among other things, that which is produced, a thing that results from any action, process or effort, a product, a product of human activity or effort. The word manufacture has also not been defined in the Act. The expression manufacture is generally understood to mean bringing into existence a new substance and does not mean merely to produce some change in a substance. <u>*Sec.32A(2)(b)(iii).*</u>**[201ITR17]**

2. The word production has a wider connotation than the word manufacture.While every manufacture can be characterised as production, every production need not amount to manufacture.The word production or produce when used in juxtaposition with the word manufacture takes in bringing into existence new goods by a process which may or may not amount to manufacture. **[B.S.Bajaj and Sons-CIT,P&H.,**Sec.80HH,80J.**135CTR491,222 ITR418,136 Taxation 523,91Taxman295]**

3. The word production has a wider connotation than the word manufacture.While every manufacture can be characterised as production, every production need not amount to manufacture.The expressions manufacture and produce are normally associated with movables articles and goods but they are never employed to denote construction activity of the nature involved in construction of a dam or a building. The word article is not defined in the Income-tax Act, 1961,or the Income-tax Rules,1962.It must,therefore,be understood in its normal connotation, i.e., the sense in which it is understood in the commercial world.It is equally well-settled that a word takes its colour from its context.**[CIT-Ceo Tech Foundations and Constructions,Ker.,**Sec.80-I.**158CTR586, 241ITR90,128Taxman253]**

4. Extraction and processing of mineral ore amounts to production within the meaning of the word in section 32A(2)(b)(iii) of the Income-tax Act,1961. **[CIT - Sesa Goa Ltd.,SC,** Sec 32.**192CTR557,271ITR331]**

5. The word production has a wider connotation than the word manufacture. While every manufacture can be characterised as production, every production need not amount to manufacture. **[CIT- Jansons and Co.,All.,** Sec.35.**202 CTR528,283ITR175,152 Taxman59]**

6. The expression production has a wider meaning than the word manufacture. The word production includes the activity of manufacturing by applying human endeavour on some existing raw material. **[Arihant Tiles & Marbles P. Ltd. - ITO,Raj.,** Sec.80-IB.**295ITR148]**

7. The activity of cutting, polishing and sizing granites would be covered within the meaning of the expression production. **[Fateh Granite Pvt. Ltd., Bom.,**Sec.10B.**222CTR,638,314ITR32]**

8. The word production used in sections 80HH and 80-I cannot derive its colour from the word manufacture.The word new is not used in the meaning of production.The important test which distinguishes production from manufacture is that the word production is wider than the word manufacture. **[Vijay Ship Breaking Corporation-CIT,SC,**Sec.80HH,80I.**219CTR639,314 ITR309,175Taxman77]**

9. The word production or produce when used in juxtaposition with the word manufacture takes in bringing into existence new goods by a process, which may or may not amount to manufacture.It also takes in all the by-products,intermediate products and residual products, which emerge in the course of manufacture of goods. **[India Cine Agencies-CIT, SC,*Sec.80HH,80-I.* 220CTR223,308 ITR98,175Taxman361]**

Production and manufacturing process **Cal.** **Simplex Concrete Piles (India) Pvt. Ltd. - CIT**

.... production and manufacturing process means putting in raw materials and emergence of a new thing. Sec.32A.**[206CTR219, 286ITR470]**

Production of mineral **AP.** **CIT - Singareni Collieries Co. Ltd.**

The expression produce used in section 32A of the Income-tax Act, 1961, must be understood in its normal connotation and according to commercial usage. The word production has a wider connotation than the word manufacture.While every manufacture can be characterised as production, every production need not amount to manufacture. The activity of winning or excavating the coal from the mines can be aptly described as production activity.The expression production of mineral is used in the allied provision of the Act (section 35E) and this is a definite pointer that Parliament employed the expression production to the minerals extracted from underneath the surface. Sec.32A **[221ITR48]**

Production or produce **Mad.** **CIT-Madurai Pandian Engineering Corp. Ltd.**

The word production or produce has been used in section 80HH of the Income-tax Act, 1961,in juxtaposition with the word manufacture and it would take in bringing into existence new goods by a process which may or may not amount to manufacture. The article referred to in the section, therefore, has reference to new articles and brought into existence by a process of manufacture or by any other mode, which can be regarded as production. The resultant article whether it is by manufacture or by way of production must be a new article. The term new is not found in the section. It must be held to be implicit in the word manufacture. Having regard to the fact that the word production or the word produce is used in juxtaposition with the word manufacture these terms also must be regarded as referring to production, which brings into existence a new article. *Sec.80HH.***[239ITR375]**

Profession SC CIT - Manmohan Das

A profession involves the idea of an occupation requiring either purely intellectual skill or if any manual skill,as in painting and sculpture or surgery, skill controlled by the intellectual skill of the operator as distinguished from an occupation which is substantially the production or sale or arrangements for the production or sale of commodities. *Sec.2(36).***[59ITR699]**

Profession Raj. CIT - Bhagwan Broker Agency

The word profession has been defined in section 2(36) to include vocation, which refers to a way of living and not necessarily a course of activity indulged in,for earning one's livelihood or making any income.Business is a term with a wider import than profession. All professions are businesses,but all businesses are not professions.There should be some special qualification of a person apart from skill and ability, which is required in carrying on any activity which could be considered as profession. This could be by having education in a particular system either in a college or university or it may be even by experience.In the case of a broker, the activities are carried on either under a written agreement or even verbal agreement in respect of different constituents and the activities, therefore, would amount to business. *Sec.2(36).***[212ITR133]**

2. The distinguishing feature of a profession is the possession by the practitioner of the profession of specialised knowledge involving intellectual skill and higher education and learning. The services rendered by a professional while practising the profession, are services for which he has been trained. The practice of a profession cannot be regarded as a commercial activity though the practice is not without compensation or profit. The compensation earned by the practitioner of a profession is by reason of the personal qualification possessed by him.........assistance rendered by the clearing and shipping agent to those who import or export, by attending to the documentation and ensuring the clearance of goods cannot be regarded as a profession based on intellectual attainment or personal service rendered on account of possession of specialised skill and knowledge based on higher learning and intellectual skill.

[CIT - International Clearing and Shipping Agency,Mad., *Sec.2.*158CTR672, 241 ITR172,118Taxman730]

Profession / vocation All. Addl.CIT -Ram Kripal Tripathi

The expression profession involves the idea of an occupation requiring purely intellectual skill or manual skill controlled by the intellectual skill of the operator, as distinguished from an occupation or business which is substantially the production or sale, or arrangements for the production or sale,of commodities.Profession is a word of wide import and includes vocation which is only a way of living and a person can have more than one vocation, and the vocation need not be for livelihood nor for making any income nor need it involve systematic and organised activity. *Sec.2(24)(va)r.w.s. 28(iv).***[17CTR79,125ITR408]**

Profession, vocation or occupation All. B. Malick - CIT

Section 4(3)(vii) of the Act obviously refers to an exercise of something more than and distinct from the faculties of the assessee. The use of faculties of the assessee must be so directed and bear such a character as to give the assessee a profession, vocation, or occupation. An activity of the assessee, before the assessee has actually acquired a profession, vocation, or occupation, either by a habitual pursuit of the activity or by engaging in it as a result of a design to pursue an occupation, cannot be considered the exercise of a profession or occupation. The activity of the assessee, however, disorganised or irregular or desultory, must assume or acquire the form of or flow from an occupation before the resulting income becomes taxable under the provisions of section 4(3)(vii) of the Act. Whether it has assumed that character or not will be a question of fact, which could only be determined by taking the intention with which,the number of times on which, and the whole set of attendant circumstances in which an activity, which may produce some monetary gain, is carried on. In every case, however, it is the exercise of a profession, vocation or occupation by an assessee which has to be established and not the mere use of the faculties or energies of an assessee which may result in an income. *Sec.10(3).* **[67ITR616]**

Professional income Guj. CIT-Dr.K.K.Shah and Dr. (Mrs.) Haribala K. Shah

Income from nursing home run by the doctors who themselves treat their own patients as an integral part of the profession must be treated as professional income. Sec.2.r.w.s.28,64(1).
[28CTR126,135ITR146,9Taxman131]

Profit Bom. Devkaran Nanjee Banking Co. Ltd. - CEPT

The word profits used in rule 5 of Schedule II must be construed to mean statutory profits as defined in section 2(19) of the Act.It cannot mean book profits or actual profits. Excess Profits Tax Act.
[18ITR47]

2....the expression profits or gains means profits or gains not merely from the particular business in respect of which depreciation is claimed, nor profits or gains from any business conducted by the assessee, but the profits or gains which may accrue or arise to the assessee under any of the head referred to in section 6. **[Ambika Silk Mills Co. Ltd.-CIT,Bom.,**Misc.**AIR1952Bom483, 22ITR58]**

3. The words profit and pains used in Section 23-A(1) Income Tax Act 1922 cannot be limited only to the profits and against and gains of a company carrying on business. The provisions of the section are applicable to companies whose only source of income is property assessable under Section 9. **[Kilburn Properties Ltd.-CIT,Cal.,** Sec.22.r.w.s.104 **AIR1949Cal.333,17ITR134]**

4. The word profit in Rule 6 has the same meaning as the word profit in section 2(6)and section.10(7)and does not include technical surplus of mutual insurance societies except the case of a mutual life insurance society where this surplus is expressly brought by the clear language of Rules2&3. **[Calcutta Hospital and Nursing Home Benefit Association –CIT,Cal.,**Sec.2(24).**AIR1963Cal.598, 47ITR247]**

5. For the purpose of income-tax, a transaction must be looked at from a commercial point of view and in trying to determine whether a certain transaction resulted in profits,we must come to a conclusion that the transaction resulted in real profits, profits which from the commercial point of view meant a gain to the person who entered into the transaction,not profits from any narrow, technical or legalistic point of view. **[CIT-Mugneeram Bangur and Co.,Cal.,**Misc.**47ITR565]**

6. Where a number of persons combine together and contribute to a common fund for the financing of some venture or object and in this respect have no dealings or relations with any outside body, then any surplus returned to those persons cannot be regarded in any sense as profit. There must be complete identity between the contributors and the participators. If these requirements are fulfilled, it is immaterial what particular form the association takes. Trading between persons associating together in this way does not give rise to profits which are chargeable to tax. **[CIT-Bankipur Club Ltd.,SC,*Misc.*AIR1997SC2312,140CTR102,226 ITR97,138Taxation720,92Taxman278]**

7. The word profit in sub-sections (1) and (3)(a) and (b) of section 80HHC means a positive profit. In other words, if there is a loss then no deduction would be available under sub-section (1) or sub-section (3)(a) or sub-section (3)(b). In arriving at the figure of positive profit, both the profits and the losses will have to be considered. If the net figure is a positive profit then the assessee will be entitled to deduction ; if the net figure is a loss then the assessee will not be entitled to deduction. **[IPCA Laboratory Ltd. -DCIT,SC,*Sec.80HHC.*187 CTR513,266ITR521,135Taxman 594]**

8. The word profit has to be given its natural meaning. The profit mentioned in the proviso is clearly a profit and not loss. When there is no difficulty in understanding the word used in the statute, it is not necessary to find out the nature of the provisions. **[CIT-A.M. Moosa, Bharath Sea Foods,Ker.,*Sec.80HHC.*191CTR441,272 ITR29,141Taxman99]**

9. The word profit in section 80HHC(1) and (3) of the Income-tax Act,1961, means a positive profit. A plain reading of sub-section (3) (c) shows that profits from such exports has to be profits from exports of self- manufactured goods plus profits from exports of trading goods.The profit is to be calculated in the manner laid down in sub-section (3)(c)(i) and (ii). The opening words profits derived from such exports together with the word and clearly indicate that the profits have to be calculated by counting both the exports. Deduction can be permitted only if there is a positive profit in the exports of both self- manufactured goods as well as trading goods. If there is a loss in either of the two then that loss has to be taken into account for the purposes of computing the profits. **[CIT-Sahara India Financial,Del.,*Sec.80HHC.*294ITR1,163 Taxman 741]**

10. The word profit implies a comparison between the state of business at two specific dates, usually separated by an interval of twelve months. Stock-in-trade is an asset : it is a trading asset. Therefore, the concept of profits and gains made by a business during the year can only materialize where a comparison of the assets of the business at two different dates are taken into account. **[CIT- Woodward Governor India P.Ltd.,SC,**<u>*Sec.145.*</u>**223CTR1, 312ITR254,179 Taxman326]**

11. The word profit in sub-sections (1) and (3)(a) and (b) of section 80 HHC means a positive profit.CircularNo.636,dated August 31,1992 nowhere provided for negative figures. **[Universal Cold StorageLtd.-DCIT,Mad.,**<u>*Sec.80HHC.*</u> **326ITR533]**

Profit in lieu of salary **AP.** **M.Krishna Murthy - CIT**

The receipt which is in the nature of addition to the remuneration of the employee and not a casual and non- recurring receipt, which is exempt from tax under section 10(3). It is a profit in lieu of salary within the meaning of section 17(3)(ii). <u>*Sec.17.*</u>**[75CTR36,152ITR 163]**

Profit or loss of a capital nature **Cal.** **Union Carbide India Ltd. - CIT**

Where profit or loss arises to an assessee on account of appreciation or depreciation in the value of foreign currency held by it, on conversion into another currency, such profit or loss would ordinarily be trading profit or loss if the foreign currency is held by the assessee on revenue account or as a trading asset or as part of circulating capital embarked in the business.But, if on the other hand,the foreign currency is held as a capital asset or as fixed capital, such profit or loss would be of a capital nature.

<u>*Sec.33.r.w.s.43.*</u>**[19CTR77,130ITR351,59Taxation71]**

Profits **Ran.** **CIT - Bengalee Urban Co-operative Credit Society Ltd.**

Profits are the surplus by which the receipts from the trade or business exceed the expenditure necessary for the purpose of earning those profits. <u>*Misc.*</u>**[2ITR121]**

2. Section 84 of the Income-tax Act,1961, is similar to section 15C of the Indian Income-tax Act,1922. Rule 19 of the Income-tax Rules, 1962,seeks to give effect to the provision of section 84 and so the term profits in rule 19(5) must have the same meaning as in section 84.The term profits has to be construed to mean taxable profits, i.e., profits arrived at in accordance with the provisions of the

Act,i.e.,after deducting the depreciation permissible under the Act, and, therefore, the amount of depreciation cannot be added back for the purposes of determining the profit under rule 19(5). **[Jaipur Udhyog Ltd.-CIT,Raj.,** *Sec.84* **198ITR282]**

3. The expression profits used in the section connotes positive profit. It is a profit earned from the business of export of goods alone which can be the subject matter of exemption there under. A fortiori, if a profit is not earned, the question of claiming exemption under that section would not arise. **[ITO-Induflex Products Pvt. Ltd.,SC,***Sec.80HHC(1).***199CTR712,280ITR1,149Taxman687]**

Profits and gains of business or profession **Guj.** **CIT-Alchemic Pvt. Ltd.**

If what is received either by way of benefit or perquisite is money, there is no question of considering the value of such monetary benefit or perquisite under clause (iv) of s. 28 and including the value of such benefit or perquisite under the head Profits and gains of business or profession. It is only if the benefit or perquisite is not in cash or money that section 28(iv) would apply and the question of including the value of such benefit or perquisite as income from business would ever arise. *Sec.28(iv).***[20CTR83,130ITR168]**

2. The expressions used in sub-section (3) of section 80HHD such as business and profits and gains of business or profession are expressions used in a larger context and not a limited context. Therefore, all receipts from the business activity of running a hotel necessarily go into constituting the total receipts which is in the denominator portion of the formula. .....The formula in terms of sub-section (3) of section 80HHD which has to be worked out, based on the language of sub-section (3) of section 80HHD and the profits which can claim benefit in section 80HHD(1) being such profits attributable to the earnings from services provided by the assessee to foreign tourists and being the amount which bears to the profits of the business the same proportion as receipts specified in sub-section (2) of section 80HHD, etc., one cannot either give a restrictive meaning or understand the words profits of the business of the assessee as profits attributable to only the recognised units of the assessee in the hotel business. .... Even in section 80HHD the reference is to the business of the assessee which is the business as a whole and as one unit and not by sub-dividing the total income of the business into unit-wise total income and then arriving at the unit-wise overall profit proportionate profit and then adding up the same for arriving at the benefit under section 80HHD. That is neither envisaged under the statutory scheme nor is indicated in the language of the section. **CIT-ITC Hotels Ltd., Kar.** *Sec.80HHD.*[**344ITR680]**

Profits from business Bom. CIT - Cilag Ltd.

The patent of a patentee or the secret knowledge possessed by a person of a process are both capital assets and both these assets possessed by the possessors thereof are capable of being employed in business in either of two ways, viz., that he can himself exploit the patent or the secret knowledge or process in business for profit or can grant a licence to others to do so on payment of royalty. The profit, which he would derive by exercising the invention or process or the profit he derives from the royalty are profits from business, though parting of the patent or process in favour of a third party may, in certain circumstances, amount to a disposal of a capital asset and not its employment in trade. What is required to be considered is whether, on the facts and the circumstances of a given case, the licence granted to make use of the patents or the secret information to another was in the nature of parting of an asset or only in the nature of employing it in trade. *Misc.*[**70ITR760]**

Profits in lieu of salary Kar. CIT - P. Surendra Prabhu

Profits in lieu of salary would mean a gain or advantage that the assessee would receive instead of or in place of salary. The Legislature again uses the expression includes immediately after the expression profits in lieu of salary to include any amount of compensation due or received by an assessee from his employer or former employer at or in connection with the termination of his employment or the modification of the terms and conditions relating thereto. *Sec.17.r.w.s.10.*[**198CTR209,279ITR402,149Taxman82]**

Profits whichever is greater Bom. Zenith Assurance Co. Ltd. - CIT

The expression profits and gains in rule 2 of the Schedule to the Indian Income-tax Act, 1922, covers not only profits and gains of the business but also losses of that business. Rule 2 plus profits and gains of life insurance business on the same footing as the losses of life insurance business and the mode of computation for both is the same.It is either under sub-clause (a) or sub-clause (b), whichever is the greater.The expression whichever is the greater is used by the

Legislature in order that advantage should ensue to the revenue by computing the profits and gains or losses under sub-clause (a) or sub-clause (b).The option is left to the taxing department and it can avail itself of the option from the point of view of benefit to the revenue. Sec.44C.**[26ITR256]**

Proper service of notice Del. R.L.Narang - CIT

Where a notice is served through a process server and there is neither a service report nor any evidence to show the identity of persons on whom the service was effected, there is no proper service. Sec.282.**[136ITR108]**

Property Lah. Trustees of the Tribune

The word property in section 4(3) of the Income-tax Act does not bear the restricted meaning that it bears in section 9 of the Act but includes also securities, business or a share in a business. Sec.5.**[3ITR246]**

2. The word property as used in sub-section (1) of section 42 means something tangible but it is not confined to immovable property or to buildings or land appertaining thereto. The word is used as an ordinary English word to be taken in its usual signification and there is nothing in the sub-section to exclude from its scope any of the six clauses of income mentioned in section 6 of the Act.**[CIT-Currimbhoy Ebrahim & Sons Ltd.,PC,**Sec.9.**3ITR395]**

3. Property is a term of the widest import, and subject to any limitation or qualification which the context might require, it signifies every possible interest which a person can acquire, hold and enjoy. Business would undoubtedly be property, unless there is something to the contrary in the enactment. There is nothing in section 4(3)(i) of the Income-tax Act,1922,which restricts in any manner the normal and accepted meaning of the word property and excludes business from its connotation.Business would, therefore, be property for the purposes of section 4(3)(i). The managing agency of a company is business and is property for the purposes of section 4(3)(i).

[CIT-J.K.Trust,SC,Sec.5.**AIR1957SC816,32ITR535]**

4. Property is a term of widest import and, subject to any limitation or qualification which the context may require,signifies every possible interest which a person can acquire, hold and enjoy. **[Dharma Vijaya Agency- CIT, Bom.,**Sec.5.**AIR1960Bom380,38 ITR392]**

5. Property is a term of the widest import and subject to any limitation which the context may require, it signifies every possible interest which a person can hold or enjoy. Therefore, excluding the exceptions enumerated in the Act, all movable and immovable property, no matter of what description, is an asset for purposes of the Wealth-tax Act.Ordinarily, property includes the right to transfer it but because of the peculiar incidents of a property or because of statutory or contractual restrictions that right may be abridged or excluded altogether. Nonetheless,what remains would still be property.Section 7 is merely concerned with the mode of valuing an asset and does not indicate what is an asset for purposes of charge to wealth-tax.The right to remuneration granted to a mutawalli under a wakf deed is not transferable.Nevertheless,it is an asset under section 2(e) and assessable as such to wealth-tax. **[CWT-Rani Kaniz Abid, All.,** *Sec.2(e).WT.Act.***93ITR332]**

6. Under section 2(14) of the Income-tax Act,1961, a capital asset means property of any kind held by an assessee, whether or not connected with his business or profession. The word property used in section 2(14) of the Act is a word of the widest amplitude and the definition has re-emphasised this by the use of the words of any kind. Any right which can be called property will be included in the definition of capital asset. A contract for sale of land is capable of specific performance.It is also assignable.Therefore,a right to obtain conveyance of immovable property is clearly property as contemplated by section 2(14)

[CIT-Tata Services Ltd.,Bom., *Sec.2(14).***122ITR594]**

7. It is true that the expression property must bear a comprehensive sense; but there must be a right, present or contingent, before it can be said that the assessee has an interest in property.

[CWT- Arvind Narottam,SC,*Sec.2(m),21(1).WT.Act.***72CTR94,173 ITR479,39Taxman368]**

8. ...the word property is used in the Indian Penal Code in a much wider sense than the expression movable property. The word property in a particular section covers only that type of property with respect to which the offence contemplated in that section can be committed. It is not necessary that a false pretence should be made in express words, but may be inferred from circumstances, including the conduct of the accused.

[M.K. Mathivathanan - ITO,Mad., *IPC.r.w.s.276C,277.***194ITR503]**

9. The meaning of property is well-settled that it is a term of the widest import and, subject to any limitation which the context may require, it signifies every possible interest which a person can hold or enjoy. **[CWT-Vidur V.Patel, Bom.,*Sec.2(e).WT.Act*124 CTR 343,215ITR30,127Taxation243,79Taxman289]**

10. Property is a term of the widest import and subject to any limitation which the context may require, it signifies every possible interest which a person clearly holds and enjoys. There is no justification or reason to give any restricted meaning to the word asset as defined by section 2(e) of the Wealth-tax Act,1957, when the language employed shows that it was intended to include property of every description. Merely because an item of expenditure is 100 per cent. allowable under the Income-tax Act while computing the income that does not automatically mean that such item ceases to be an asset or that it has no value. It will still remain an asset and will have some market value which in appropriate cases may be brought to tax under the provisions of the Wealth-tax Act. **[CWT-Narendra Kumar Gupta,All.,*Sec.2(e)WT.Act.*274ITR553]**

11. The expression property used in section 11 of the Act is of wide amplitude and it includes the business undertaking itself. It includes immovable and movable property.

[CIT-Krishi Utpadan Mandi Samiti,All.,*Sec.11.*231CTR505,331 ITR 154,186 Taxman460]

Property is in the occupation of the owner .... Cal. Calcutta Stock Exchange Association Ltd. - CIT

The expression where property is in the occupation of the owner for his our residence in the said proviso is intended to apply only to dwelling houses of human beings. It is not applicable to fictional persons such as limited companies. *Sec.22.* **[3ITR105]**

Property of any kind AAR Foster's Australia Ltd.

Property of any kind undoubtedly includes intellectual property which is but a species of intangible property. Trade mark, brand, goodwill, technical know-how relating to the manufacture of goods would all qualify to be treated as capital assets within the meaning of section 2(14) of the Act.

Sec.2(14). **[217CTR,21,302ITR289,170Taxman341]**

Property of which he is owner PC CIT-Dewan Bahadur Dewan Krishna Kishore

The words property of which he is the owner in section 9 of the Act cannot be read as meaning of which annual value he is the owner.

Sec.22. **[9ITR695]**

Provided credit facility **All.** **Addl.CIT- U.P.Co-operative Cane Union**

The expression providing credit facilities in section 81(i)(a) of the Income-tax Act 1961 takes its colour from the preceding expression viz. business of banking. In order that banking or providing of credit facilities may constitute a business it is necessary that these activities must be the chief source of income. A trader may earn interest from his customers to whom goods are supplied on credit but that is only incidental to the main business he carries on. The intention of the Legislature is to grant exemption to the income of co-operative societies engaged in the business enumerated in clauses (a) to (f) of section 81(i). These are the only kinds of business which are exempt. To hold that a co-operative society which sells goods on credit to its members is engaged in the business of providing credit facilities would amount to extending the exemption to businesses other than those mentioned in clauses (a) to (f) of section 81(i) of the Act. *Sec.80P(2).* **[114ITR70]**

Providing credit facilities **All.** **CIT- U.P. Co-op. Cane Union Federation Ltd.**

The expression providing credit facilities in section 80P(2)(a)(i) of the Income-tax Act,1961, would comprehend the business of lending money on interest.It would also comprehend the business of lending services on profit for guaranteeing payments, because guaranteeing payment is as much a part of banking business for affording credit facility as advancing loans. *Sec.80P(2).* **[9CTR160,122ITR913]**

2.The expression providing credit facilities in section 80P(2)(a)(i) cannot include the mere sale of goods on credit by an out-and-out consumer co-operative society. **[Rodier Mill Employees' Co-op. Stores Ltd.-CIT, Mad.,** *Sec.80P(2).* **135ITR355]**

3. In order to constitute the business of banking, it is necessary that these activities must be the main source of income of the society. It is equally true that the words which are used in one particular clause have to be given the same colour and understood as forming part of one genus of which they may be different species.The words providing credit facilities would, therefore, have to be interpreted to comprehend the business of lending services of credit facilities in connection with the business of banking. **[CIT-Co-op. Supply and Commission Shop Ltd.,Raj.,** *Sec.80P(2)(a).***204ITR713]**

4. The words providing credit facilities, occurring in section 80P (2) (a)(i) of the Income-tax Act, 1961, should be construed as similar to, or akin to the carrying on the business of banking, the preceding clause in the same subsection. The words providing credit facilities to its members means providing credit by way of loans and not selling goods on credit. **[Kerala Co-op. Consumers' Federation Ltd.-CIT,Ker.,** *Sec.80P(2).***67CTR73,170 ITR455,35Taxma441]**

Provision **Del. CIT - Orissa Cement Ltd.**

In the definition of provision enunciated by the Supreme Court in Metal Box Co.'s case 73 ITR 53 viz., that if any amount is retained by way of providing for any known liability of which the amount cannot be determined with substantial accuracy it would amount to a provision and not reserve, the expression known liability, in the context of the Super Profits Tax Act, should be taken to refer only to a liability existing on the relevant date and not a future liability. So long as the assessee has not estimated the current liability and set it apart or charged it to the current year and claimed a deduction therefore for purposes of income-tax,it cannot be said that a provision has been made by it towards an existing liability.

*Super Profits Tax Act.***[124ITR251]**

2. Though the term provision is defined positively by specifying what it means, the definition of reserve is negative in form and not exhaustive in the sense that it only specifies certain amounts which are not to be included in the term reserve. There could be no dispute about the principle that if provision for a known or existing liability is made in excess of the amount that would be reasonably necessary for the purpose, the excess shall have to be treated as a reserve and, therefore, would be includible in the capital computation. The question whether the concerned amounts in fact constituted reserves or not will have to be decided by having regard to the true nature and character of the sums so appropriated depending on the surrounding circumstances particularly the intention with which and the purpose for which such appropriations had been made. **[CIT - Gramophone Co. of India Ltd., Cal.,** *Sec.80O.***52CTR316, 162 ITR725,26Taxman250]**

3. A provision is a liability which can be measured only by using a substantial degree of estimation.A provision is recognized when : (a) an enterprise has a present obligation as a result of a past event; (b) it is probable that an outflow of resources will be required to settle the obligation,and (c) a reliable estimate can be made of the amount of the obligation.If these conditions are not met, no provision can be recognized. **[CIT - Hewlett Packard India (P.) Ltd.,SC,** *Sec. 37.*__223CTR,425,314ITR62,180Taxman422]__

Provision made by the assessee **SC** **Shree Sajjan Mills Ltd. - CIT**

The expression provision made by the assessee is not used in any artificial sense,e.g., of setting apart specifically by the assessee for meeting the liability for gratuity his in account books but in him in its ordinary sense. *Sec.40A(7)r.w.s.28,37.* **[49CTR193,156ITR585,23Taxman37]**

Public **Bom.** **Raghuvanshi Mills Ltd. - CIT**

The expression public in the Explanation to section 23A(1) of the Indian Income-tax Act,1922, is used in contradistinction to the directors and it cannot be given its ordinary natural meaning. The whole object of the third proviso to section 23A(1) is that there must be voting power exercised which must be independent of the control of the directors. If members of the public who are shareholders are under the control of the directors and if their voting power is controlled by the directors and if the votes cast by them are not their own votes but in substance the votes of the directors, then for the purpose of the third proviso the shares in effect are not held by the public at all but are held by the directors.In India, unlike in England,there is no statutory presumption as to control and therefore in each particular case it must be found as a fact that a director exercises de facto control over a shareholder. If that is found as a fact, then that particular shareholder for the purposes of section 23A will not be considered as a member of the public.

*Sec.104,109.***[24ITR338]**

2. The word public is used in contradistinction to one or more persons who act in unison and among whom the voting power constitutes a block.If such a block exists and possesses more than seventy- five per cent. of the voting power, the company cannot be said to be one in which the public are substantially interested. Such a group may be formed by the directors of a company acting in concert, or by some directors acting in concert with others, or even by some shareholder or shareholders, none of whom may be a director.The test is first to find out whether there is an individual or a group which controls the voting power as a block. If there be such a block,the shares held by it cannot be said to be unconditionally and beneficially held by members of the public. In the category of shares held by the public, only those shares can be counted which are unconditionally and beneficially held by the public, or, in other words, which are uncontrolled by the group which controls the affairs of the company. The group itself may be composed of directors or their nominees or relations in different combinations, but none can be said to belong to that group, be he a director or a relative,unless he does not hold the shares unconditionally and beneficially for himself. It is only such a person, who can fall properly outside the word public. **[Raghuvanshi Mills Ltd. - CIT,SC,***Sec.104,109.***41ITR613]**

3. The word public has not been defined in the statute but is of a very wide connotation and sweep and capable of taking in even an incorporated company if the context in which the said expression is used warrants such connotation. The expression public occurring in sub-clause (i) of clause (b) of the Explanation to section 23A will include companies other than companies to which section 23A applies and hence will take in a company in which the public are substantially interested. Further, in view of the two decisions of the Supreme Court in Shree Changdeo Sugar Mills Ltd.vs.CIT(41 ITR 667) and Pilani Investment Corporation Ltd. vs. CIT (89 ITR 53) when a company in which the public are substantially interested has been allotted unconditionally, or has acquired unconditionally, or was throughout the previous year beneficially holding not less than 50 per cent. of the voting power, it can be said that the public themselves were allotted or acquired the said voting power and they beneficially held the same throughout the previous year.

[Yercaud Coffee Curing Works Ltd.-CIT,Mad., *Sec.104,109.***111ITR787]**

4. The word public has not been defined under the Income-tax Act. The word public is used in contradistinction to one or more persons who act in unison and among whom the voting power constitutes a block.Thus,when a shareholder is constrained to transfer his shares to another shareholder, he is not freely transferring his shares to a member of the public. If a shareholder is permitted to transfer his shares to a person of his choice who is not a shareholder of the company without any unreasonable restriction, then alone can it be said that the shares are freely transferable to members of the public. **[CIT- Light Publication Ltd., Guj.,** *Sec.2(18).Companies Act.***167CTR450,251ITR120,119Taxman1106]**

Public are substantially interested SC CIT - Amrutanjan Ltd.

Where there is no individual member or a group of members acting in concert holding fifty-one per cent. or more of the voting power, which controls the working of a company,it is from its very nature a company in which there is no controlling member or group, and therefore,the public are substantially interested. Sec.104,109. **[53ITR218]**

Public interest Guj. Wood Polymer Limited and Bengal Hotels Pvt. Ltd.

If the only purpose discernible behind the amalgamation is defeating tax by creating a paper company and transferring an asset to such company which can without consequence be amalgamated with another company to whom the capital asset was to be transferred so that on amalgamation it may pass on to the amalgamating company, it would distinctly appear that the provision for such a scheme of amalgamation was utilised for the avowed object of defeating tax.It is true that the parties may so arrange their affairs that it may amount to avoidance of tax liability and not evasion of tax; law frowns upon tax evasion and not on tax avoidance.But such a benefit cannot be permitted to be enjoyed when it could not be done without the aid of the court. The court is charged with a duty, before it finally permits dissolution of the transferor- company by dissolving it without winding up, to ascertain whether its affairs have been carried on, not only in a manner not prejudicial to its members but in even public interest. The expression public interest must take its colour and content from the context in which it is used. The context in which the expression public interest is used, enables the court to find out why the transferor company came into existence, for what purpose it was set up, who were its promoters, who were controlling it, what object was sought to be achieved through creation of the transferor company and why it was being dissolved by merging it with another company. That is the colour and content of the expression public interest as used in the second proviso to section 394(1) of the Act which have to be enquired into. If the only purpose appears to be to acquire certain capital asset through the intermediary of the transferor- company created for that very purpose to meet the requirement of law, and in the process to defeat tax liability which would otherwise arise, it could not be said that the affairs of the transferor- company sought to be amalgamated, created for the sole purpose of facilitating transfer of capital asset through its medium, have not been carried on in a manner prejudicial to public interest. Public interest looms large in this background and the machinery of judicial process is sought to be utilised for defeating public interest and the court would not lend its assistance to defeat public interest.The court would, therefore,not sanction the scheme of amalgamation. *CompaniesAct.r.w.s.45,47.* **[6CTR431,109ITR177]**

Public Limited Co. Mad. Amrutanjan Ltd. - CIT

A public limited company would not come within section 23A of the Income-tax Act, if any one of the two conditions were satisfied: (1) that no member of the company possessed more than 50 per cent. of the voting power, and (2) that even if a member of the company possessed more than 50 per cent. of the voting power, the public held at least 25 per cent. of the voting power. The term public meant all members of the company other than the person having control,i.e.,more than 50 per cent.of the voting power. *Sec.104,109.* **[41ITR21]**

Public policy Kar. CIT - Bangalore Arrack Co.

Public policy is the principle which declares that no man can lawfully do that which has a tendency to be injurious to public welfare and public policy comprehends only the protection and promotion of public welfare.The question whether elimination of competition in trading is opposed to public policy cannot be considered in isolation de hors the nature of the trade,privilege or the right with respect to which an auction is held. *Sec.37.* **[201ITR25]**

Public servant Bom. Emperor - Osman Chotani

The public servant referred to in Section 54 of the Income-tax Act is a public servant to whom disclosure has been made under the Income-tax Act. *Sec.137.* **[10ITR429]**

Purchase Bom. Vidarbha Co- op. Marketing Society Ltd. - CIT

The first requirement before exemption from tax can be claimed under clause (d) of section 81(i) is that the co-operative society must be engaged in the purchase of the specified articles. Secondly, those articles must be intended for agriculture and, thirdly, the purchase must be for the purpose of supplying them to its members. The provision clearly implies that there must be purchase and the purchase of the articles must be for the purpose of supplying them to the members.It is only when the articles are purchased at a lesser cost and sold at a higher cost that a question of profit arises. It is this profit which is sought to be exempted under section 81(i). While dealing with a taxing statute which deals with income from business, the word purchase will, therefore, have to be construed in the commercial sense. In the commercial sense, a transaction of purchase is a part of a transaction of sale. A transaction of sale can never be complete unless there is a transfer of property from the seller to the buyer and the buyer who is the purchaser, must, therefore, acquire the property before he can claim to have purchased the property. *Sec.80P.* **[36CTR400,156ITR422]**

2. The word purchase occurring in section 54(1) of the Act had to be given its common meaning, viz.,buy for a price or equivalent of price by payment in kind or adjustment towards a debt or for other monetary consideration. **[Balraj - CIT, Del.,** *Sec.137.* **173CTR452,254ITR22,123Taxman290]**

Purpose Mad. CIT - A.M.M.Mohammad Ibrahim Sahib

The word purpose of section 44D(3)(a) signifies an intention or design to achieve a particular result, namely, the avoidance of liability to Taxation. Where the purpose is shown to be other than the avoidance of liability to Taxation the exemption given by the section would apply.The mere fact that the transfer results in the avoidance of the tax liability cannot mean that there was an intention to avoid such liability. *Sec.6.* **[45ITR166]**

Put to use Mad Siv Industries Ltd. - DCIT

Section 32(1) of the Income-tax Act,1961, uses the expression put to use.In the absence of a definition of a word or an expression, which requires construction, the usual course to be adopted is to assign the meaning given to the word or expression in the legal dictionary. If the expression put to use is given the meaning as defined in the dictionary,the block of assets acquired by the assessee during the previous year is applied or employed for the purpose of business of the assessee in that previous year for 180 days that would make the assessee eligible for full depreciation. The machinery cannot be used on all days and every day, right through the year from the date on which it was first put to use. There may be normal working hours even during a day. There may be holidays intervening the 180 days. *Sec.32.* **[306ITR114]**

Q

Qualification **Bom. Dr. J.M.Mokashi - CIT**

The word qualification simpliciter is a word of very wide import and, in the absence of any qualifying words or expression, conveys the idea of any quality which makes a man fit for any job or any activity in life.But the word qualification in the proviso to section 64(1)(ii) is qualified by the words technical or professional. In that view of the matter, its broad meaning will not be relevant for the purpose of section 64(1)(ii).The word profession does not take within its ambit any and every activity or employment undertaken by a person for his livelihood. The word qualification occurring in section 64(1)(ii) must be such which makes a person eligible for technical or professional work.A person can, therefore, be said to be in possession of requisite technical qualification when,by virtue there of,he is eligible to perform that function.Similarly, professional qualification must mean qualification which is necessary for carrying on the particular profession. A spouse, well-versed in law and experienced in the working of the legal profession, cannot be said to be in possession of professional qualification for carrying on the legal profession if he or she does not possess the requisite degree or diploma.Payments made to a spouse in such a case for any legal services cannot be brought within the purview of the proviso to section 64(1)(ii) by reference to the words knowledge and experience occurring in the latter part thereof. Therefore, the possession of technical or professional qualification is a condition precedent on fulfilment of which that part of the income which falls in the second part of the proviso is excluded from the operation of the clubbing provision.

Sec.64(1).[**115CTR73,207ITR252,72 Taxman98**]

Question **AAR Morgan Stanley & Co.**

Having regard to the context of the first proviso to section 245R(2) the term question would mean a point of investigation or theme of enquiry. <u>*Sec.245.*</u> **[201CTR67,284ITR260,152Taxman1]**

Question of law arising out of such order **SC** **CIT - Indian Molasses Co. P. Ltd.**

The expression question of law arising out of such order in section 66(1) is not restricted to take in only those questions which have been expressly argued and decided by the Tribunal. If a question of law is raised before the Tribunal, even if an aspect of that question is not raised, that aspect may be urged before the High Court.

<u>*Sec.256.*</u>**[78ITR474]**

Quoted share **Mad.** **CWT - A.S. Ananth**

Quoted share is defined in rule 2(9) of Schedule III to the Wealth-tax Act,1957, which Schedule sets out the rules for determining the value of assets. Quoted share or quoted debenture in relation to any equity share or a preference share or, as the case may be, a debenture is defined as meaning a share or debenture quoted on any recognised stock exchange with regularity from time to time, where the quotations of such shares or debentures are based on current transactions made in the ordinary course of business. The only mode of valuation of these shares therefore is the rate quoted at the exchange.

<u>*WT.Act.Sch.III.r.2*</u> **[181CTR505,261ITR763,131Taxman342]**

R

Rate **SC** **Sundaram & Co. - CIT**

By the use of the expression Rate a relation between the taxable income and the tax charged is intended, but the relation need not be of the nature of proportion or fraction.The expression rate is often used in the sense of a standard or measure.

<u>*Sec.147.*</u> **[AIR1968SC124,66ITR604]**

Raw material **SC** **Collector of Central Excise-Ballarpur Industries Ltd.**

The expression raw material is not a defined term. The meaning to be given to it is the ordinary and well- accepted connotation in the common parlance of those who deal with the matter.....The ingredients used in the chemical technology of manufacture of any end-product might comprise, amongst others, those which may retain their dominant individual identity and character throughout the process and also in the end- product; those which, as a result of interaction with other chemicals or ingredients, might themselves undergo chemical or qualitative changes and in such altered form find themselves in the end-product; those which, like catalytic agents,while influencing and accelerating the chemical reactions, however, may themselves remain uninfluenced and unaltered and remain independent of and outside the end- products and those which might be burnt up or consumed in the chemical reactions. To determine whether the ingredients of the last mentioned class qualify themselves as and are eligible to be called raw material for the end- product, one of the valid tests could be that the ingredient should be so essential for the chemical processes culminating in the emergence of the desired end product that having regard to its importance in and indispensability for the process, it could be said that its very consumption on burning-up is its quality and value as raw material. In such a case, the relevant test is not its absence in the end- product, but the dependence of the end-product for its essential presence at the delivery end of the process. The ingredient goes into the making of the end- product in the sense that, in its absence,the presence of the end- product, as such, is rendered impossible. This quality should coalesce with the requirement that its utilisation is in the manufacturing process as distinct from the manufacturing apparatus.....For an item of material used in the manufacture of goods to qualify as raw material, it need not necessarily and in all cases go into, and be found, in the end-product, or endure, either in its original or in an altered form, as a composite element of the end-product. _Central Excise Rules,_ **[186ITR244]**

Ready to commence Guj. Hotel Alankar - CIT

When a business is established and is ready to commence business then it can be said of that business that it is set up.The words ready to commence would not necessarily mean that all the integrated activities are fully carried out and/or wholly completed.The requirement is also complied with in a given case where an assessee had undertaken the first of the kind of integrated activities of which the business is overall comprised of. The question whether a business has been set up or not is always a question of fact which has to be decided on the facts and in the circumstances of each case, subject to the broad guidelines provided by different decisions in that behalf. _Misc._**[22CTR252,133ITR866,6Taxman183]**

Real income Bom. Western India Oil Distributing Co.Ltd.- CIT

Whether accrual of income has taken place or not must be judged on the principles of the real income theory. After accrual,non-charging of tax on the same because of certain conduct based on the ipse dixit of a particular assessee cannot be accepted.In determining the question whether it is hypothetical income or whether real income has materialised or not,various factors will have to be taken into account. It would be difficult and improper to extend the concept of real income to all cases depending upon the ipse dixit of the assessee.What has really accrued to the assessee has to be found out and what has accrued must be considered from the point of view of real income taking the probability or improbability of realisation in a realistic manner, but once accrual takes place, on the conduct of the parties subsequent to the year of closing, an income which has accrued cannot be made no income. *Sec.5.*[**116CTR296,206ITR359,121Taxation334,73Taxman565]**

2. The Income-tax is a tax on real income, i.e., the profits arrived at on commercial principles subject to the provisions of the Act. **[Southern Technologies Ltd.-JCIT,SC,***Misc.r.w.s.2(24).***228CTR 440,320ITR577,187Taxman346]**

Real income theory SC State Bank of Travancore - CIT

The following propositions emerge in relation to the theory of real income: (1) It is the income which has really accrued or arisen to the assessee that is taxable. Whether the income has really accrued or arisen to the assessee must be judged in the light of the reality of the situation. (2) The concept of real income would apply where there has been a surrender of income which in theory may have accrued but in the reality of the situation, no income had resulted because the income did not really accrue. (3) Where a debt has become bad,deduction in compliance with the provisions of the Act should be claimed and allowed. (4) Where the Act applies,the concept of real income should not be so read as to defeat the provisions of the Act. (5) If there is any diversion of income at source under any statute or by overriding title, then there is no income to the assessee. (6)The conduct of the parties in treating the income in a particular manner is material evidence of the fact whether income has accrued or not.(7)Mere improbability of recovery, where the conduct of the assessee is unequivocal,cannot be treated as evidence of the fact that income has not resulted or accrued to the assessee. After debiting the debtor's account and not reversing that entry but taking the interest merely to a suspense account cannot be such evidence to show that no real income has accrued to the assessee or has been treated as such by the assessee. (8) The concept of real income is certainly applicable in judging whether there has been income or not but, in every case, it must be applied with care and within well-recognised limits. *Misc. r.w.s.36.* **[50CTR290,158ITR102,80Taxation37,24Taxman337]**

Reason to believe Pun. R.S.Chiranji Lal and Sons - CIT

The words reason to believe suggest something more than the satisfaction of the Income-tax Officer. The expression suggests reasonable grounds on which the Income-tax Officer may take action. Power under this section cannot be exercised on mere rumours or suspicion. *Sec.147.***[36ITR407]**

2. The expression reason to believe in section 34 of the Income-tax Act does not mean a purely subjective satisfaction on the part of the Income-tax Officer. That belief must be held in good faith; it cannot be merely a pretence. Consequently, it is open to the court to examine the question whether the reasons for the belief have a rational connection or a relevant bearing to the formation of the belief and are not extraneous or irrelevant to the purpose of the section. **[Y.Rajan -ITO,AP.,*Sec.147.*77ITR839]**

3. The words reason to believe suggest that the belief must be that of an honest and reasonable person based upon reasonable grounds and that the Income-tax Officer may act on direct or circumstantial evidence but not on mere suspicion, gossip or rumour. The Income-tax Officer would be acting without jurisdiction if the reason for his belief that the conditions are satisfied does not exist or is not material or relevant to the belief required by the section. The court can always examine this aspect though the declaration or sufficiency of the reasons for the belief cannot be investigated by the court.

[Sheo Nath Singh - AAC, SC, *Sec.147.*1CTR484,82ITR147]

4. The words reason to believe in section 147(b) of Income-tax Act, 1961, which states the conditions to be satisfied before the Income-tax Officer resorts to reassessment, suggest that the belief must be that of an honest and reasonable person based upon reasonable grounds.The Income-tax Officer may act on direct or circumstantial evidence but not on mere suspicion, gossip or rumour. The Income-tax Officer would be acting without jurisdiction if the reason for his belief that the conditions are satisfied does not exist or is not material or relevant to the belief required by the section and the court can always examine this aspect though the declaration or sufficiency of the reasons for the belief cannot be investigated by the court. If there was no new material before the Income-tax Officer, unknown to him previously, which could provide reasons for his belief under section 147(b) when he issued the notice of reassessment, it would be a case of mere change of opinion which cannot be a ground for reopening the assessment. **Purushottam DasBangur-ITO,Raj.,*Sec.147.*14CTR161,126ITR580]**

5. Under the Act of 1922,the only circumstance in which the Commissioner could authorise search of any building or Inspecting Assistant Commissioner under section 37(2) was when he had reason to believe that books of account or other documents which may be found would be useful in proceedings pending under the Act. ...... Section 132 was amended in 1964, 1965 and 1975 and in 1975 section 132A was added,which widened the scope of the

provision. The widening was, however, in respect of items in respect of which authorisation could be issued to seize them. It did not expand the power in respect of the ground or material on which it could be exercised.The conditions precedent for action under section 132 or section 132A are: (i) information in the possession of the Director of Inspection or Commissioner; (ii) in consequence of which he should have reason to believe;(iii) that any person is in possession of any money, bullion, jewellery or other valuable article or thing; (iv) and such jewellery,bullion,etc.,must represent either wholly or partly undisclosed income. If any of these conditions is not fulfilled the officer concerned can have no jurisdiction to proceed.The information mentioned in condition (i) must relate to conditions (iii) and (iv) so as to empower the authority concerned to reach the conclusion contemplated in condition (ii). Reason to believe has been considered to be the most salutary safeguard on the exercise of power by the officer concerned. It is made up of two words reason and to believe.The word reason means cause or justification and the word believe means to accept as true or have faith in it. Before the officer has faith or accepts a fact to exist there must be a justification for it. The belief may not be open to scrutiny as it is the final conclusion arrived at by the officer concerned as a result of mental exercise made by him on the information received. But the reason due to which the decision is reached can always be examined.When it is said that the reason to believe is not open to scrutiny, what is meant is that the satisfaction arrived at by the officer concerned is immune from challenge,but, where the satisfaction is not based on any material or it cannot withstand the test of reason, which is an integral part of it, then it falls through and the court is empowered to strike it down. **[Ganga Prasad Maheshwari-CIT,All.,**<u>*Sec.132.*</u>**21CTR83,139ITR1043]**

6. The expression used is reason to believe and not reason to suspect. Action can be taken only when there is an honest and reasonable presumption based on reasonable grounds and not mere guess, gossip, suspicion or rumour. There should be a direct link or nexus with the information or material with the officer and the formation of the belief as to the escapement of net wealth in a particular assessment year. **[Lokendra Singh Rathore - WTO, MP.,**<u>*Sec.17.WT.Act.*</u>**155 ITR629]**

7. The word reason in the phrase reason to believe would mean cause or justification.If the Assessing Officer has a cause or justification to think or suppose that income had escaped assessment, he can be said to have a reason to believe that such income had escaped assessment. The words reason to believe cannot mean that the Assessing Officer should have finally ascertained the facts by legal evidence. Unless the ground or the material on which his belief is based, is found to be so irrational as not to be worthy of being called a reason by any honest man, his conclusion that it constitutes a sufficient reason, cannot be overridden. If the Assessing Officer honestly comes to a conclusion that a mistake has been made, it matters nothing so far as his jurisdiction to initiate the proceedings under section 147 is concerned,that he may have come to an erroneous conclusion whether on law or on facts. The court will not in exercise of its extraordinary jurisdiction under the Constitution, examine the sufficiency of the reason which led the Assessing Officer to believe that the income had escaped assessment. **[Praful Chunilal Patel -M.J.Makwana,ACIT,Guj.,** *Sec. 147.***148CTR62,236ITR832]**

8. The expression reason to believe refers to the belief which prompts the Assessing Officer to apply section 147 of the Income-tax Act,1961, to a particular case. It will depend on the facts of each case.The belief must be of an honest and reasonable person based on reasonable grounds. The Assessing Officer is required to act, not on mere suspicion, but on direct or circumstantial evidence. The expression reason to believe does not mean a subjective satisfaction on the part of the Assessing Officer. **[IPCA Laboratories Ltd.-Gajanand Meena, DCIT,Bom.,***Sec.147.***170CTR585,251ITR420]**

9. The expression reason to believe in common parlance would mean to have sufficient cause to believe. It is not synonymous with subjective satisfaction as it would be open to the court to examine the matter whether reason to believe has a rational connection or a relevant bearing to the formation of the belief. The belief must be held in good faith and cannot be merely pretence. It must be stronger than mere satisfaction. The belief entertained by the Assessing Officer must not be arbitrary or irrational but should be reasonable and in other words it must be based on reasons which are relevant and material. Adequacy or sufficiency of reasons can hardly be investigated by the court. Reasonableness, relevancy and good faith are factors which should be demonstrated in the order itself or at best the record in support thereof. Reason to suspect is not reason to believe.They are terms of apparent definition and cannot be permitted to be interchanged in operation of fiscal provisions. Reasons are the link between the material on which the conclusions are based and they alone can exhibit how the mind is applied to the subject- matter for arriving at a decision. Rational nexus between the facts considered and conclusions reached is the main consideration. Reason to believe must relate to the standards of belief of a reasonable man and not to a prejudicial or biased mind.

[Sony India Ltd.-CIT,Del.,*Sec.147.*196 CTR81,276ITR278,146 Taxman98]

10. The expression reason to believe in section 132 of the Income-tax Act,1961,means a genuine satisfaction arrived at upon an honest and reasonable evaluation of information coming to the authority. There must be a reasonable nexus between the satisfaction and the situation contemplated in any of the clauses (a), (b) and (c) of section 132(1). The meaning of the expression reason to believe is stronger than satisfaction. There should be reasons to believe and such reasons to believe must be on the basis of information which is in the possession of the concerned officer. The information on the basis of which the concerned officer forms reason to believe should be in the possession of the concerned officer.The expression possession means hold as own property, actual holding or occupancy.

[Sikkim Subba Associates -Union of India, Sik., *Sec.132.*276ITR456]

11. The words reason to believe suggest that the belief must be that of an honest and reasonable person based on reasonable grounds, relevant, and available materials on record.

[Vinbros and Co.-ITO, Mad., *Sec.147.*206CTR371,286ITR 439]

12. Reason to believe cannot mean that the Assessing Officer should have finally ascertained the facts by legal evidence. **[BPL LTD. - DCGT, Kar., *Sec.16.G.T.Act.*208CTR346,293ITR321,161 Taxman64,TaxLR405]**

13. The expression reason to believein section 147 would mean cause or justification. If the Assessing Officer has cause or justification to know or suppose that income had escaped assessment, he can be said to have reason to believe that income had escaped assessment.The expression cannot be read to mean that the Assessing Officer should have finally ascertained the fact by legal evidence or conclusion.What is required is reason to believe but not the established fact of escapement of income. At the stage of issue of notice, the only question is whether there was relevant material on which a reasonable person could have formed the requisite belief. Whether material would conclusively prove escapement of income is not the concern at that stage. This is so because the formation of the belief is within the realm of the subjective satisfaction of the Assessing Officer. **[ACIT - Rajesh Jhaveri Stock Brokers P. Ltd.,SC,*Sec.147.*210CTR30,291ITR500]**

Reasonable Pun S. Ranghbir Singh Sandhawalla - CIT

The expression reasonable means rational according to the dictates of reason and not excessive or immoderate.An act is reasonable when it is comfortable or agreeable to reason, having regard to the facts of the particular controversy. *Misc.*[**AIR1958Pun250,34ITR 719]**

2. A thing is generally considered reasonable if it is not actuated by bad faith, dishonesty or false grounds.No hard and fast rule can be laid down for governing or deciding when a certain ground would be considered reasonable and when it would not be so held. Each case has to be decided on its own merits and facts by the authority. If the decision is based on relevant grounds by the authority which has power to condone the default,the appellate court cannot interfere with the same. **[CWT - S.L.Khunnah, All.,** *Sec.18(1)(a).WT.Act.***79 CTR 229,180ITR340,46Taxman240]**

Reasonable cause Raj. Addl.CIT - Mohammed and Sons

Penalty can be imposed under section 271(1)(a) of the Income-tax Act,1961, for failure to furnish returns in time without reasonable cause. Before a cause can be said to be reasonable, it must be found as a fact that a particular cause operated upon the mind of the assessee which prevented him from filing the return in time. The onus is on the assessee to show that he was prevented by sufficient cause from complying with the statutory requirement of filing the return in time. *Sec.271(1)(a).*[**43CTR114,154ITR220]**

2. The reasonable cause has to be explained by the assessee and the cause should be so reasonable that a person instructed under law should believe the explanation as plausible. If the explanation is supported by evidence,then the veracity of such evidence is to be examined and, in case of doubt, the benefit has to be given to the assessee. **[CWT-Taranath Tondon, Raj.,** *Sec.18(1)(a).WT.Act.* **[203ITR871]**

3. The words reasonable cause have not been defined under the Act but they could receive the same interpretation which is given to the expression sufficient cause. Therefore, in the context of a penalty provision, the words reasonable cause would mean a cause which is beyond the control of the assessee. Reasonable cause means cause which prevents a reasonable man of ordinary prudence acting under normal circumstances,without negligence or inaction or want of bona fides from furnishing the return in time.

[CWT- Jagdish Prasad Choudhary,Pat.,*Sec.18(1)(a).WT.Act.* **123CTR200,211 ITR 472, 80Taxman 42]**

4. The words reasonable cause in section 273B of the Income-tax Act,1961, must necessarily have a relation to the failure on the part of the assessee to comply with the requirement of the law which he had failed to comply with.In the case of delay in compliance, the cause shown must be for the whole of the period of the delay and not merely for a part thereof. If the cause shown is such as to explain the delay as a whole and constitute a good reason for the non- compliance,no penalty would be leviable. However, in cases where the cause shown is such as to explain a part of the delay, or the cause shown is only to mitigate the gravity of the non-compliance,such a cause cannot be extrapolated and treated as being good cause for the whole of the period of the delay in its entirety.

[Kalakrithi-Income-tax Officer,Mad., *Sec.273B.***73CTR523, 125Taxman97]**

5. Reasonable cause as applied to human action,is that which would constrain a person of average intelligence and ordinary prudence. It means an honest belief founded upon reasonable grounds, of the existence of a state of circumstances, which assuming them to be true, would reasonably lead an ordinary, prudent and cautious man, placed in the position of the person concerned, to come to the conclusion that the same was the right thing to do.The cause shown has to be considered and only if it is found to be frivolous, without substance or foundation, would the prescribed consequences follow. **[Woodward Governor India P. Ltd.-CIT,Del.,***Sec.271C.r.w.s. 273B.* **168CTR394,253ITR745,118 Taxman433]**

Reasons beyond his control **All.** **Mehra International - CIT**

The expression reasons beyond his control is not defined in the Act. It should be interpreted keeping in view the context in which is appears.The intention of the Legislature in enacting the aforesaid provision should also be kept in mind. The purposive interpretation of a statute is one of the recognised principles of interpretation of statute. *Sec.80HHC.***[195CTR368,273ITR8,143Taxman206]**

Reasons to believe Del. CIT - Smt. Shaila Agarwal

The reasons to believe would mean cause or justification of the Assessing Officer to believe that the income has escaped assessment and not that the Assessing Officer should have finally ascertained the fact by legal evidence or reached a conclusion, as this is determined and decided in the assessment order,which is the final stage before the Assessing Officer. *Sec.147* **[338ITR51]**

Reassessment All. Jawahar Lal Mani Ram - CIT

The word reassessment is not confined to a fresh assessment being made of an income in the hands of the same person but is wide enough to include a fresh assessment of an income in hands different from those in which it had been assessed originally....... Whether an assessment in pursuance of a direction of the Tribunal be treated as a separate class of assessment or an assessment under section 23 or 34,the fresh assessments were not barred by limitation and were valid even though they were made after the expiry of four years from the end of the various assessment years. *Sec.143,147.* **[48ITR837]**

Rebate Bom. Harihar Cotton Pressing Factory - CIT

Rebate is a remission or a payment back and of the nature of a deduction from the gross amount.It is sometimes spoken of as a discount or a drawback.The dictionary meaning of the term includes a refund to the purchaser of a thing or commodity of a portion of the price paid by him. It is not confined to a transaction of sale and includes any deduction or discount from a stipulated payment, charge or rate.It need not necessarily be taken out in advance of payment but may be handed back to the payer after he has paid the stipulated sum.The repayment need not be immediate. It can be made later and in case of persons who have continuous dealings with one another it is nothing unusual to do so.

*Misc.***[39ITR594]**

Receipt All. CIT - Gulab Chand

The word receipt must be understood as synonymous with income. The expression income has been defined in clause (24) of section 2 of the Income-tax Act,1961, to include capital gains.

*Sec.24(2).***[192ITR495]**

Received **Bom.** **CIT - Bai Navajbai N. Gamadia**

Received in section 4(3) means not only actual receipt but also what is deemed to be received under the Act. <u>*Sec.5.*</u>**16ITR109]**

2. The word received in the fourth proviso to section 10(2)(vii) does not mean receivable. If what is not a trading receipt is taxed as a result of a fiction, the scope of the same ought to be strictly limited. The assessee is not liable to pay tax on amounts deemed to be profits under the fourth proviso to section 10(2)(vii) if they have not been actually received by him, though the assessee keeps his accounts on the mercantile system and the amounts are credited in his accounts. **[CIT - Moon Mills Ltd., Cal.,** <u>*Sec.41.*</u>**46ITR771]**

Recognised provident fund **AP.** **CIT- Indocean Engineer P. Ltd.**

The expression recognised provident fund as defined in section 2 (38) of the Income-tax Act,1961, contemplates two categories of provident funds- (i) a provident fund which has been recognised by the Chief Commissioner or Commissioner by applying the rules contained in Part A of the Fourth Schedule to the Act; and (ii) a provident fund established under the scheme framed under the Employees' Provident Funds Act.

<u>*Sec2(38).*</u>**[146CTR161,225ITR201,143Taxation12,92Taxman476]**

Reconstruction **Del.** **CIT - Hindustan General Industries Ltd.**

It is not every alteration in the mode, method or scope of the activities of a business and it is not every transfer of assets from one unit to another that will involve reconstruction. The expression is no doubt very wide but it does not take in a case of a company setting up or establishing a totally independent and viable industrial unit for carrying on the same or similar business even though it might be so set up by way of expanding the already existing business. The emphasis in section 84 is not on business but on undertaking. The exemption is granted to new undertakings and the essence of the exemption is that it is a new industrial unit that is established and that it is not merely a rehash of an already existing unit. <u>*Sec.84.*</u>**[23CTR73,137ITR851]**

Reconstruction **Ker.** **Kerala State Cashew Dev. Corporation - CIT**

Reconstruction is the rejuvenation or rehabilitation of an existing undertaking. The original business or undertaking continues to exist without its identity being lost. <u>*Sec.80J*</u> **[205ITR19]**

Reconstruction of business	**Bom.**	**CIT- Gaekwar Foam & Rubber Co. Ltd.**

The reconstruction of a business or an industrial undertaking must necessarily involve the concept that the original business or undertaking is not to cease functioning, and its identity is not to be lost or abandoned.The concept essentially rests on changes but the changes must be constructive and not destructive. The underlying idea of a reconstruction is of a business already in existence: there must be a continuation of the activities and business of the same industrial undertaking. The undertaking must continue to carry on the same business though in some altered or varied form. If the alterations and changes are substantial, there would be little scope for describing what emerges as a reconstruction of the business. For instance, if the ownership of a business or an undertaking changes hands not ostensibly but in reality and effectively, that would not be reconstruction.Or, if the very nature of the business is changed, that again would not be reconstruction. On the other hand, reorganisation of the business on sounder lines or alterations in the mode or method or scope of the activities of the business or in its personnel or infusion of new blood in the management or control of the business which may even be by some changes in the constitution of persons interested in the undertaking would be no more than reconstruction of the business if it is substantially the same business carried on by substantially the same persons. In these matters, we have to look at the substance of the transaction and not the form. If, looking at the substance of the transaction, it is a sale, then the concept of reconstruction must be ruled out for in such a case there is no scope for speaking about any reconstruction of an existing business. <u>*Sec.80J.*</u>**[35ITR662]**

2. In the reconstruction of a business, as in the reconstruction of a company, there is an element of transfer of assets and of some change, however, partial or restricted it may be, of ownership of the assets. The transfer, however, need not be of all the assets. It is nonetheless imperative that there should be continuity and preservation of the old undertaking though in an altered form. The concept of reconstruction of business would not be attracted when a company which is already running are industrial unit sets up another industrial unit. The new industrial unit would not lose its separate and independent identity even though it has been set up by a company which is already running an industrial unit before the setting up of the new unit.**[CIT-Ganga Sugar Corporation Ltd.,Del.,** *Sec.80J.***92ITR173]**

3. In order to attract sub-section (4)(i) of section 80J,there must be transfer of any assets of the old business to the new industrial undertaking to resuscitate the undertaking. It must be a new one in the sense that new machinery and plant are erected to produce the same or distinct goods by preservation of the old one to continue the old undertaking in the altered form so that the persons carrying on the business will substantially do the same or similar business. It is only in such a case that it can be considered to be a reconstruction of business. **[CIT - U Foam Private Limited, AP.,** *Sec.80J.***63CTR204,167ITR586,31Taxman491]**

Reconstruction of a business already ... Cal. CIT-Textile Machinery Corporation

The change of producing one's own goods systematically used in the existing business instead of buying them from outside would only be a reconstruction of a business already in existence.

Sec.80J. **[80ITR428]**

Record SC Maharana Mills (Pvt.) Ltd. - ITO

The record contemplated by section 35 does not mean only the order of assessment but it comprise all proceedings on which the assessment order is based and the Income-tax Officer is entitled for the purpose of exercising his jurisdiction under section 35 to look into the whole evidence and the law applicable to ascertain whether there was an error.If he doubts the written down value of the previous year it is open to him to check up the previous calculations and if he finds any mistake it is open to him to make fresh calculations in accordance with the law applicable including the rules made thereunder.If, for instance, the Income-tax Officer finds that in an earlier assessment year there was an apparent arithmetical mistake in the account of the written down value of the properties of the assessee which resulted in a corresponding mistake in the assessment of the relevant assessment year he can take the corrected figure for the purposes of the assessment and it cannot be said that the mistake was not apparent from the record.A fortiori if he discovers that the very basis of the different earlier assessments was erroneous because of an initial mistake in determining the written down value it cannot be said that this would not be a mistake apparent from the record. And if in order to determine the correct written down value the Income-tax Officer makes correct calculations,it cannot be said that that is not rectifying a mistake apparent from the record but is dehors it.

*Sec.154.***[36ITR350]**

2. The power of rectification is confined to mistakes apparent from the record. The record is not confined to the order of assessment. It comprises all proceedings on which the assessment order is based and the Income-tax Officer is entitled to look into the whole evidence and the law applicable to ascertain whether there was an error and if he doubts the written down value of the previous year, it is open to him to check up the previous calculations and if he finds any mistake, it is open to him to make fresh calculations in accordance with the law applicable including the rules made there-under.If he discovers that the very basis of the different earlier assessments was erroneous because of an initial mistake in determining the written down value,it cannot be said that this mistake would not be a mistake apparent from the record and if in order to determine the correct written down value, the Income-tax Officer makes correct calculations, it cannot be said that that is not rectifying a mistake apparent from the record but is dehors it. The limit to which the Income-tax Officer can go does not stop at the written down value of the previous year but extends to the figure of the original cost. **[Bihar State Road Transport Corporation-CIT,Pat.,***Sec.263.* **55CTR270,162ITR114]**

3. The record contemplated in section 263(1) of the Income-tax Act, 1961,does not mean only the order of assessment but it comprises all proceedings on which the assessment is based.The Commiss-ioner is entitled for the purpose of exercising his revisional jurisdiction to look into the whole evidence.The expression record as used in section 263 of the Act is comprehensive enough to include the whole record of evidence on which the original assessment order was based. All proceedings which constitute evidence on which the assessment order is based must normally be regarded as part of the record.

[CIT-S.M.Oil Extraction Pvt. Ltd.,Cal.,
*Sec.263.***190ITR404,61Taxman205]**

4. Record includes all records relating to any proceeding under the Act available at the time of examination by CIT.Valuation report forms part of the record though received after completion of assessment.The word record after the amendment of Section 25(2) by the Finance Act, 1989, includes all records relating to any proceedings which are available at the time of his examination.

[CWT-Smt.B. Indira Devi, SC, *Sec.25.WT.Act.*208ITR26]

5. It has been provided that record shall include and shall be deemed always to have included all records relating to any proceeding under the Act available at the time of examination by the Commissioner.It cannot be said that the correct and settled legal position, with respect to the meaning of the word record till June 1,1988, is that it meant the record which was available to the Income-tax Officer at the time of passing of the assessment order. Such a narrow interpretation of the word record is not justified in view of the object of the provision and the nature and scope of the power conferred upon the Commissioner.The revisional power conferred on the Commissioner under section 263 is of wide amplitude.It enables the Commissioner to call for and examine the record of any proceeding under the Act. It empowers the Commissioner to make or cause to be made such enquiry as he deems necessary in order to find out if any order passed by the Assessing Officer is erroneous in so far as it is prejudicial to the interests of the Revenue.After examining the record and after making or causing to be made an enquiry, if he considers the order to be erroneous, then he can pass the order thereon as the circumstances of the case justify.Obviously, as a result of the enquiry he may come into possession of new material and he would be entitled to take that new material into account. If the material, which was not available to the Income-tax Officer when he made the assessment could thus be taken into consideration by the Commissioner after holding an enquiry, there is no reason why the material which had already come on record though subsequent to the making of the assessment, cannot be taken into consideration by him. Moreover, in view of the clear words used in clause (b) of the Explanation to section 263(1), it has to be held that while calling for and examining the record of any proceeding under section 263(1),it is and it was open to the Commissioner not only to consider the record of that proceeding but also the record relating to that proceeding available to him at the time of examination.

[CIT-Shree Manjunathesware Packing Products and Champhor Works,SC,*Sec.263(1).*143 CTR406,231ITR53,96 Taxman1]

6. The record contemplated by section 154 of the Income-tax Act, 1961,does not mean only the order of assessment, but it comprises all proceedings on which the assessment order is passed and the Income-tax Officer is entitled for the purpose of exercising his jurisdiction under section 154 of the Act, to look into the whole evidence and the law applicable to ascertain whether there was an error.Sections 147 and 154 of the Act are mutually exclusive in their operation and in a given case, where the statutory requirements are satisfied,the Income-tax Officer can have recourse to either of the two provisions and it cannot be said that they are overlapping and either of the sections at the choice of the assessing authority would not bar the officer to have recourse to the other provision of law. **[India Forge and Drop Stampings Ltd.-CIT,Mad.,***Sec. 154.* **146CTR28,233ITR112]**

7. The ambit of the word record in clause (b) of the Explanation to Sec. 263(1) has been clarified to mean that it shall include and shall be deemed always to have included all records relating to any proceeding under the Act available at the time of examination by the CIT.That it has now to be taken as settled law after the decision of the Supreme Court in South India Steel Rolling Mills v. CIT [224 ITR 654], that the material which came to light,after the order of the ITO but before the revisional power was exercised,could be taken into consideration for exercise of revisional power under Sec. 263(1) **[CIT-K.C.RangaiahandCo.,AP.,***Sec.263.* **146CTR694,230 ITR385,101Taxman110, 147Taxation103]**

8. The Finance Act,1988, inserted an Explanation to sub-section (2) of section 24 of the Gift-tax Act,1958. The Explanation declares that the expression record includes and shall be deemed always to have included all records relating to any proceeding under the Act available at the time of examination by the Commissioner. **[CGT-O.E.ArumughaMudaliar,Mad.,***Sec.24.G.T.Act.* **237ITR789, 154Taxation467]**

9. The term record is not defined in the section or in the definition section of the Act. For determining the true scope of this provision and the meaning to be properly assigned to the term record it is necessary to keep in view the object of the provision and the nature of the power conferred on the authorities under that provision. The absence of definition cannot have the consequence of limiting its meaning to a very narrow and limited sphere of the record of the original proceedings alone.The object with which power is conferred by section 154 is as stated in the marginal heading rectification of mistake.The principal condition for exercising the power under section 154 is the existence of a mistake in the record. The mistake is not to be a mistake which requires in- depth probing to discover but is a mistake which is apparent from the record. The power conferred by this provision is only to enable the authorities to rectify the apparent mistakes in the record. The record referred to is the record which the authorities are required to examine for the purpose of rectifying mistakes in the orders mentioned in sub-clauses(a), (b) and (c) of section 154(1). The section does not either expressly or implicitly require that the authorities exercising power under this provision should limit their attention only to the order sought to be rectified. The requirement that the mistakes in the record be apparent does not imply that no other relevant document should be looked into. If in the light of other legally valid orders it is found that the original order contains mistakes which are apparent, the rectification of such mistakes is not barred under section 154. It is neither necessary nor possible to set out exhaustively all the material that can possibly be regarded as forming part of the record for the purpose of examination under section 154(1). **[CIT-M.R.M. Plantations(P.)Ltd.,Mad.,***Sec.154.***153CTR71,240ITR660,102 Taxman1]**

10. The term record is not confined to the material available to the Assessing Officer. It would include material available on record at the time of examination by the Commissioner of Wealth-tax. The valuation report received after completion of the assessment forms part of the assessment record and the Commissioner can consider the report for the purpose of revision of the assessment orders under section 25 of the Wealth-tax Act. **[CWT-P.P.Ghosh,Cal.,** *Sec.25.WT.Act.***161CTR365,244ITR574,115Taxman606]**

11. The term record in section 25(2) of the Wealth-tax Act,1957, has been defined in Explanation,clause (b),as including records relating to any proceeding under the Act at the time of examination by the Commissioner. **[CWT-S.L.Chitale,Mad.,***Sec.25.WT.Act.***252ITR 586,121Taxman459]**

12. The record would include all records relating to any proceedings under the concerned direct tax laws available at the time of examination by the Commissioner.The Legislature again stepped in through the Finance Act,1989,so as to clarify that the provisions of the Explanation to section 263(1) of the Income-tax Act,1961, shall be deemed to have always been in existence. **[CIT-Vallabhdas Vithaldas,Guj.,***Sec.263(1).***172CTR728,253ITR 543,123 Taxman 110]**

13. The expression record in section 263 means the record available at the time of examination by the CIT and is not confined merely to the material available to the Income-tax Officer.**[CIT-Export House,P&H.,***Sec.263(1).***175CTR137,256ITR603,122Taxman879]**

14. The record . . . . . does not merely mean the assessment order. The return, the things which accompanied the return are also part of the record and if there has been omission on the part of the Assessing Officer to take note of the contents of that record, while making his order, the mistake in the assessment can be regarded as apparent. **[Annamallais Agencies -CIT,Mad.,**_Sec.154._ **189CTR73, 260ITR478,134Taxman]**

15. ...the clear words used in clause (b) of the Explanation to section 263(1) of the Income-tax Act,1961,while calling for and examining the record of any proceeding under section 263(1), it is and it was open to the Commissioner not only to consider the record of that proceeding but also the record relating to that proceeding available to him at the time of examination. **[CIT - Prakashwati, All.,** _Sec.263._**276ITR575,144Taxman313]**

16. The provisions of section 263(1) of the Income-tax Act, 1961, and section25(2)of the Wealth-tax Act,1957,are analogous.While calling for and examining the record of any proceeding under section 25(2) it is open to the Commissioner not only to consider the record of that proceeding but also the record relating to that proceeding available to him at the time of examination.**[CWT-Smt. Phoolwati Agarwal,All.,**_Sec.25(2).WT.Act._ **276ITR623,145Taxman436]**

17. Under Explanation (b) to section 263(1) of the Income-tax Act, 1961,the word record shall include and shall always be deemed to have included all the records, relating to any proceedings under the Act,which are available at the time of examination by the Commissioner.The words shall be deemed always would signify that, even in cases where the orders of revision were passed prior to the amendment, records would include those available at the time of examination by the Commissioner and not merely those records which were available before the assessing authority or those referred to in the order of assessment. Parliament has chosen to insert Explanation (b) to section 263(1), by the Finance Act, 1988, with effect from June 1,1988..... ....Since the power of revision under section 263(1) is required to be exercised in cases where it is prejudicial to the interests of the Revenue, the power of the Commissioner is not limited only to the material which was available before the Assessing Officer and, in order to protect the interests of the Revenue, the Commissioner is entitled to examine any other records which are available at the time of examination by him and to take into consideration even those events which arose subsequent to the order of assessment.**[M.S.Raju -DCIT,AP.,** _Sec.263(1)._ **298ITR373**_Sec.263(1)._**298ITR373]**

18. A reading of sub-section (2) of section 25 of the Wealth-tax Act, 1957, makes it clear that the basis of jurisdiction to revise an order is not the examination of records simpliciter or the examination of the record of the assessee: the requirement is to examine the record of any proceedings under this Act in relation to any order passed therein. Therefore, the first limitation envisaged under sub-section (2) of section 25 of the Act is that the Commissioner cannot travel beyond the record of the proceedings in which the order has been made. **[CWT-Amichand C.Shah(HUF),Guj.,** *Sec.25.W.T.Act.* **131CTR471,218 ITR 659,85Taxman272]**

Reduce Ker. Kerala State Cashew Dev. Corp. Ltd. - ITO

Reduce means to diminish or decrease and, waive means to refrain from insisting on, or claiming or demanding or to relinquish or forgo or give up,etc. *Sec.139(8)r.w.s.117.***[35CTR391,145ITR266, 14Taxman451]**

Reduced by the amount of tax deductible Guj. CIT - Ranoli Investment P. Ltd.

The words reduced by the amount of tax deductible which appear in sub-section (5) of section 215 also occur in clause (iii) of section 209 (1)(a) dealing with computation of advance tax. The amount of tax deductible in accordance with section 194A would obviously mean the tax as was required to be deducted in respect of the interest income at the time of credit to the account of the payee or payment whichever is earlier. *Sec.215,209.***[146CTR745,235ITR433]**

Regular Assessment Mad. M.Rm.M.M.N.Natarajan Chettiar - ITO

The words regular assessment which occur in sub-section (6) of section 18A of the Income-tax Act must bear the same meaning attributed to them in sub-section (5) of that section. *Sec.214.* **[42ITR29]**

2. If the provisions of section 18A(9) are read in the context of the last portion of section 34(1) it is clear that the expression regular assessment occurring in section 18A(9) would cover reassessment proceedings. Penalty can be initiated by issuing a notice under section 18A(9)(a) read with section 28(3) in the course of reassessment proceedings. In section 18A(9) it is a question of levying penalty on the assessee for having furnished estimates of the tax payable by him which he knew or had reason to believe to be untrue which aspect in a given case could be brought home to the assessee only after proceedings under section 34(1) are commenced. The nature of the subject-matter dealt with by section 18A(9) and the purpose served thereby necessitate the adoption of a different meaning of the expression regular assessment occurring therein than the meaning attributed to similar expression while considering the provisions of section 18A(5) and (6). Regular assessment in section 18A(9) would, therefore, cover cases of reassessment undertaken under section 34(1).

[Deviprasad Kejriwal - CIT, Bom., *Sec.143.*102ITR180]

3. The obligation under section 214 is to pay interest on the amount by which the advance tax paid exceeds the tax determined on regular assessment. An order which is made by the Income-tax Officer to give effect to the order of the Appellate Assistant Commissioner is an order of assessment under section 143. If that is the position, then, in view of section 2(40) of the Act, the regular assessment as contemplated by section 214(1) should be the assessment made by the Income-tax Officer initially or the first assessment made by the Income-tax Officer if there is no appeal there from, but in a case where there is an appeal, the order passed by the Income-tax Officer finally to give effect to the direction, if any, of the appellate authority. Having regard to the scheme of the Act and the context in which the expression has been used, regular assessment under section 214 would include in the particular facts and circumstances of the case an assessment made by the Income-tax Officer pursuant to the direction of the Appellate Assistant Commissioner.

[Chloride India Ltd.-CIT,Cal., *Sec. 214.*106ITR38]

4. There is nothing in the context of section 214 of the Income-tax Act,1961, which requires the expression regular assessment not being understood as the first or original assessment according to the definition in section 2(40) of the Act.

[Lala Laxmipat Singhania-CIT,All., *Sec.214.*2CTR160,110ITR289]

5. Regular assessment, as defined in section 2(40) of the Act,means the assessment made under section 143 or section 144.It does not, therefore, include an order made under section 215 or section 216 or section 217. **[CIT- Rajkuverba Dowager Maharani Saheb of Gondal,Kar.,*Sec.217.r.w.s. 263.*115ITR301]**

6. That the expression regular assessment in section 214 should be construed as referring only to the first or initial regular assessment and not to subsequent modifications thereof.**[National Agr. Coop. Marketing Federation of India-Unionof India,Del.,**<u>*Sec.214.*</u> **130ITR928,6Taxman123**<u>*Sec.214.*</u>**130ITR928,6Taxman123]**

7. The words regular assessment used in section 214 refer to the assessment made by the Income-tax Officer under section 143 or section 144 on the first occasion and do not include the consequential order passed by him as a result of an order in appeal.

[Nizam's Religious Endowment Trust(HEH)-ITO,AP.,<u>*Sec.214.*</u> **15CTR239,131ITR239,2Taxman110]**

8. The expression regular assessment as defined by section 2(40) means assessment made under section 143 or section 144 in contradistinction to assessments under section 140A, a provisional assessment before its abolition in and an assessment/reassessment under section 147.When the expression regular assessment is defined to mean only assessments made under certain sections of the Act,it would be beyond the jurisdiction of any court to give it a wider and comprehensive meaning so as to include all assessments done under the Act, whether under section 143/144 or completed after issue of notice under section 148. **[Charles D'souza - CIT,Kar.,**<u>*Sec.2(40).*</u>**147ITR694]**

9. When the first assessment of the Income-tax Officer is final, that is the regular assessment for purposes of section 214(1). **[Bardolia TextileMills-ITO,Guj.,**<u>*Sec.214.*</u>**151ITR389]**

10. ......The context and setting in which the term regular assessment appears in section 214 only warrants the conclusion that it connotes the assessment made by the Income-tax Officer for the first time when he is bound to determine the sum payable by the assessee or refundable to him on adjustment of the advance tax paid before such assessment. **[CIT-G.B.Transports,Ker.,** <u>*Sec.214.*</u>**155ITR 548]**

11. The expression regular assessment has been defined in section 2 (40) as an assessment made under section 143 or 144. The definition contained in section 2 prevails unless the context requires otherwise.Only orders passed by the Income-tax Officer under sections 143 and 144 could be considered as regular assessments within the meaning of section 2(40) of the Act and it is not possible to expand the scope of the expression regular assessment to include other orders.

[CIT-PadmaTimberDeopt,AP.,<u>*Sec.2(40).*</u>**67CTR109, 169 ITR646, 35Taxman117]**

12. The expression regular assessment occurring in section 273 has been defined by section 2(40) of the Act to mean, unless the context otherwise requires, the assessment made under section 143 or section 144 of the Act.The clause used in section 273 of the Act in the course of any proceedings in connection with the regular assessment for the assessment year is wide enough to include within its ambit proceedings under section 155 of the Act for amendment of the assessment made under section 143 or section 144 of the Act, which is the regular assessment, as defined by section 2(40) of the Act, in contradistinction to a self-assessment under section 140A, a provisional assessment under section 141(before its deletion in the year) and an assessment or reassessment under section 147 of the Act. **[CIT-MamtaTiwari,MP.,**<u>*Sec.273A.r.w.s.2(40).*</u>**66CTR83,171 ITR59,35Taxman237]**

13. The words regular assessment occurring in section 214 of the Income-tax Act,1961,refer only to the original assessment or the first assessment made by the Income-tax Officer and not to any order which may be passed pursuant to an appellate order.

[Warner Hindustan Ltd.-CIT,AP.,<u>*Sec.214.*</u>**171ITR224,36Taxman106]**

14. The words regular assessment used in section 214 of the Income-tax Act,1961,do not connote the first assessment only. Where the entire assessment itself has been set aside by the CIT under section 263 of the Act, it cannot be the intention of the Legislature that interest payable under section 214 of the Act alone will remain intact while the assessment itself is non-existent in the eye of law. Unless interest is quantified, no payment can be made. Such quantification can only be made after assessment is made and the tax payable is determined. It cannot be the intention of the Legislature to enjoin the Central Government to pay interest under section 214 even in the absence of an order of assessment.

[Calcutta Electric Supply Corporation Ltd.-CIT,Cal., <u>*Sec.214.*</u> **179 ITR 580]**

15. In the Income-tax Act, 1961, regular assessment has been defined by referring to the assessment made under section 143 or section 144. The legislative intent was not to confine the meaning of regular assessment to the initial or first assessment. Hence regular assessment as contemplated under section 214(1) should be the assessment made by the Income-tax Officer initially or the first assessment made by the Income-tax Officer if there is no appeal there from.But, in a case where there is an appeal, the order passed by the Income-tax Officer finally to give effect to the direction, if any, of the appellate authority should be the regular assessment. **[CIT - Chloride India Ltd.,Cal.,** *Sec.214.* **186ITR217]**

16. A perusal of sections 147 and 148 makes it clear that at any rate an assessment for the first time made by resort to section 147 is a regular assessment. Section 148 enjoins the Income-tax Officer before making an assessment under section 147 to serve a notice to the assessee, containing all or any of the requirements which may be included in a notice under section 139(2). The further provision in that section is very significant which provides that the aforesaid notice has to be treated as if it is a notice under section 139(2) and that all the provisions of the Act shall apply to the subsequent procedure and the final assessment. In other words, the notice issued under section 148 has to be deemed to be a notice under section 139(2) and if the other provisions of the Act have to be applied, an assessment in pursuance of that can be made only under section 143 or section 144. There is no other provision by which the Income-tax Officer is authorised to make an order of assessment under the Act. The provisions contained in section 140A also give an indication that an assessment made in pursuance of a notice under section 148 is a regular assessment under section 143 or section 144, for section 140A(2) provides that any admitted tax paid in pursuance of section 140A(1) shall be deemed to have been paid towards the regular assessment under section 143 or section 144. It is pertinent to note that section 140A(1) deals with a return required to be furnished under section 139 or section 148. That makes the provision clear that an assessment made under section 147 also will be a regular assessment under section 143 or section 144. This is also made explicit by sub-section (6) of section 215 added with effect from April 1,1985. Even in the absence of sub-section (6) of section 215, such assessments will be regular assessments. Sub-section (6) of section 215 is only a clarification of the earlier law as different High Courts have expressed different opinions on the question and by the inclusion of that sub-section alone it cannot be said that for the assessment year in question such assessments cannot be treated as regular assessments. Any observation made in a decision has to be understood in the context in which it is made. **[Lally Jacob -ITO, Ker.,** *Sec.147.* **197ITR439]**

17. Under section 214 of the Income-tax Act, 1961, regular assessment includes the final effect-giving assessment order.There is nothing contrary to this to be found in the definition of regular assessment which is at clause (40) of section 2 of the Act. In fact regular assessment is defined with reference to section 143, and, on the view that the assessment under section 143(3) still continues when the Income-tax Officer gives effect to the appellate decision, regular assessment would clearly also include,by the definition clause, this effect-giving order.**[Calcutta Electric Supply Corp. (India)Ltd.-ITO,Cal.,**_Sec.214.r.w.s.2(40), 143._**197ITR 563]**

18. The words regular assessment in section 214 of the Income-tax Act,1961, would include not only the order of assessment made by the Income-tax Officer initially if the matter rests at that stage, but also any order passed by the Income-tax Officer finally to give effect to the directions, if any, of the appellate or revisional authority or competent court.

[Super Spinning Mills Ltd.-CIT,Mad.,_Sec.214._**199ITR832]**

19. Ordinarily, the term regular assessment would cover all assessments which can be made under section 143 in respect of a given assessment year.Where as a result of any direction given by the AAC or the Tribunal the Income-tax Officer is required to pass a fresh assessment order, it would also be a regular assessment under section 143.

[CIT-Shambhu Investment,Bom., _Sec.143._**203 ITR773]**

20. The first order of assessment under section 143 or 144 of the Income-tax Act, 1961, will be the regular assessment. Any order passed in consequence of a finding or a direction given by a higher authority will not be a regular assessment. The higher authority may direct the Income-tax Officer to modify his order. The higher authority may also set aside the order and direct the Income-tax Officer to pass a fresh order of assessment. In neither of these two cases will the consequential order be a regular assessment order. It will be an assessment made pursuant to a direction given by a higher authority. The date up to which interest under section 214 will be paid is the date of the regular assessment, i.e., the date of the first order of assessment under section 143. The quantum of interest, however, may be varied as a result of subsequent orders modifying the assessment pursuant to the direction of any higher authority. **[CESC Ltd.-CIT, SC,** _Sec.143._**150CTR607,233ITR50]**

21. It is clear from sub-section (1) of section 215 of the Income-tax Act,1961, that the liability to pay interest arises if, in any financial year,the advance tax paid by the assessee falls short of seventy- five per cent. of the assessed tax. In that event, interest is payable on the deficient amount from the first day of April of the next financial year up to the date of regular assessment. Assessed tax has been defined in sub-section (5) to mean the tax determined on the basis of the regular assessment.Thus,both for the purposes of determining the liability of the assessee to pay interest under this section and also the date up to which interest is to be charged, the material assessment is the regular assessment. Regular assessment means and refers to the original assessment made under section 143 or 144 of the Act. **[CIT-Harihar Jethalal Jariwalla,Bom.,** *Sec. 143.***147CTR381,234ITR1,98Taxman404]**

22. Any assessment made for the first time by resort to section 147 will also be a regular assessment for the purpose of section 217 of the Act. **[CIT-K.P. Baburaj, Ker.,** *Sec.147.***148CTR215, 234ITR718]**

23. The expression regular assessment has been used in the provisions of sections 215 and 214 of the Income-tax Act, 1961, which are pari materia so far as the interest for the period is concerned, that is from the first day of April next following the said financial year up to the date of regular assessment.A reassessment under section 147 is not a regular assessment for the purpose of section 215. **[CIT-Mahalaxmi Rice Mill,Cal.,***Sec.147.***168CTR194,249 ITR 456,118Taxman663]**

24. Regular assessment means the original assessment made under section 143 or 144 of the Income-tax Act, 1961. If any consequential order has to be passed by the Income-tax Officer to give effect to an order passed by the higher authority that consequential order cannot be treated as regular assessment nor can the date of the consequential order be treated as the date of the regular assessment. Where no regular assessment has been made under section 143 or 144 in the first instance, the assessment under sections 147 and 148 is also to be treated as regular assessment.

[CIT-Ghewar ChandSoni,Raj.,*Sec.143.***177CTR499,263ITR650,132Taxman915**

25. In section 214 of the Income-tax Act, 1961, the words regular assessment have been used.The expression clearly means and refers to the original assessment made under sections 143 and 144. It does not speak about any assessment after the year of regular assessment. The words regular assessment in section 214 are clear. Section 214 contains unmistakable and irrefutable indications that regular assessment therein means the original assessment alone. The amendments made to section 214 from time to time also go to indicate that regular assessment was used in the sense of the first assessment. **[CIT - Udhoji Shri Krishandas,MP.,** *Sec.214.***186CTR26,268ITR244,136Taxman465]**

26. .....the words regular assessment in section 215 mean only the first order of regular assessment and not the last operative order of regular assessment at any given point of time passed in appellate or revisional proceedings. **[CIT - Carona Sahu Co. Ltd., Bom.,** *Sec.215.***38CTR219,146ITR452,16Taxman32]**

Reheard Pat. CWT- Jagdish Prasad Choudhary

Reasonable opportunity of hearing mentioned in section 18(2) means a reasonable opportunity of oral hearing and similarly reheard mentioned in section 39 would mean reasonable opportunity of reheard.This construction flows from a combined reading of the safeguards given in section 18(2) read with section 39 of the Act.In other words, if the legislative intent of giving a reasonable opportunity of hearing is confined merely to a consideration of the written representation without an oral hearing, there would be no meaning in giving an opportunity of reopening of the case or an opportunity of being reheard before the succeeding officer and in that case the succeeding officer could continue the proceeding on the basis of the materials on record and there would be no purpose in giving a right to the assessee to demand a reopening of the proceedings or demand a rehearing. The legislative intent behind giving to the assessee a right of reopening of the proceeding or a right of rehearing becomes meaningful only if the right of hearing given in section 18(2) of the said Act is construed as a right of oral and personal hearing. *Sec.39.r.w.s.18.WT.Act.* **[211ITR472]**

Regulation and control SC State of West Bengal - Kesoram Industries Ltd.

It is well settled that the power of regulation and control is separate and distinct from the power of Taxation.A power to regulate, develop or control would not include within its ken a power to levy tax or fee except when it is regulatory. *Constitution of India.* **[187CTR219,266ITR721]**

Relative Cal. Bhagwati Trading Co. Ltd. - CIT

Relative has been defined as meaning husband, wife, lineal ascendant or descendant,brothers and sisters.Therefore,the persons must be related to one another by one of the relationships indicated in the definition. The provision requires not only that one must be related to the other, but also that the other must be related to the next person. Each must be related to the other in view of the expression one another. If there are two brothers the wife of one of them would not be related to the other in terms of this definition. The expression relative has been defined and it is not proper to construe the relationship by some other notion because there cannot be two different meanings to the same expression.

Sec.2(18),2(41.)r.w.s.104. **[109ITR353]**

2. Clause (ii) of Explanation 1 to section 2(18) of the Income-tax Act, 1961, is categorical and definite that unless two or more persons are relatives of one another, they cannot be treated as a single person. The expression relative as appearing in section 2(41) is in relation to an individual and the persons who are included in the category of relatives have been spelt out in the provision itself, i.e.,husband, wife, brother or sister or any lineal ascendant or descendant of that individual.The expression that individual is the key to the intention of the Legislature. It has link with the expression an individual which appears in the earlier part of the provision. Therefore, while husband, wife,brother and sisters or lineal ascendant of an individual is concerned, merely because another person who is relative being one of the indicated categories, relatives of that person are not covered by the expression relative under section 2 (41),so far as the other individual is concerned. **[Addl.CIT-Motor and General Finance Ltd.,Del.,***Sec.2(18),2(41)***163CTR92,246 ITR471,112Taxman183]**

Religious community **Guj.** **Shantagauri Ramniklal Trust - CIT**

In order to reach a conclusion whether a community is a religious community, there must be something to show that the community denotes abiding by a particular faith to be considered as a member of that community, or its identification is on account of practicing or following customs or practices. Merely by adhering to distinct cultural practices in its social interactions may not be sufficient to treat a community, as a religious community.The question also requires to be considered whether a caste in order to be excluded from the benefit of registration must be also a religious caste or a part of a religious community as an identifiable mark. Without any such material, it is not possible to reach a concrete conclusion.

Sec.12A. **[153CTR145,239ITR528]**

Relinquishment **Bom.** **CIT - Rasiklal Maneklal (HUF)**

In a transaction of relinquishment the interest of a person in a property is either given up, abandoned or surrendered; but the property in which interest is relinquished continues to exist and the property continues to be owned by some person or persons after the transaction of relinquishment. Sec.45.**[95ITR656]**

2.A relinquishment takes place when the owner withdraws himself from the property and abandons his rights thereto. It presumes that the property continues to exist after the relinquishment.Where, upon amalgamation, the company in which the assessee holds shares stands dissolved, there is no relinquishment by the assessee. **[CIT - Rasiklal Maneklal (HUF),SC,** Sec.45**77CTR31,177ITR 198,42Taxman259]**

Remand **Mad.** **M.R.M.Periannan Chettiar - CIT**

A remand implies a termination of the proceedings, so far as the appellate authority is concerned, and no question would arise about the continuation of the appeal thereafter. *Misc.***[39ITR159]**

Remission of the liability **Kar.** **Liquidator, Mysore Agencies Pvt. Ltd.- CIT**

The remission of the liability arises when the creditor voluntarily gives up the claim.The cessation of such liability arises only when it ceases to exist in the eye of law for all intents and purposes.When a debt becomes barred by time the creditor would not be able to recover the amount by enforcing his right in a court.But the right will not come to an end nor does the liability cease. *Sec.41.* **[114ITR853]**

Remittance **Mad.** **S.Kanagasabapathi Pillai - CIT**

Remittance is an elastic word and is not a term of art. It is not merely where cash or value in kind physically passes that there is remittance. Even where there is transfer by credit entries in the books of account without actual cost being handled and the credit entry so transferred in such that the person is whose favour it is made has a right to operate upon and draw the amount, there is remittance. *Misc.***[67ITR735]**

Remuneration Guj. Alembic Glass Industries Ltd. - CIT

Remuneration would include all that is quantifiable in money and paid to a person for his services or work. It will be noticed from the provisions of section 309(2) of the Companies Act,1956, dealing with the topic of remuneration of directors, that a director may receive remuneration by way of a fee for each meeting of the board, or a committee there of, attended by him. Thus, the fee which a director receives for attending any board or a committee meeting thereof is described by the statute itself as his remuneration. The provisions of sub-section (2) of section 198 of the Companies Act, 1956, under which fees are to be excluded while computing the maximum managerial remuneration to be given to its directors cannot control the provisions of the Income-tax Act, 1961, which operate in an independent and separate field.Even though the remuneration payable by way of a fee to a director for attending a board or committee meetings may not be included while calculating the overall maximum managerial remuneration payable to the directors under section 309 of the Companies Act,1956, it cannot be said that such remuneration would not be taxable under the Income-tax Act. Such fees would fall within the meaning of the word remuneration under section 40(c)(i) of the Income-tax Act, 1961.

<u>*Sec.40(c)(i).*</u>**[112CTR6,205ITR200,69Taxman165]**

2. The word remuneration is defined in section 198 of the Companies Act and it includes any expenditure incurred by the company in providing any other benefit or amenity free of charge or at concessional rate or any expenditure incurred in respect of any obligation or service which, but for such expenditure by the company, would have been incurred by any of the persons specified in sub-section (1) of section 198, i.e., directors and manager of the company. This is why when expenditure incurred by the managing director on medical treatment is to be paid by the company, it has to be approved by the Central Government under section 310 of the Act.

[Ambica Mills Ltd.-CIT,Guj.,<u>*CompaniesAct.*</u>**146CTR195,23 ITR583,99Taxman341]**

Rendering technical services SC Continental Construction Ltd.- CIT

In order to say that a person is rendering technical services to another, it is not necessary that the services should be rendered by the former personally and not through the medium of others.

Sec.80HHB[**AIR1992SC803,101CTR386195ITR81,107Taxation 132,60Taxman429]**

Renewal Cal. CIT - Borhat Tea Co. Ltd.

The word renewal must be interpreted in its popular sense. In common parlance renewal does not mean extension, it means repetition of the earlier contract. In that sense every repeat deposit is a new deposit. The word renewed or renewal, as applied to promissory notes in commercial and legal parlance, means something more than the substitution of another obligation for the old one. It means re-establishment of a particular contract for another period of time, to restore to its former condition an obligation on which the time of payment has been extended. A deposit is repayable either on the maturity date fixed therefor or according to the terms of the agreement relating to the demand on the making of which the deposit will become repayable. It, therefore, logically follows that the renewal of a deposit is a fresh contract of deposit because the first deposit falling repayable has to be construed as having been repaid and renewal is recycling the fund in a fresh deposit. It cannot be said that renewal is a continuation of the contract of the first deposit. It is more so when the renewal is effected by execution of a fresh contract under the terms and conditions of the deposit rules. *Misc.*[**203ITR987]**

Renovate P&H. CIT - Porrits and Spencer (A) Ltd.

Renovate in its ordinary sense means to restore to life, to a former state (as of freshness, soundness, purity or newness of appearance). The process of renovation involves expense. But it does not mean acquisition of a new asset. The amounts spent on renovation cannot be the sole measure for deciding the nature of expense. Various relevant factors have to be taken into account. The basic test is - has a new asset come into being. *Sec.37.*
[176CTR333,257ITR49,124Taxman155]

Rent Cal. Smt. Bishaka Sarkar - Union of India

The term rent has been defined in the Explanation to section 194- I of the Income-tax Act, 1961, as any payment, by whatever name called,under any lease, sub-lease, tenancy or any other agreement or arrangement for the use of any land or building (including factory building), together with furniture, fittings and the land appertaining thereto, whether or not such building is owned by the payee. The definition for the purpose of this Act of the nomenclature rent as expounded in the Explanation column of the section itself, amply reveals that the same is projected as the generic term which includes within its ambit payment made on any account whatsoever for occupation of a tenanted portion. It appears to be a composite concept. Once the rent is comprehended as a composite concept then it is not capable of being fragmented. It is comprehended in the section itself that the person who is responsible for such payment to the landlord is required to deduct at source (a) 15 per cent. if the payee is an individual or a Hindu undivided family; and (b) 20 per cent. in other cases. *Sec.194I.* **[139CTR,219ITR327]**

2. Explanation (i) to section 194-I of the Income-tax Act,1961, states that rent means any payment, by whatever name called, under any lease, sub-lease, tenancy or any other agreement or arrangement for the use of any land or any building (including factory building), together with furniture, fittings and the land appurtenant thereto, whether or not such building is owned by the payee. A perusal of the provision shows that the word rent, as defined has a wider meaning than rent in common parlance. **[United Airlines - CIT,AAR,** *Sec.194I.* **287ITR281,157Taxman368]**

Rent and premium Del Bharat Steel Tubes Ltd. - CIT

Section 105 of the Transfer of Property Act, 1882, makes a distinction between rent and premium payable under a lease. When the interest of the lessor is parted with for a price, the price paid is premium or salami. But the periodical payments made for the continuous enjoyment of the benefits under the lease are in the nature of rent. *Transfer of Property Act.*
[170CTR442,252ITR622,119 Taxman6]

Rental /business income Cal. CIT - Shambhu Investment Pvt. Ltd.

Merely because income is attached to any immovable property, that cannot be the sole factor for assessment of such income as income from property.If the main intention of the assessee is to let out the property or any portion thereof the income must be considered as rental income or income from property whereas if the primary object is to exploit the immovable property by way of complex commercial activities,in that event it must be held as business income. *Sec.22.r.w.s.28.***[168CTR237,249ITR47,116Taxman795]**

Repair **Bom.** **New Shorrock Spinning & Manufacturing Co. Ltd. -CIT**

The expression repairs must be understood in contradistinction to renewal or restoration. The test that has to be applied is that as a result of the expenditure which is claimed as an expenditure for repairs what is really being done is to preserve and maintain an already existing asset. The object of such expenditure is not to bring a new asset into existence, nor is its object the obtaining of a new or fresh advantage. *Sec.31.***[30ITR338]**

2. Renewal is a repair if it is only restoration by renewal or replacement of subsidiary parts of a whole. If, on the other hand, it amounts to a reconstruction of the entirety or of substantially the whole of the subject-matter it is not a repair but a reconstruction. The test, therefore, which ecides the question whether a thing is a repair or not is to see whether the act actually done is one which in substance is a replacement of defective parts or a replacement of the entirety or a substantial part of the subject-matter. Replacements can be of two different types. There can be replacements which amount to repair and there can also be replacements which cannot be treated as repair. **[Simbholi Sugar Mills Ltd.-CIT,All.,***Sec.37.***45ITR125]**

3. Repair is restoration by renewal or replacement of subsidiary parts of a whole. Renewal,as distinguished from repairs, is reconstruction of the entirety, meaning by the entirety not necessarily the whole but substantially the whole subject-matter under discussion. **[Sir Shadi Lal and Sons,Shamli-CIT,All.,** *Sec.24(1)(i)(b).***92ITR453]**

4. Repair means restoration by replacement of subsidiary parts or the whole. Current repair does not mean petty repair. Repairs may be small or major. If it is a major repair, it may involve considerable amount of money. But the amount of money spent alone cannot be a factor to determine whether the expenditure falls under current repairs or not. It is the nature of the repairs carried out by the assessee that matters for grant of deduction. **[Nathmal Bankatlal Parikh and Co.-CIT,AP.,***Sec.31.***122ITR168,3Taxman97]**

5. The idea of repair may include replacement or even a renewal. But the converse may not be true. All replacements or renewals need not necessarily be repairs. In the case of a building, restoration of stability or safety of a subordinate or subsidiary part of it or any portion of it can be considered as repair while the reconstruction of the entirety of the subject matter may not be so regarded. **[Sir Shadi Lal and Sons-CIT,SC,** *Sec.24(1)(a)(b).***67CTR1,169ITR 510,35Taxman400]**

6. Repair implies the existence of a thing which has malfunctioned and can be set right by effecting repairs which may involve replacement of some parts, thereby making the thing as efficient as it was before or as close to it as possible. After repair the thing to which the repair was carried out continues to be available for use. Replacement is different from repair. Replacement implies the removal or discarding of the thing that was in use, by a different or new thing capable of performing the same function with the same or greater efficiency. The replacement of a section in a series of machines which are interconnected, in a segment of the production process which together form an integrated whole may in some circumstances, be regarded as amounting to repair when without such replacement that unit in that segment will not function. That logic cannot be extended to the entire manufacturing facility from the stage of raw material to the delivery of the final finished product. In order to constitute current repairs the expenditure must have been incurred to preserve and maintain an already existing asset, and the object of the expenditure must not be to bring a new asset into existence or for obtaining a new advantage. It is difficult to accept that for the purpose of determining the allow ability of expenditure under the head Repairs the entire productive apparatus of a manufacturing company be treated as one single asset and wholesale replacement of complete identifiable and distinct parts be regarded as a repair effected to the production facility as a whole. While judging a claim for deduction under the Income-tax Act the scope of the term repair cannot be stretched beyond all recognition. **[CIT-Madras Cements Ltd.,Mad.,***Sec.31.***175CTR551, 255ITR243,170Taxation520,123Taxman412]**

7. In the case of repair or replacement of certain parts to set right the equipment may be current repairs. The expression repair in ordinary connotation means a restoration by renewal, or replacement of subsidiary parts of a whole and it is also understood that it should be allowed to make good the effect of the equipment. When we say to make good the effect of the equipment, it should be understood to mean that with the repair the defective condition of the equipment is cleared and the equipment is restored to the position to be used as it was a new one. If the equipment as a whole is to be replaced, it would mean that it is not repaired and it may be a case of replacement. **[DCIT-S.T.N.Textiles Ltd.,Ker.,** *Sec.31(i).***175CTR543,257ITR161,131Taxman73]**

8. Repair means restore (building, machine, garment, tissue, strength, etc.) to good condition, renovate or mend by replacing or refixing parts or compensating loss or exhaustion. **[CIT-Udaipur Mineral Developement Syndicate Pvt. Ltd., Raj.,***Sec.37.***203ITR556]**

Repair and renew   Guj.   Addl. CIT - Desai Bros.

Repair and renew are not words expressive of a clear contract. Repair always involves renewal; renewal of a part. Repair is restoration by renewal or replacement of subsidiary parts of a whole. In finding out whether expenses are for current repairs deductible under section 31 of the Income-tax Act, 1961, the real test is the aim and object of the expenses. If the expenses are incurred in bringing into existence a new asset or in achieving an advantage or benefit of an enduring nature or in substantially replacing plant or machinery, the expenditure would not be allowable. If the expenditure is incurred in the continuous process of use or employment of plant or machinery it would be deductible. *Sec.31.* **[108ITR14]**

Repairs   Kar.   CIT - Motor Industries Co. Ltd.

....the term repairs has a special meaning. What is necessary for the upkeep and maintenance of a building which necessarily includes periodical color/white washing and painting can be treated as repairs. But, amounts lavishly or even unnecessarily spent just to satisfy the ego or the eccentricities of an employee cannot be treated as amounts spent on repairs. *Sec.37.***[173ITR374]**

2. There is a clear distinction between the expression repairs and the expression current repairs. The word repairs is much wider than the expression current repairs. The expression current repairs is much more restricted than the word repairs because the latter is qualified by the word current. There is a distinction between expenses incurred by a tenant for repairs of the premises and the expenses incurred by a person who is not a tenant towards current repairs to the premises. **[CIT-Hi Line Pens Pvt.Ltd.,Del.,** *Sec.30(a)(i).***219CTR397, 306 ITR 182,175Taxman132]**

Repairs and maintenance **Gau.** **George Williamson (Assam)Ltd.-CIT**

In section 43 of the Income-tax Act, 1961, the expression motor vehicle has not been specifically mentioned, but the expression plant includes motor vehicle. Section 31 speaks about amounts spent on repairs, whereas section 37(3A) speaks about running and maintenance.The expressions repairs and maintenance are two different expressions.The expression repair presupposes certain injury or partial destruction. But the expression maintenance does not do so. It means to keep a particular thing in its similar state. The Legislature being fully aware of the difference of expressions, dealt with the expenses on repairs in section 31 and expenses for running and maintenance in section 37(3A) and(3B).*Sec.37(3A)(3B).* **[139CTR194,223ITR203,91Taxman308]**

Repairs/maintenance **Ker.** **CIT - Travancore Cements Ltd.**

The expenditure on repairs contemplated under section 31 is entirely different from the expenditure towards maintenance contemplated in section 37(3B). From the dictionary meaning of these two expressions it is very clear that the expression repair pre supposes certain injury or partial destruction.But the expression maintenance does not do so. It means to keep a particular thing in its similar state. Hence, the expenditure on repairs dealt with under section 31 of the Act is entirely different from the expenditure on maintenance covered by sub-sections (3A) and (3B) of section 37.*Sec.37(3B).***[157CTR395,240ITR816,108Taxman313]**

Replacement **Mad.** **CIT - Madras Cements Ltd.**

Replacement is different from repair.Replacement implies the removal or discarding of the thing that was in use, by a different or new thing capable of performing the same function with the same or greater efficiency. The replacement of a section in a series of machines which are interconnected, in a segment of the production process which together form an integrated whole may in some circumstances, be regarded as amounting to repair when without such replacement that unit in that segment will not function. That logic cannot be extended to the entire manufacturing facility from the stage of raw material to the delivery of the final finished product. *Sec.31.***[175CTR551,255ITR243,123Taxman412]**

Representative assessee/ Individual **Mad.** **CIT-Venu Suresh Sanjay Trust**

A plain reading of sections 160 to 162 of the Income-tax Act,1961, would go to show that a representative assessee has either to be an individual or an artificial juridical person,who is also equated with an individual. Under section 160(1)(iv) it is the trustee,who is the representative assessee. Sec.160-62.**[138CTR97,221ITR649 90Taxman486]**

Representative-assessees **MP.** **CIT - G.B.J.Seth**

In the case of representative- assessees they take their status from the beneficiaries they represent and it is wholly immaterial whether there is one representative- assessee or there are two or more of them representing the same beneficial interest or interests. In the case of executors, under section 168 of the Income-tax Act, 1961, if there is only one executor, he will be assessed as if the executor were an individual or if there are more executors than one, then, as if the executors were an association of persons. But this difference is of no consequence as the assessee or assessees are assessed as representing the estate of the deceased. Sec.168. **[133ITR192,6Taxman318]**

Requisition **All.** **Chandra Prakash Agrawal - ACIT**

The word requisition means taking of actual possession. The requisition is complete only when the seized books of account and other documents, which have been requisitioned, have been delivered to the requisitioning authority. The provisions of Chapter XIV-B of the Act would come into play only when the books of account or other documents or assets are actually received by the Assessing Officer pursuant to the requisition made under section 132A. Sec.132A.**[203CTR125,287ITR172,154Taxman372]**

Reserve **Mad** **CIT - Vasantha Mills Ltd.**

One ordinary natural meaning of the word reserve is to set apart for some purpose or with some end in view to keep for some use.Misc. **[AIR1957Mad326,32ITR237]**

2. A reserve is by its very nature a fund which is created and maintained for the purpose of being drawn up in future, and nothing can be reserved unless it has been reserved or laid by or stored for use or application in a future contingency which is anticipated as certain or likely. Payments which are made in discharge of a present liability in the course of the year to which the balance- sheet and profit and loss account relate cannot be said to be made by way of creating a reserve; they are clearly of the nature of expenditure, although it may not be expenditure in the Income-tax sense allowable in an assessment. A reserve is created only out of the whole or a part of the surplus profits as they are found to be in the hands of the company at the end of the year.**[Indian Steel and Wire Products Ltd.-CIT,Cal.,*Misc.* 33ITR579]**

3. A reserve can only mean profit earned by a company and not distributed as dividend to the shareholders but kept back by the directors for any purpose to which it may be put in future. **[First National City Bank-ITC,SC,*Misc.*AIR1961SC1393,42ITR 17]**

4. In its ordinary meaning the expression reserve means something specially kept apart for future use or for a specific occasion. Earned surplus, an account in which the net profits less the appropriations are added year after year, represents reserve. **[ITC-Standard Vaccum,SC, *Misc.*AIR1966SC1393,58ITR685]**

5. A reserve may be a general reserve or a specific reserve, but in order to constitute a reserve there must be a clear indication to show that it was a reserve either of the one or the other kind.A mass of undistributed profits is not a reserve,even though it is shown in the balance-sheet as a reserve. **[Aluminium Industries Ltd.-CIT,Ker., *Misc.*68ITR125]**

6. In its ordinary meaning, the expression reserve meant something specifically kept apart for future use or for a specific occasion. **[CIT - Mafatlal Chandulal & Co. Ltd.,Guj., *Misc.*107ITR489]**

7. In computing the capital of a company for the purpose of surtax, the true nature and character of a sum designated as a reserve is to be determined with reference to the substance of the matter.A reserve created in regard to a payment to be made on account of liabilities which have already arisen cannot be properly termed as reserve. The accounts of a company may be made up for a year up to a particular date at a later point of time.A company can, similarly, finalise its appropriations for various purposes including reserves and provisions at a later date with retrospective effect. A company can take advantage of retrospective effect of its determination of appropriations but it cannot then contend that,by being retrospective, the nature of the appropriation will change.**[Braithwaite and Co.(India)Ltd.-CIT,Cal.,**<u>*Super profit tax Act.*</u>**111ITR825]**

8. The true nature and character of a sum disputed as reserve has to be determined with reference to the substance of the matter. In case there is any conflict between the expressions reserveand provision, they should be understood in the ordinary sense in contradistinction to the sense in which these words are understood in company jurisprudence. ..... An item which could not be ascertained with any substantial accuracy, by following any normal. **[CIT-Karam Chand Thapar and Bros.,Cal.,**<u>*Companies(Profits) Surtax Act.*</u>**19 CTR49,131ITR175]**

9. The expression reservehas not been defined in the Super Profits Tax Act, 1963, or the C.(P.) S.T.Act, 1964. The dictionaries do not make any distinction between the two concepts reserve and provision while giving their primary meanings, whereas in the context of those Acts a clear distinction between the two is implied. Though the expression reserveis not defined, since it occurs in taxing statutes applicable to companies only and to no other assessable entities, the expression has to be understood in its popular sense, that is to say, the sense or meaning that is attributed to it by men of business; trade and commerce and by persons interested in or dealing with companies. Therefore, the meanings attached to the words reserves and provisions in the Companies Act, 1956, dealing with the preparation of the balance-sheet and the profit and loss account would govern their construction for the purposes of the two enactments. The broad distinction between the two is that whereas a provisionis a charge against the profits to be taken into account against gross receipts in the profit and loss account, a reserve is an appropriation of profits, the asset or assets by which it is represented being retained to form part of the capital employed in the business. **[VazirSultan TobaccoCo.Ltd.-CIT,SC,**<u>*Companies(Profits)Surtax Act.*</u> **25CTR 186,132ITR559,7Taxman28]**

10. In order to create a reserve there must be a positive and overt act on the part of the company and in the case of a secret reserve there cannot be any overt act of creating a reserve by segregating or setting apart of profits appropriated for a given purpose. Consequently, a mere excess provision for depreciation allowance or writing down of the trade investments on the assets side of the balance-sheet cannot be considered to be a reserve. **[CIT-Peirce Leslie and Co. Ltd.,Mad.,**<u>*Companies(Profits) Surtax Act.*</u> **38CTR 309, 147ITR157,16 Taxman8]**

11. Reserves are amounts set aside out of profits and other surpluses, not designed to meet a particular liability, contingency, commitment or diminution in the value of assets known to exist at the date of the balance-sheet. **[CIT-Industrial Credit and Dev. Syndicate Ltd., Kar.,**<u>*Companies (Profits) Surtax Act.*</u>**177ITR51]**

12. A mass of undistributed profits cannot automatically become a reserve and somebody possessing the requisite authority must clearly indicate that a portion thereof has been earmarked or separated from the general mass of profits with a view to constituting it either as a general reserve or as a specific reserve. The surrounding circumstances should make it apparent that the amount earmarked or set apart is in fact a reserve to be utilised in future for a specific purpose and on a specific occasion.A clear conduct on the part of the directors in setting apart a sum from out of the mass of undistributed profits avowedly for the purpose of distribution as dividend in the same year would run counter to any intention of making that amount a reserve. Thus, any amount set apart for diminution in the value of the assets which is a known liability can only be taken to be a provision and not a reserve. **[CIT - Gordon Woodroffe and Co. (Madras) Pvt. Ltd., Mad.,** <u>*Companies (Profits) Surtax Act.*</u>**183ITR465]**

Reserve / provision **Mad.** **CIT-Crompton Engineering Co.(Madras) Ltd.**

There is a distinction between a reserve and a provision. A reserve is a standby created out of profits of a business to meet unknown and unforeseen contingencies and does not go out of the business but is retained in the business while a provision is a charge against profits created to meet known and foreseeable liabilities but whose exact timing and quantification alone are uncertain at the moment...........Gratuity for employees is a definite liability of the business. Although, when exactly it would have to be paid actually is dependent upon several contingencies, it is not impossible to provide for the liability so as to meet the situation as and when it arises and such a provision is a definite liability provided it is done on scientific principles. In such a case the provision made will be a charge against the profits of the year in which the provision is made and, as it is a provision, it cannot be a reserve....While an ad hoc provision or an actual provision made on the basis of an actuarial valuation can be considered only as a provision, any amount in excess of the actual amount obtained on an actuarial valuation will really constitute a reserve. <u>*Super Profits Tax Act.*</u> **[31CTR347,140ITR320]**

2. The question whether a particular amount kept apart by the assessee for a particular purpose is a provision or a reserve will not depend upon the nomenclature given by the assessee in his books of account. What is shown in the books as a provision may, in reality, turn out to be a reserve and what is shown as a reserve, may, in fact, be a provision. The answer depends upon the true nature and character of the provision. A particular amount set apart by the assessee can be termed as a provision if it is so kept apart for providing for any known or existing liability.But if such provision is made in excess of the amount that would be reasonably necessary for such purpose, the excess would be treated as a reserve.... that a provision for doubtful debts or doubtful advances is nothing but a reserve. **[Glaxo Laboratories (India) Ltd. - CIT, Bom.,** <u>*Sec.36.*</u>**120CTR310,206 ITR 335,121Taxation440]**

Reserves Bom. CIT -Century Spinning and Manufacturing Co.

The expression reserves cannot be given any technical meaning but must be construed in its plain natural meaning-A reserve can only mean profit earned by a company and not distributed as dividend to the share holders but kept back by the directors for any purpose to which it may be put in future. <u>*Misc.*</u>**[AIR1951Bom420,20ITR260]**

2. In the contemplation of the Explanation,reserves are one of the constituent or component parts of the capital of the company as defined in the proviso. The mere fact that a mass of undistributed profits is carried over by a company from one year to another is not sufficient to make such a mass reserve but it is further necessary that it must be clearly allotted to a reserves, whether general or specific. **[CIT-Braithwaite&Co.,Cal.,** <u>*Super Profits Tax Act.*</u> **111ITR729]**

3. The expression reserves has not been defined in the Act. It has, therefore, to be understood in the manner in which it is used in commercial practice. As observed by the Supreme Court in Metal Box Company of India Ltd. v. Their Workmen (73 ITR 53) it is an amount which is set apart to meet a contingency which is not known at the time when the balance-sheet is prepared. It need not always be created out of profits. **[Addl. CIT- Indian Telephone Industries, Kar.,** <u>*Companies (Profits)Surtax Act.*</u>**10CTR44, 118ITR291]**

Residence **Guj.** **Harsutrai J. Raval - CIT**

The word residence has a variety of meanings, according to the statute in which it is used and position in which it is found.The word residence refers to a place of permanent and not merely temporary abode. The word permanent here, is a relative term and is not synonymous with everlasting. A stop gap arrangement for stay in a house property not intended to be purchased or constructed for permanent residence, that is as an abode where the assessee would ordinarily dwell, cannot deny the entitlement to the benefit of the provisions of section 54(1). *Sec.54(1).* **[174CTR540,255ITR315,122Taxman165]**

Residence of Hindu Undivided Families **SC** **V.Vr.N.M. Subbayya Chettiar - CIT**

The test for deciding the residence of Hindu undivided families laid down in section 4A(b) of the Income-tax Act, 1922, is based very largely on the rule which has been applied in England to cases of corporations. In order to bring the case under the exception, the court has to ask whether the seat of the direction and control of the affairs of the family is inside or outside British India.*Sec.6.***[19 ITR 168]**

Resident **Bom.** **Dhirajlal Haridas-CIT**

A plain reading of section 6(1)(b) of the Act of 1961 shows that what it requires, in order that the assessee may be considered as a resident in India, is that the assessee must have maintained or caused to be maintained for him a dwelling place in India for the period stated in the said clause and also resided in India for a period of 30 days or more during the relevant previous year. It is not sufficient in order to satisfy this test that there is a dwelling place in the taxable territories in which the assessee goes and lives. There must be in him a right to live in such a dwelling place maintained for him, because without that right, it could not be said that he has either maintained a dwelling place or a dwelling place has been maintained for him. The volition on the part of the assessee is essential. Whether the assessee maintains a dwelling place or causes it to be maintained,the maintenance of the dwelling place has to be at his instance or behest and,when it is maintained by someone other than the assessee, it has to be for the assessee or for his benefit. <u>*Sec.6(1)(b).*</u> **[138ITR570]**

Resident but not ordinarily resident/ Ordinarily resident **Guj.** **Pradip J. Mehta - CIT**

Section 6(6) of the Income-tax Act,1961, does not define ordinarily resident in India, but describes not ordinarily resident in India. It resorts to the concept of resident in India for which the criteria are laid down in section 6(1) of the Act. Ordinarily resident for the purposes of Income-tax connotes residence in a place with some degree of continuity and apart from accidental or temporary absences. When an individual has been a resident in India for nine out of ten preceding years, then in order to escape tax on his foreign income, he must not have been in India for seven hundred and thirty days or more in the aggregate during the preceding seven years.The test is one of presence and not absence from India and the length of presence will determine when an individual is not ordinarily resident in India. In order that an individual is not an ordinarily resident, he should satisfy one of the two conditions laid down in section 6(6)(a) of the Act, the first condition is that he should not be resident in India in all the nine out of ten years preceding the accounting year and the second condition is that he should not have during the seven years preceding that year, been in India for a total period of seven hundred and thirty or more days. <u>*Sec.6(1).*</u>**[175CTR394,256ITR647,169Taxation668,123Taxman1118]**

Resident of India **AAR** **Anurag Chaudhary**

The net effect of section 6(1) of the Income-tax Act,1961, read with the Explanation,is that an individual who had left India for employment outside India should be treated as a resident only if he was in India during the relevant period/year for 182 days or more. In other words, if an individual had spent less than 182 days in India during the previous year and was outside India for the purpose of employment, then, regardless of his being in India for 365 days or more during the four preceding previous years, he cannot be treated as a resident of India. If the applicant was not present in India for more than 365 days in the four preceding years, then clause (a) of sub-section (1) of section 6 would apply, and it requires stay of 182 days or more in India to be treated as a resident. <u>*Sec.6(1).*</u>**[322ITR293]**

2. In order to hold an individual a resident in India under section 6(1)(c), the said individual having been in India for a period of 365 days or more within 4 years preceding the previous year should also remain for a period of 60 days or more in that previous year. Explanation (b) to section 6(1)(c) stipulates that in the case of a citizen of India who being outside India comes on a visit to India in any previous year, instead of 60 days of stay in India, such stay should be for 182 days. **[ADIT (AIU)-Apparasu Ravi,Mad.,** *Sec.6(1).***332ITR497]**

Residential accommodation **Ker** **CIT - Hotel Amritha (P.) Ltd.**

The expression residential accommodation has not been defined in the Act. The question whether a hotel can be considered to be residential accommodation has to be approached and considered with special reference to the nature of the hotel, and the extent and type of residential accommodation provided thereby, the nature of the premises where the plant or machinery has been installed and whether the premises, vis-a-vis the machinery installed, could be passed as office premises or residential accommodation. *Sec.33*. **[11CTR20,122ITR296]**

Residing **Gau.** **CCIT - Neepco Tribal Employees' Welfare Association**

Though the word residing read in isolation may not necessarily imply an act of permanent residence, it has to be construed in the context in which it appears as well as the legislative intent conveyed thereby. The meaning to be accorded to the expression has to be necessarily that the underlying purpose of exemption is not only in the interest of members of the Scheduled Tribes of the area but also of the locality in which they reside and the situs of source of their income rests. The chargeability of income is dependent upon the locality of accrual or receipt thereof and the scheme and pattern of the Act acknowledge relevance of the territorial nexus with the assessable income. *Sec.10(26).***[290ITR481]**

Residing in any area specified **Gau.** **Pradip Kr. Taye**

The expression residing in any area specified occurring under section 10(26) cannot be given a narrow and restricted meaning to imply that the members of a scheduled tribe migrating from their place of origin, which happens to fall in one of the areas specified in sub-section (26), to another area although once again falling within the areas specified in the sub-section, would not get the benefit of the exemption under section 10(26). The expression residing in any area specified must be interpreted in the context of sub-section (26).

Sec.10(26).[**320ITR29]**

Retained profits Bom. CIT - Hindustan Antibiotics Ltd.

A company,which accumulates and retains, for the purposes of its future expansion, large amounts out of the profits earned during the year, can be said to have accumulated the amounts for purposes of its business. These are what are known in business parlance as retained profits. Retained profits may be retained for the purpose either of meeting and reducing the present liabilities or for meeting the working capital required in prasenti or for reduction of loans and borrowings or for the future expansion of the company. Regarded in any, of these ways such retained amounts must be treated as moneys required for the purpose of the business of the company. *Sec.80J.*[**26CTR129,137ITR42]**

Retirement Bom. CIT - D.P Malhotra.

Retirement is a word of wide import. In the context of employment, it means conclusion of a career. One of the meanings of the word retire is resign. Both retirement and resignation result in the conclusion of the service career. In fact, resignation from service is also one of the modes of retirement from service. Resignation is a voluntary act of the employee to retire from service. Once an employee resigns, his service stands terminated from the date on which his letter of resignation is accepted by the appropriate authority, unless there is any law or statutory rule governing the conditions of service to the contrary. In other words, on acceptance of resignation, the employee stands retired from service.The word retirement has not been used in clause (10AA) in the restricted sense to mean retirement on superannuation. On the other hand, it is clear from the language of clause (10AA) itself that it has been used in the widest possible terms to mean and include all cases of retirement, whether on superannuation or otherwise.What is relevant is retirement-how it took place is immaterial for the purpose of this clause. It is therefore, clear that if on retirement.

Sec.10AA. **[229ITR394,98Taxman110]**

Retransfer or reassume **Bom.** **Ramji Keshavji - CIT**

The words retransfer or reassume necessarily involve a second transaction of a later time. *Sec.64.* **[13ITR105]**

Retrospective **Gau.** **Ashutosh Banik - CIT**

A statute is retrospective when it takes away or impairs any vested right acquired under the existing laws, or creates a new obligation, or imposes a new duty, or attaches a new disability in respect of transactions or considerations already past. But a statute is not retrospective because a part of the requisites for its action is drawn from a time antecedent to its passing. *Misc.* **[20 CTR101, 132ITR544]**

Return **Bom.** **Princess Maheshwari Devi of Pratapgarh - CIT**

The word return could not be interpreted as meaning only a return for labour or skill employed or capital invested. Such a definition of return would be too narrow and would exclude the case of voluntary payments, when it is the settled position in law that in some cases even voluntary payments could be regarded as income. *Sec.2(24).* **[33CTR117,147ITR258,12Taxman220]**

Revenue expenditure **SC** **Assam Bengal Cement Co. - CIT**

If the expenditure is made not for the purpose of bringing into existence any asset or advantage but for running the business or working it with a view to produce profits it is a revenue expenditure. *Sec.37.* **[AIR1955SC89,27ITR34]**

2. If a trader or businessman buys a quantity of goods which forms his stock-in-trade (and this he may do in various ways), it is a part of his revenue expenditure. In such a case, he is buying stock-in-trade in order to carry on his business, and looking at it from a commercial point of view, the expenditure incurred is one solely for the purposes of his business. **[CIT - H.Dear and Co. Ltd.,Cal.,** *Sec.37.***52ITR65]**

3. A payment made with a view to obtain the benefit of technical assistance for running the assessee's business more efficiently so as to earn more profits and not by way of transfer of fruits of research once and for all, can be treated as an item of revenue expenditure. **[CIT-Alembic Glass Industries Ltd.,Guj.,***Sec. 37(1).***71ITR752]**

4. Ordinarily,any expenditure which is not of capital nature incurred in the course of business or trade is treated as revenue expenditure. **[Nippon Electronics (P.) Ltd.-CIT,Kar.,** *Sec.37.***116ITR231]**

5. If an expenditure is intrinsically connected with the running of a company and if it had not been incurred the company would not be able to operate in the field of its business, the expenditure would be relatable to revenue. **[CIT - Utkal Machineries Ltd.,Ori.,** *Sec.37(1).***131ITR650,6Taxman288]**

6. To decide whether all expenditure incurred in connection with the profit earning apparatus would be revenue expenditure, the proper approach should be what was the predominant purpose or intention of incurring the expenditure. If the predominant purpose of incurring the expenditure was the carrying on of the business, the incidental advantage of that expenditure cannot affect its revenue character. **[CIT-Karanpura Development Co. Ltd.,Cal.,** *Sec.37.***31CTR170,144ITR538,69Taxation108]**

Revenue/capital expenditure **P&H.** **Saraswati Industrial Syndicate Ltd.- CIT**

One of the guidelines for distinguishing revenue expenditure from capital expenditure is that if the expenditure is incurred for obtaining an advantage of enduring benefit it would be capital expenditure but the test of enduring benefit is not a certain or conclusive test and it cannot be applied blindly and mechanically with regard to the particular facts and circumstances of a given case. It is not every advantage of enduring nature acquired by an assessee that would bring his case within the said concept.What is material to consider is the nature of the advantage in a commercial sense and it is only where the advantage is on the capital filed that the expenditure would be disallowable on an application of this test. If the advantage consists merely in facilitating the assessee's trading operations or enabling the management and conduct of the assessee's business to be carried on more efficiently or more profitably while leaving the fixed capital untouched,the expenditure would be on revenue account. *Sec.31(1)(ii) 37.r.w.s.37*
[26CTR304,137ITR886,62Taxation54]

Right to receive **Mad.** **S. Khader Mohideen - ITO**

The words right to receive in section 44AC(1)(b) of the Income-tax Act,1961,have to be construed having regard to the object of the provision and its place in the scheme of the Act. The provision is a special provision for computing the profits and gains from the business of trading in certain goods. One of the items of the goods specified is forest produce. The transfer of title of the forest produce from one person to another in the course of trade is, therefore, to be governed by this provision. It covers not only the actual sale of the goods but also the right to receive the goods of the nature specified in the table including forest produce. The word receive therefore, in the context has to be construed as including the gathering of the forest produce from the forest, as immediately after such gathering it is to be appropriated by the person who has been given the right to gather, as his own property.The words right to receive have been used in the statutory provision to indicate the right to become the owner of the goods and not merely an instance of one person delivering goods to another without any intent to transfer the ownership therein.What is material is the transfer of title.The auction of the right to gather the forest produce is, therefore, an auction in relation to the right to receive forest produce by going into the forest and gathering the produce. *Sec.44AC.* **[167CTR444,248ITR554,102Taxman]**

Road transport vehicle **Cal.** **CIT- Birla Jute and Industries Ltd.**

. . . road transport vehicle is defined in the Road Transport Corporations Act,1964 ,to mean a service to carry passengers or goods or both by road in vehicles for hire or reward. It clearly means a service by road, whereas vehicle is defined in the Act to mean a vehicle capable of being used for road transport and includes tram car, trolley vehicle or trailer. It does not include rail. The stress is on road and includes tram car but not rail or train. If a statute in an inclusive definition refers to one item distinct from the other and omits that other item, then the intention of the Legislature appears to be clear. While including one by special reference without referring to the other the omission is deliberate. *Sec.32A. r.w.r.5.***[180CTR399,260ITR55,133Taxman337]**

2. The expression road transport vehicles is not defined in the Income-tax Act.In the absence of a statutory definition, the meaning in common parlance has to be adopted.Road transport vehicles would mean vehicles which are used for transportation of goods or passengers on roads. A tractor is not basically used for transportation of goods or passengers on roads. It is machinery used for tilling agricultural land. People can ride on a tractor but that is not its main purpose. Hence investment allowance can be claimed on tractors. **[CIT-U.P. Paschimi Kshetriya Vikas Nigam Ltd.,All.,**<u>*Sec.32A.*</u> **188CTR40,264ITR273,137Taxman378]**

Royalty Guj. CIT - Ahmedabad Manu. and Calico Printing Co.

In the case of secret processes,patents, special inventions,when right of exploitation is given by the owner of the inventions, patents, etc., to a third party instead of outright sale, then for the right to exploit these inventions, secret processes, some amount may be paid and the amount paid may be correlated to the extent of the exploitation. It is in this sense that licence agreements for the exploitation of patents, inventions, etc., are being entered into in modern commercial world and as part of such agreements, even knowledge derived from his own experience and technical know-how for the most economical and efficient user of the patents, inventions,etc., are parted with by the licensor to the licensee. Payments of this kind are known as royalties. This is also evident from several double Taxation avoidance agreements between the Govt. of India and foreign countries such as Sweden in which the term Royalty has been defined. That such payments are royalties is also evident from the definition of the word Royalty in section 9 (1) (vi),Explanation 2,which was subsequently introduced by the Finance Act,1976,with effect from June 1,1976. <u>*Sec.9(1)(vi).*</u> **[139ITR806]**

2. The dictionary meaning of the term royalty makes it clear that the said term connotes payments periodic or at a time for user by one person of certain exclusive rights belonging to another person. The examples of such exclusive rights are rights in the nature of a patent mineral rights or rights in respect of publications.It is possible for a person carrying out operations of manufacture and production of a particular produce to acquire specialised knowledge in respect of such manufacture and production which is not generally available. A person having such specialised knowledge can claim exclusive right to the same as long as he chooses not to make such specialised knowledge public. It is also conceivable that such a person can exploit and utilise such specialised knowledge in the same way as a person holding a patent or owning a mineral right or having the copyright of a publication to allow a limited user of such specialised knowledge to others in confidence against payment.There is no reason why payment for the user of such specialised knowledge though not protected by a patent should not be treated as royalty or in the nature of royalty. **[N.V.Philips - CIT,Cal.,**<u>*Sec.9.*</u>**65CTR103,172ITR521]**

3. A plain reading of section 9(1)(vi)(b) of the Income-tax Act,1961, shows that any amount paid by a person who is a resident by way of royalty will fall within the phrase income deemed to accrue or arise in India. The expression royalty is defined in Explanation 2 of the clause. The Explanation defines royalty to mean consideration including any lump sum consideration paid for imparting any information concerning the working of, or the use of, a patent, invention, model, design, secret formula or process or trade mark or similar property and also the imparting of any information concerning technical,industrial, commercial or scientific knowledge,experience or skill.A lump sum consideration paid which is in the nature of income chargeable under the head Capital gains, is excluded from the meaning of royalty. **[CIT-Klayman Porcelains Ltd.,AP.,** <u>*Sec.9(1) (vi).*</u>**229ITR735,96Taxman221]**

4. ....the term royalty normally connotes the payment made by a person who has exclusive right over a thing for allowing another to make use of that thing which may be either physical or intellectual property or thing. The exclusivity of the right in relation to the thing for which royalty is paid should be with the grantor of that right. Mere passing of information concerning the design of a machine which is tailor-made to meet the requirement of a buyer does not by itself amount to transfer of any right of exclusive user, so as to render the payment made therefor being regarded as royalty. **[CIT-Neyveli Lignite Corporation Ltd.,Mad.,**<u>*Sec.9.*</u>**162CTR 206,243ITR459]**

5. The term royalty is assigned a specific meaning in Explanation 2 to clause (vi) of section 9(1) of the Act. Sub-clause (i) of the Explanation deals with a situation where rights in intellectual property are transferred including the granting of a licence.What sub-clause (i) envisages is the transfer of rights in respect of property and not transfer of right in the property. In the first category, the rights are purchased which enable use of those rights, while in the second category, no purchase is involved, only a right to use has been granted.Thus, the definition of the term royalty in respect of the copyright, literary, artistic or scientific work, patent, invention, process, etc. does not extend to the outright purchase of the right to use an asset. In the case of royalty, the ownership on the property or right remains with the owner and the transferee is permitted to use the right in respect of such property. **[Asia Satellite Telecommunications Co. Ltd.-DIT,Del.,***Sec.5. r.w.s.9(1),90.***238CTR233,332 ITR340,197Taxman263]**

6. Under the DTAA with the U. S. A., the necessary ingredient to be satisfied to find out whether the payment would amount to royalty is that it is received as a consideration for the use of, or the right to use,any copyright of literary, artistic or scientific work. **CIT - Samsung Electronics Co. Ltd., Kar.** *DTAA-USA* **[245CTR481, 345ITR494,203Taxman477]**

S

Salami SC Maharaja Chintamani Saran Nath Sah Deo - CIT

Salami is a single payment made for the acquisition of the right of the lessor by the lessee to enjoy the benefits granted to him by the lease. <u>Sec.4.</u>**[82ITR464]**

Salaries Raj. CIT - Pramod Kumar Jain

Section 15 of the Income-tax Act,1961, has provided for the income chargeable to Income-tax under the head Salaries. The provisions of section 15 of the Income-tax Act, 1961, contemplate the jural relationship of employer and employee.In the case of an employment the existence of a relationship of master and servant has to be established. A firm is not a legal person and has no legal existence apart from its partners.Though under the Income-tax law,it is a unit of assessment by virtue of the special provisions it cannot be considered that the firm is the employer of its partner.There cannot be a contract of service, in strict law,between a firm and one of its partners.Payment of salary to a partner represents a special share of the profits. Salary paid to a partner retains the same character of the income of the firm as profit. Explanation 2 added by the Finance Act,1992 from April 1,1993, in section 15 makes it clear that the salary received by a partner of a firm shall not be regarded as salary. <u>Sec.15.</u>**[125CTR154,216ITR 598,80Taxman333]**

Salary Sin. CIT - Mills Store Co.

Before a payment from one party to the other can be regarded as salary within the meaning of section 7 of the Act some relationship of employer and employee or of principal and agent is essential. A restrictive covenant, whereby a person undertakes for consideration to abstain from doing a particular act or from following a particular course of conduct, is something quite outside an ordinary contract of employment and the word salary cannot be held to include the consideration of such a contract. *Sec.15.* **[9ITR642]**

2. Conceptually there is no difference between salary and wages, both being a recompense for work done or services rendered, though ordinarily the former expression is used in connection with services of non-manual type while the latter is used in connection with manual service.The expression wages does not imply that the compensation is to be determined solely upon the basis of time spent in service; it may be determined by the work done; it could be estimated in either way. If conceptually salary and wages mean one and the same thing then salary could take the form of payment by reference to the time factor or by the job done. In fact, in the case of salary, the recompense could be determined wholly on the basis of time spent on service or wholly by the work done or partly by the time spent in service and partly by the work done. In other words, whatever be the basis on which such recompense is determined it would all be salary.If under the terms of the contract of employment remuneration or recompense for the services rendered by the employee is determined at a fixed percentage of turnover achieved by him, then such remuneration or recompense will partake of the character of salary, the percentage basis being the measure of the salary and, therefore, such remuneration or recompense must fall within the expression salary as defined in rule 2(h) of Part A of the Fourth Schedule to the Act.

[GestetnerDuplicatorsPvt.Ltd.-CIT,SC,*Sec.17(1)(iv),r.w.s.40(c)(iii),Sch.IV,* **8CTR371,117ITR1,1Taxman1]**

3. The word salary in its widest coverage would definitely include all those payments which an employer makes to his employees and the payment of bonus is consistently taken to be part of salary. **CIT - New India Industries Ltd. Guj.** *Sec.17* **[201ITR208]**

4. There is no general definition of the word salary applicable for all the provisions of the Income-tax Act. The Explanation to section 10 (10)states that, In this clause and in clause (10AA) salary shall have the meaning assigned to it in clause (h) of rule 2 of Part A of the Fourth Schedule to the Income-tax Act. Clause (h) of rule 2 of Part A of the Fourth Schedule reads that salary includes dearness allowance, if the terms of employment so provide, but excludes all other allowances and perquisites. There is no ambiguity whatever in the definition contained in the above clause, and it is the said definition which should be applied while construing section 10(10) and section 10(10AA). **[K.Gopalakrishnan -CBDT, Mad., <u>*Sec.17.r.w.s.(10AA).*</u>206ITR183]**

5. Section 17(1) of the Income-tax Act,1961, provides that for the purposes of sections 15 and 16, salary includes any fees, commissions,perquisites or profits in lieu of or in addition to any salary or wages.Salary means to recompense, reward;to pay for something done.If under the terms of the contract of employment remuneration or recompense for the services rendered by the employee is determined at a fixed percentage of the turnover achieved by him then such remuneration or recompense will partake of the character of salary, the percentage basis being the measure of the salary.

[CIT - Govind Chandra Pani CIT - Govind ChandraPani,Ori., <u>*Sec.17(1).*</u>126CTR 359,213ITR783, 83Taxman364]

6. ....The definition of income in section 2(24) of the Income-tax Act, 1961,though inclusive, includes also the receipts which would be income in their normal sense and it is not a mere catalogue of receipts which otherwise would not be income.Under the head Salaries, to be chargeable income, it should not only be income, but it should also be income of the nature indicated in section 15 read with section 17, which defines salary.Under section 17(1)(iv), salary would include any fees, commissions, perquisites or profits in lieu of or in addition to any salary or wages. Any special allowance or benefit other than perquisite which are treated as income by virtue of section 2(24) have not been separately mentioned in the definition of salary and can be classified as salary only if they are covered by the expression profits in lieu of or in addition to salary or wages. The use of the word profit in the expression profits in lieu of or in addition to salary or wages, itself involves a deduction of expenses properly incurred in realising the proceeds out of which the profit arises.Section 2(24) of the Act speaks of benefits and allowances granted to the assessee to meet the expenses for the performance of duty and not the expenses incurred in the performa- nce of duty. To the extent that such amount granted to the assessee for meeting his expenses wholly necessarily and exclusively for the purpose of his duties are not actually spent by him, that portion of such income included in section 2(24) will result in profits to the assessee in addition to his salary or wages. Certain types of expenses falling within this artificial definition of income which are incurred out of the amount so granted are specifically exempted from being included in the total income by virtue of section 10(14). The type of such incurred allowances or benefits granted to the assessee if not exempted under section 10(14) would not thereby automatically become his salary. The receipt has to fall under the head Salaries in order to become chargeable to tax. The deductions contemplated under section 16 of the Act are to be made while computing the income chargeable under the head Salaries and these statutory deductions are quite distinct from the removal of expenses incurred for the purpose from the receipts for working out the profits in lieu of or in addition to salary under clause (iv) of section 17(1).

[CIT-Kiranbhai H. Shelat, Guj.,<u>*Sec.17.r.w.s.10(14).*</u>**147CTR43, 235ITR635,99Taxman63]**

7. A contract of employment necessarily involves the rendering of services by the employee to the employer and the use of his skill, energy and time for the benefit of the business in which he is employed. If the employer chooses to remunerate those services by adopting different measures for different aspects of the services received from the employee, the payments nevertheless retain the character of compensation to the employee for the skill, labour and the time put in by the employee for the benefit of the employer. The definition of salary in section 17 of the Act is couched in wide terms to take into account the remuneration paid to the employee, whether it is labelled as salary or otherwise by the employer. It is not the label given to the payment that is determinative of the question as to whether it is salary.The definition of salary in section 17(1) of the Act is an inclusive definition.Sub-clause (iv) of section 17(1) of the Act makes a specific reference to commission paid in lieu of, or in addition to, any salary or wages.**[CIT -T.Abdul Wahid andCo.,Mad.,**<u>*Sec.17.*</u>**243ITR467,125 Taxman702]**

8. The definition of salary under section 15 of the Income-tax Act, 1961, is so wide and is only an inclusive one taking in all receipts from the employer in the form of wages, commission, bonus, profits in lieu of or in addition to salary, etc.Therefore, any payment by the employer to the employee towards consideration for services rendered in the course of employment comes within the description of salary which includes perquisites as well.

[CIT -T.K.Ginarajan, Development Officer, LIC of India,Ker., <u>*Sec.15.*</u>**253ITR463,120 Taxman 646]**

9. The definition of salary in rule 3 of the Income-tax Rules,1962, is an inclusive one and,therefore,it is not restricted to what is included in the definition.The device of inclusive definition is employed by the Legislature with a view to enlarge the meaning of the ordinary words and hence the rule of interpretation of such definition adopted by the courts is to read the word defined so as to enlarge its meaning and not to restrict it to the words included in its inclusive part unless the context otherwise requires.The Income-tax paid by the employer on behalf of the employee as a part of salary of the assessee. In a word salary would in its natural import comprehend within it taxes paid on behalf of the employees. **[MitsubishiCorporation-CIT,Del.,***Sec.17.r.w.r.-3.***337ITR498, 200Taxman372]**

10. For the purpose of rule 3 of the Income-tax Rules, 1962, the term salary would not include the value of perquisites specified in section 17(2) of the Income-tax Act,1961.Perquisites and profits in lieu of or in addition to salary and wages have been treated as a part of salary.... The term wages or salary may acquire different interpretations and meanings depending upon the legislative intent and the definition clause in the enactment in question. Rule 3 specifically draws and makes a distinction between perquisites covered by section 17(2) in contradistinction to the word wages, etc., mentioned and treated as a part of salary in other clauses of section 17(1). What is covered by section 17(2) has to be excluded from salary for the purpose of ascertaining the perquisite value of rent free accommodation in view of the specific language of rule 3, Explanation clause (vi)(d). **CIT - Telsuo Mitera, Del.** *Sec.17(2).* **[345ITR256]**

Salary/fee P&H. CIT - Dr. Mrs. Usha Verma

According to Corpus Juris Secundum the word salary is usually applied to the reward paid to a public officer for the performance of his official duties.. It is paid at stated intervals. Under the Income-tax Act,1961, it is not merely defined to mean the compensation for services rendered but by providing an inclusive definition the scope of the provision has been widened. The legislation does not confine salarywithin the narrow limit of compensation for services rendered during the subsistence of a relationship of employer and employee but even includes the benefits which may become available at the end of that relationship.In clause (iv) of section 17 of the Income-tax Act it has been provided that even fees, commissions, perquisites or profits which are paid to a person in lieu of or in addition to any salary or wages shall be included in income taxable under section 15. According to Corpus Juris Secundum fee in a generic sense, implies compensation or salary; but if used in its narrow, distinctive sense it signifies the compensation for particular acts or services rendered in the line of official duties. It has been defined as a charge fixed by law for the services of a public officer, or for the use of a privilege under the control of the Government; a charge for services ; a charge or emolument . . .This meaning conforms to the provisions of section 17 of the Act and it is in consonance with the broad concept of salary as compensation for services rendered. *Sec. 15.r.w.s.17.***[172CTR98,254ITR404,120Taxman738]**

Salary, Profit SC Karamchari Union -Union of India

.....A reading of clause (1) of section 17 of the Act makes it abundantly clear that the word salary is given an exhaustive meaning as stated in sub-clauses (i) to (vii). The inclusive definition of the word salary given in section 17 provides that apart from salary received by the employee, it includes wages, any annuity or pension,any gratuity, any fees, commissions, perquisites or profits in lieu of or in addition to, any salary or wages, any advance of salary,any payment received by an employee in respect of any period of leave not availed of by him and other payments mentioned in sub-clauses (va), (vi) and (vii). These sub-clauses (i) to (vii) of clause (1) indicate that the Legislature intended to include in salary, the specified or named amount paid to the employee in respect of the services rendered by him. Sub-clause (iv) of clause (1) provides for inclusion of four types of payments in the word salary- (i) fees, (ii) commissions, (iii) perquisites, and (iv) profits in lieu of, or in addition to, salary. In common parlance fees, commissions, perquisites or payments of profits in lieu of salary may not be considered to be salary. But by this inclusive definition it has been provided so.Clause (3) provides for an inclusive definition of the phrase profits in lieu of salary.....The Income-tax Act is a self-contained code and the taxability of the receipt of any amount or allowance is to be determined on the basis of the meaning given to the words or phrases in the Act. Section 2(24) of the Act gives a wide and inclusive definition to the word income. Similarly, for levying tax on salary income, an exhaustive definition is given under section 17, which includes perquisites and profits in lieu of salary. The only exclusion provided under clause (3) is any payment referable to clauses (10), (10A), (10B), (11), (12), (13) or (13A) of section 10.In view of this specific inclusion and exclusion in the meaning of the word income and salary, the payment received by an employee has no connection with the profits of the employer. The word profits is used only to convey any advantage or gain by receipt of any payment by the employee.Webster's Comprehensive Dictionary gives the meaning of the word profit, inter alia, as advantage or benefit.Applying the general meaning of the word profits and considering the dictionary meaning given to it under sections 17(1)(iv) and 17(3)(ii), it can be said that advantage in terms of payment of money received by the employee from the employer in relation or in addition to any salary or wages would be covered by the inclusive definition of the word salary. Because of the inclusive meaning given to the phrase, profits in lieu of salary would include any payment due to or received by an assessee from an employer, even though it has no connection with the profits of the employer. It is true that the Legislature might have avoided giving an inclusive meaning to the word salary by stating that any payment received by the employee from an employer would be considered to be salary except the payments which are excluded by section 17(3)(ii), i.e., those referable to clauses (10),(10A),(10B), (11), (12),(13) or (13A) of section 10. However, it is for the Legislature to decide the same.This would not mean that by giving an exhaustive and inclusive meaning, the word profits can be given a meaning only when it pertains to sharing of profits by the employer. For the assessee, the receipt of such amount would be a profit, gain or advantage in addition to salary, even though it is not named as salary.Therefore,the word profits in the context is required to be understood as a gain or advantage to the assessee. May be that to the extent that Government or statutory corpor-ations do pay something less than what is required to be reimb-ursed, the receipt of city compensatory allowance cannot be termed as profit in common parlance.However, for income, salary, and its taxability under the Act, the meaning given by the Legislature is to be taken into consideration,as for that purpose,it is a complete code. *Sec.17 (1).* **[159CTR148,243ITR143,109Taxman1]**

Sale **Cal.** **Calcutta Electric Supply Corp. Ltd. - CIT**

The ordinary meaning of the word sale is a transaction entered into voluntarily between two persons known as the buyer and the seller by which the buyer acquires property of the seller for an agreed consideration known as a price. *Misc.* **[19ITR406]**

2.The word sale had become a word of well- recognised legal import at the time when it was introduced in the Act, and in that sense a sale could be made only by a registered instrument in the case of immovable property of the value of Rs.100 and upwards.

[CIT-Meatles Ltd.,Del.,*Transfer of Property Act.* **84 ITR37]**

3. The term sale means the transfer of a property for a price, and partition is, on principle and authority,not a transfer of the property but is merely a change in the mode of enjoyment. **[CIT - Govindlal Mathurbhai Oza, Guj.,** *Sec.171.***,138ITR711]**

Sale and barter SC CIT - Motors and General Stores (P.) Ltd.

Sale is a transfer of property in goods or of the ownership in immovable property for a money consideration. But in exchange there is a reciprocal transfer of interest in immovable property, a corresponding transfer of interest in movable property being denoted by the word barter.The difference between a sale and an exchange is this, that in the former the price is paid in money, whilst in the latter it is paid in goods by way of barter. *Sec.41(2).* **[66ITR692]**

Sale and sold Ker. CIT - Nataraj Motor Service

The expressions sale and sold are not defined in the Act but are used in their ordinary meaning. Sale according to its ordinary meaning is a transfer of property for a price. *Sec.32*.**[86ITR109]**

Sale or disposal SC Goodyear India Ltd. - State of Haryana

The mere despatch of goods by a manufacturer to his own branches outside the State does not in any way amount to a sale or disposal of goods as such; the consignment or despatch of goods is neither a sale nor a purchase. *Constitution of India,Art.269(1)(h)***[188ITR402]**

Sale or Transfer Bom. Provident Investment Co. Ltd. - CIT

A sale or transfer presupposes the existence of the property which is sold or transferred.It presupposes the transfer from one person to another of the right in property. *Sec.45.***[AIR1954Bom95,24ITR33]**

Sale proceeds . .receivable by the assessee SC Sea Pearl Industries- CIT

The phrase sale proceeds . . . receivable by the assessee in section 80HHC cannot be construed to mean sale proceeds ultimately received. *Sec80HHC.***[165CTR395,247ITR578,114Taxman618]**

Sales promotion Cal. CIT - Hindusthan Motors Ltd.

Sales promotion is a phrase of wide amplitude. In fact, advertisement and publicity may well come within the ambit of the phrase sales promotion. By using the phrase advertisement, publicity and sales promotion in section 37(3A) of the Act, the Legislature clearly indicated that a certain type of expenditure will not be allowed as deduction beyond the limit specified in that section. Sales promotion must be construed ejusdem generis with the earlier two expressions advertisement and publicity.It appears from the dictionary meaning of the word promote that costs incurred to sell goods simpliciter will not come within the meaning of the word promotion. In order to sell the goods manufactured, selling costs have to be incurred. The seller may engage middlemen or wholesalers to sell the goods; commission and brokerage will have to be paid to them for this purpose. These expenditures will have to be incurred for the purpose of selling the goods. But every type of expenditure incurred in connection with sale of goods will not come within the phrase advertisement, publicity and sales promotion. Selling costs may include many types of expenditures. Only those expenditures which are of the nature of sales promotion such as fashion shows, consumer gift offers, etc., will come within the mischief of section 37(3A) of the Act. The position has been made clear by the circular issued by the Central Board of Direct Taxes in this connection after the introduction of section 37(3A) in the statute by Circular No. 240 dated May 17. Sec.37(3A).**[63CTR110,192ITR619]**

Sales promotion Cal. CIT - The Statesman Ltd.

The expression sales promotion used in section 37(3B) of the Income-tax Act,1961, though one of wide amplitude, is not defined. It has, therefore, to be understood in its meaning in the setting in which it occurs. Sales promotion necessarily involves an element of advertisement and publicity. A manufacturer of a product may intend to further the popularity or sales by publishing and advertising or by several other modes, but the cost incurred to sell the product will not come within the purview of sales promotion. The expression sales promotion is preceded by the words advertisement and publicity in clause (i) of sub-section (3B) of section 37. Here, the legal maxim, ejusdem generis, is of aid. The maxim serves to restrict the meaning of a general word to things or matters of the same genus as the preceding particular words.Where the statute imposes restriction on advertisement, publicity and sales promotion,the expression sales promotion cannot include the selling expenses incurred in the ordinary course of the business.It only restricts such expenses as are of like nature as advertisement and publicity...The provision cannot be construed as restricting anything and everything connected with sale, such as transportation of goods for sale, collection of sale proceeds and so on. Sec.37 (3B).[198ITR582]

2.The expression sales promotion in section 37(3A) of the Income-tax Act, 1961,though of wide amplitude and undefined,is to be understood in its meaning in the setting in which it occurs. Sales promotion necessarily involves an element of advertisement and publicity.A manufacturer of a product may intend to further the popularity of his product by advertising or by several other like promotional methods. But the cost incurred in the ordinary course of sale of the product will not come within the purview of the expression sales promotion. **[CIT-Bata India Ltd.,Cal.,** *Sec.37(3A).***120CTR139,201ITR884]**

3. The term sales promotion is not to be confused with the sales actually effected. While sales promotion means the measures taken by the assessee to promote generally the sales of the products manufactured by it, or dealt with by it, individual sales made in the normal course of business on commercial terms either directly to the customer, or through its wholesale and other dealers to whom, under the terms of trade,discounts and commissions are allowed, cannot be regarded as sales promotion.**[CIT-Tuticorin Alkali Chemicals and Fertilisers,Mad.,***Sec.37 (3A).***261ITR80,136 Taxman625]**

Same business Del. Narain Singh - CIT

Determination of the nature of business depends more on the amend rather than the means and the method of procuring it cannot determine the nature of the business. The test of interlacing or dovetailing are to be applied and considered for finding out whether or not as a result of closure of one activity, the same business was continued. *Sec.73.* **[AIR1967Del75,66ITR341]**

2. The question whether two or more lines of a business may be regarded as the same business or different businesses depends not upon the special methods prescribed by the Income-tax Act for computation of the taxable income, but upon the nature of the businesses, the nature of their coorganisation, management, source of the capital fund utilised, method of book-keeping and a host of other related circumstances which stamp them as the same or distinct and the test is whether there was any interconnection, any interlacing, any interdependence and any unity at all embracing the two businesses.The existence of common management, common business organisation, common administration, common fund and a common place of business furnishes such interconnection, interlacing, interdependence and unity. If one business cannot conveniently be carried on after the closure of the other, there would be a strong indication that the two businesses constitute the same business, but no decisive inference may be drawn from the fact that, after the closure of one business, the other may conveniently be carried on. **[CIT-R.M. Maruthai Naidu and Sons,Mad.,*Sec.28.*93CTR265, 192ITR666]**

Satisfaction Kar. BPL Ltd.- DCGT

Satisfaction is necessary in terms of the material available on record.Formation of belief must be in good faith and it should not be a mere pretence. *Sec.16.G.T.Act.***[208CTR346,293ITR321,161 Taxman64]**

2. The word satisfaction has not been defined in the Act. The satisfaction by its very nature must precede before the papers/ documents are sent by the Assessing Officer of the person searched to the Assessing Officer of the third person. Mere use or mention of the word satisfaction in the order/note will not meet the requirement of the concept of satisfaction as used in section 158BD. The satisfaction has to be in writing and can be gathered from the assessment order, if it is so mentioned/ recorded, or from any other order, note or record maintained by the Assessing Officer of the person searched. The word satisfaction refers to the state of mind of the Assessing Officer of the person searched, which gets reflected in a tangible shape/ form when it is reduced into writing. It is the conclusion drawn or the finding recorded on the foundation of the material available.The Assessing Officer is satisfied when he makes up his mind or reaches a clear conclusion when he takes a prima facie view that the material available establishes undisclosed income of a third party.The Assessing Officer must reach a clear conclusion that good grounds exist for the Assessing Officer of the third person to initiate proceedings as material before him shows or would establish undisclosed income of a third person. **[CIT – Radhey Shyam Bansal,Del.,*Sec.158BD.*243CTR375,337ITR217]**

Satisfied Guj. CIT - Parmanand M. Patel

The term satisfied means make up one's mind not troubled by doubt or reach a clear conclusion on the evidence before the authority. The provision does not empower any other authority under the Act to

record satisfaction : the only authorities empowered are the AO., the AAC or the CIT (Appeals).Even if the CIT is superior to the authorities mentioned in the administrative hierarchy, when it comes to discharging the functions in a quasi-judicial capacity, one has to proceed strictly in accordance with the requirement of the provision.The CIT is not empowered to record satisfaction by Sec. 271(1) of the Act, and if he is not entitled to do so on his own,he cannot do it by directing the assessing authority. *Sec271(1).*
198CTR641,278ITR3,149Taxman403]

Scholarship Mad. CIT - V.K.Balachandran

Section 10 sets down various items of income which are wholly outside the purview of taxable total income.The opening words of the section show very clearly that the items listed there under are undoubtedly receipts of an income-character, but they are nevertheless to be excluded from the computation of taxable income. Scholarship to meet the cost of education falling under clause (16) has been included in section 10, not because scholarship bears an income-character. By scholarship, as ordinarily understood, we mean anything which makes education free of charge, or at a concessional rate of fees. In sectio 10(16), however, scholarship is not used in that sense of something in education opportunity which is given free. The basis postulate of a scholarship in clause (16) is that it is an income receipt.Nevertheless it is excluded from the total income by being brought under section 10.The view of the Income-tax statute of a scholarship, therefore, differs from the popular, or dictionary, view of a scholarship.Whereas under the popular view,scholarship is education made available gratis,the sense in which the same expression is used in the I.T. Act is positive payment made to a scholar for pursuit of his education.The considerations which make up the concept of a scholarship for meeting the cost of educationin section 10(16) are that the payment is intended to be an income receipt in the hands of the scholar and that whatever is paid is intended to meet the cost of education of the recipient. Since the purpose of the payment is to meet the cost of education,the question whether the quantum of payment is adequate or inadequate, or, is or is not in excess of the requirements are all beside the point.It is enough if the whole object of the payment is to meet the cost of education of a person and no further enquiry is called for in order to exclude the amount from the taxable income under section 10(16). If the payment is only for the cost of education, the fact that the recipient does not spend the whole of the amount or saves something out of it or utilises it for other purposes would not detract from the character of the payment being one for scholarship. Sec.10(16).[147ITR4]

Science **Del.** **E.P.W. Da Costa - Union of India**

The word science is a very general word. Statistical tables complied after analysing masses of numerical data can be said to be commercial or scientific knowledge. Sec.80-O.[121ITR751]

Search **P&H.** **CIT - Ramesh Chander**

..... The word search is used because, in the nature of things, in a given case, when section 132 of the Act applies, the articles to be searched and seized have been concealed by the person concerned from the Income-tax department for the purpose of assessment of income-tax. The word search has been used in this sense and not in any other sense. Sec.132.[93ITR450]

2. The word search in section 132(1)(c)(i), considering the object and scope of the section, should not be given a far too technical meaning. The word search has varied meanings and it should be given the general meanings, to look for or seek which are well-known meanings attributable to the word. **[Assainar - ITO, Ker.,** Sec.132(1).**101ITR854]**

Search initiated **Kar.** **CIT - Wipro Finance Ltd.**

The expression search initiatedhad to be interpreted to mean the commencement and conducting of the initial search, i.e., the first search in the case. Sec.132.**[218CTR386,323ITR467,176Taxman 233]**

Security **All.** **CWT - Janki Kishori Devi**

The word security occurring in clause (xxii) of section 5(1) of the Wealth-tax Act,1957, has not been defined in the Act. In construing the term, those provisions of law which specifically regulate the various types of Government securities can be considered. In order to remove the difficulties in the working of the Indian Securities Act,1920,a comprehensive enactment, namely,the Public Debt Act,1944, was enacted.Under section 2 of the Public Debt Act, the expression Government in relation to any Government security means the Central or State Government issuing the security.

Sec.5(1).WT.Act.**[192ITR229,59Taxman206]**

Seizure **P&H.** **Ramesh Chander - CIT**

Seizure is an expression which implies a forcible exaction or taking possession from either the owner or one who has the possession and who is unwilling to part with possession. Sec.132.**[93ITR244]**

2. Though the word seizure normally implies taking possession forcibly, it also means taking possession under the authority of law. That the person who is in possession may not object to the handing over is no criterion for holding that such an authority is not necessary to deprive a person of his right to be in possession. Even if such other department is holding the property in its own right and in enforcement of any other provision of law still, on the authority of a warrant, section 132(3) could be invoked and such other department prohibited from handing over the same to any other person when the purpose for which they were holding the same was no longer available.Hence it is not correct to state that because the goods are known to be in a particular place and in the custody of another department of the Government the power under section 132 cannot be invoked. **[Gulab and Co.-Supdt. of Central Excise(Preventive),Trichy, Mad.,** Sec.132.**98ITR581]**

Seller **Pat.** **State of Bihar - CIT**

The word seller as referred to in section 44AC of the Income-tax Act,1961, has to be read not only in the context of the word buyer but also upon taking into consideration the purpose for which the said provision as also section 206C had been inserted by Parliament. Sec.44AC.**[202ITR535]**

Separate business **Mad.** **Ginning and Oil Mills - CIT**

If in the course of carrying on any business,speculative transactions are entered into along with non-speculative transactions, by reason of the statutory fiction in Explanation 2 to section 28 of the Income-tax Act,1961, the speculative transactions will constitute a separate business.In other words, where a trader carries on a regular business in spot dealings in a commodity and also indulges in speculative transactions in the same commodity, while in point of fact the business of the assessee may be regarded as an integrated business combining in itself not only spot transactions but also forward dealings,by virtue of Explanation 2 to section 28, the speculative transactions must be separated and regarded as constituting a separate business. <u>*Sec.28.*</u>**[133ITR85]**

Servant /agent SC Ram Prashad - CIT

There is no doubt that for ascertaining whether a person is a servant or an agent, a rough and ready test is whether, under the terms of his employment, the employer exercises a supervisory control in respect of the work entrusted to him. A servant acts under the direct control and supervision of his master. An agent, on the other hand, in the exercise of his work is not subject to the direct control or supervision of the principal, though he is bound to exercise his authority in accordance with all lawful orders and instructions which may be given to him from time to time by his principal. But this test is not universal in its application and does not determine in every case, having regard to the nature of employment, that he is a servant. <u>*Misc.*</u>**[86ITR122]**

Serve All. Sri Niwas - ITO

The retention of the word serve in sub-section (1) by the Legislature even though section 34 was amended several times leads also to the inference that the Legislature did not intend that serve must be read as equivalent to issue.Under section 63 of the Income-tax Act, a notice to an association of persons had to be addressed to the principal officer thereof. A notice served on the son of one of the members of the association would not be a proper service of notice on the association. <u>*Sec.282.*</u>**[30ITR381]**

Served with notice Lah. Lala Har Kishen Das - CIT

The words served with notice in section 66(2) of the Indian Income-tax Act,1922 .....do not mean served with a written notice of such order or served with a copy of the detailed order giving reasons.

<u>*Sec.282.*</u>**[2 ITR 484]**

Service Pat. Associated Cement Co. Ltd. - CIT

.....loading of packed cement into trucks and wagons is a work as it is commonly understood and the resultant thing is the accomplishment of the loading. In a very broad sense, a work done by one person is service rendered to another and one of the dictionary meanings of the word service is work. *Sec.194C.* **[11CTR211,120ITR444,55Taxation139]**

Service of notice **Guj.** **Shanabhai P. Patel - R.K.Upadhyaya, ITO**

The words service of notice or issuance of notice in section 34 have no fixed connotation but are inter-changeable. The same meaning should be given to the words issue of notice in section 148 and service of notice in section 149. Under the Act of 1961 also there are no two distinct and separate stages of issue of notice and service of notice.Notice of reassessment is issued to the assessee when it is served on him. A notice of reassessment issued against the assessee before limitation but served on the assessee after limitation would be without jurisdiction, void and ineffective. *Sec.148.***[96ITR141]**

Services **AAR** **Geofizyka Torun Sp. Z. O. O**

That there was no compelling reason to assign a narrow and restricted meaning to the expression services in section 44BB by resorting to Explanation 2 to section 9(1)(vii) and confine it to services other than technical, consultancy or managerial services. In the absence of any words of limitation or exclusion, the word servicesin section 44BB had to be understood in its plain and ordinary sense.The word services followed by an expansive phrase in connection with were relatable to prospecting for and exploration of mineral oils.That meant that all services associated with prospecting for and exploration activities were brought within the scope and reach of section 44BB. *Sec.44BB.***[320ITR268,186 Taxman213]**

Set up **Mad.** **Ramaraju Surgical Cotton Mills Ltd.- CWT**

Unless a factory is erected and the plant and machinery are installed therein, it cannot be said to have been set up. The resolutions of the board of directors, the orders placed for purchasing the machinery and the licence obtained from the Government for constructing the factory, are merely initial stages towards the setting up,however,necessary and essential they may be to further the achievement of the end.It is not, however, the actual functioning of the factory or its going into production that can alone be called setting up of the factory. The setting up is a stage anterior to the commencement of the factory.The proper meaning of the expression set up would be ready to commence business. *Sec.5(1)(xv).WT.Act.* **[46ITR820]**

2. The word set up in the principal clause of section 5(1)(xxi) is equivalent to the word established,but operations for establishment cannot be equated with the establishment of the unit itself or its setting up. The applicability of the proviso has, therefore, to be decided by finding out when the company commenced operations for establishment of the unit, which operations must be antecedent to the actual date on which the unit is held to have been set up for the purpose of the principal clause.**[CWT - Ramaraju Surgical Cotton Mills Ltd., SC,** *Sec.5(1)WT.Act*.**63ITR478]**

3. The words set up in the context in which they are used refer to an industrial undertaking which commenced manufacture in the year in which the claim for deduction is made or had commenced such manufacture within a period of three years preceding the year of claim. An industrial undertaking which had been set up long prior, and which had commenced manufacture several years earlier, clearly cannot be regarded as an industrial undertaking set up in the year in which the ownership changes. It is not the year in which the legal entity that owns the undertaking is formed that matters; what is material is the year in which the industrial undertaking is set up.Every change in the ownership of that undertaking does not result in the same industrial undertaking being set up over and over again. **[Mad.,CIT-Ponds India Ltd.,** *Sec.37(3D).***174CTR235, 253 ITR686,122Taxman706]**

4. The words used in section 37(3D) of the Income-tax Act,1961, are has set up an industrial undertaking. On a plain reading, these words show the requirement that an industrial undertaking must be set up. The words set up as defined in various dictionaries means established and organised. The words set up are not qualified by the word new.Hence,the benefit of section 37(3D) would also be available to the assessee,which was an industrial establishment already set up but had launched a new product. The requirement of the section was not that of setting up of a new undertaking but would be satisfied where the industry has undertaken the production of a new article. **[CIT-Agra Beverages Corporation Pvt.Ltd.,All.,***Sec.37(3D).***219CTR662, 300ITR286]**

Setting up of a business Guj. CIT - Sarabhai Sons Pvt. Ltd.

There is a clear distinction between commencing a business and setting it up.For the purpose of section 3(1)(d) what is required to be considered is the setting up of a business.When a business is established and is ready to start business it can be said to be set up. The business must be put into such a shape that it can start functioning as a business or a manufacturing organisation.

Sec.3(1)(d). **[90ITR318]**

Setting up of business **Bom.** **CIT -Forging and Stamping Pvt. Ltd.**

When a business is established and is ready to commence business, then it can be said of that business that it is set up. In the case of a manufacturing company, the installation and erection of machinery is not sufficient by itself and till some end product is or can be obtained, it cannot be said that the company is ready to commence production and it is the readiness to commence production which is equivalent to the setting up of the business. *Sec.28.* **[119ITR616]**

Settled **Mad.** **R.Chinnaswami Chettiar - CIT**

There is no basis for construing the word settled in the definition of the term speculative transaction in section 43(5) of the Income-tax Act,1961,as restricted to the settlement of a contract before its breach.Though on the breach of a contract, the cause of action for the claim of damages is based on its breach, it does not mean that when the claim is settled, the settlement is not of the contract itself but of the breach alone. The rights and liabilities of the parties flow from the contract which was broken.The emphasis in section 43(5) of the Income-tax Act,1961,is on the words periodically or ultimately settled otherwise than by actual delivery.The word settled is used in this part of the section without any restriction as to whether it was before or after breach of a contract. Whether the settlement was before or after breach of the contract is immaterial if actual delivery of the goods is absent. Even where the contract is highly speculative and amounts to a wagering contract, if it is settled by actual delivery, it will not be a speculative transaction for the purpose of section 43(5). The section dispenses with all other tests except whether there was actual delivery or transfer of the commodity when the contract is settled. *Sec.43(5).* **[96ITR353]**

Settlement **SC** **CIT - Anjum M.H. Ghaswala**

Though the term settlement may have a very wide dictionary meaning and, in the absence of a statutory definition, generally the word settlement could give the Settlement Commission sufficient power to arrive at a settlement which it deems fit, when the statute qualifies the expression with the mandatory words in accordance with the provisions of this Act, the width of the term settlement becomes subject to the mandate found in the section, which would mean that while the Commission has sufficient elbow-room in assessing the income of the applicant under section 245D(4) it cannot make any order with a term of settlement which would be in conflict with the mandatory provisions of the Act. <u>Sec.245D.</u>**[171 CTR1,252ITR1,119Taxman352]**

Settlement of a contract and the settlement of claims **AP.** **CIT -Andhra Oil and Fertilisers Co.**

There is a clear distinction between the settlement of a contract and the settlement of claims arising out of a breach of contract. While the former would be a speculative transaction, the later cannot be. Where the contract is settled by payment of difference before the occurrence of the due date, it is tantamount to a settlement of contract within the meaning of section 43(5) of the Income-tax Act, 1961.However, where the payment is effected to satisfy the claims arising out of a breach of contract, there is no settlement of contract as contemplated under section 43(5) and as such they cannot be termed as speculative transactions within the meaning of section 43(5). <u>Sec.43(5).</u>**[34CTR104,143ITR661,69Taxation219,13Taxman64]**

Settlement of a contract and the settlement of a dispute **Raj.** **CIT - Dina Lal Gupta**

A distinction has to be drawn between a settlement of a contract and the settlement of a dispute arising out of a breach of a contract. A contract is said to be settled if instead of effecting the delivery or transfer of the commodity envisaged by the contract, the promisee in terms of section 63 of the Indian Contract Act, 1872, accepts instead of it any satisfaction which he thinks fit. In other words, the contact is said to be settled when it is either performed or the promisee dispenses with or remits, wholly or in part, the performance of the promise made to him or accepts instead of it any satisfaction which he thinks fit. In cases where there is a breach of the contract and by virtue of section 73 of the Contract Act, the party suffering by such breach becomes entitled to receive from the party who has broken the contract, compensation for any loss or damage caused to him thereby, the settlement of such a dispute with regard to damages for the breach of the contract cannot be regarded as a settlement of the contract. What is settled by the award of damages is the dispute arising between the parties as a result of the breach of the contract. This would mean that a contract can be said to be settled before its breach and in cases where there has been a breach of the contract and any settlement takes place between the parties to the contract with regard to compensation or damages, it cannot be regarded as the settlement of the contract but it is a settlement of the dispute with regard to damages on account of the breach of the contract. For the purpose of deciding whether a particular transaction is a speculative transa-ction under section 43(5) of the Income-tax Act,1961, the transaction falling in the first category, namely, where there is a settlement of the contract, can be regarded as a speculative transaction and a transaction falling in the second category, namely, where there is a breach of the contract, it cannot be regarded as a speculative transaction. <u>Sec.43(5).</u>**[170ITR583, 33Taxman129]**

Settlement of a contract and the settlement of a..... Raj. CIT - Ganesh Das Ram Swaroop Kakani

A distinction is to be drawn between a settlement of a contract and the settlement of a dispute arising out of the breach of a contract. A contract is said to be settled if, instead of effecting delivery or transfer of the commodity envisaged by the contract, the promisee, in terms of section 63 of the Contract Act, accepts, instead of it, any satisfaction which he thinks fit. In other words, the contract is said to be settled when it is either performed or the promisee dispenses with or remits, wholly or in part, the performance of the promise made to him or accepts, instead of it, any satisfaction which he thinks fit. In cases where there is a breach of the contract and by virtue of section 73 of the Contract Act, the party suffering by such breach becomes entitled to receive from the party who has broken the contract, compensation for any loss or damage caused to him thereby, the settlement of such a dispute with regard to damages for the breach of contract cannot be regarded as a settlement of the contract. What is settled by the award of damages and the acceptance of the same by the aggrieved party is the dispute arising between the parties as a result of breach of the contract. This would mean that a contract can be said to be settled before its breach and in cases where there has been a breach of the contract and any settlement takes place between the parties to the contract with regard to compensation or damages, it cannot be regarded as a settlement of the contract but it is a settlement of the dispute with regard to damages on account of a breach of the contract.For the purpose of deciding as to whether a particular transaction is a speculative transaction under section 43(5) of the Act, a transaction falling in the first category, viz., where there is a settlement of the contract, can be regarded as a speculative transaction and a transaction falling in the second category, viz., where there is a breach of the contract and the dispute with regard to damages or compensation for the breach of the contract is settled, it cannot be regarded as a speculative transaction. <u>*Sec.43(5).*</u>**[181ITR93]**

Settlement of cases **All.** **Smt. Neeru Agarwal - Union of India**

The very use of the words settlement of cases is indicative of the fact that the provision has been made to settle the case in its entirety for ever and leave no issue open for subsequent decision. <u>*Sec.245I.*</u> **[231CTR153,330ITR422,187Taxman198]**

Shall **Ker.** **CIT - Malayalam Plantations Ltd.**

....the term shall need not in all contexts, circumstances and situations be treated as indicating a mandatory rule is a proposition by now well settled. That, in every case, where the question is one of consequence of non- observance of a provision, the court has to decide the legislative intent, and to decide this, the court has to consider not only the actual words used but the scheme of the statute, the intended benefit to the public of what is enjoined by the provisions and the material danger to the public by the contravention of the scheme, was observed by the Supreme Court in Banwarilal Agarwalla-State of Bihar.The examination of the provision, therefore, must be with a view to ascertain the intent of the legislature where it is the provision of a statute that arises for consideration and that of the rule making authority when it is a rule that calls for consideration. <u>*Sec.33A r.8.*</u>**[103ITR835]**

2. The question whether a provision is mandatory or directory depends upon the intent of the Legislature and not upon the language in which the intent is clothed. The court must carefully get into the underlying idea of the provisions and ascertain the purpose to be achieved notwithstanding the text of the provision. The use of the word shall does not conclude the matter. The word is no doubt obligatory in its ordinary import. But it need not be given that connotation in each and every case, and a provision can be interpreted as directory despite the use of the word shall, with reference to the context, the subject matter and the object of the statute in question. **[CIT-AntonyMendez,Ker.,**<u>*Sec.7.WT.Act.*</u>**191ITR346]**

3. Where the question is whether a provision which uses the word shall is mandatory, one of the principles is to find out the consequences of reading the provision as mandatory and, if the consequence is to render the statutory provision harsh, onerous or arbitrary, such a reading should be avoided. Another principle is to read down a statutory provision, so as to make it a valid provision and prevent its nullification as unconstitutional; the third principle applicable is to read the provision in consonance with the object and scheme of the statute and thus limit the operation of the particular provision to effectuate the said statutory object. **[CWT-S. Jindal,Kar.,**<u>*Sec.7.WT.Act.*</u>**100CTR217,194ITR 539]**

4. The construction of the expression shall depends on the provisions of a particular Act, the setting in which the direction is given the consequences that would flow from the infringement of the direction and such other considerations. Having regard to the other provisions of the Act regarding filing of the return or revised return or rectifying the defects in the return, the provisions of section 12A of the Income-tax Act,1961, are directory in the sense that the Assessing Officer is not powerless to allow an assessee to file the audit report, if not filed along with the return, any time before the completion of the assessment. **[CIT -Rai Bahadur Bissesswarlal Motilal Malwasie Trust, Cal.,**<u>*Sec.12A.*</u> **195ITR825]**

5. Rules made under a statute must be treated for all purposes of construction or obligation exactly as if they are in the Act and are to be of the same effect as if contained in the Act, and are to be judicially noticed for all purposes of construction or obligation. While the expression shall need not always be indicative of the mandatory nature of a provision and the word may need not always be indicative of its directory nature, ordinarily the word shall may be taken to be indicative of the intention of the Legislature or the rule-making authority to make the provision mandatory.The intention has to be gathered on the basis of the general purpose of the statute, the purpose of the particular provision,the inter-relationship between the two,the language used and the consequences of interpreting the provision to be mandatory or directory and the mischief, if any, sought to be avoided. **[Ajay Kumar Saharia-CWT, Gau.,**<u>*Sec.7.WT.Act.*</u>**113CTR44, 206ITR98]**

6.It is true that the word shall has been used in connection with the issuance of an intimation but it is well-established that the construction of the expression shall depends upon the provisions of the Act,the setting in which the direction is given and the consequences that would follow from the infringement of the direction and other such considerations.The context in which the word shall has been used in section 143(2) has to be read in the background of the proviso to the section and that is that where there is no scope for any adjustments in terms of the proviso, there would be no scope for sending any intimation. **[Modern Fibotex India Ltd.-DCIT,Cal.,**<u>*Sec.143.*</u>**125CTR323,212ITR496]**

7. Section 201(1A) of the Income-tax Act,1961, makes it clear that the levy of interest is mandatory.It is true that use of the expression shall is not always determinative of the fact whether a provision is directory or mandatory in nature, but the context in which expression shall is used in section 201(1A) makes it clear that the levy is mandatory.The purpose of the levy is to claim compensation on the amount which ought to have been deducted and deposited and has not been done.The ultimate liability for tax being not there (since the firm which received the interest from the assessee had paid tax on such interest) did not dilute the requirements for the non-compliance of which interest is levied under section 201(1A Section 201(1A) of the Income-tax Act,1961, makes it clear that the levy of interest is mandatory. It is true that use of the expression shall is not always determinative of the fact whether a provision is directory or mandatory in nature, but the context in which expression shall is used in section 201(1A) makes it clear that the levy is mandatory. The purpose of the levy is to claim compensation on the amount which ought to have been deducted and deposited and has not been done. The ultimate liability for tax being not there (since the firm which received the interest from the assessee had paid tax on such interest) did not dilute the requirements for the non- compliance of which interest is levied under section 201(1A). **[CIT - Dhanalakshmy Weaving Works,Ker.,** <u>*Sec.201(1A).*</u>**160CTR374,245ITR13, 109Taxman395]**

9. The expression shall used in sections 234A,234B and 234C cannot be construed as may.Prior to the Finance Act,1987,the corresponding sections pertaining to imposition of interest used the expression may ; but the change brought about by the Finance Act, 1987, is a clear indication that the intention of the Legislature was to make the collection of statutory interest mandatory.That expression is used deliberately. **[CIT-Anjum M.H.Ghaswala,SC,** <u>*Sec.234.*</u>**170 CTR424,252ITR1,119Taxman352]**

10. The word shall is ordinarily mandatory, but it is sometimes not so, if the context or intention otherwise demands. When a statute uses the word shall, prima facie, it is mandatory, but the court may ascertain the real intention of the Legislature by carefully attending to the real scope of the statute. ....There are several provisions where the Legislature has made a distinction between interest payable and penalty imposable.Under section 201(1A) of the Income-tax Act,1961,the levy of interest is a compensatory measure for withholding tax which ought to have gone to the exchequer.The provision makes it clear that the levy is mandatory. It is true that the use of the expression shall is not always determinative of the fact whether a provision is directory or mandatory in nature.But the context in which the expression shall is used in section 201(1A) makes it unambiguously clear that the levy is mandatory. **[CIT -Prem Nath Motors(Pvt.)Ltd., Del.,** *Sec.201(1A).***170CTR424,253ITR705]**

11. The use of the word shall in a statutory provision,though generally taken in a mandatory sense, does not necessarily mean that in every case it shall have that effect, that is to say,unless the words of the statute are punctiliously followed,the proceeding or the outcome of the proceeding would be invalid.On the other hand, it is not always correct to say that where the word may has been used,the statute is only permissive or directory in the sense that non-compliance with those provisions will not render the proceedings invalid. **[CITPunjabFinancialCorporation,P&H.,** *Sec.32AB.***172CTR561,254ITR6,121Taxman656**

12. Word shall in a statutory provision though generally taken in a mandatory sense, not necessarily have that effect in every case. **[CIT - Magnum Export (P) Ltd.,Cal.,** *Sec.80HHC.***183CTR75, 262ITR10,130Taxman702]**

13. The expression shall used in the section cannot be considered as may.The intention of the Legislature by bringing the provisions of section 234C by the Amending Act,1987, was to make the collection of statutory interest mandatory.To judge the constitutionality of the provision, the generality of its provisions are to be looked into and not by the freaks and exceptions. **[Mrs. Prabha Lal - CIT,Pat.,** *Sec.234.***190CTR99,269ITR212,137Taxman277]**

Shall not be allowed   Kar.   Addl. CIT - India Tin Industries P. Ltd.

The expression shall not be allowed in section 34(3)(a) of the Act is a clear indication of the intention of the Legislature that compliance with the conditions specified therein is mandatory and has, therefore, to be fulfilled.The conditions laid down are: (i) the amount specified therein has to be debited to the profit and loss account of the relevant previous year and credited to the reserve account, and (ii) this reserve fund so created has to be used by the assessee in the manner and for the period specified therein.In order to attract the provisions of section 154 of the Act,there must be a mistake and it must be a mistake apparent from the record. Sec.34. **[57CTR70,166ITR454,29Taxman128]**

Shall Presume SC P. R. Metrani - CIT

Shall presume leaves no option with the court not to make the presumption.The court is bound to take the fact as proved until evidence is given to disprove it. In this sense such presumption is alsorebuttable. Misc.**[203CTR290,287ITR209,155Taxman186]**

Share SC CIT - Standard Vacuum Oil Co. Ltd.

A share is not a sum of money, but represents an interest measured by a sum of money and made up of diverse rights contained in the contract evidenced by the articles of association of company.

Sec.187.**[AIR1966SC1393,58ITR685]**

2. The word share in the expression definite shares in section 41(1) of the Indian Income-tax Act,1922, includes not only a definite fraction or proportion of the income, but also a definite portion or part of the income, e.g., a fixed sum of money out of the income.

[Bankim Ch. Datta-CIT,Cal., Sec.164.**62ITR239]**

3. Share is a right to receive a proportion of the profits of the company and it is assessed on winding up and all other benefits to membership combine the obligation to contribute to its liabilities, all measured by a certain sum of money which is the nominal value of the share, and all subject to control by the regulations of the company. Therefore, a share can be either in the first phase or stage or in the second phase or stage. It remains either in its shell as a part of the capital or resides in a shareholder.It cannot lie suspended in any intermediate phase or stage.Hence it is necessary to find out the modus operandi of the transit from one phase or stage to another to appreciate the meaning of the word issue, which ordinarily means sending out or putting out.Section 30 of the Companies Act furnishes the modus operandi or mechanism for the transformation and ultimately the completion of the transit. Therefore,though the subscription to the memorandum or an agreement to take a share may keep the share in preparedness by fulfilling the preliminary requirements for its exit and transit to the owner, it still remains in the womb or shell of the company, though it may be then in an animated condition. The exit and transit, however, takes place when the entry in the register is made. Hence.....issue of the share takes place when entry of the name of the subscriber or the successful offer or is made in the register of members. It follows from the foregoing observation that a share is issued when it finds an owner......The word share has more than one meaning in common parlance. The Indian Companies Act defines shares: Share means under section 2(16) share in the share capital of the company and includes stock except when a distinction between stock and shares is expressed or implied.Therefore the statutory meaning of share covers the three phases of the share, share when it is a part of the share capital still remaining unexploited by the company, share when it is exploited by the company finding a share-holder and lastly when the share is converted into stock.The first phase arises because under the company law Every company limited by shares has nominal or authorised or registered share capital. This capital is one of the essential features in the company's constitution. It is to be mentioned in the memorandum of association and the capital so mentioned is to be divided into shares of a fixed amount. The capital is usually fixed at some round figures according to the requirements of the company assessed by the promoters of the company. .... first part of the definition of the word share in the Companies Act refers to the share in this limited sense when the share is still in the womb of the company or in the shell of the company and has no shareholder.The second phase arises when it attracts section 28 of the Indian Companies Act. Section 28 of the Indian Companies Act is as follows: (1) The shares or other interests of any member in a company shall be movable property, transferable in the manner provided by the articles of the company.(2) Each share in a company having a share capital shall be distinguished by its appropriate number. Therefore, the share when it becomes associated with a member becomes a movable property. It is however not movable property whose transfer is solely regulated by the Sale of Goods Act. Its transfer is also governed by the Companies Act and/or articles of the company. Each share again bears a distinguishing number. It may be noticed that certificate of share is not the shares or a share.**[Shree Gopal Paper Mills Ltd.-CIT,Cal.,**Sec.2,28.Companies Act.**64ITR233]**

Share holder **Bom.** **Shree Shakti Mills Ltd. - CIT**

The shareholder mentioned in section 18(5) of the Indian Income-tax Act is the person who owns certain shares and who is shown as a shareholder in the register of the company and it is only the shareholder of a company to whom dividends are paid who is entitled to the procedure of processing permissible under sections 16(2) and 18(5). *Misc.*[16ITR187]

2.The shareholder mentioned in section18(5)of the Indian Income-tax Act, 1922, is a person who owns shares and who is shown as a shareholder in the register of the company.A person who had purchased certain shares but did not get his name registered in the books of the company is not entitled to the benefits of section 16(2) and 18(5) in respect of the income from dividend on these shares because he is not registered as a shareholder........ The shareholder mentioned in section 18(5) of the Indian Income-tax Act, 1922, is a person who owns shares and who is shown as a shareholder in the register of the company.A person who had purchased certain shares but did not get his name registered in the books of the company is not entitled to the benefits of section 16(2) and 18(5) in respect of the income from dividend on these shares because he is not registered as a shareholder. **[Jaluram Bhikulal Firm of Itwara-CIT,Nag.,*Misc*.22ITR490]**

3. A person who has purchased shares in a company under a blank transfer and in whose name the shares have not been registered in the books of the company is not a shareholder in respect of such shares within the meaning of section 18(5) of the Income-tax Act, notwithstanding his equitable right to the dividend on such shares, and is not, therefore, entitled to have this dividend income gross up under section 16(2) of the Act by the addition of the Income-tax paid by the company in respect of those shares,and claim credit for the tax deducted at source,under section 18(5) of the Act. **[Howrah Trading Co. Ltd.-CIT,SC,*Misc.*36ITR215]**

4. The expression shareholder means a shareholder registered in the books of the company.A Hindu undivided family can be a registered shareholder of the company. **[CIT-Shakuntala,SC, *Sec.2(22)(e).*43ITR352]**

5.The term share holder means a shre-holder whose name is recorded in the books of the company. **[Rameshwarlal-CIT,Ass., *Sec.2(22)(e).*58ITR271]**

6. Section 2(6A)(e) of the Indian Income-tax Act,1922, must receive a strict construction.When the section speaks of shareholder,it refers to the registered shareholder and not to the beneficial owner. **[CIT - Sarathy Mudaliar, SC,*Sec.2(22)(e).*83ITR170]**

7. Shareholder mean only a registered shareholder.Only the person whose name is entered in the register of shareholders of the company as the holder of the shares who can be said to be a shareholder. **[Rameshwarlal Samwarmal-CIT,SC,** <u>*Sec.2(22)(e).*</u>**AIR1980SC372,14CTR372,122ITR1,3Taxman1]**

Shares are not beneficially held by the persons **Bom.** **Zycus Infotech P. Ltd. - CIT**

In Explanation 1 to sub-section (9) inserted with effect from April 1, 2000,the present tense is used with an injunction that the shares are not beneficially held by the persons who hold the shares in company.The present tense cannot be assumed to describe the status of the shareholder as the owner, but the status of the shares which are beneficially held. On this interpretation the language of the section can only be understood to describe the date on which the undertaking was set upas applicable only for those who are setting up the undertaking after the new provision, so that in case of others, the date has to be understood at best, as on April 1,200, the date on which the law was brought in the statute. <u>*Sec.10A.*</u> **[235CTR113, 331ITR 72,196Taxman594]**

Shift **SC** **East India Hotels Ltd. - CIT**

The expression shift is not defined in the Income-tax Act, 1961. The expression as defined in section 2(r) of the Factories Act,1948, is relevant and is with reference to the workers,not the establishment. <u>*FactoriesAct.*</u>**[136CTR246,223ITR1,135Taxation481,89Taxman 395]**

Ship **Mad.** **Chola Fish and Farms P. Ltd. - CIT**

The word ship is not defined in the Income-tax Act, 1961. Normally, in such circumstances, the word should not be understood in any technical sense, but it has to be construed in the popular sense. The word will have to be understood in the context in which it occurs in the statute,the purpose of the provision and other relevant considerations.For the purpose of section 32 of the Income-tax Act, 1961, which provides for depreciation in respect of certain classes of vessels,Part I of Appendix I to the Income-tax Rules,describes ship as a class of vessel and under the heading of Ships, four different classes of ships are mentioned: (1) Ocean-going ships (i) Fishing vessels with wooden hull; (ii) Other ships, (2) Vessels ordinarily operating on inland waters (i) Speed boats; (ii) Other vessels. It is

evident from the description of ship that trawlers would come within the meaning of the word ship.Except that section 33 is applicable to priority industries whereas depreciation is allowed in respect of every category and not restricted to priority industries there is no distinction between sections 32 and 33 in so far as the meaning of the word ship is concerned.Trawlers are covered by the expression ship in section 33. Sec.33.
[132CTR358,217ITR609]

2. ...the word ship has to be interpreted as per the natural meaning which it conveys.Section 3(55) of the General Clauses Act, 1897, gives an inclusive definition of the word ship which is as follows :Ship shall include every description of vessel used in navigation not exclusively propelled by oars. Thus ship means all types of vessels which have one common factor or feature and this common factor or feature is that it is used in navigation.The most important and significant word,therefore, is navigation, since navigation is the crux of the whole thing. The meaning of the verb navigate is given as to sail over or up or down (sea, river). Hence, conveying goods, men and material from one place to another is the sine qua non for being qualified as a ship.Pontoons are not used to go from one place to another place alone.Their main use is that they serve the purpose of a platform, on which the work of constructing bridges can be undertaken.Their usefulness commences after they reach that workspot. In contradistinction to the pontoon, for other vessels, their main and indeed the only use is to transport across the water. These vessels navigate from one point to some other point across the water.Thus, they navigate in the fullest sense of the word. A pontoon, on the other hand, is nothing but a platform for work. Just because it floats on water, it cannot be described as a ship. Pontoons are platforms used for working in the water and nothing else. Many things float in the water, but just because they float, they cannot be described as ships.A contrary legislative intent cannot be attributed or inferred, and such a broad meaning cannot be given to the word ship so as to include pontoons within its sweep. **[Gammon India Ltd. – CIT, Bom.,** *General Clauses Act r.w.s.33.*
228ITR691]

3. ..the word ship has not been defined in the Income-tax Act,1961. However, in the table of rates at which depreciation is admissible, the meaning of ship has been given in the Appendix to the Income-tax Rules,, where ship is defined to include : 1. Ocean-going ships(i) Fishing vessels with wooden hull; (ii) Dredgers, tugs, barges, survey launches and other similar ships used mainly for dredging purposes; (iii) Other vessels; and (2) Vessels ordinarily operating on inland waters (i) Speed boats; (ii) Other vessels. This definition makes it clear that ships would include tugs. However, pontoons have not been specifically included within the meaning of ship. However, having regard to the types of vessels that are entitled to depreciation like dredgers,tugs, barges, etc, an expanded meaning has to be given to ships, and pontoons will come within the expanded meaning. According to the Concise Oxford Dictionary, edition, a pontoon is a flat-bottomed boat used as ferry boat or to carry lifting-gear, etc. Having regard to the width of the meaning given to the word ships in the depreciation table of the Income-tax Rules, 1962, it is clear that a flat-bottomed boat used as ferry-boat will clearly come within the description vessels ordinarily operating on inland waters.Therefore, pontoons and tugs are ships.

[CIT - Digvijay Cement Co. Ltd., SC,<u>*Sec.33.rule5.*</u>**150CTR206, 232 ITR709,100Taxman374]**

Similar association Guj. Sports Club of Gujarat Ltd. - CIT

Section 44A of the Income-tax Act,1961, which begins with a non obstante clause applies to any trade, professional or similar association and but for the use of the words other than an association or institution referred to in clause (23A) of section 10, even the professional associations referred to in section 10(23A) would have derived the advantage of section 44A.The intention of the Legislature was to give the benefit of section 44A to all professional associations other than those referred to in section 10(23A). They are clearly words of limitation not intended to enlarge the scope of the expression similar association beyond its meaning in section 28(iii). <u>*Sec.44A.*</u>**[67CTR233,171ITR504,37Taxman38]**

Similarly apportioned SC CIT-T.V.Sundaram Iyengar & Sons (P.) Ltd.

The words similarly apportioned in Explanation 2 to section 23A mean apportioned with reference to the amounts of profits and gains attributable to the two parts of the company's business. Similarly apportioned means simply similarly split up. The dividends have to be split up in the same ratio as the industrial and non-industrial profits bear to each other after the total profit is split into two parts, industrial and non- industrial. It is not open to the company to split up and apportion the dividends to the profits of the two segments in such manner as it finds convenient or thinks fit.

<u>*Sec.104,109.*</u> **[101ITR764]**

Six per cent. per annum Mad. CIT - Simpson & Co.

The words six per cent. per annum are ordinarily applied to calculation of interest and in similar contexts but the words per annum could be inappropriate in a taxing statute levying tax on the income earned during the previous year which is not necessarily a period of twelve months though it would ordinarily be a period of twelve months. The words per annum in section 84 of the Income-tax Act, 1961, have been added only to ensure that the assessee would get for each of the five years during which the relief under section 84 is available, six per cent. on the capital employed. These words cannot be understood as indicating any broken period during which the assets were used. Sec.84.[122ITR283]

Smallness of profit made Cal. Indra Singh and Sons Ltd. - CIT

Where with respect to a company to which section 23A of the Income-tax Act, 1922, would otherwise apply, the Income-tax Officer has to consider whether having regard to the smallness of profit made the payment of a dividend or a larger dividend than that declared would be otherwise unreasonable, in judging the reasonableness of the company's action in distributing dividend at the rate at which it actually made the distribution or in making no distribution at all, the Income-tax Officer must pay regard to the accountable profits of the company actually at its disposal, which alone are contemplated by the words profit made and not profits, of which it might be deemed to be possessed composed partly of the accounting profits and partly of notional income, coming in either as disallowed items of expenditure or as income computed on some artificial basis.The evil which the section aims at checking is unreasonable withholding of profits from distribution as dividend in spite of money for distribution being available.At the point of time when the Income-tax Officer has to consider under section 23A(1) whether the failure or omission to distribute as dividend at least sixty per cent. of the assessable income was caused by the smallness of the profit made,what he has to pay regard to is the profit which was available to the company for distribution as dividend or, in other words, the size of the distributable fund which the company had in its hands. Dividend is distributable only out of revenue profits,but all profits derived from whatever source contribute to the fund out of which the dividend may be declared. The Income-tax Officer has to consider the whole fund available for distribution and not merely a portion of that fund contributed by a particular source.The profit contemplated is not limited to profit from business assessable under section 10 but it also includes income from interest on securities assessable under section 8, income from property assessable under section 9, and income from dividend assessable under section 12. *Sec.104,109.*[33ITR341]

Small-scale industrial undertaking **Ker.** **CIT-Travancore Mats and Mattings Co.**

The definition of small- scale industrial undertaking in clause (2) of the Explanation below sub-section (2) of section 32A, which is applicable for purposes of section 35B(1A), it is with regard to the machinery and plant installed that the value should not exceed Rs. 10 lakhs. The importance of the requirement of installation is made more specific by the words appearing in the brackets to exclude tools, jigs, dies and moulds. *Sec.32A.*[**142CTR445,229ITR93,92 Taxman 10**]

Social worker **Mad.** **CIT - P.S. Chelladurai**

.......a social worker who spends his entire time on social work can be considered to be a bread-winner of his family because it costs someone or some institution or some group of people to maintain them and the amount expended must be equated to their real income.The fact that the social worker simply appropriates to himself the material requisites for the well being of himself and his family without any account does not mean that he has no income at all to report. Whatever money had gone into the maintenance and upkeep of the assessee and his family can only be attributed to the receipts from his profession or occupation as a social worker.

Misc.[**145ITR139**]

Sold **Bom.** **Cement Agencies Ltd. - CIT**

..though under Expln.(2)to sub-section (1) of section 32,the term sold included, inter alia, a compulsory acquisition under any law for the time being in force, yet confiscation meant appropriation to public treasury by way of penalty or seizure as if by authority.

Sec.32.[**32CTR99,146ITR136,12Taxman110**]

Sole selling agent All. CIT - Arkay Wires P. Ltd.

The phrase sole selling agent has not been defined under the Companies Act,1956.In common parlance and in ordinary sense it would mean the exclusive and sole right to sell all the products of the principal to the exclusion of all others. *Companies Act.w.r.s. Sec.37.*[193CTR697,277ITR225]

Solely AP. Raasi Cements Ltd.-CIT(No.2)

...the expression solely in clause (iv) of section 32(1) indicates that such buildings should be occupied by the employees drawing a salary of Rs 10,000 per annum or less. *Sec.32(1).*[275ITR582]

Source Cal. Performing Right Society Ltd. - CIT

Under section 9(1) of the Income-tax Act, 1961, all income accruing or arising from any source of income in India is deemed to accrue or arise in India. The word source does not mean any legal concept, but refers to that which a practical man would regard as a real source of income. *Sec.9.*[93ITR44]

2. Source means not a legal concept but something which a practical man would regard as a real source of income. The assessee may have separate sources of income.All taxable income must necessarily have a definite source. **[Sterling Foods-CIT,Kar., *Sec.14.*150ITR292]**

3. Source means not a legal concept but which a practical man would regard as a real source of income.Therefore, to find out whether any particular income is in respect of a source different from the one for which the previous year has already been opted for by the assessee, a more practical approach has to be made.

[CIT-B. Murali Mohan Rao, AP.,*Sec.14.* 124CTR198,216ITR166, 81Taxman116]

Source of income SC CIT - Lady Kanchanbai

Source of income is defined in Act 2 ,1886. Source mean not a legal concept but which a practical man would regard as a real source of income. *Sec.14.*[AIR1970SC69,77ITR123]

2. A source of income may be described as the spring or fount from which a clearly defined channel of income flows. It is that which by its nature and incidents constitutes a distinct and separate origin of income capable of consideration as such in isolation from other sources of income and which by the manner of dealing adopted by the assessee can be treated so. **[Seth Shiv Prasad-CIT,All.,** *Sec.14.***84ITR15]**

3. The term source of income, as understood in the Income-tax Act is to identify or classify income so as to determine under which head, out of the various heads of income referred to in section 14 of the Act, it would fall for the purposes of computation of the total income for charging Income-tax thereon. **[Sheraton Appearels-ACIT,Bom.,***Sec.14.***175CTR651,256ITR20,123Taxman 238]**

Specify Cal. CIT - Shyam Narain Mehrotra

The word specify has not been defined either in the Income-tax Act or in the Rules. So, the word specify must be interpreted reasonably and fairly.In fact, section 26A of the Indian Income-tax Act does not indicate that the specification of shares of the partners is for distribution of profits and losses.But that is normally implied. Having regard to this, the word specify under section 26A of the Indian Income-tax Act and rule 2 means mentioning or describing or defining in detail but does not mean expressly set out in fractional or other shares. *Sec.185.***[122ITR362]**

Specifying the individual share of the partners Bom. Parekh Wadilal Jiwanbhai,

The words specifying the individual shares of the partners in section 26A of the Income-tax Act meant expressly and definitely mentioning the individual shares of the partners in the profits of the firm.Section 13 of the Partnership Act, which provided that there being no agreement to the contrary in the deed the partners were entitled to share equally in the profits, could not be called in aid for the purposes of section 26A. *Sec.185.***[42ITR266]**

Speculation business AP. CIT-Puttaiah Seshaiah and Co.

The definition of speculation business makes it clear that the intention of the parties is made immaterial and irrelevant. If a contract is settled otherwise than by the actual delivery or transfer of the commodity or scrips it will be a speculative transaction notwithstanding the fact that at the time the transaction was entered into the parties intended to effect delivery of the commodity or scrips, as the case may be. Explanation 2 to section 28, section 43(5) and section 73 are parts of a single scheme. Parliament has first defined what is a speculative transaction and in doing so, they have totally excluded from consideration the intention of the parties.Having so defined speculative transactions, Parliament provided that if these transactions are business transactions they must be treated as a separate and distinct business and then declared that the losses arising from speculative business can be set off only against the profits of any other speculative business. *Sec. 43(5).r.w.s.73.***[37CTR69,146ITR168,15Taxman47]**

2. Expression used in section 73(1) is speculation business. If speculation business is to be understood in the normal commercial parlance, the ingredients are totally different and a speculative transactionartificially defined in section 43(5) of the Act does not amount to speculation businessas is commercially understood. Explanation 2 to section 28 does not also have the effect of deeming every speculative transaction as speculation business.The term business connotes some real, substantial and systematic or organised course of activity or conduct with a set purpose. Even a single or isolated transaction can, in a conceivable case, amount to business.The question whether an assessee is carrying on speculation business can be determined by the application of recognised tests, viz.,whether the speculative transactions carried on by the assessee connote some real, substantial and systematic or organised course of activity or conduct with a set purpose.In the application of this test, it can conceivably be found that even a single or isolated transaction constitutes speculation business. **[CIT-Sri Venkateswara Rice and Oil Mills, AP.,***Sec.43(5).r.w.s. 73.***154ITR756]**

3. The question whether the purchase and sale of shares in a given case will amount to a business activity has to be judged by the ordinary yardstick of deciding whether an activity is business activity or not. If it is found that the activity of the assessee is of such a nature as to constitute a business but the transaction in question was settled otherwise than by actual delivery or transfer, then such a transaction will have to be regarded as speculation business. A speculative transaction would amount to a speculation business if it fulfilled the definition of businessunder section 2(13) of the Income-tax Act,1961,or if it amounted to an adventure in the nature of trade. The object of Explanation 2 to section 28 is not to substitute a different definition of businessfrom that contained in section 2(13). The object is merely to declare that speculation business is to be treated as distinct and separate from any other business.Much significance could not be attached to the use of the expression speculative transactions, in the plural, for the words in the singular include the plural and vice versa. A solitary speculative transaction by an assessee will not prevent the transaction from being speculative business.If the transaction amounted to an adventure in the nature of trade, then the profits of such trade will have to be treated as business income. But, if the transaction was settled without actual delivery of goods, then the adventure in the nature of trade will be an adventure in speculative business. Speculative business is only a species of general business where contracts for purchase and sale of commodities are settled without actual delivery of the commodities. **[CIT-Ganga Prasad Birla (HUF),Cal.,**Sec. 43(5).**199ITR173]**

Speculation transaction/ business **Bom.** **CIT - Kamani Tubes Ltd.**

There is a perceptible difference between a speculation transaction and speculation business.An isolated transaction of settlement of a contract otherwise than by actual delivery of the goods might amount to speculative transaction within the meaning of section 43(5) but in the absence of something more to show that the nature of the transactions was such as to constitute a business, it cannot be termed as speculation businesswhich has been treated as distinct and separate from other business. Sec.43(5). **[207ITR298, 121Taxation285,75Taxman55]**

Speculative transaction **Cal.** **Hoosen Kasam Dada (India) Ltd. - CIT**

The definition contained in Explanation 2 of section 24 has severely restricted the meaning of the expression speculative transaction and in a sense simplified it for the purposes of computation of income-tax. Where there is no delivery under a settlement contract, it is a speculative transaction.On the other hand, however speculative the transaction might be,if there is delivery,it cannot be considered as a speculative transaction for the purposes of section 24. Sec.43(5).**/[52ITR171]**

2. The definition of speculative transaction in section 43(5) of the Income-tax Act,1961, gives a simple test for deciding for the purpose of Income-tax what a speculative transaction means. If a contract for sale or purchase is ultimately settled and no actual delivery of the goods was effected under the settlement then it is a speculative transaction. The requirement under section 30 of the Contract Act of the existence of the intention of the parties even at the time of the original contract not to give or take delivery of the goods to make it a speculative or wagering transaction is dispensed with for the purpose of income-tax. If the actual delivery of the goods is not given under the settlement of the contract, then the intention of the parties at the time of the contract is immaterial. The necessity of entering into such a contract or the reason for settling the same by payment of the difference in price is also immaterial. If the settlement of the contract was otherwise than by actual delivery of the goods then section 43(5) of the Income-tax Act,1961, is attracted and the section is not qualified in any manner except as regards delivery of the goods in pursuance of the settlement. **[A.Muthukumara Pillai-CIT,Mad.,** _Sec.43(5)._ **96ITR557,36 Taxation26]**

3. In order that a transaction may fall within the scope of the expression speculative transaction, it must be a transaction in which a contract for purchase or sale of any commodity, including stocks and shares, is periodically or ultimately settled otherwise than by the actual delivery or transfer of the commodity or scrips. A contract can be settled only during the subsistence of the contract. If a breach occurs by the non-performance of the contract by actual delivery or transfer of the commodity or scrips and, thereafter, the parties to the contract settle the amount of damages by paying the difference between the contract price and the market price on the due date of performance, that would not amount in law to settling a contract.What is settled in such a case is settling the damages consequent on the breach. **[Bhandari Rajmal Kushalraj–CIT,Mys.,**_Sec.43(5)._**96ITR401]**

4. A breach of the contracts puts an end to the contract and creates a liability in damages. Section 43(5) of the Income-tax Act,1961, does not say that a settlement of a claim for damages arising out of a breach of contract will be a speculative transaction. A contract settled means contract settled before breach.After breach of contract, the cause of action is no longer based on the contract itself but on its breach.A settlement of a claim for damages arising out of a breach of contract cannot, therefore, be a speculative transaction under section 43(5). **[CIT-Ramjeewan Sarawgee & Sons,Cal.,** _Sec.43(5)._**107ITR845]**

5. ...... the legislature, while defining a speculative transaction in Explanation 2 to section 24(1) of the Indian Income-tax Act, 1922, used words which people conversant with speculative transactions and stock exchange employ.In a stock exchange, there are certain days marked out for settlement, called clearing days. On the clearing date, the settlement is effected with reference to the price fixed by the stock exchange. That is why the expression setting day is understood to mean the day on which transactions for the account are made up on the stock exchange. By the use of the expression settled in the definition of a speculative transaction, the legislature meant to deal with the performance of a contract by the parties and not with its non- performance. If the contract is performed by actual delivery of the contracted commodity, such a transaction falls outside the purview of a speculative transaction, as defined by Explanation 2 to section 24(1) of the Act. If the contract is, however, performed otherwise than by actual delivery or transfer of the contracted commodity, then the contract is so settled as to constitute it a speculative transaction as defined by the aforesaid provision. Therefore, the expression contract settled, occurring in Explanation 2 to section 24(1) of the Act, has to be understood as contract performed.It cannot be held to cover a case where, in consequence of non- performance of a contract, parties resolve their disputes arising out of the breach of contract. Such a settlement, in the sense of arranging matters in dispute, as ordinarily understood would not make it a settlement of the contract in the sense in which the expression contract settled is used in Explanation 2 to section 24(1) of the Act. **[Thakurlal Shivprakash Poddar-CIT,MP.,** *Sec.43(5).***116ITR190]**

6. For Income-tax purposes,speculative transaction means what the definition of the expression in Explanation 2 to section 24(1) of the Indian Income-tax Act, 1922, says.Whether a transaction is speculative in the general sense or under the Contract Act is not relevant for the purpose of the Explanation.Thus, a transaction which is otherwise speculative, would not be a speculative transaction within the meaning of Explanation 2 .....if actual delivery of the commodity takes place and, on the other hand, a transaction which is not otherwise speculative in nature may be required to be considered speculative according to Explanation 2 if there is no actual delivery of the commodity. **[Seksaria Riswan Sugar Factory Ltd.-CIT,Bom.,***Sec.43(5)***. 11CTR 274,121ITR196]**.

7. The artificial definition of speculative transaction in section 43(5) renders every transaction, in which a contract for the purchase or sale of any commodity, including stocks and shares, is periodically or ultimately settled otherwise than by the actual delivery or transfer of the commodity or scrips, a speculative transaction. A contract is speculative if it is settled without actual delivery. If the contract is broken, i.e., for any reason one party is unable to give delivery or the other is unable to take delivery, it is a case of breach of the contract. It depends,therefore, on the facts and circumstances of the case as to whether there has actually been a breach of the contract or a settlement of the contract. A breach takes place on account of repudiation of the contract or failure to perform it or a contract may be terminated through frustration, impossibility or through any other cause which ends the contract under the Contract Act or the Sale of Goods Act. When the obligation to supply or take delivery, as the case may be, comes to an end by operation of law, it does not make the transaction speculative. However, if the parties mutually settle the contract without any intervening circumstances, then it may be a speculative transaction if all the other conditions of section 43(5) are satisfied. The material question is: Why was the contract settled ?If it was settled by mutual consent to avoid delivery, then it would be speculative. If it was settled because of inability of the assessee to supply or on account of the fact that it did not have the necessary resources to give the delivery, then it would be a breach of contract. **[CIT-Sri Venkateswara Rice & Oil Mills,AP.,***Sec.43(5).***154ITR756]**

8. A speculative transaction as contemplated by section 43(5) of the Income-tax Act,1961, should fulfil the following four essential conditions:(1) the contract should be for purchase or sale;(2) the purchase or sale should be of stocks or shares or commodity;(3) periodical or ultimate settlement of the contract, and(4) settlement to be otherwise than by actual delivery or transfer.Payment of compensation on account of breach of contract does not amount to a loss in a speculative transaction. After the breach of contract has taken place, the party receives by way of settlement the damages suffered by it on account of the breach of the contract. In these circumstances, the nature of the transaction cannot be said to be a speculative transaction under the Act. The words contract settled used in section 43(5) refer to the settlement of the contract prior to its breach and once there is a breach of the contract, then the question of settlement of the contract, as such, does not arise. **[CIT-Rajasthan Wool Agencies,Raj.,***Sec.43(5).***160ITR358]**

9. The provisions of the Explanation to section 73 have to be contrasted with the provision of section 43(5),which defines speculative transaction to mean a transaction in which a contract for the purchase or sale of any commodity, including any stocks and shares, is periodically or ultimately settled otherwise than by the actual delivery or transfer of the commodity or scrips. .... For the purpose of setting off and carrying forward of loss, the buying and selling of shares by certain companies are regarded by the statute as speculation business, even though the transaction of purchase and sale was followed up by delivery of scrips and as such cannot be treated as speculative transaction as defined in section 43(5). ....The opening words of the Explanation to section 73 are where any part of the business of a company. Any is a word which excludes limitation or qualification. A restricted meaning should not be given to the phrase any part of the business.The object of Circular No.204,dated July24,1976, is to curb devices to manipulate and reduce the taxable income of a company under the management of a controlling group of persons. But the circular has clearly stated in paragraph 19.1 that the business of purchase and sale of shares by companies which are not investment or banking companies or companies carrying on business of granting loans and advances will be treated on the same footing as speculation business. The phrase in the Explanation to section 73 to the extent to which the business consisted of purchase and sale of such shares also does not indicate that the Legislature had several other actual and existing non-speculative activities of business in mind. It merely indicates that the business activity which consists of purchase and sale of shares will be treated as speculation business. If the entire business activity of a company consists of purchase and sale of shares of other companies, then the entire business will be treated as speculation business. But, if, apart from purchase and sale of shares, the company has other business activities, then those other activities will not be treated as speculation business. **[CIT-Arvind Investments Ltd.,Cal.,***Sec.43(5).r.w.s.73.***94CTR263,192ITR365, 58Taxman216]**

10. Speculative transaction, as defined, means a transaction in which a contract for purchase or sale of any commodity is periodically or ultimately settled otherwise than by actual delivery. The contract has to be an enforceable contract and not an unenforceable one by reason of any taint or illegality resulting in its invalidity. If an assessee has carried on speculative business which is partly legal and partly illegal, the legal business is deemed to be distinct and separate from the illegal business.

[CIT- Shivlal Dhirajlal,Guj.,*Sec.43(5).***193ITR196,104Taxation 405,63 Taxman130]**

11. A bare reading of section 43(5) of the Income-tax Act,1961, makes it clear that though a transaction in which a contract for the purchase or sale of a commodity is settled otherwise than by actual delivery or transfer of the commodity..... speculative transaction has been defined as certain species of such transactions entered into under specific circumstances have been taken out of the definition by virtue of the three provisos thereto. Proviso (a) clearly states that for the purposes of this clause a contract in respect of raw materials or merchandise entered into by a person in the course of his manufacturing or merchanting business to guard against loss through future price fluctuations in respect of his contracts for actual delivery of goods manufactured by him or merchandise sold by him shall not be deemed to be a speculative transaction.This proviso is not confined to contracts in respect of raw materials entered into by persons in the course of their manufacturing business.It applies equally to cases of persons carrying on manufacturing as well as persons carrying on business of selling goods. There is no scope to confine it to manufacturers. **[CIT-Ramachandra Shivnarain, Bom.,**<u>*Sec.43(5).*</u>**201ITR 862]**

12. A transaction cannot be described as a speculative transaction within the meaning of section 43(5) of the Income-tax Act,1961, where there is a breach of the contract and on a dispute between the parties damages are awarded as compensation. **[CIT- Kamani Tubes Ltd., Bom.,** <u>*Sec.43(5).*</u>**207ITR298,121Taxation285,75 Taxman55]**

13. A transaction cannot be described as a speculative transaction within the meaning of sub-section (5) of section 43 of the Income-tax Act,1961,where there is a breach of the contract and on a dispute between the parties damages are awarded as compensation.

[CIT-Panachand Khemchand,Guj.,<u>*Sec.43(5).*</u>**210ITR1053]**

14. ....reading of the definition of speculative transaction in section 43(5) of the Income-tax Act,1961, shows that where a contract for the purchase or sale of any commodity, which includes stocks and shares, is settled, periodically or ultimately,otherwise than by way of actual delivery or transfer of the commodity or scrips, it would be treated as a speculative transaction. There are various modes of entering into business transactions.For example, forward contracts, hedging contracts, etc., are contracts where the goods do not pass from the seller to the purchaser but only an account of the sale is maintained on purchase and sale of the goods and such transactions are settled periodically or ultimately without involving the delivery or transfer of the commodity or the scrips. They are dealt with for the purposes of setting off and carrying forward of losses. The effect of treating such transactions as speculative transactions is that loss suffered in such transactions can only be set off against the income earned from speculative business and not against income from other business as contemplated in section 73 of the Act.Any trader in commodities generally intends to buy and sell the commodity with a stipulation that the delivery of the commodity by the seller to the purchaser is an essential ingredient of that transaction. Under the Sale of Goods Act such delivery can be made either by physical delivery of the commodity directly to the purchaser or to the carrier for him or by transferring the document of title to the commodity. In the expression otherwise than by actual delivery or by transfer of the commodity in section 43(5), these two modes of delivery must be contemplated.Taking this expression along with the previous phrase settled periodically or ultimately, it would mean that in such a settlement of the contract delivery of the commodity was never contemplated at all. Where the nature of transaction entered into between the parties is such that under the contract the seller parts with possession of the goods and while the goods were in transit the buyer enters into a subsequent transaction, then so far as the first seller and the first buyer are concerned, there would be actual delivery of the goods and the transaction cannot be a speculative transaction.

[CIT-LakshminarayanaTradingCo.,AP.,<u>*Sec.43(5).*</u>**134CTR499, 219ITR90,82Taxman301]**

15. The definition of the expression speculative transaction in section 43(5) which defines it to mean a transaction in which a contract for the purchase or sale of any commodity, including stocks and shares, is periodically or ultimately settled otherwise than by the actual delivery or transfer of the commodity or scrips,cannot be read into the provisions of section73 having regard to the plain meaning of the Explanation to section73. **[CIT-Lokmat Newspapers P.Ltd.,Bom.,**<u>*Sec.43(5).*</u>**230CTR521,322ITR43]**

16. A transaction cannot be described as a speculative transaction within the meaning of section 43(5) of the Income-tax Act,1961, where there is a breach of contract and on a dispute between the parties damages are awarded as compensation,e.g.,by an arbitration award. What is really settled by the award of such damages and their acceptance by the aggrieved party is the dispute between the parties. Section 43(5), however, speaks of a settlement of the contract and a contract is settled when it is either performed or the promisee dispenses with or remits, wholly or in part, the performance of the promise made to him or accepts, instead of it, any satisfaction which he thinks fit. A contract can be said to be settled if instead of effecting the delivery or transfer of the commodity envisaged by the contract, the promisee, in terms of section 63 of the Indian Contract Act, 1872, accepts any satisfaction which he thinks fit. It is quite another matter when instead of such acceptance the parties raise a dispute and no agreement can be reached for a discharge of the contract. There is a breach of the contract and by virtue of section 73 of the Indian Contract Act, 1872, the party suffering by such breach becomes entitled to receive from the party who broke the contract compensation for any loss or damage caused to him thereby. The award of damages for breach of a contract is not the same thing as a party to the contract accepting satisfaction of the contract otherwise than in accordance with the original terms thereof. What is really settled by the award of such damages and their acceptance by the aggrieved party is the dispute between the parties.......The word settled or settlement in connection with the contract has not been defined in the Income-tax Act or in the Contract Act or in the Sale of Goods Act, or in any other statute. The proper meaning to be given to the words contract settled in the definition clause would be a contract determined or concluded or disposed of. By the use of the expression settled, what is intended to be dealt with is a case of performance of contract and not non-performance. **[CIT- Hans Machoo and Co., Del.,**<u>*Sec.43(5)*</u>**164CTR93,247ITR 79,113Taxman427]**

17. ....any speculative transaction cannot be treated as speculative unless such a transaction takes place as a part of business activity of the assessee concerned. It is not necessary that the assessee must carry on speculative business exclusively-it can be one of the businesses but this must be distinct and separate from any other business.The definition of speculative transaction in section 43 (5) demands that there must be actual delivery or transfer of the commodity or scrip.Only if it is by way of a paper transaction it can be said to be a speculative transaction. ...However, part of the business of a company other than a few category of companies as mentioned therein,shall be treated as speculation business,if such part consists of the purchase and sale of shares,not other category of speculative dealings. **[Paharpur Cooling Towers Ltd.-CIT, Cal.,**<u>*Sec.43(5)r.w.s73.*</u> **239CTR394, 338ITR295, 198Taxman83]**

Spending **SC** **General Insurance Corporation of India-CIT**

Spending in the sense of paying out or away of money is the primary meaning of expenditure.Expenditure is what is paid out or away and is something which is gone irretrievably.

<u>*Sec.44*</u>.**[AIR1999SC3759,156CTR425,240ITR139,153Taxation 398,106 Taxman389]**

Splitting or reconstruction of .. Guj. Nagardas Bechardas and Brothers P. Ltd. - CIT

The expressions new industrial undertaking and splitting up, or reconstruction of a business already in existence in section 84 (now section 80J) must be understood in a broad commercial sense from a common sense point of view. Interpreting these expressions the purpose of the provision, i.e., to encourage the setting up of new industries must be borne in mind. A comparison between the cost of establishment and the totality of turnover of the new undertaking with that of the previously running business may assume importance in some cases but does not provide a proper guideline in every case. Though every case must depend upon its own peculiar facts, a broadly correct solution can be arrived at by asking two questions:(1) Whether there was any activity in the existing business which could have been reconstructed and could have assumed a new form on such reconstruction; and (2) Whether the nature of the business which has assumed a new form as a result of the change which is introduced is the same as the nature of the business which was existing before the change was introduced. If either of these two questions is answered in the negative, it would mean, prima facie,that there is no reconstruction which would attract clause (i) of section 84(2) [now section 80J(4)(i)] of the Act.

Sec.80J.**[104ITR255]**

2. The expression splitting up of the business already in existence in section 84 of the Income-tax Act,1961, indicates a case where the integrity of a business earlier in existence is broken up and different sections of the activities previously conducted are carried on independently. **[CIT-Hindustan General Industries Ltd., Del.,**Sec.84**23CTR73,137ITR851]**

3. The expression new industrial undertaking and splitting up or reconstruction of business already in existence in section 15C must be understood in a broad commercial sense from a common sense point of view. In interpreting the expression, the purpose of the provision, i.e., to encourage the setting up of new industries, must be borne in mind. Reconstruction of the business or splitting up of a business already in existence must be in relation to the new industrial undertaking.Further, the new industrial undertaking must not be formed by transferring building, plant or machinery of the existing business. Sub-section (1) requires separate capital but not new or different capital.But whether a new industrial undertaking is entitled to exemption under section 15C must depend upon the facts of each case. **[CIT-Orient Paper Mills Ltd., Cal.,** <u>*Sec.80J.*</u>**94ITR73]**

Standing to the credit of the share premium account **SC** **CIT - Allahabad Bank Ltd.**

The words standing to the credit of the share premium account do not mean that the share premium shall be maintained in a separate account apart from the reserve. <u>*Misc.*</u>**[73ITR745]**

Stock **Gau.** **CIT - Hardware Exchange**

Stock means the goods and wares of a merchant or tradesman kept for sale and traffic and, in a larger sense, the capital of a merchant or other person including his merchandise, money and credits, or, in other words, the entire property employed in business.A careful consideration of the meaning of the two expressions, expenditure and stock-in-trade makes it abundantly clear that payments made for purchase of stock-in-trade cannot be termed as expenditure. Money does not go irretrievably in such a case.What is acquired by such payments, namely,for stock-in-trade,forms part of the business assets of the assessee. Such assets are sold in due course and the money paid is ordinarily recovered with some profit. Stock-in-trade which remains unsold at the end of the year is reflected as assets in the balance-sheet. <u>*Sec.40A.*</u>**[95CTR183,190ITR61]**

Subject of an appeal **Mad.** **A.V.Sreenivasalu Naidu - CIT**

An order is made the subject of an appeal within the meaning of clause (c) of the first proviso to sub-section (2) of section 33A only when it is the subject- matter of an effective appeal. If an appeal to the Appellate Tribunal under section 33 is not admitted and is disposed of on the ground that it was filed after the prescribed time the order cannot be said to be the subject of an appeal.

<u>*Sec.264.*</u>**[16ITR341]**

2. An order can be said to be the subject of an appeal within the meaning of section 264(4) of the Income-tax Act,1961, only if the appeal was considered and disposed of on the merits; an appeal which was dismissed as withdrawn at the instance of the assessee or where an appeal was dismissed on the ground that the same was incompetent or where an appeal was dismissed as barred by limitation or where the appeal was dismissed for non- payment of undisputed tax as a condition precedent for entertaining the appeal cannot be said to have been made subject of an appeal. Subject of an appeal means subject of an effective appeal. **[C.C.Jayaram,Ker.,** *Sec.264.* **207ITR662]**

Subject of dealing in any stock exchange **Cal.** **Star Company Ltd. - CIT**

The expression subject of dealing in any stock exchange emphasises that the shares are actually sold and purchased in the open market by members of the public. Quotations in the stock exchange bulletin are prima facie evidence of dealings in shares in the stock exchange, but it can be established by other evidence that in fact there were no such dealings. *Misc.* **[58ITR149]**

Subject to **Bom.** **CIT - F.Y.Khambaty**

The proviso to clause (c) of sub-section (1) of section 5 of the Income-tax Act,1961, provides that in the case of a person not ordinarily resident in India,the income which accrues or arises to him outside India shall not be included in his total income, unless it is derived from a business controlled in or a profession set up in India. Similarly,the plain language of sub-section (2) indicates that the total income of a non-resident includes only the income received or deemed to be received by him in India or income accruing or arising or deemed to accrue or arise to him in India during the relevant accounting year.The expression subject to used in the opening portion of both sub-sections (1) and (2) of section 5 has to be read keeping in mind that section 5 is intended to explain the scope of total income.Therefore, what the use of the said expression shows is that in considering what is total income under section 5, one has to exclude such income as is excluded from the scope of total income by reason of any other provision of the Income-tax Act and not that the other provisions of the Income-tax Act override the provisions of section5.

Sec.5. **[50CTR275,159ITR203,80Taxation135,24Taxman29]**

Subject to any rules made in this behalf **Ker.** **CIT- Antony Mendez**

There is nothing in the language of section 7(1) of the Wealth-tax Act, 1957, requiring the valuation of any particular asset to be done in a particular manner.What is contemplated is that the value determined should approximate to the market price on the valuation date.The expression subject to any rules made in this behalfin section 7(1) does not make the rules determinative of the mode of estimation of the value.The rules are intended to carry out the purposes of the Act.The rules can only guide the Wealth-tax Officer in forming his opinion on the value of the assets as contemplated by section 7(1), namely, the market price as on the valuation date.The rules can at best lay down only guidelines to enable the Wealth-tax Officer to determine the value of the asset, in the performance of his functions under section 7(1). When the substantive provision in section 7(1) itself has not created any mandatory or fictional mode of valuing the asset, the rule cannot be read as imposing, for fixation of the market price, such a mandate or creating a fiction. *Sec.7.WT.Act.* **[191ITR346]**

Subsidiary company Bom. Petrosil Oil Company Ltd. - CIT

The expression subsidiary company has not been defined in the Income-tax Act.The expression subsidiary company has been defined in section 4 of the Companies Act.Under the Companies Act, a subsidiary company of a subsidiary is also deemed to be a subsidiary of another company of which such company is a subsidiary.Once we import the definition of the expression subsidiary company appearing in section 4(1) of the Companies Act to find out the true meaning of the word subsidiary company in clause (b) of section 108 of the Income-tax Act, it will have to be read in the context of the requirements of clause (b) of section 108. In order to get the benefit of the lower rate of Income-tax prescribed in clause (1) of sub-paragraph I of Paragraph F of the First Schedule to the Finance (No. 2) Act,1971, which is applicable to domestic companies, every company, whether holding company as well as the subsidiary or subsidiary of a subsidiary company must be a domestic company and should also meet the requirements of section 2(6)(b) of the Finance Act read with section 108 of the Income-tax Act.In order to get the benefit of the lower rates of Income-tax applicable to a company in which the public are substantially interested in the capacity of a subsidiary of a company in which the public are substantially interested by reference to clause (b) of section 108 of the Act, it is necessary that the parent company or the holding company should also be a domestic company......Section 108(b) of the Act is satisfied only if the entire share capital of such subsidiary is held by the company falling under section 108(a) of the Act. Section 4 of the Companies Act,1956, defines the expression holding company and subsidiary company for purposes of the Companies Act only. The artificial extension of the meaning of the word subsidiary company by the

Companies Act is for purposes of the Companies Act only and it cannot be imported in section 108(b) of the Income-tax Act. A sub-subsidiary company does not fall under section 108(b) of the Act.....Section 108(b) of the Act does not state that the said section shall be applicable only to subsidiary companies of which the holding company is a domestic company. *Companies Act.r.w.s.108.* **[155CTR445,236ITR220]**

Subsidy Cal. Kesoram Industries and Cotton Mills Ltd. - CIT

Subsidy is a financial aid by the Government towards expenses of an undertaking or institution held to be of public utility. *Misc.* **[191ITR518]**

2. The word subsidy connotes aid, particularly financial, given by the Government to entrepreneurs, industries, etc., or to keep down the price of a commodity, by making available to the consumer goods at a lower cost. The predominant object is generally to set up new industries and expansion of industries in certain backward and specified areas. The scheme for subsidy may provide assistance to entrepreneurs for purposes of industrial development and expansion in a variety of ways. **[Srinivas Industries-CIT,Mad.,** *Sec.43(1).***188ITR22]**

Substantial SC Raghuvanshi Mills Ltd. - CIT

The Explanation to section 23A(1) of the Income-tax Act (prior to its amendment in 1955),which provides when a company shall be deemed to be a company in which the public are substantially interested, lays down, among the tests, the minimum interest which can be called substantial by saying that shares of the company carrying not less than 25 per cent., of the voting power must be allotted unconditionally to,or acquired unconditionally by, the public and they must be beneficially held by the public.The essence of the Explanation lies not in the percentage which only shows the limit of the minimum holding by the public, but lies in the words unconditionally and beneficially.These words underline the fact that no person who holds a share or shares not for his own benefit, but for the benefit of another and who does not exercise freely his voting power, can be said to belong to that body, which is designated public.

*Sec.2(22)(e).***[41ITR613]**

Substantial contribution Del. CIT - Charat Ram Foundation

The expression substantial contribution may mean something quite different to a person who is affluent from what it may mean to a person who is not so. But the language of a taxing statute does not recognise such a differentiation. *Sec.13(3)(b).* **[168CTR261,250ITR64,116Taxman255]**

Substantial interest Bom. Dr. J.M.Mokashi - CIT

...the expression substantial interest appearing in section 64(1)(ii) read with Explanation 2(ii)referred only to a proportion of the whole interest and not the whole interest and as such section 64(1)(ii) had no application to a proprietary concern in which the assessee had 100 per cent.interest. *Sec.64(1)(ii).* **[115CTR73,207ITR252, 72Taxman98]**

Substantial part of the business Bom. CIT - Parle Plastics Ltd.

The expression used under clause (ii) of section 2(22)(e) is substantial part of the business. The expression substantial part does not connote an idea of being the major part or the part that constitutes majority of the whole.If the Legislature intended that a particular minimum percentage of the business of a lending company should come from the business of lending, the Legislature could have specifically provided for that percentage while drafting clause (ii) of section 2(22)(e) of the Act. The Legislature had deliberately used the word substantial instead of using the word major and/or specifying any percentage of the business or profit to be coming from the lending business of the lending company for the purpose of clause (ii) of section 2(22)(e). Any business of a company which the company does not regard as small, trivial, or inconsequential as compared to the whole of the business is substantial business.Various factors and circumstances would be required to be looked into while considering whether a part of the business of a company is its substantial business. Sometimes a portion which contributes a substantial part of the turnover, though it contributes a relatively small portion of the profit, would be a substantial part of the business. Similarly, a portion which is relatively small as compared to the total turnover, but generates a large portion, say more than 50 per cent. of the total profit of the company would also be a substantial part of its business. Percentage of turnover in relation to the whole as also the percentage of the profit in relation to the whole and sometimes even percentage of manpower used for a particular part of the business in relation to the total manpower or working force of the company would be required to be taken into consideration.Employees of a company are now called its human resources and, therefore, the percentage of human resources used by the company for carrying on a particular division of business may also be required to be taken into consid-eration while considering whether a particular business forms a substantial part of its business.Undisputedly, the capital employed by a company for carrying on a particular division of its business as compared to the total capital employed by it would also be relevant while considering whether the part of the business of the company constitutes substantial part of the business of the company.

*Sec.2 (22)(e).***[236CTR382,332ITR63,196Taxman462]**

Substantial question of law Del. Mahavir Woollen Mills - CIT

The words substantial question of law have not been defined. But the expression has acquired a definite connotation through a catena of judicial pronouncements. Usually five tests are used to determine whether a substantial question of law is involved.They are as follows: (1) whether, directly or indirectly, it affects substantial rights of the parties, or (2) the question is of general public importance, or (3) whether it is an open question in the sense that the issue has not been settled by pronouncement of the Supreme Court or Privy Council or by the Federal Court, or (4) the issue is not free from difficulty, and (5)it calls for a discussion for alternative view.A question of fact becomes a question of law, if the finding is either without any evidence or material, or if the finding is contrary to the evidence, or is perverse or there is no direct nexus between the conclusion of fact and the primary fact upon which that conclusion is based.But, it is not possible to turn a mere question of fact into a question of law by seeking to find whether as a matter of law the authority came to a correct conclusion upon a matter of fact.

*Sec.260A.***[162CTR267,245ITR297,111Taxman566]**

2. A question of fact becomes a question of law if the finding was either without evidence or material or if the finding was contrary to the evidence or was perverse or there was no direct nexus between conclusions of fact and the primary fact upon which the conclusion was based.But it is not possible to turn a mere question of fact into a question of law by asking whether as a matter of law the authority came to a correct conclusion on a matter of fact.

[ChamanLal-CIT,Del.,*<u>Sec.260A.</u>*164CTR192,245ITR730,112 Taxman291]

3. The words substantial question of law mean a question of law which affects the substance of the case, e.g., money liability, and not merely those questions which the court thinks to be somehow specially serious. **[CIT- Agarwal Hardware Works Pvt. Ltd.,Cal., *<u>Sec.260A.</u>*248ITR155,117Taxman249]**

4. The expression substantial question of law is not defined. The following five tests have been used to determine whether a substantial question of law is involved. They are: (a) whether directly or indirectly it affects substantial rights of parties; (b) the question is of general public importance; (c) whether it is an open question in the sense that the issue has not been settled by pronouncement of the Supreme Court or the Privy Council or by the Federal Court; (d) the issue is not free from difficulty; (e) if it calls for a discussion for alternative view......Section 260A is analogous to the provisions of section 100 of the Civil Procedure Code, 1908. Under section 100 of the Code, a second appeal could be entertained only when a substantial question of law is involved. A question of fact becomes a question of law if the finding is either without any evidence or material,or if the finding is contrary to the evidence or is perverse or there is no direct nexus between the conclusion of fact and the primary facts upon which that conclusion is based. But it is not possible to turn a mere question of fact into a question of law by asking whether as a matter of law the authority came to a correct conclusion upon a matter of a fact. **[Bhagat Construction Co.(P.)Ltd.- CIT, Del., *<u>Sec.260A.</u>*250ITR291,114Taxman606]**

5. A point of law which admits of no two opinions may be a proposition of law but cannot be a substantial question of law. To be substantial, a question of law must be debatable,not previously settled by law of the land or a binding precedent, and must have a material bearing on the decision of the case, if answered either way, in so far as the rights of the parties before it are concerned. To be a question of law involved in the case there must be first a foundation for it laid in the pleadings and the question should emerge from the sustainable findings of fact arrived at by the court of facts and it must be necessary to decide that question of law for a just and proper decision of the case.Any entirely new point raised for the first time before the High Court is not a question involved in the case unless it goes to the root of the matter......The word substantial as qualifying question of law, means having substance, essential, real, of sound worth, important or considerable. It is to be understood as something in contradistinction with technical, of no substance or consequence, or academic merely. The expression substantial question of law has not been suffixed by the words of general importance as has been done in other provisions such as section 109 of the Code of Civil Procedure or article 133(1)(a) of the Constitution of India. The substantial question of law, on which a second appeal shall be heard, need not necessarily be a substantial question of law of general importance.[**Santosh Hazari-Purushottam Tiwari, SC,** *Constitution of India.***170CTR160, 251ITR84]**

6. Though the expression substantial question of law has not been defined in the Act or in any of the statutes where this expression appears, e.g., section 100 of the Civil Procedure Code, 1908, the true meaning or connotation of this expression is now well settled by various judicial pronouncements. There is a difference between a question of law and a substantial question of law.

[Nek Ram Sharma and Co.-ITAT,J&K.,*Sec.260A.*
262ITR692, 115 Taxman636]

Succession **Mad.** **Jupudi Kesava Rao - CIT**

The word succession as used in section 26(2) of the Indian Income-tax Act connotes a transfer of ownership and the person who succeeds another must have by such succession become the owner of the business which his predecessor was carrying on and which he, after the succession, carries on in such capacity. Consequently, there is no succession within the meaning of section 26(2) where the business of a joint Hindu family devolves on a co- parcener by survivorship under the Hindu Law. *Sec.170.***[3ITR339]**

2.In order that there may be succession ....three conditions have to be satisfied:-(i) change of ownership; (ii) continuity in the integrity of the business; and (iii) identity and continuity of the business in the hands of the person succeeded and the person succeeding.

[B. Koteswara Rao-CIT,AP.*Sec.170.* **46 ITR882]**

3. For the purposes of income-tax,a firm is a distinct entity from the persons who compose it and there is a succession when a business carried on by a person is transferred to a firm, even though the transferor continues as a partner of the firm; and the assessee was accordingly entitled to relief under section 25(4) of the Act.The view that in all cases where the exercise of a profession or vocation depends on the skill or ability of the person who carries it on, the profession or vocation could not be succeeded to, is fundamentally opposed to the provisions of section 25(4) of the Act.

[M.Subbaraya Iyer-CIT,Mad.,*Sec.170.*44ITR801]

4. Succession involves change of ownership; that is, the transferor goes out and the transferee comes in; it connotes that the whole business is transferred;it also implies that substantially the identity and the continuity of the business are preserved. If there is a transfer of a business, any arrangement between the transferor and the transferee in respect of some of the assets and liabilities not with a view to enable the transferor to run a part of the business transferred but to enable the transferee to run the business unhampered by the load of debts or for any other appropriate collateral purpose cannot detract from the totality of the succession.....The expression succession has acquired a somewhat artificial meaning. The tests of change of ownership, integrity, identity and continuity of a business have to be satisfied before it can be said that a person succeeded to the business of another.**[CIT-K.H.Chambers,SC.*Sec.170.*55ITR674]**

5. Succession implies that there is an end of an entity carrying on the business, and its place has been taken by an entirely new entity to run in continuity and as a going concern, the same business. Substantial identity and continuity of the business must be preserved. The tests of change of ownership, integrity, identity and continuity of a business have to be satisfied before it can be said that a person succeeded to the business of another

[Oriental Fire and General Insurance Co. Ltd.-CIT, Del., *Sec.170.*162CTR2,244ITR631, 11 Taxman 571]

Such **Gau.** **CIT - Bimal Auto Agency**

Prior to the amendment by the Finance Act, 2002 ,section 158BB authorised the Assessing Officer to make an assessment of undisclosed income on the basis of evidence found as a result of search and such other materials or information as may be available with the Assessing Officer.The use of the word such points out that such materials or information must have some connection with the search and not constitute independent materials, i.e., independent of the search. *Sec.158BB.***[314ITR191]**

Such assets **Ker.** **M.G.Kollankulam - CIT**

The words such assets occurring in sub-clauses (i) and (ii) of section 4 connote assets which have been mentioned earlier in the section, viz.,properties falling within the definition of assets which, on the valuation date, are held by the spouse or minor child of the assessee.If the properties held by the spouse or minor child are assets under section 2(1)(e) of the Act on the valuation date and if they are found to have been transferred by the assessee to the spouse or minor child directly or indirectly otherwise than for adequate consideration, the requirements of sub-clauses (i) and (ii) are satisfied.It is not necessary that the properties concerned should have been assets as defined in the Act on the date the transfers were effected by the assessee. *Sec.4.WT.Act.***[7CTR16,115 ITR160]**

Such capital **Cal.** **Shewkissen Bhatter - CIT**

The expression such capital in the last part of section 9(1)(iv) of the Indian Income-tax Act,1922,which provides for the allowance of interest on borrowed capital refers to capital employed for the acquisition or construction or repair or renewal or reconstruction of the property and it is this capital which is borrowed capital indicated by the expression such capital.It does not include interest on that capital or interest on interest. *Sec.24.***[74ITR 331]**

Such income **Mad.** **S.A.S.Marimuthu Nadar - CIT**

The expression such income in section 2(6AA) of the Income-tax Act means the income on which the assessee has to be assessed to tax, and includes the income of any other person if that income has to be treated as the assessee's income for the purposes of Taxation, under any of the provisions of the Act. *Sec.2(45).* **[30ITR670]**

2. The words such income in the latter part of section 2(6AA) of the Indian Income-tax Act, 1922, refer to the whole definition of earned income given by the Act before it says what is to be included in it, and mean earned income determined in the same manner in which that income is to be determined under the earlier part of the definition. **[CIT-S.A.S.MarimuthuNadar,SC.,***Sec.2(45)***44ITR1]**

Such orders as it thinks fit **Mad.** **T.M.S.Mohamed Abdul Kader-CGT**

The term such orders as it thinks fit in section 23(5) of the Gift-tax Act, 1958, is sufficiently wide to include a power to direct the lower authorities to collect fresh evidence and submit findings on such evidence, if necessary, pending the disposal of the appeal.The power is not restricted to making an order of remand for fresh disposal by the lower court. Sec.23.G.T.Act.**[70ITR237]**

Such other material or...as are available with A.O. **Mad.** **CIT- G.K. Senniappan**

The expression such other materials or information as are available with the Assessing Officerin section 158BB of the Income-tax Act, 1961,cannot be bisected or taken in isolation for the purpose of computation.Such other materials or information as are available with the Assessing Officer should be relatable to such evidence.The word such used as a prefix to the word evidence assumes much significance in this provision as it indicates only the evidence found as a result of search or requisition of books of account or other documents, at the time of search. Any other material cannot form the basis for computation of undisclosed income of the block period. Sec.158BB.**[284ITR220]**

Such person **Kar.** **Srimathi Indira -ITO**

Whenever a person proposes to transfer, assign, limit or extinguish his right, title or interest in any property the value of which exceeds Rs. 50,000 which is compulsorily registrable under the Indian Registration Act, 1908, the Registering Authority is prohibited from registering such document unless that person produces before him a certificate issued by the concerned Income-tax Officer certifying that taxes due by him under that Act and other Acts referred to in that section, if any, have been paid or satisfactory arrangements for their payment have been made. The words such person occurring in the section refer to the transferor or the person creating interest in favour of another.In the context, the words such person cannot be read as any and every person comprehending both the transferor and the transferee.The section requires the Income-tax Officer to issue a certificate only to the person that proposes to transfer or create an interest in favour of another and to no other person. If the Legislature in its wisdom creates a right on the transferor only, the court cannot create that right in favour of the transferee also. Any such attempt will really amount to legislation in the guise of interpretation, which is impermissible. Sec.230A.**[150ITR351]**

2. The provisions of section 153A of the Income-tax Act,1961, make it clear that only in the case of a person on whom a search is initiated under section 132 or books of account or other documents or any assets are requisitioned under section 132A after March 31,, the Assessing Officer shall after issuing notice assess or reassess the total income of such person for six assessment years immediately preceding the assessment year relevant to the previous year in which such search is conducted or requisition made. The legislative intent is clear from the use of the expression such person in clause (a) of section 153A.The expression clearly relates to a person in respect of whom search under section 132 has been initiated as section 153A itself provides.Thus to exercise powers under section 153A in the case of a person the mandatory requirement is that there must be initiation of a search as contemplated under section 132 or requisition under section 132A in respect of such person. The word person appearing in section 132 and in section 153A is one and the same person. Thus the person, in respect of whom search under section 132 is initiated, must be the same person against whom notice under section 153A is to be issued for making assessment/reassessment under that section. **[Smt. Vijay Dhir -DIT,All.,**<u>*Sec.153A.*</u>**[335ITR112]**

Such profits Cal. CIT - Belliss and Morcom (I.) Ltd.

The legislative intent in enacting section 80-I of the Income-tax Act, 1961 was to give relief to certain priority industries and develop them.The expression where the gross total income includes any profits and gains attributable to any priority industry is descriptive of the types of income or profits which will be entitled to relief and the expression such profits must relate to the quality of the profits, that is to say, profits of the priority industry. It would be improper to interpret the section in such a manner as would run contrary to the legislative intent. The expression such profits in section 80B cannot be understood in the light of sub-section (5) of section 80B which enjoins deduction of carried forward depreciation losses from non-priority industries.Hence,the deduction under section 80-I should be given on the profits of the priority industry without deducting there from any loss arising in any other business activity under section 70 or section 71. <u>*Sec.80I.*</u>**[26CTR76,136ITR481]**

Such sum Bom. Loyal Motor Service Co. Ltd.- CIT

Section 10(2)(x) of the Indian Income-tax Act must be strictly construed in favour of the subject. The sum excepted under the expression such sum in this clause must be the same sum as is described by the expression any sum paid as bonus or commission. It refers to the quantum and not the character of the payment. *Sec. 40c.* **[14ITR647]**

Sufficient cause **Del.** **Surinder Kumar Boveja - CWT**

In order to get condonation of delay in filing of an appeal, a party has to show sufficient cause. Sufficient cause means a cause beyond the control of the party, e.g.,a mistake made in good faith in respect of exercise of due care and attention. *Misc.* **[204CTR173,287ITR52,155Taxman311]**

Sum **Ori.** **CIT - Aloo Supply Co.**

The word sum has no statutory definition and must have the common parlance meaning. While legislating,Parliament tries to convey its intention through express words.It is one of the well-settled rules of interpretation that where a word used in a statute carries more than one meaning, that meaning which makes the provision workable and is nearest to the legislative intention, has to be adopted. The word sum in section 40A(3), second proviso,of the Income-tax Act, 1961, is used only to indicate an amount of money and does not refer to the totality of the expenditure. *Sec.40A(3).* **[121ITR680]**

Sums paid as donations **Bom.** **CIT - Associated Cement Co. Ltd.**

In order that an assessee may be entitled to the rebate under section 15B in respect of sums paid as donations,the donations need not be in the shape of actual cash. *Sec.80G.* **[68ITR478]**

Sums paid by the assessee **SC** **H.H.Sri Rama Verma - CIT**

The context in which the expression sums paid by the assessee has been used makes the legislative intent clear that it refers to the amount of money paid by the assessee as donation. The Act provides for assessment of tax on the income derived by an assessee during the assessment year, and the income relates to the amount of money earned or received by an assessee.Therefore, for purposes of claiming deduction under section 80G(2)(a), the donation must be a sum of money paid by the assessee.The plain meaning of the words used in the section does not contemplate donations in kind.

Donations may be made by supplying goods of various kinds including building, vehicle or any other tangible property but such donations, though convertible in terms of money, do not fall within the scope of section 80G(2)(a) and will not entitle an assessee to deduction.Donation of shares of a company does not amount to payment of any amount though the shares, on their sale, may be converted into money, and the donation so made does not fall within the ambit of section 80G(2)(a).Since the expression and language used in section 80G(2)(a) is plain and clear, it is not open to the courts to enlarge the scope by its interpretative process founded on the basis of the object and purpose underlying the provision for granting relief to an assessee. *Sec.80G.*[**187ITR308**]

Superannuation **All.** **CIT - Ashok Kumar Dixit**

The superannuation is of an employee's severance of the relationship of contract of employment between employer and employee.After attaining a specified age, an employee ceases to be an employee ipso facto in pursuance of the terms and conditions governing the employment. *Sec.10(10AA),17(1)(va).* **[134CTR105,223ITR209,88Taxman99]**

Survey **Mad.** **Madras Fertilisers Ltd. - CIT**

The word survey occurring in section 35D(2)(a)(iii) of the Act should be understood in the context of the nature of the business carried on by the assessee.There is no point in attributing a mere dictionary meaning to this word. The word survey does not mean a mere onlooking or overlooking of what happened, but would also include attracting customers to a particular spot demonstrating to them the utility and value of the assessee's products and studying there from the business possibilities or determining the action necessary to extend the business.

Sec.35D(2)(a).[**209ITR174**]

T

Taken into custody **MP. Samta Construction Co.-Pawan Kumar Sharma, DDIT (Invs.)**

The expression taken into custody is different from the expression in custody.The words taken into custody when given the natural and objective meaning do convey that possession of the asset is taken over by an authority without the consent/volition of the person who is in possession or control of the same. The authority or officer who has taken into custody gains control over it. A person from whose custody it is taken cannot get it back as and when he desires. The officer or authority exercises the lawful control over the asset and it is taken into custody with a purpose. Quite apart from the above, the provision also stipulates that the officer or authority must have power or jurisdiction under any other law for the time being in force. On a plain reading of the provision it is clear that power must be vested in the authority or the officer under some law to take the asset into custody. While analysing the concept taken into custody it is also appropriate to scan the provision in sub-section (2) of section 132A. It postulates that when a requisition is made, the officer or authority shall deliver the books of account, other documents or assets to the requisitioning officer...A banker cannot take into custody a draft from the possession or control of a person. It is the owner or the title holder or the person in possession who voluntarily parts with it for carrying out some transaction with the bank. It is presented before the bank for definite purposes. At any point of time the person who has presented the draft can withdraw or bring it back.The bank has no power to retain it under any provision of law unless the person concerned gives his consent. It is well known that in the case of a fixed deposit, it cannot be adjusted in respect of a loan unless there is an agreement to that effect, as they form two different transactions.Thus, the banker has no control or authority over the draft.It only accepts the instrument for a limited purpose. When it is in the custody of the banker, the banker retains it on behalf of the customer. The banker cannot utilise the draft for any purpose other than what has been desired by the customer.The bank draft when presented for clearing by the customer to the bank cannot be said to have been taken into custody by the bank to attract the applicability of section 132A.

*Sec.132.***[155CTR405,244ITR845,107Taxman198]**

Tax SC Union of India - Harbhajan Singh Dhillon

The requisites of a tax under entry 49 of List II are: (a) it must be a tax on units, that is, lands and buildings separately as units; (b) the tax cannot be a tax on totality, i.e., it is not a composite tax on the value of all lands and buildings; (c) the tax is not concerned with the division of interest in the buildings or lands. In other words, it is not concerned whether one person owns or occupies the land or building or two or more persons own or occupy it.

*Constitution of India.***[83ITR582]**

2. Tax includes any levy. **[Shyam Sunder - Gift Tax Officer,All.,** *Constitution of India.***AIR1967All.19,66ITR74]**

3. The word tax in section 221 includes advance tax and the failure to pay advance tax amounts to default for which penalty can be levied. **[CIT-Sreerama and Co.,AP.,***Sec.221.***101ITR531]**

4. Royalty is not a tax. **[State of West Bengal - Kesoram Industries Ltd, SC,** *Constitution of India.***187CTR219, 266ITR721]**

5. In the generic sense,tax, toll, subsidies, etc., are manifestations of the exercise of the taxing power.The primary purpose of a taxing statute is the collection of revenue.Tax is levied as a part of the common burden.The basis of a tax is the principle of ability or capacity. In the case of a tax, there is no identification of a specific benefit and even if such identification is there, it is not capable of direct measurement. In the case of a tax, a particular advantage, if it exists at all, is incidental to the State's action.It is assessed on certain elements of business, such as, manufacture, purchase, sale, consumption, use or capital, but its payment is not a condition precedent.It is not a term or condition of licence. A fee is generally a term of licence.A tax is a payment where the special benefit, if any, is converted into common burden.On the other hand, a fee is based on the principle of equivalence.This principle is the converse of the principle of ability to pay.In the case of a fee or compensatory tax, the principle of equivalence applies.The basis of a fee or a compensatory tax is the same.The main basis of a fee or compensatory tax is the quantifiable and measurable benefit. In the case of a tax, even if there is a benefit, the same is incidental to the

Government action and even if such benefit results from the Government action,the same is not measurable.Under the principle of equivalence,as applicable to a fee or a compensatory tax, there is an indication of a quantifiable data,namely,a benefit which is measurable.A tax can be progressive. **[Jindal Stainless Ltd.-State of Haryana,SC,**<u>*Constitution of India.*</u>**283ITR1]**

6. The expression tax has been defined in section 2(43) of the Income-tax Act,1961 and under the definition tax does not include interest which has been independently referred to under section 2 (28)(a) of the Act. While considering the language of a section, the courts are bound to look at the definitions in the legislation for the purpose of interpreting and construing the expressions and words under the Act, the object being to avoid conflict and have a harmonious interpretation unless the context otherwise requires. For the purpose of section 249(4) of the Act, the deposit of tax, which is a condition precedent,does not include interest under sections 234B and 234C of the Act. **[Manoj Kumar Beriwal,Bom.,** <u>*Sec.234.*</u>**217CTR,407,316ITR218]**

7. The phrase tax as contemplated under section 179 of the Income-tax Act,1961,does not include penalty and interest in so far as the directors of the company are concerned. **[H.Ebrahim-DCIT,Ker.,** <u>*Sec.179.*</u>**332ITR122]**

Tax due **SC** **Harshad Shantilal Mehta - Custodian**

Tax due usually refers to an ascertained liability. However, the meaning of the words taxes due will ultimately depend upon the context in which these words are used. . . . In the Special Court (Trial of Offences Relating to Transactions in Securities) Act, 1992, the words taxes due occur in a section dealing with distribution of property.At this stage, the taxes due have to be actually paid out. Therefore,the phrase taxes due cannot refer merely to a liability created by the charging section to pay the tax under the relevant law. <u>*Sec.2(43).*</u> **[231ITR871,99Taxman16]**

Tax outstanding **SC** **CWT - Kantilal Manilal**

Section 2(m)(iii)(a) of the Wealth-tax Act,1957, comes into play only after a demand of payment of tax has been made. The clause speaks of tax outstanding in consequence of an order passed under the relevant taxing statute.Tax becomes payable in consequence of such an order when a notice of demand is served on the assessee. Where the notice of demand is served on the assessee subsequent to the relevant valuation date, it cannot be said that on the valuation date the amount is outstanding and, in such a case, a material requirement of section 2(m)(iii)(a) is not satisfied and, therefore, that provision cannot be invoked by the Department to deny deduction of such an amount of tax as a debt in the computation of the net wealth of the assessee.Parliament intended, in enacting section 2(m)(iii)(a), that if the amount of tax payable in consequence of an order was challenged by the assessee as not being payable by him by recourse to any of the statutory remedies, by way of an appeal or other proceedings, such claim to deduction would be barred. Plainly, in order to give effect to that intent, it is immaterial whether the statutory remedy is being availed of on the valuation date or has been taken thereafter.A challenge by the assessee that the amount outstanding is not payable by him is sufficient to bar his claim to deduction of the amount whether the challenge is subsisting on the valuation date or is initiated after the valuation date has passed.In order to invoke the bar prescribed by section 2(m)(iii)(a), it is necessary for the Revenue to establish that both the above requirements are satisfied, that is to say, that an amount of the tax is outstanding on the valuation date and further that the amount is claimed by the assessee in appeal as not being payable by him. *Sec.2(m).WT.Act.***[152ITR447]**

Tax paid **Raj. Addl. CIT - Farasol Ltd.**

The object of section 40(a)(i) of the Income-tax Act,1961, is to protect the interests of the Revenue by ensuring that in respect of interest chargeable under the Act and payable outside India, the tax payable by the non-resident is either paid or deducted in cases where the non-resident does not have any agent in India from whom it can be recovered.From this point of view, it is immaterial whether the Revenue has received payment of the tax due either by a voluntary act on the part of the assessee or by initiation of the recovery proceedings against the assessee. It may also be observed that, under the Act, an involuntary payment of tax, whether by way of deduction at source or by way of recovery under the provisions of the Act, is regarded as tax paid. A perusal of sections 199 and 237 makes this clear. Hence, the word paid as used in sub-clause (i) of clause (a) of section 40 covers not only voluntary payments but also payments made otherwise,i.e.,the amount recovered from the assessee under the provisions of the Act. *Sec.40(a).***[163ITR364, 22Taxman418]**

Tax payable Mad. Chengalvaroya Chettiar - CIT

The words tax payable in the said section mean tax which the firm or partnership would be liable to pay if it had not been discontinued, tax either found to be due already or that will be found to be due in the future.

Sec.187.[5ITR70]

2. The expression tax payable in section 280ZB of the Income-tax Act, 1961, cannot mean the tax payable on self-assessment under section 140A or provisional assessment under section 141: it can only mean the tax payable upon regular assessment by the Income-tax Officer for the base year or the succeeding base year. Entitlement to the grant of a tax credit certificate accrues only upon such regular assessment and the right to adjustment of any existing tax liability or to refund accrues only after the grant of the tax credit certificate. A manufacturing company is not, therefore, entitled, while paying tax upon self-assessment for any assessment year subsequent to the base year or the succeeding base year, to adjust the amount of tax credit on the basis of its liability for that earlier year unless regular assessment for that year has been made.

[Orissa Cement Ltd.-CBDT,Del.,*Sec.280ZB.*84ITR451]

Tax so assessed MP. Kaushal Construction Co. - CIT

In the case of reassessment following an assessment, the quantification of tax cannot be on the income which is found to have escaped assessment because the escaped income has to be added to the income earlier determined for finding out the total tax liability, as the rate of tax is affected by the increase in the total income. So, at the stage of assessment of tax, even in cases under section 147, the tax is assessed not only on the income that had escaped assessment but on the total income. Out of the tax so assessed any tax that had already been paid would be deducted and the tax payable would be the tax assessed minus the tax already paid. But the assessed tax even in cases of reassessment would be that tax which is assessed on the total income including the income that had escaped assessment. This is clear from the Explanation to section 271(1) which says that assessed tax means the tax as reduced by the sum, if any, deducted at source under Chapter XVII-B or paid in advance under Chapter XVII-C.If the intention of the law-makers was that assessed tax should mean the difference between the tax assessed at the stage of reassessment and the tax originally assessed, the explanation would have said so.

<u>*Sec.271.*</u>**[15CTR146,126ITR466]**

Taxable receipts SC Elel Hotels and Investments Ltd.

The taxable receipts as defined in the statute cannot be held to fall outside such a wider connotation of incomein the wider constitutional meaning and sense of the term as understood in entry 82, ListI. <u>*HotelReceiptsTaxAct*</u>**[77CTR168,178ITR140,44Taxman304]**

Taxi All. CIT - Jagran Prakashan

That the word taxi denotes only passenger vehicles and not goods vehicles. <u>*Sec.37.*</u>**[296ITR627]**

Technical know- how Del. Lurgi India Co. P. Ltd. - CBDT

The meaning of technical know-how is the knowledge which would enable a company receiving such know-how to do the project.

<u>*Sec.80MM.*</u>**[121ITR287]**

2. There is no ambiguity in the Explanation, which defines the words technical know-how, in section 35AB of the Income-tax Act, 1961. The definition is very specific. It means industrial information or technique which is likely to assist in the manufacture or processing of goods or in the working of a mine, oil well or other sources of mineral deposits. **[CIT-Frontline Software and Services P.Ltd.,MP.,**<u>*Sec.35AB.*</u> **301ITR396]**

Technical or professional qualification Mad. CIT-Smt. R. Bharathi

The words technical or professional qualification do not necessarily connote a qualification conferred by a recognised university after examining the candidate who has undergone a course of study in the technical subject or a course of study preparing him for a profession like law, accountancy, etc. The term qualification must be given a wide meaning as referring to the qualities which are required to be possessed by a person performing the work that he does, so long as that work is capable of being regarded as technical or professional. The word professional is a term capable of very broad meaning and would encompass varieties of occupation, although the term is also capable of being given a limited meaning where the context so requires. In the context in which the words professional qualifications are used in the Act, it is not possible to hold that Parliament intended to confine the scope of the proviso only to the professions such as medicine, law,engineering or accountancy. A large number of occupations that are being practised and which form a source of livelihood are capable of being regarded as professions, so long as they require a degree of skill. The degree of skill required is a matter for examination in each case. A person having skill, experience and competence in that line of work can be regarded as professionally qualified for the purpose of section 64 (1) (ii), proviso.

Sec.64. **[158CTR127,240ITR697,107Taxman93]**

Technical service **Del.** **Bharti Cellular Ltd**

The expression technical service would have reference to only technical service rendered by a human. It would not include any service provided by machines or robots.

Sec.9(1)(vii.)r.w.s.194J.
[220CTR258,319ITR139,175Taxman573]

2. Technical services in section 80-O would not include commercial services or managerial services. Managerial service may be professional, service like legal or medical service, but it would not be technical service like engineering service. The main reason why the word technical in section 80-O cannot be given a wider meaning to include managerial or commercial is that the performance of managerial or commercial services by an Indian company for a foreign enterprise would amount to virtually managing or running the foreign company and remuneration obtained by the running or managing of a foreign company would be in the nature of profits, while section 80-0 deliberately restricts itself to income by way of royalty, commission or fees and excludes other types of remuneration.

[J.K.(Bombay)Ltd.-CBDT,Del.,*Sec.80-O.***118 ITR312,1Taxman537]**

3. The format or the method of technical services may vary depending upon the nature of the work undertaken or entrusted. It is for the parties to agree upon what should be the nature of technical assistance or service to be rendered. In a given case, supply of design and drawing could be in the nature of technical services. Supply of design and drawing cannot, in all circumstances, be treated as cost of plant and machinery. Supply of technical services in the nature of printed documents by way of design, chart or drawing, depending upon the nature of the services rendered, can be treated as technical service.

[AEG Aktiengesllschaft -CIT, Kar.,*Sec.9.* **188CTR497, 267ITR209, 137Taxman1]**

4. Explanation 2 to section 9 of the Income-tax Act has clearly defined what is technical services. Technical services have been defined as managerial, technical or consultancy services including the provision of services of technical or other personnel.

[Goa CarbonLtd.-V.M.Muthuramalingam,Bom., *Sec.9.***251ITR348]**

Technician AAR John A. Sayre - CIT

Under the Explanation to section 10(5B) a technician may also mean person who has specialised knowledge and experience in, inter alia, construction or manufacturing operations. In order to succeed on this point, the applicant has to establish that not only has he specialised knowledge and expertise in constructional and manufacturing operations but he was also employed in India for this purpose. On the question whether the applicant was eligible for the benefit of section 10(5B) of the Income-tax Act, 1961, the authority on the stated facts ruled.

Sec.10(5B). **[151CTR651,236ITR652,103Taxman78]**

2. The definition of the expression technician in section 10(5B) includes a person having specialised knowledge and experience in constructional or manufacturing operation.

[P.No.20of,AAR., *Sec.10(5B).r.w.s.245Q.***153CTR119,237ITR382, 115Taxation113,104Taxman168]**

3. .....technician has been defined as a person having specialised knowledge and experience in constructional or manufacturing operations . . . It may be that in the course of marketing, the company would have to install the machinery at the working place of the customer but the installation of a machinery which had been manufactured abroad in the laboratory or in the office of a customer would not amount to constructional activity either by the company or by the individual employee of the company. Affixation of an instrument at a place where it is expected to function cannot amount to construction. **[P.36 of 1998,AAR.,** *Sec.10(5B).* **158CTR364, 242ITR698,109Taxman39]**

Termination of service Ker. State Bank of Travancore - CBDT

The word termination of service is used in section 17(3)(i) in a liberal sense so as to take all categories of cases such as voluntary retirement, superannuation, compulsory retirement, resignation, dismissal and so on. *Sec.17.***[282ITR587,151Taxman133]**

Terms SC CIT - Anjum M.H. Ghaswala

The expression terms used in sub-section (6) does not contemplate any power to waive or reduce tax, penalty or interest : all that the expression means is that the Commission can stipulate the conditions of payment like installments, last date for payment, etc. Beyond that, sub-section (6) does not authorise waiver or reduction of tax, penalty or interest settled under sub-section (4). *Sec.245D.r.w.s.234.* **[171CTR1,252ITR1,119Taxman352]**

Territorial waters, Continental shelf, EEZ Mad. CIT - Ronald William Trikard

Under the provisions of article 297 of the Constitution all lands, minerals and other things of value underlying the ocean within the territorial waters or the continental shelf or the exclusive economic zone of India vest in the Union. Although the Constitution does not itself define the terms territorial waters, continental shelf and exclusive economic zone, clause (3) of article 297 states that their limits shall be such as may be specified by Parliament. In 1976 , Parliament implemented the amendments introduced to the Constitution of India in the years and by passing the Territorial Waters, Continental Shelf,Exclusive Economic Zone and Other Maritime Zones Act,1976 . A careful reading of sections 3, 6 and 7 of the 1976Act shows that territorial waters,the sea bed and sub-soil underlying therein and the air space over such territorial waters form part of the territory of India and that the sovereignty of India extends over such areas.However, the position is different in the case of the continental shelf and the exclusive economic zone of India. The continental shelf of India comprises the sea bed beyond territorial waters to a distance of 200 nautical miles.The exclusive economic zone represents the sea or waters over the continental shelf. *Constitution of India.***[215ITR638]**

Textiles **SC** **CIT - Sundaram Spinning Mills**

The word textiles in it is not used in isolation but is stretched by bringing in more in its company through the following words including those dyed, printed or otherwise processed made wholly or mainly of cotton including cotton yarn, hosiery and rope.Thus textiles as is understood at common parlance or as is understood in its natural sense which is limited, is not indicated here. The Legislature has deliberately widened its sphere for the purpose of giving larger benefit to other items included in it by extending it to include even cotton yarn, hosiery and rope to be understood as textiles.It is always open for a Legislature to stretch or shrink or to give an artificial projection to any word including one used for goods,to make it more meaningful to subserve the objectives it intends to achieve. That is why this inclusive clause brings in more goods, which may not strictly come within the field of such goods. This is in order to give them similar benefit or to make them equally treated. Similarly, hosiery and rope could not but for their inclusion under this item have been classified as textiles. Similarly may be cotton yarn. It is true that manufacture of cotton yarn is a stage earlier than manufacture of textiles as understood commonly. In fact, cotton is the first stage, next comes cotton yarn which finally produces textiles.But here the Legislature intended to give the higher rate of initial depreciation even to the manufacture of goods which commonly understood could not have been included as textiles.So, this entry has to be interpreted to subserve the intended objective of the Legislature. *Sec.32.r.w.Sch.IX.***[241ITR350, 108Taxman337]**

Textiles made wholly or mainly of cotton **Mad.** **Premier Mills Ltd.- CIT**

The very wording of item 32, Fifth Schedule to the Income-tax Act, 1961,clearly shows that the expression textiles made wholly or mainly of cotton is meant to be exhaustive and not expansive. It is not possible to effect a division of item 32 into four categories. A reading of items 32 and 33 would clearly show that Parliament kept in mind a clear distinction between textiles made wholly or mainly of cotton including cotton yarn and textiles made wholly or mainly of jute including jute twine. Hence, the words cotton yarn found in item 32 have to be read in a restrictive sense and cannot be applied to any yarn made out of material other than cotton such as staple fibre. *Sec.33.r.w.Sch.***[152ITR457]**

The and such Mad. CIT - Bosotto Brothers Ltd.

The use of the definite article the,and the word such in section 10 makes it clear that the business contemplated in clause (vi) of the section is the business carried on by the assessee, and not by any person or person other than the assessee, and what it is in respect of a building belonging to the assessee, and used for his business, that he claim a depreciation allowance. <u>*Sec.32.*</u>**[8ITR41]**

The A.O.is of the opinion Mad. CIT - NEPC India Ltd.

The words in section 40A(2)(a) the Assessing Officer is of the opinion indicate that the opinion must be formed by the Assessing Officer and it is of course, implicit that the opinion must be an honest opinion,having been formed based on the circumstances available before him.There should be some material available for the Assessing Officer for invoking section 40A(2)(a) to initiate action to disallow or refuse to deduct the excessive or unreasonable expenditure mentioned thereunder. ...... such decision of the Assessing Officer should be based on reasons well-founded, which are judiciously acceptable. <u>*Sec.40A(2).*</u>**[303ITR271]**

The amount of tax, if any, payable by him SC CIT-Vegetable Products Ltd.

Tax payable is not the same thing as tax assessed. The tax payable is the amount for which a demand notice is issued under section 156 of the Income-tax Act,1961.In determining the tax payable, the tax already paid has to be deducted. Hence, the expression the amount of tax, if any, payable by him in the earlier part of section 271(1)(a)(i) refers to the tax payable under a notice of demand. The words the tax in the latter part of the provision can only refer to the tax, if any, payable by the assessee mentioned in the earlier part of sec.271(1)(a)(i). <u>*Sec.271(1)(a).*</u> **[88ITR192]**

The amount of the interest paid in respect of capital Mad. K.Somasundaram & Brothers - CIT

Section 36(1)(iii) of the Income-tax Act,1961, refers to the amount of the interest paid in respect of capital borrowed for the purposes of the business or profession.It is implicit in this provision that the capital so borrowed should not only be invested in the business, but that the amount borrowed should continue to remain in the business. ......This provision cannot be construed as enabling an assessee to burden the business with interest even while taking the amount initially borrowed for the business, but subsequently taken out of the business by diverting it as interest- free loans to relatives ofthepartners. *Sec.36(1)(iii).***[153CTR153,238ITR939,153Taxation 560]**

The application is in order All. Hajie Saeed and Sons - CIT

The expression the application is in order in rule 6A relates to the form of the application and not to the correctness of the statement made therein. *Sec.184.***[15ITR51]**

..the assessee offers no explanation SC CIT - P. Mohanakala

The expression the assessee offers no explanation means the assessee offers no proper,reasonable and acceptable explanation as regards the sums found credited in the books maintained by the assessee. *Sec.68.***[210CTR20,291ITR278,161Taxman169]**

The date of assessment Bom. Sarangpur Cotton Manufacturing Co. Ltd. - CIT

The date of the assessment under section 23 up to which interest is payable to an assessee on the advance tax paid by him under section 18A, refers to the date of the assessment under section 23 in fact.The mere fact that an assessment made under section 23 was set aside on appeal by the Appellate Assistant Commissioner, and a fresh assessment was made at a subsequent date, would not entitle the assessee to interest on the advance tax up to the date of the subsequent assessment. *Sec.143.***[31ITR698]**

2. The plain meaning of the words the date of assessment occurring in section 73A(b) is the date on which assessment is made by the primary authority under section 58 of the Act and not the date of disposal of the appeal, revision,etc. **[Merla Sitarama Prasad - ACED,AP.,***Estate Duty Act.* **80ITR672]**

The number of days on which the factory or SC South India Viscose Ltd. - CIT

In Part I of Appendix I to the Income-tax Rules,1962,relating to extra shift depreciation allowance the words the number of days for which the concern worked double shift or triple shift, as the case may be as well as the words the number of days on which the factory or concern actually worked during the previous year in clauses (a) and (b) clearly indicate that for the purpose of calculating the extra shift depreciation allowance allowable under Part I of Appendix I to the Income-tax Rules, the number of days on which the concern as a whole actually worked double shift and triple shift has to be determined and it is not necessary to see whether any particular item of machinery or plant had actually been used in double shift or triple shift on the days on which the concern had worked in double shift or triple shift. All that is excluded from extra shift depreciation allowance are the items of machinery or plant against which the letters N.E.S.A. were inscribed in sub-item (ii) of the Table in Part I of Appendix I to the Rules and certain specified items of machinery or plant to which the general rate of depreciation of 10 per cent. is applicable.

Sec.32.**[AIR 1991 SC 2940,114CTR374,227ITR286,139Taxation 688, 93Taxman 371]**

The person liable to penalty Gau. CIT - Maskara Tea Estate

The expression the person liable to penalty in section 271(2) refers to persons who are liable to penalty under section 271(1)(a) read with section 271(1)(i). The expression refers to the persons who are liable to penalty under sections 271(1)(a)/271(1)(i) and no other person.Those exempted from liability to penalty cannot be described as the person liable to penalty under section 271(2) of the Act. If a person liable to penalty under section 271(1)(i) happens to be a registered firm its case squarely falls under section 271(2). A registered firm is denuded of the advantage of registration when it is liable to penalty under section 271(1). It takes the colour of an unregistered firm and the penalty imposable shall be the same as would be imposable on that firm if that firm were an unregistered firm.The penalty imposable shall be in the manner and method provided in section 271(1)(i). Sec.271.
[21CTR47,130ITR955,6Taxman191]

The public Ker. CIT-Aspinwall and Co. Ltd.

A company in which the public are substantially interested is included in the expression the public.When two categories of persons, namely, a director or a company to which section 2(18) (b)(B)(i) does not apply,are said to be not included in the expression the public under sub-clause (d) there of, every other category of persons must, by implication, be deemed to be included by the expression the public.Otherwise, the exclusion of certain persons or companies alone becomes meaningless.Thus understood, a company in which the public are substantially interested clearly comes within the expression the public in section. Sec.2(18).**[98ITR291]**

The same business All. Shadi Ram Ganga Prasad-CIT

The same business in section 24(2) means a business that is carried forward as a whole,without keeping any distinction inter se between the several kinds of business that an assessee may be carrying on. Sec.73.**[68ITR701]**

The same business, profession or vocation Cal. CIT - Jiwanmal Bhuturia

The words the same business, profession or vocationin section 24(2) could not be held to cover more than one business or more than one profession or vocation. Sec.73.**[19ITR444]**

The subject of an appeal Bom. Jagmohandas Gokaldas -CWT

An order can be said to be the subject of an appeal within the meaning of proviso (b)to sub-section (1) of section 25 of the Wealth-tax Act (which prohibits the Commissioner from revising an order where the order is the subject of an appeal to the Assistant Commissioner or the Appellate Tribunal), only if the Appellate Court has considered the order on the merits. Sec.25WT.Act. **[50ITR578]**

2. An order is made the subject of an appeal within the meaning of clause (c) of the first proviso to sub-section (2) of section 33A only when it is the subject- matter of an effective appeal. If an appeal to the Appellate Tribunal under section 33 is not admitted and is disposed of on the ground that it was filed after the prescribed time, the order cannot be said to have been made the subject of an appeal, and the Commissioner can entertain an application for revision under section 33A. **[Krishna Flour Mills Ltd.-CIT, Mys.,** Sec.264.**55ITR259]**

The tax, if any,payable by him Del. CIT - Hindustan Industrial Corporation

The words the tax, if any, payable by him,as used in clause (i) of section 271(1) means the tax which has been assessed on the assessee and is chargeable and not the residue of the tax to be paid by him after making adjustment of the tax already deposited by him. Sec.271(1). **[86ITR657]**

The whole of the amount of profits and gains of business **SC** **Totgar's Co-operative Sale Society Ltd - ITO**

The words the whole of the amount of profits and gains of business in section 80P(2)of the Income-tax Act,1961,emphasise that the income in respect of which deduction is sought by a co-operative society must constitute the operational income and not the other income which accrues to the society. Sec80(P).**[229CTR209, 322ITR283,188Taxman282]**

Thenceforward **Mad.** **G.Chandrasekhara Reddi-CIT**

Section 61(3) of the Indian Income-tax Act, 1922, does not by itself prevent the Income-tax authorities (including the Central Board of Revenue)from permitting a person against whom they had passed an order under that provision to appear before the authorities on behalf of the assessees at a later date, provided, of course, there was nothing in the original order itself,which precluded them from so doing.An order in terms of the language in section 61(3) without mentioning the period of disqualification would not be such an order as to preclude reconsideration on a subsequent date. The word thenceforward in that sub-section-prima facie-fixes only one limit,namely, the limit from which the order begins to have effect. By itself it does not necessarily fix the other limit.Thenceforward cannot be read as thenceforward and forever. Sec.288.**[19ITR616]**

Thereon **Ker.** **Mrs. Lucy Kochuvareed - CWT**

The expression thereon restricts the jurisdiction of the Tribunal to the subject matter of the appeal, and the words pass such order as the Tribunal thinks fit include all the powers (except possibly the power of enhancement),which are conferred on the Appellate Assistant Commissioner. Sec.254.**[183CTR308,263ITR215,131 Taxman159]**

2. It is evident from the provisions of section 254 of the Income-tax Act, 1961,and rule 11 of the Income-tax (Appellate Tribunal) Rules, 1963, that the power of the Tribunal while dealing with an appeal is in the widest possible terms and the word thereon occurring in section 254 of the Act, restricts the jurisdiction of the Tribunal to the subject matter of the appeal only. The Tribunal therefore has jurisdiction to go into every aspect of the assessment proceedings before the taxing authorities and into the question whether such assessment was made in accordance with law provided a ground is taken before the Tribunal in that respect or an additional ground by way of amendment is allowed by the Tribunal.The Tribunal also has the jurisdiction to examine a question of law which arose from the facts as found by the taxing authorities which has a bearing on the tax liability of the assessee. **[The Assam Tribune-CIT, Gau.,** *Sec.254.***285ITR452]**

Thing **Raj.** **CIT - Trinity Hospital**

The dictionary meaning of the word thing is a product of work or activity; useful or appropriate object. *Sec.32.***[131CTR328,225ITR 178,87Taxman127]**

Through or from any money lent at interest **Bom.** **Salzgitter Industrie Bau Gmbh**

Section 9(1)(i) of the Income-tax Act, 1961, brings within the ambit of the charging section income accruing or arising whether directly or indirectly in five situations. One such situation is where income is accruing or arising, whether directly or indirectly, through or from any money lent at interest and brought into the taxable territories in cash or in kind.Admittedly, the situation contemplated herein is that the income arises outside India but by the statute, it is deemed to accrue or arise within the taxable territory. However, the deeming provision is not absolute. It is applicable only if the conditions laid down therein are satisfied.On the face of it, this part of the clause can conveniently be divided into three parts, namely: (1) income accrues or arises, whether directly or indirectly; (ii) through or from any money lent at interest; and (iii) brought into India in cash or in kind. The incident of bringing money into India in cash or in kind to the knowledge of the lender or borrower is an integral part of the transaction. *Sec.9(1).***[82CTR284,184ITR7]**

To account **Bom.** **Sheraton Appearels - ACIT**

.... to account means to reckon, and it is difficult to conceive of any accounting which does not involve either additions or subtractions or both of these operations of arithmetic. <u>*Sec.271(1).*</u>**[256ITR20]**

To assess SC ITO - K.N. Guruswamy

In the normal sense to assess means to fix the amount of tax or to determine such amount. The process of reassessment is to the same purpose and is included in the connotation of the term assessment. Section 34 of the Indian Income-tax Act contemplates different cases in which the power to assess escaped income has been given. Where there has been no assessment at all, the term assessment may be appropriate and where there was assessment at too low a rate or with unjustified exemptions, the term reassessment may be appropriate,and it may have been necessary to use the two different terms to cover with clarity the different case dealt with in the section.But this does not mean that the two terms should be treated as mutually exclusive or that the word assessment should be given a restricted meaning. <u>*Sec.147.r.w.s.2*</u>**[34ITR601]**

To inform Ker. United Mercantile Co. Ltd. - CIT

To inform means to impart knowledge and a detail available to the Income-tax Officer in the papers filed before him does not by its mere availability become an item of information.It is trasmited into an item of information in his possession only if,and only when, its existence is realised and its implications are recognised. <u>*Sec.147.*</u> **[64ITR218]**

To the extent SC CED - Parvati Ammal

The words to the extent in section 10 cannot that, if the donee does not assume immediate bona fide possession and enjoyment of a part or fraction of the gifted property and thenceforward retain it to the entire exclusion of the donor or of any benefit to him by contract or otherwise, it shall be that part or fraction of the gifted property which shall be deemed to pass on the death of the donor. Those words thus seek to restrict the liability to pay estate duty in respect of only the aforesaid part or fraction of the property. <u>*Estate Duty Act.*</u>**[97ITR621]**

to the extent to which such income is applied Del. DIT(Exemp.) -National Association of Software and Services Companies

That the contention that the words to the extent to which such income is applied to such purposes in India appearing in section 11(1)(a) of the Act only require that the charitable purposes should be confined to India and that the application of the income of the trust to the execution of such purposes can be outside India, is opposed to the natural and grammatical meaning that can be ascribed to the words. The word applied is a verb used in the past tense. The requirement of the provision is that the income of the trust should be applied not only to charitable purposes, but also applied in India to such purposes. Therefore, on a proper interpretation of section 11(1)(a) of the Act, the amount spent by the assessee-trust in Germany could not be considered as application of the income of the trust in India for charitable purposes. *Sec.11(1)(a).* **[345ITR362]**

Too low a rate SC Sundaram and Co. (Pvt.) Ltd. - CIT

The word rate in the expression too low a rate in section 34(1)(b) of the Indian Income-tax Act, 1922, does not mean a fraction of the total income.By the use of the word rate in the context in which it occurs, undoubtedly a relation between the taxable income and the tax charged is intended.But the relation need not be of the nature of proportion or fraction.The word rate is often used in the sense of a standard or measure.Provided, the tax is computable by the application of a prescribed standard or measure, though not directly related to taxable income, it may be called tax computed at a certain rate.The rebate of tax and the reduction of such rebate are essentially matters of measure or standards of rate. *Sec.147.* **[66ITR604]**

Tools and implements of agriculture Cal. Kanan Devan Hills Produce Co. Ltd. - CWT

Tools and implements of agriculture include not only all instruments used in ancient times but also those used in modern times,such as bullock-drawn ploughs,manually operated spades, shovels and sickles as well as power-driven tractors, earth movers, mechanism for irrigating fields and lifting water by electrically operated water pumps from deep tube wells,mechanical sprayers and insecticidal instruments,etc.If electrical transformers and switch gears supply energy for raising agricultural produce,they will be implements used for raising agricultural produce.

Sec.5(1)(ix)WT.Act. **[67ITR 823]**

Total income JK. CIT - Mohd. Amin Tyamboo

Total income as defined in section 2(45) means the total amount of income of as assessee from all sources, including (i) income,other than those mentioned in Chap. III, chargeable to tax under the Act; (ii) incomes chargeable to tax by way of deduction at source under section 198; (iii) incomes deemed to be those of the assessee by force of certain provisions including section 64 of the Act; and (iv) incomes on which no tax is leviable but which are nevertheless included for rate purposes, but after making deductions by way of permissible expenditure under Chap.VI-A and section 280-O.

<u>*Sec.2(45).*</u>**[14CTR190,125ITR375]**

2. The expression total income referred to in the Explanation to section 271(1)(c) refers to the total income disclosed by the assessee in his original return filed under section 139(1).Whatever is disclosed subsequently by way of filing a revised return may relate back and become part of the total income provided it is covered by section 139(5). In a case where the disclosure is not bona fide, as for example, where the revised return is filed on coming to know that the Income-tax Officer was investigating certain transactions, although it may not be necessarily fraudulent or wilful,the difference of 80 per cent.has to be reckoned with reference to the income as shown in the return filed under section 139(1).

[Mohammad Ibrahim Azimulla-CIT,All.,<u>*Sec 271(1)(c).r.w.s.139.*</u> **131ITR680]**

3. Section 5 of the Income-tax Act,1961,lays down that all the income, profits or gains which had accrued or arisen to the assessee or been received by or on behalf of the assessee would be liable to be included in his total income.This provision is subject to the other provisions of the Act and, therefore, if the income of any person is declared by section 64 to be includible in computing the total income of the assessee such income would form part of the total income exigible to tax under section 4 of the Act.

[Smt.Radha Gajapathi Raju-CIT, Mad.<u>*Sec.5.r.w.s.64.*</u>**87CTR 113,190ITR144,100Taxation 252, 55 Taxman87]**.

4. It is not possible to read the expression total income used in section 36(1)(viii) to mean total income computed in accordance with the provisions contained in sections 30 to 43A, as provided in section 29,so as also to take into account deductions admissible under section 36(1)(viii). Since section 36(1)(viii) itself provides that the total income for the purpose of the said provision is total income before the deductions under Chapter VIA,it would mean that, for the purpose of working out deduction under that provision, the total income would be the total income before deductions (1) under Chapter VIA, and (2) under that provision [section 36(1)(viii)]. What was implicit in section 36(1)(viii) has now been made explicit by the subsequent amendment of the said clause (viii) with effect from April1,1985. **[CIT-Gujarat State Finance Corporation,Guj.,** *Sec.36(1)(viii).***196 ITR 822]**

5. Under section 5 of the Income-tax Act,1961,total income includes all income from whatever sources it is derived as provided therein...The total income from different sources is to be computed in the manner laid down in the Act. In the computation of the total income of the assessee under the head Profits and gains of business,the total income has to be arrived at after considering the profit and loss from different heads of income.Total income cannot be computed without adjusting the profit and loss from different heads.Irrespective of adjustments under sections 70,71 and72(1) income from all sources accruing to the assessee in the year has necessarily to go into the calculation of the total income.Section 72 provides for carried forward loss only when such loss cannot be set off against income under any other head.For calculating the total income of the assessee, profits and losses from different sources of income are required to be taken into consideration.Under the Act, there is no provision which gives an option to the assessee to show the profit as income from one source and carry forward the loss from another source of income to the next year. **[CIT - Milling Trading Co.Pvt.Ltd.,Guj.,** *Sec.5.r.w.s.70-72.***[211ITR690]**

6. Section 5(1) of the Income-tax Act, 1961, deals with the scope of total income. It states, inter alia, that subject to the provisions of the Act,the total income of any previous year of a person who is a resident includes all income from whatever source derived which is received or is deemed to be received in India or which accrues to him in India or which accrues to him outside India during such year. In other words, the total income of a resident Indian shall, inter alia, include even income which accrues to him outside India. On the other hand, section 5(2) indicates the meaning of accrual of income.It states, inter alia, that the total income of any previous year of a non-resident shall include all income from whatever source derived which is received by him in India or which accrues to him in India.In other words,broadly, in the case of a resident Indian all income which accrues to him whether in or outside India is taxable whereas in the case of a non-resident only income which accrues to him in India or which is received by him in India is taxable. **[CIT - Avtar Singh Wadhwan, Bom.,** Sec.5(1).**[165CTR 414,247ITR260,115Taxman536]**

7. The definition of the term total income involves two ingredients : firstly,the income must consist of the total amount of income referred to in section 5; and, secondly, it must be computed in the manner laid down in the Income-tax Act.Therefore, the manner of computation laid down in the Income-tax Act forms an integral part of the definition of total income. The correct method of approach is to treat nothing as being charged to tax until by the process laid down by the Act the status of income, profits and gains emerges. The tax is on income and not on gross receipts.

[CIT- Williamson Fin. Services,SC,Sec.2 (1A),(2)(45)r.w.s.5.**213 CTR,612,297ITR17,165Taxman638]**

8. The expression total income is defined in section 2(45) of the Act to mean the total amount of the income referred to in section 5, computed in the manner laid down in the Act. It would thus be clear that for clause (c) of section 253(6) to apply the total income assessed has to be in excess of Rs.2,00,000. The use of the words more than would also indicate that it has to be a positive figure in excess of Rs. 2,00,000.What has to be determined is the purpose for deciding the quantum of fee that has to be paid and what is the total income that is computed by the Assessing Officer for the year to which the appeal relates.It is that figure that determines the quantum of fee payable. **[Gilbs Computer Ltd., Bom.,** Sec.253(6)r.w.s.2(45).**226CTR,19,317ITR159]**

Total turnover Raj. CIT - Sharda Gum and Chemicals

The total turnover refers to turnover of business and cannot be turnover of sale and purchase of goods.There is no justification for importing the definition of the State Sales Tax Act into the Income-tax Act to assign a limited meaning to the expression turnover.The meaning of any expression used in a Central enactment operating throughout the territory of India cannot be interpreted differently for its implementation in different States depending on the expression defined in a particular State for a very different object.

Sec.80HHC.**[288ITR116]**

Total turnover of the business Mad. CIT - Madras Motors Ltd.

The words total turnover of the business would be controlled by and have to be read in the colour of the opening clause. The sub-section has been created only to see the ratio of the income out of the export to the total income out of the business in respect of those goods because of the obvious difficulty of segregating the profits earned out of export alone.The total turnover of the business would contemplate only the business regarding such goods part of which are exported and the others are not so exported. Hence, it is impermissible to apply the section even to goods which are outside the limits of clause (a) of sub-section (2). Sec.80HHC.**[174CTR221, 257ITR60,122Taxman516]**

Towards purchase Bom. CIT - Dr. Laxmichand Narpal Nagda

Taking into consideration the letter as well as the spirit of section 54 of the Income-tax Act,1961,and the word towards used before the word purchase in sub-section (2) of section 54,it is clear that the said word is not used in the sense of legal transfer and, therefore, the holding of a legal title within a period of one year is not a condition precedent for attracting section 54 of the Act. Sec.54(2). **[211ITR804]**

Trade SC State of Punjab - Bajaj Electricals Ltd.

Trade in its primary meaning is the exchanging of goods for goods or goods for money; in its secondary meaning it is repeated activity in the nature of business carried on with a profit motive, the activity being manual or mercantile, as distinguished from the liberal arts or learned professions or agriculture.The question whether trade is carried on by a personal at a given place must be determined on a consideration of all the circumstances. No test or set of tests which is or are decisive for all cases can be evolved for determining whether a person carries on trade at a particular place. The question, though one of mixed law and fact, must in each case be determined on a consideration of the nature of the trade, the various steps taken for carrying on the trade and other relevant facts. Sec.2(13).**[70ITR730]**

2. Trade means that particular business activity where the person engaged in the profession buys or sells.All businesses may be carried on for the purpose of earning profit but only that particular kind of business where the business man buys and sells a commodity can be designated as trade. **[H.Mohmed & Co. – CIT,Guj.,**Sec.2(13).**107ITR637]**

3. Trade means a business which a person has learnt or carries on for procuring subsistence or profit; occupation or employment. The commission received in the real estate transaction was,therefore, income arising out of business.**[CIT - Assam Hard Board Ltd., Gau.,** _Sec.2(13)._**141CTR187,224ITR318]**

Trading loss SC Ramchandar Shivnarayan - CIT

A businessman has to keep moneys either when he gets it as sale proceeds of the stock-in-trade or for disbursement to meet the business expenses or for purchasing stock-in-trade and if he loses such money in the ordinary course of business,the loss is a deductible trading loss.It is immaterial whether the money is a part of the stock-in-trade, such as, of a banking company or a money-lender,or is directly connected with other business operations.The risk is inherent in the carrying on of the business and is either directly connected with it or incidental to it. _Sec.28._ **[7CTR5,111ITR263,50Taxation1]**

Trading receipt Bom. M.R.Goyal - CIT

Consideration received by a businessman for transferring the benefits of a contract entered into in the ordinary course of business, though the particular contract is of great magnitude, is a trading receipt and not a capital receipt.Moneys so received do not stand on the same footing as consideration received for transferring managing agency rights or selling agency rights. _Sec.28._**[46ITR 1095]**

2. The true character of the receipt must be judged with reference to the reasons for the collection and the liability for meeting which, the collection was made.When the liability is a statutory liability, which the assessee was required to meet and for meeting which it was by the statutes or authorities permitted to collect the amount required from its customers, the true character of the collection is a trading receipt.By calling a portion of the amount deposit, it cannot be said that the assessee had constituted itself a trustee,and therefore,the amounts received were not required to be regarded as part of its trading receipt. **[CIT - A.R.A.S.P.V.and PV, Mad.,** _Sec.28._**182CTR524,263ITR616, 129Taxman391]**

Transacting business AAR Fidelity Northstar Fund

The words transact business and the transaction of business ..... postulate transaction of sale and purchase, they do not refer as such to trading activity.Transacting business is different and distinguishable from carrying on business.Transact business is a general term referring to carrying on all types of activities whereas business transaction refers only to commercial trading activities. *Sec.14.r.w.s.115AD .***[207CTR9,288ITR641,158Taxman372]**

Transaction **Mad. G.S.Ramaswamier & Sons - CIT**

The word transaction has a very wide meaning. It can be applied to any particular act done in the carrying on of a business; but one of its meanings is the carrying on or completion of an action or a course of action. *Excess Profits Tax Act.* **[13ITR24]**

2. A transaction entered into by a person, should be an act which is bilateral or multilateral in character,and not a mere unilateral action which is all that occurs when a coparcener throws his self-acquisitions into the hotchpot of his joint family. A transaction by a person must be a transaction with some other person; to interpret the words enter into a transaction as if they had the same meaning as do an act or abstain from doing an act gives no real effect to the words enter and transaction. **[P.K.Subramania Iyer -CGT,Ker.,** *Sec.2(xii),2(xxiv)(d).G.T.Act.* **67ITR612]**

3. Throwing of self-acquired property into joint family property amounts to transfer of property and does not cease to be a transaction within the meaning of section 2(xxiv)(d) of the Gift-tax Act,simply because it is unilateral in nature. **[G.V.Krishna Rao - GTO (First Additional),AP.,** *Sec.2(xxiv)G.T.Act.* **70ITR 812]**

4. The word transaction in clause (d) of section 2(xxiv) takes its colour from the main clause; it must be a transfer of property in some way.Section 2(xxiv) deals with transfer of properties in various ways and not any other transactions. The words disposition, conveyance, assignment, settlement, delivery and payment, are all used to indicate some of the modes of transfer of property. **[CGT - N.S.Getti Chettiar,SC,** *Sec.2(xxiv)G.T.Act.* **82ITR599]**

5. The expression transaction referred to in clause (d) of section 2 (xxiv) takes its colour from the main clause. It must be a transfer of property in some way. A partition in a Hindu undivided family does not effect any transfer as generally understood in law and it cannot beconsidered to be a disposition,conveyance,assignment,settlement, delivery,payment or other alienation of property within the meaning of these words in section2(xxiv)and,therefore,it does not constitute a gift under section 2(xii). **[VidyawatiDeviRathi -CGT, Raj.,** *Sec.2(xxiv)G.T.Act.* **64CTR241,169ITR708,35Taxman242]**

6. The word transaction in clause (d) of section 2(xxiv) of the Gift-

tax Act,1958, takes its colour from the main clause ;it must be a transfer of property in some way.The words disposition, conveyance, assignment, settlement, delivery and payment are all used to indicate some kind of transfer of property.**[Jagatram Ahuja -CGT, SC,**<u>*Sec.2(xxiv)G.T.Act.*</u>**164CTR1,246ITR609,113Taxman459]**

7. For clause (v) of section 2(47) to be applicable there must be a transaction under which the possession of immovable property is allowed to be taken or allowed to be retained.The Legislature advisedly referred to any transaction with a view to emphasize that it is not the factum of entering into agreement or formation of contract that matters, but it is the distinct transaction that gives rise to the event of allowing the contractee to enter into possession that matters. That transaction is identifiable by the terms of the agreement itself and it takes place within the framework of the agreement.What is contemplated by section 2(47)(v) is a transaction which has a direct and immediate bearing on allowing possession to be taken in part performance of the contract of transfer. It is at that point of time that the deemed transfer takes place. Though entering into the agreement/contract might be a transaction in a broad sense,yet when the agreement envisages an event or an act on the happening or doing of which alone possession is allowed to be taken in part performance of the contract, the transaction of the nature contemplated by clause (v) cannot be said to have occurred before thatdate. **[Jasbir Singh Sarkaria,AAR,**<u>*Sec.2(47).*</u>**[212CTR107, 294ITR196,164Taxman108]**

8. A lease of a building or property is a transfer within the meaning of the Transfer of Property Act, 1882. **Auto Sales Properties v. CIT, All.**<u>*Sec.64.*</u>**[294ITR507]**

Transaction entered into SC Goli Eswariah - CGT

The transaction entered into by one person with another, contemplated by clause (d) of section 2(xxiv) of the Gift-tax Act, 1958, cannot apply to a unilateral act. The act must be one to which two or more persons are parties.The declaration by a coparcener whereby he impresses the character of joint family property on his self-acquired property does not fall within clause (d) of section 2(xxiv). <u>*Sec.2(xxiv)G.T.Act.*</u>**[76ITR675]**

Transfer Mad. Wilfred Pereira Ltd. - CIT

The word transfer in section 12B includes both a transfer by act of parties and a transfer by operation of law. <u>*Sec.2(47).*</u>**[53ITR747]**

2. The word transfer has been used in the strict sense and not in the sense of including every means by which the property may be passed from one to another partition of the joint Hindu family even after the assessee threw his self-acquired property into the hotchpot of family property,is not transfer of property under S. 16 (3)(a)(iii). **[CIT-Keshavlal,SC,*Sec.64.*AIR1965SC866,55ITR637]**

3. Section 2(47) defines transfers in relation to a capital asset. This definition gives an artificially extended meaning to the term by including within its scope and ambit two kinds of transactions which would not ordinarily constitute transfer in the accepted connotation of that word, namely, relinquishment of the capital asset and extinguishment of any rights in it.But, even in this artificially extended sense, there is no transfer of interest in the partnership assets involved when a partner retires from the partnership.The interest of a partner in a partnership is not interest in any specific item of the partnership property. It is a right to obtain his share of profits from time to time during the subsistence of the partnership and on dissolution of the partnership or on his retirement from the partnership to get the value of his share in the net partnership assets which remain after satisfying the debts and liabilities of the partnership.When,therefore, a partner retires from a partnership and the amount of his share in the net partnership assets after deduction of liabilities and prior charges is determined on taking accounts on the footing of notional sale of the partnership assets and given to him, what he receives is his share in the partnership and not any consideration for transfer of his interest in the partnership to the continuing partners. His share in the partnership is worked out by taking accounts in the manner prescribed by the relevant provisions of the partnership law and it is this, namely, his share in the partnership which he receives in terms of money. There is in this transaction no element of transfer of interest in the partnership assets by the retiring partner to the continuing partners. The transfer of a capital asset in order to attract capital gains tax must be one as a result of which consideration is received by the assessee or accrues to the assessee. When a partner retires from a partnership what he receives is his share in the partnership which is worked out and realised and does not represent consideration received by him as a result of the extinguishment of his interest in the partnership assets. **[CIT - MohanbhaiPamabhai,Guj.,*Sec.2(47).*91ITR393]**

4. If an existing title in a capital asset is extinguished and a new one created,there is within the meaning of section 12B(1) of the Indian Income-tax Act, 1922 (as reintroduced by the Finance (No. 3)

Act) a transfer of a capital asset.The fact that the divestiture of title takes place under a law relating to compulsory acquisition of property would make no difference.The word transfer is comprehensive and is regarded generally as comprehending within its scope transfers both voluntary and involuntary. In the absence of a distinct genus or category in section 12B(1),no presumption can arise that the word transfer in section 12B(1) must be construed in the sense of a voluntary transfer, because the words sale, exchange and relinquishment are in the normal acceptation of those terms voluntary acts.The words sale, exchange, relinquishment and transfer must be given their plain and actual meanings and there is no justification for restricting the wide comprehension of the word transfer to voluntary transfers by the application of the ejusdem generis rule.The deletion of the clause in the third proviso to section 12B(1) as originally introduced in, which carved out an exception in relation to transfers of capital assets by reason of compulsory acquisition, also reflects indelibly the true legislative intent, viz., that transfers of capital assets by reason of compulsory acquisition are comprehended within the meaning of the word transfer.

[Mangalore Electric Supply Co. Ltd.- CIT,SC,*Sec.2(47)* .AIR 1978SC1272,7CTR61,113ITR655]

5. The word transfer occurring in sections 60, 61 and 62 is not limited to its connotation and meaning given in the Transfer of Property Act, 1882, or other Acts. Section 63(b) defines transfer as including any settlement, trust, covenant, agreement or arrangement. The word arrangement has not been used as a term of art but has been used in a business sense. The factum of advance carrying or not carrying any interest is not decisive of the fact as to whether the advance qualified to be called a loan or not.What is decisive is the circumstance as to whether the advance was made to meet some genuine need for loan. **[S.P.Jaiswal-CIT,P&H.,*Sec.60,63(b).* [130ITR643,2Taxman313]**

6. The expression transfer in section 45 of the Income-tax Act,1961, will have to be read in the light of its definition in section 2(47). A combined reading of section 45 with section 2(47) shows that any profits or gains arising from the transfer of a capital asset,which expression includes sale, exchange or relinquishment of the asset or the extinguishment of rights therein,will be chargeable to tax as capital gains.The conditions precedent for the operation of section 45 are: (i) There should be a transfer of a capital asset, and (ii) As a result of such transfer, gains should have accrued or arisen to the transferor. **[CIT-KartikeyV.Sarabhai,Guj.,*Sec.45r.w.s.2(47).* 131ITR42]**

7. The word transfer in section 84(2)(ii) of the Income-tax Act,1961, cannot be restricted to a case where the full rights of ownership were transferred and that it would include a transfer also of some limited right or interest in or to the property and, therefore, where a transfer is effected by the creation of a lease in a building in favour of the new business or the person carrying on the new business, it would amount to a transfer within the meaning of that expression. It is not necessary that the building transferred to the new-industrial undertaking must have been previously used by the assessee himself in any other business and a building earlier used for business by a stranger would come within the ambit of section 84(2)(ii).

[CIT-Indian Expanded MetalsP.Ltd.,Bom.<u>*Sec.84(2).*</u>**21CTR143, 134ITR483,]**

8. The expression transfer occurring in section 155(5) has not been defined.However,in legal parlance,it bears a wide connotation. Even the transfer of a mere fractional interest in property would come within the conception of transfer of property. There is an indication in section 155(5) itself that Parliament intended the term transfer to be understood in the widest sense possible. This is seen from the expression sold or otherwise transferred. The transfers, as it were, are divided into two categories, sales and non-sales. The expression otherwise exhausts all the categories of transfers other than sales which are transfers of a kind for consideration.The expression otherwise occurring in a combination of words has sometimes been regarded as indicating the application of the ejusdem generis rule. The subject and context of section 155(5) clearly point to the intention of Parliament that the machinery which has obtained a grant of development rebate by reason of its having come under the ownership of the assessee should continue to remain in the same ownership and should not be parted with by him for a period of at least eight years from the date of installation.In this context, therefore, any parting with that asset would involve a breach of the statutory condition, subject to which alone development rebate is originally granted. It stands to reason,therefore, that the expression otherwise transferred must be given such wide amplitude of meaning as is consistent with its ordinary connotation.There can be no warrant for cutting down that meaning, to any extent. **[Baldevji-CIT,Mad.,**<u>*Sec.155.*</u>**40CTR120,156ITR776]**

9. The contribution of joint stock by a partner to the firm would amount to a transfer within the meaning of section 2(47) of the Income-tax Act,1961,and gains arising there from would be liable to tax as capital gains. **[CIT-Ramanbhai B.Amin,Guj.,** <u>*Sec.2(47)r.w.s.45.*</u>**163ITR125,24Taxman639]**

10. The word transfer has been given an artificial definition in section 2(47) of the Income-tax Act, 1961, different from its meaning in the ordinary popular and natural sense. The words extinguishment of rights therein have a wider meaning than mere destruction as used in section 41(2) of the Act. These words have very wide amplitude covering every possible transaction and situation which results in the destruction, annihilation, extinction, termination, cessation or cancellation of any bundle of rights, either qualitative or quantitative,that the assessee had in the capital asset comprising either movable or immovable property.The word transfer spoken of in respect of extinguishment of any rights therein clearly includes a case where there is extinguishment of the capital asset itself. A right in any property could be extinguished even when the property ceases to exist.It cannot be doubted that when a capital asset ceases, the right is also extinguished therein.When an assessee receives money from an insurance company as compensation for the extinguishment of his capital asset, he receives that money in lieu of the capital asset and not in lieu of the premia paid to the insurance company.This amounts to a transfer within the meaning of section 2(47) and the amount received would be liable to be taxed as capital gains. **[CIT-J.K.Cotton Spinning & Weaving Mills Co. Ltd.,All.,** *Sec.2(47).***[164ITR18]**

11. The transfer contemplated under section 2(47) of the Income-tax Act, 1961, envisages, no doubt, sale, exchange or relinquishment of the asset, etc. But mere conversion of one currency into another currency cannot be considered as exchange. The exchange in the context must mean transfer of one capital asset for another capital asset.Like a sale, it requires two persons. There cannot be a sale or exchange to oneself. **[Jayakumari and Dilharkumari-CIT,Kar.,** *Sec.2 (47)r.w.s.45.***165ITR787]**

12. The definition of transfer is an inclusive definition. It specifically refers to sale, exchange, relinquishment,extinguishment or compulsory acquisition.These five categories are specified by way of illustration or by way of abundant caution and not to exclude other categories which naturally come within the expression transfer.The expression must be read widely and not narrowly. The definition denotes extension and cannot be treated as restricted. The expression otherwise transferred occurring in sections 34(3)(b) and 155(5) is wide enough to include a transfer where under the right to exclusive possession and enjoyment stood transferred, albeit subject to a right of reversion in favour of the transferor.

[BlueBayFisheries(P.)Ltd.-CIT,Ker.,*Sec.2(47)r.w.s.34(3)(b)& 155(5).***62CTR66,166ITR1,31Taxman393]**

13. Section 2(47) of the Income-tax Act,1961, applies to transfer in relation to a capital asset. It is not possible to accept that the definition clause, i.e., section 2(47), does not or will not apply while considering the word transfer as used in section 34(3)(b). The contribution by a partner of a capital asset to the firm as capital amounts to transfer within the meaning of section 34(3)(b). **[D.S.Joshi-CIT,Bom.,***Sec.2(47)r.w.s.34.***95CTR5,191ITR302,57 Taxman18]**

14. Section 2(47) of the Act states that, in relation to a capital asset, transfer includes the sale, exchange or relinquishment of the asset or the extinguishment of any rights therein or the compulsory acquisition thereof under any law. This is an inclusive definition, which should not ordinarily bear a restrictive or limited construction.The five categories of transfer mentioned therein are by way of illustration or for abundant caution,and were not intended to exclude other categories which naturally come within the expression transfer. The expression has to be read broadly and not narrowly. The transfer contemplated by section 80J4(ii) comprehends leases as well (except to the extent excluded by the parenthesis therein).There is nothing in the Explanation to section 80J to exclude the value of building or machinery forming part of the lease in computing the value of the building, machinery or plant previously used for any purpose. Even the parenthesis which appears in clause (ii) of sub-section (4) of section 80J not being a building taken on rent or lease does not find a place in the Explanation. Therefore, the value of the building,machinery and plant leased out has to be taken into consideration for the purposes of the Explanation.

[Kerala State Cashew Dev.Corp.-CIT,Ker., *Sec.80J.***205ITR19]**

15. The word transfer in section 2(47) of the Income-tax Act,1961, has a very wide meaning. Its meaning cannot be narrowed down by referring to the provisions of other statutes which are quite different and applicable in different contexts. **[CIT-Tata Iron and Steel Co. Ltd., Bom.,** *Sec.2(47).***113CTR95,206 ITR196,71 Taxman188]**

16. The expression transfer is used in varying senses in different statutes, depending on the context. In a broad sense, it will include acquisition of an asset by a person from whatever source. But the scheme of section 80J of the Income-tax Act,1961 indicates that what is being aimed at is to prevent exemption to those industrial undertakings which are formed by the splitting up or by reconstruction or by transfer to a new business,of plant or machinery of the old business.Transfer, in this context,must mean a transfer of plant or machinery which is essential for the formation of the new industrial undertaking and that must again mean a transfer to the new business of the transferee of any machinery used by the said transferee in his old business. **[CIT- Dandeli Ferro Alloys Pvt. Ltd.,Bom.,*Sec.80J.*129CTR44, 212ITR1,89 Taxman292]**

17. The definition of transfer in section 2(47) is an inclusive one and does not exclude the contextual or the ordinary meaning of the word,transfer. The words otherwise transferred occurring in section 34(3)(b) should bear an appropriate meaning, in the context of the main provision, section 33(1)(a) of the Act. Section 34(3)(b) is closely linked to section 33(1)(a) of the Act. Keeping in view the purpose for which the relief by way of development rebate is afforded under section 33(1)(a) of the Act, in cases where the mach-inery or plant is not wholly used by the assessee for the purpose of business carried on by him, for the specified period, and such user is given over to another,it can be stated that the machinery or plant is otherwise transferred by the assessee to another person.

[CIT-Narang Dairy Products,SC,*Sec.2(47)r.w.s.33.*133CTR65, 219 ITR478,85 Taxman375]

18. Section 2(47)of the Income-tax Act,1961, defines transfer in relation to a capital asset.It is an inclusive definition which, inter alia, provides that relinquishment of an asset or extinguishment of any right therein amounts to a transfer of a capital asset. It is not necessary for a capital gain to arise, that there must be a sale of a capital asset. Sale is only one of the modes of transfer envisaged by section 2(47)of the Act. Relinquishment of the asset or extingui-shment of any right in it, which may not amount to a sale, can also be considered as a transfer and any profit or gain which arises from the transfer of a capital asset is liable to be taxed under section 45. A company, under section 100(1)(c) of the Companies Act,1956, has a right to reduce the share capital and one of the modes which can be adopted is to reduce the face value of the preference shares. Section 87(2)(c) of the Companies Act, inter alia, provides that where the holder of any preference share has a right to vote on any resolution in accordance with the provisions of this sub-section, his voting right on a poll, as the holder of such share, shall, subject to the provisions of section 89 and sub-section (2) of section 92, be in the same proportion as the capital paid up in respect of the prefere-nce share bears to the total paid-up equity capital of the company. Hence,when as a result of the reducing of the face value of the share, the share capital is reduced,the right of the preference share-holder to the dividend on his share capital and the right to share in the distribution of the net assets upon liquidation is extinguished proportionately to the extent of reduction in the capital. Such reduction of the right in the capital asset would clearly amount to a transfer within the meaning of that expression in section 2(47) of the Income-tax Act,1961. **[Kartikeya V.Sarabhai -CIT,SC,** *Sec.2(47).***228ITR163,94Taxman164]**

19. The definition of transfer in section 2(47) clearly contemplates the extinguishment of rights in a capital asset distinct from and independent of such extinguishment consequent upon the transfer thereof.It is not correct to view the expression extinguishment of any rights therein as not extending to mean the extinguishment of rights independent of or otherwise than on account of transfer. To read so is to render the expression ineffective and its use meaningless.The expression includes the extinguishment of rights in a capital asset independent of and otherwise than on account of transfer.

[CIT - Mrs. Grace Collis,SC, *Sec.2(47).***248ITR323,115Taxman 326]**

20. Transfer, as defined in section 2(47)has to be given a simple meaning,taking into consideration the object of the Act.There cannot be different criteria to ascertain the meaning of the expression in the case of sale and lease.

[CIT - Mormasji Mancharji Vaid,Guj., *Sec.2(47).***168CTR565,250ITR542]**

21. The ordinary meaning of transfer under the Transfer of Property Act,1882, read with the Registration Act,1908, is to be understood in a liberal sense for the purpose of the Income-tax Act. Anyone in possession of property in his own title,exercising such dominion over the property as would enable others being excluded there from and having the right to use and occupy the property and/or to enjoy its usufruct in his own right, would be the owner of the property though a formal deed of title may not have been executed and registered as contemplated by the Transfer of Property Act,1882, and the Registration Act,1908.The definition in the Wealth-tax Act has to be understood and considered in the context of the legal and actual ownership meaning thereby title to the property for the assessment of wealth.This definition though the words employed therein are similar and identical, cannot be brought in assistance under the Income-tax Act.

[Electro Zavod (India)Pvt.LTd.,Cal., *Sec.281B.* **278ITR187]**

22. The purpose of introducing clause (v) in conjunction with clause (vi) in section 2(47) of the Income-tax Act, 1961, defining transfer was to widen the net of Taxation of capital gains so as to include transactions that closely resembled transfers but were not treated as such under the general law. Avoidance or postponement of tax on capital gains by adopting devices such as the enjoyment of property in pursuance of revocable power of attorney or part performance of a contract of sale was sought to be arrested by introducing the two clauses, clauses (v) and (vi), in section 2(47).

[Jasbir Singh Sarkaria, AAR,*Sec.2(47).*212CTR107,294ITR196, 164 Taxman108]

Transfer of a capital asset Guj Patel Brass Works - CIT

Transfer has been defined, in relation to a capital asset, to include various contingencies stipulated by sub-clauses of section 2(47) of the Act.Sub-clause (ii) of section 2(47) of the Act states that transfer includes the extinguishment of any rights therein. That requires assigning meaning to the term therein. In other words, for the purposes of transfer of a capital asset, it would suffice if there is extinguishment of any rights in a capital asset.However, even for the purposes of extinguishment of any rights, existence of a capital asset is a must. *Sec.2(24)(vi).r.w.s.45&56.*
[205CTR139,286ITR598]

Transfer of property SC Sunil Siddharthbhai - CIT

In its general sense, the expression transfer of property connotes the passing of rights in property from one person to another. In one case, there may be a passing of the entire bundle of rights from the transferor to the transferee.In another case,the transfer may consist of one of the estates only out of all the estates comprising the totality of rights in the property. In a third case, there may be a reduction of the exclusive interest in the totality of rights of the original owner into a joint or shared interest with other persons. An exclusive interest in property is a larger interest than a share in that property.To the extent to which the exclusive interest is reduced to a shared interest, there is a transfer of interest. *Sec.45, 48.***[49CTR472,156ITR509,23Taxman14]**

Transportation Gau. A.B.C. India Ltd. - CIT

The expression transportation means the movement of goods or persons from one place to another, by a carrier.Carrier,according to the dictionary meaning, is individual or organisation engaged in transporting passengers or goods for hire. *Sec.32.***[145CTR914,226 ITR914]**

Travelling Mad. Beardsell Ltd. - CIT

The word travelling is no doubt capable of being construed in a narrow way as to limit it to the actual time spent on travel, i.e.,the time actually spent on road,rail or air while travelling from one destination to another.In the Income-tax Rules,1962, rule 6D and section 37 of the Income-tax Act,1961,however,that term has been used in a wider sense to include the entire period of absence from the headquarters including the period from the time of the departure till the time of the return, and including the time spent on actual travel and the time spent staying in hotels or elsewhere during the period when the person was not actually travelling, but remained outside the headquarters and had incurred expenditure on such stay. In other words, travelling, for the purpose of these provisions, includes the entire period of absence after the travel commences.It includes the periods spent on actual travel as also the periods spent while staying away from the headquarters before completing the route which brings the person back to the place that hestartedfrom.

*Sec.37.r.w.s.rule-6D.***[150CTR620,246ITR505, 10Taxman 65]**

Tribunal SC Engineering Mazdoor Sabha - Hind Cycles Ltd.

The tribunal as distinguished from the court,exercises judicial power and decides matters brought before it judicially or quasi-judicially,but it does not constitute a court in the technical sense.

*Misc .***[AIR1963SC874]**

Tribunals Cal. Garg Glass Tubes (P.) Ltd. - In Re.,

Apart from statutory Tribunals set up by the Legislature, authorities set up under the statute who perform quasi-judicial functions can also be termed Tribunals for the purposes of article 227 of the Constitution of India. Article 265 of the Constitution of India requires that all taxing officers and, or authorities must levy tax according to the provisions of the particular taxing statute under which they operate, and in order to levy or assess such tax, they have to act judicially. They, therefore, perform quasi- judicial functions and are Tribunals within the meaning of the term in article 227 of the Constitution. The High Court can interfere,under article 227, in order to keep such Tribunals within the bounds of their authority and to see that they do their duty in a legal manner.

Bengal Sales Tax Act.[175ITR422]

Trust for any public purpose of a **Bom.** **Trustees of Gordhandas G.Family Charity Trust - CIT**

The expression trust for any public purpose of a charitable or religious nature connotes a trust for charitable objects involving an element of public utility. Marriage expenses do not normally involve any element of public utility. *Sec.5(1)WT.Act.*[70ITR600]

Turnover **Cal.** **CIT - Chloride India Ltd.**

The general definition of the word turnover or the case law dealing with it in the Sales Tax Act cannot be imported into section 80HHC.

Sec.80HHC.[178CTR432,256ITR625,130Taxman352]

2. The word turnover is not defined in the Income-tax Act. In the absence of a definition the term has to be understood only in the popular sense as defined in the dictionaries.The dictionary meaning of the term is the total amount of business done in a given time; also the amount of goods produced and disposed of by a manufacturer; also the turning over of the capital involved in a business;also the net profit derived from a business in a given time.

[ACIT- South India Produce Company, Ker.,*Sec.80HHC.*262 ITR20]

3. The word turnover is not defined in the Act.Going by the definition in the Sales Tax Act of the word turnover it is clear that the consideration received must be for the sale of goods and it must be available with the assessee for being turned over, or in other words, it must come to the assessee's till as money belonging to him. It has got relation to the capital employed in the business ; turnover has also got relationship with profits of the business. In order that an amount can be included in the total turnover it must either be the purchase price or the sale price or something incidental to the transfer of the goods dealt with by the assessee. In other words,the turnover must relate to the purchase or the sale of the goods made by the assessee.What is includible in the total turnover is only the consideration for the transfer of goods effected by the assessee and the expenses incurred in connection with the said transfer,of course excluding the freight or insurance attributable to the transport of goods or merchandise.

[CIT- Rajendranathan Nair,Ker.,*Sec.80HHC.***187CTR201,265 ITR35,135Taxman360]**

4. One cannot interpret the words total turnover with reference to the definition of the word turnover in other laws like the Central sales tax or as defined in accounting principles.. Just as commission received by the assessee is relatable to exports and yet it cannot form part of turnover for the purposes of section 80HHC,excise duty and sales tax also cannot form part of turnover. Just as interest, commission, etc., do not emanate from the turnover so also excise duty and sales tax do not emanate from such turnover. Since excise duty and sales tax did not involve any such turnover such taxes had to be excluded. Commission, interest, rent, etc., do yield profits, but they do not partake of the character of turnover and therefore they are not includible in the total turnover. If so, excise duty and sales tax also cannot form part of the total turnover under section 80HHC(3). **[CIT-LakshmiMachineWorks,SC,***Sec.80.***210CTR1, 290ITR667,160Taxman404]**

U

Under **Guj.** **CIT - Poonjabhai Vanmalidas**

The word under occurring in section 41(4) should be given the meaning as contemplated by and hence the orders passed under section 10(2)(xi) of the Act of 1922, which was equivalent to section 36(1)(vii) of the Act of 1961, would be covered by the words of section 41(4). *Sec.41(4).***[105ITR388,44Taxation89]**

2. The only meaning of the word under in section 47(iii) can be involving or by way of : to the extent to which there is a shortfall of consideration the transfer can be said to be under or by way of a gift,the word gift being used in its ordinary meaning in common parlance and not in the sense in which it is used in the Gift-tax Act or the Transfer of Property Act. **[Addl.CIT - Mrs. Avtar Mohan Singh,Del.,** *Sec.47r.w.s.52.***136ITR645]**

Under the Act **Guj.** **Union of India - Casam Mohmad Ajam Ismail**

To give too wide a construction to the expression under the Act in section 67 might lead to the serious consequence of attributing to the Legislature, which owed its existence itself to the Constitution, the intention of affording protection to unconstitutional activities by limiting challenge to them only by resort to the special machinery provided by it in place of the normal remedies available under the Code of Civil Procedure. *Sec.293.***[62ITR367]**

Undisclosed income **Guj.** **Rushil Industries Ltd.- Harsh Prakash**

The expression undisclosed income has been defined in section 158B(b) of the Income-tax Act,1961,to include income based on an entry in the books of account or other documents which has not been or would not have been disclosed for the purposes of this Act.

The definition would also include any entries in the books of account or other documents showing concealment of the real source of income.The requirement of section 158BD of the Income-tax Act,1961,is only a prima facie satisfaction by the Assessing Officer that in the search operation there is material to show undisclosed income of a person other than the one who is searched. *Sec.158B.* **[166CTR300,251ITR608,120Taxman67]**

2. If any amount of income has not been taxed and during the course of search,some undisclosed income is found on the basis of material seized, that should be treated as undisclosed income as per the scheme of special assessment under this chapter.

[CIT-Elegant Homes Pvt. Ltd.,Raj.,*Sec.158B.***177CTR261,259ITR 232,171Taxation32,124Taxman819]**

3. The definition of undisclosed income in section 158B(b) of the Income-tax Act,1961,would show that it not only includes within its sweep property which has not been disclosed by an assessee in his/ her return but it also includes within its fold that property which would not have been disclosed for the purpose of this Act by the assessee. **[Smt. Harbans Kaur Bhatia-CIT,MP.,***Sec.158B.r.w.s. 113.***193CTR725,274ITR298]**

4. A reading of clause (b) of section 158B of the Income-tax Act, 1961, shows that undisclosed income includes any money, bullion, jewellery or other valuable article or thing or any income based on any entry in the books of account or other documents or transactions mentioned therein which represents wholly or partly, income or property which has not been disclosed or which would not have been disclosed for the purposes of the Act. Thus, a case where the income has not been disclosed for the purposes of the Act will also be a case of undisclosed income. If the assessee does not disclose any income in the return filed before the due date of filing such return under section 139(1) of the Act, notwithstanding the fact that his income is reflected in books of account at the time of search, his income is to be assessed as undisclosed income.

[Dr.Brijesh Lahoti- CIT, MP.,*Sec.158B.* **200CTR499,282ITR349, 151Taxman216]**

Unit has been set up Bom. CIT - Piem Hotel Pvt. Ltd.

A unit cannot be said to have been set up unless it is ready to discharge the function for which it is being set up. It is only when the unit has been put into such a shape that it can start functioning as a business or a manufacturing organisation that it can be said that the unit has been set up. If the unit is ready to start functioning, it does not matter that it has not actually started its business on the relevant date. Once the business is set up, expenditure incurred concerning such business can be claimed as business expenditure subject to other applicable conditions of the Act being satisfied. *Sec.28.*[**116CTR401,209ITR616,73Taxman295**]

Unless he is himself liable to pay Income-tax as an ..... Del. National Ind. Dev. Corp. Ltd.- CIT

The expression unless he is himself liable to pay Income-tax as an agent was introduced as an exception for the benefit of the person sought to be made liable for deduction under section 195(1) of the Income-tax Act,1961. It was not necessary that the status of a person paying any sum chargeable to a non-resident who was sought to be made liable for deduction of tax should first be determined under section 163 before deciding whether the other requisites under section 195(1) were or were not satisfied. *Sec.195 (1).*[**253ITR489,119Taxman721**]

Urban Consumer Co.op. Society Mys. Mysore Co-operative Society Ltd. - CIT

The Explanation to section 14(3)(iv) of the Indian Income-tax Act, 1922,and to section 81(v) (now section 80P(2)(f)) of the Income-tax Act,1961,is a complete definition of an urban consumers' co-operative society and the definition of the term in other enactments should not be read into the Income-tax Act. If a society carries on the activity of an urban consumers' co-operative society it is for that reason alone an urban consumers' co-operative society and the mere fact that it carries on some other kind of business does not make available to it an argument, that it is not an urban consumers' co-operative society.An urban consumers' co-operative society is, therefore,not entitled to the exemption contained in the above provisions notwithstanding that its operations are not confined to those mentioned in the Explanation. *Sec.80P(2)(f).*[**75ITR445**]

Use Del. E.P.W. Da Costa - Union of India

In order to qualify for a deduction under section 80-O of the Income-tax Act, 1961, he royalty, etc., received by an Indian company from a foreign enterprise must be in consideration for the supply of information for use outside India, and the information must concern industrial, commercial or scientific knowledge, experience or skill. The word use is a very general word.It is not necessary that the use to which the information is to be put must be practical, that is to say, it must result in the manufacturing or making of some concrete thing. *Sec.80-O.*[**121ITR751]**

2. The word use means the employment or using a particular article or thing for profitable purpose in relation to business or trade. In relation to a trade or business, the expression used, for the purpose of business is to be construed in consonance with the kind or character of the business or trade followed by the assessee. It cannot be interpreted without the context of the purpose of the business. While interpreting the word use for the purpose of section 32 in order to allow depreciation, the object and purpose of acquisition of the article or thing and its employment for profitable purpose or its use for taking advantage of would be material and relevant.The leasing of the article or thing would amount to employment of the article or thing for profitable use, since the advantage of the possession or acquisition of the article or thing is being obtained by reason of such leasing.Whether the rent is payable at a later date or the profit will accrue at a later point of time would not be relevant for the purpose of determining the use of the article or thing. In a business of leasing the moment the article or thing is leased out, the article or thing is put to use for the purpose of leasing or is used for leasing, the business. The rent or profit, which will accrue, will accrue on account of grant of lease from the date when the lease is granted, even though the profit or the rent may accrue at a later point of time. **[Multican Builders Ltd.-CIT,Cal.,** *Sec.32***199CTR124,278ITR142, 147Taxman103]**

Used **Bom. CIT - Viswanath Bhaskar Sathe**

The word used in section 10(2)(vi) of the Indian Income-tax Act, should be understood in a wide sense so as to embrace passive as well as active user. *Sec.32.*[**5ITR621]**

2. The word used in section 10(2)(vi) of the Indian Income-tax Act should be understood in a wide sense so as to embrace passive as well as active user. Consequently depreciation may be allowable in certain cases even though the machinery was not in use or was kept idle. **[CIT-Dalmia Cement Ltd.,Pat.,***Sec.32.***13ITR415]**

3. The expression used should have a wider meaning so as to include not only actual but also passive user and depreciation was allowable. **[CIT - India Tea and Timber Trading Co.,Gau.,** *Sec.32.***137CTR334, 221ITR857, 90 Taxman 181]**

4. The word used should be understood in a wide sense, so as to include passive as well as active user. **[CIT-Vindhyachal Distilleries Pvt. Ltd., MP.,*Sec.32.* 272ITR583]**

5. The two requirements to claim depreciation under section 32 of the Income-tax Act, 1961, (i) the asset should be owned by the assessee,and (ii) it should be used for the purposes of business or profession.The expression used has to be given a wide meaning. The expression includes passive as well as active user. **[Anil Bulk Carriers P.Ltd.-CIT,All.,*Sec.32.*194CTR226,276ITR625,142 Taxman673]**

6. ...the film rolls leased out could not be used by the lessee, even though kept ready for use,on account of strike in the film industry. Therefore,the film rolls kept under forced idleness,were in use during the entire period of the year.Consequently,the assessee, even though a passive user, was deemed to be an active user within the meaning of the word used, as the film rolls were kept ready for use. **[CIT-HeeraFin.ServicesLtd.,Mad.,*Sec.32.*212CTR,532,298 ITR245]**

Used for agricultural purposes PC Raja Mustafa Ali Khan - CIT

In order to decide the question whether land is used for agricultural purposes no assistance is to be got from the meaning ascribed to the word agriculture in other statutes. Though it must always be difficult to draw the line, yet, unless there is some measure of cultivation of the land, some expenditure of skill and labour upon it, it cannot be said to be used for agricultural purposes within the meaning of the Income-tax Act. *Sec.2(14),45.***[16ITR330]**

Used for the purpose of business SC Liquidators of Pursa Ltd. - CIT

The words used for the purposes of the business in section 10(2)(iv) of the Indian Income-tax Act,1922, mean used for the purpose of enabling the owner to carry on the business and earn profits in the business.In other words,the machinery or plant must be used for the purpose of that business which is actually carried on and the profits of which are assessable under section 10(1). *Sec.37.*

[25ITR265]

2. The expression used for the purposes of the business in section 10(2)(vii) of the Income-tax Act, 1922, does not necessarily imply that the machinery or plant referred to must have been in active use up to the date of sale. **[Niranjan Lal Ram Chandra -CIT,All.,** *Sec.32.* **49ITR177]**

3. ....the expression used for the purpose of the business (used in section 10(2)(iv) read with section 10(2)(vi) of the Indian Income-tax Act,1922),comprehended cases where the machinery was kept ready by the owner for its use in his business and the failure to use it actively in the business was not on account of its incapacity for being used for that purpose or its non-availability.And the language used in rule 8 of the Indian Income-tax Rules,1922, was not inconsistent with this interpretation. **[Capital Bus Service P. Ltd. -CIT,Del.,** *Sec.32.* **123ITR404]**

Used for the purpose of business or profession **Cal.** **Multican Builders Ltd. - CIT**

Section 32 of the Income-tax Act, 1961, uses the expression used for the purpose of business or profession. It does not qualify the word used by actual nor does it qualify the same by the phrase put to as has been used in other provisions of the 1961 Act. Therefore, the expression that has been used is to be interpreted in the ordinary grammatical sense.The word use means the employment or using a particular article or thing for profitable purpose in relation to business or trade. In relation to a trade or business, the expression used,for the purpose of business is to be construed in consonance with the kind or character of the business or trade followed by the assessee.It cannot be interpreted without the context of the purpose of the business.While interpreting the word use for the purpose of section 32 in order to allow depreciation, the object and purpose of acquisition of the article or thing and its employment for profitable purpose or its use for taking advantage of would be material and relevant.The leasing of the article or thing would amount to employment of the article or thing for profitable use, since the advantage of the possession or acquisition of the article or thing is being obtained by reason of such leasing. Whether the rent is payable at a later date or the profit will accrue at a later point of time would not be relevant for the purpose of determining the use of the article or thing. In a business of leasing the moment the article or thing is leased out, the article or thing is put to use for the purpose of leasing or is used for leasing, the business. The rent or profit, which will accrue, will accrue on account of grant of lease from the date when the lease is granted, even though the profit or the rent may accrue at a later point of time. <u>*Sec.32.*</u> **[199CTR124,278ITR142,147Taxman103]**

Used for the purposes of the business **Nag.** **Central Provinces Manganese Ore Co. Ltd. - CIT**

The expression used for the purposes of the business in section 10 (1)(iv) means used for the purposes of the business during the account year and not merely that the machinery must not have been used for other purposes, and depreciation allowance could not be allowed merely because the stocks sold during the account year were the result of the use of the machinery in previous years.<u>*Sec.32.*</u> **[5ITR734]**

2. The words used for the purposes of business are capable of larger or narrower interpretation.If the expression used is construed strictly,it can be taken as connoting or requiring the active requirement or actual working of the machinery, plant or building in the business.On the other hand,the wider meaning will include not only cases where the machinery and plant, etc., are actively employed but also cases where there is what may be described as passive user of the same in the business and they can be said to be in use when kept ready for use. **[CIT-SwarupVegetable Products IndiaLtd.,All.,<u>*Sec.32.*</u>277ITR60,145Taxman253]**

3. ......as long as the machinery was available for use, though not actually used, it fell within the expression used for the purposes of the business and the assessee could claim the benefit of depreciation.Use and discarding were not in the same field and could not stand together. ....However, a harmonious reading of the expressions used for the purposes of the business and discardedit would show that used for the purposes of the business only means that the assessee had used the machinery for the purposes of the business in earlier years. The expression used for the purposes of the business as found in section 32 when used with respect to discarded machinery would mean that the user in the business was not in the relevant financial year/previous year but in the earlier financial years. **[CIT-Yamaha Motor India Pvt. Ltd., Del.,<u>*Sec.32.*</u> 328 ITR297,193Taxman281]**

Utilised **Mad.** **CIT - M.Ct. Muthiah Chettiar Family Trust**

Though the word utilised has different shades of meaning, in the context of section 11(3)(c), it means application of income,that is, where the trustees had not applied the income for the purposes for which it was accumulated.The later amendment in section 11(3A) of the Act,which was introduced by the Taxation Laws (Amendment) Act, 1975, with effect from April 1, 1976, also gives a clue that the word,utilisation in section 11(3)(c) should be interpreted to mean application. Under section 11(3A) of the Act, it is permissible for the Income-tax Officer to change the purpose of application of income, if the trustees state before the Income-tax Officer that the income accumulated cannot be applied for the purposes due to the reasons mentioned therein. Though the word utilisation in normal connotation would connote spending of money for the purposes for which the income was accumulated,in the context of section 11(3)(c),the more appropriate meaning that can be assigned to that expression utilisation is application of the income. The section deals with three successive stages in the application of the accumulated income.The first stage is non-application of income and the second stage is conversion of the approved Government securities or deposits before the period mentioned in section 11(2) and the third stage is the non-utilisation or utilisation of the accumulated income for some other purposes either during or after the expiry of the period mentioned in section 11(2). The emphasis given under the provisions of sections 11(1) and 11(2) is application of income for charitable or religious purposes and it is only in this context, that the courts have taken the view that the handing over of the money by one trust to another trust having similar objects would amount to application of income.There are no weighty reasons to give a different meaning to the expression utilised as spent in section 11(3)(c) of the Act and it cannot be assumed that the Legislature has postulated a different test for utilisation of the accumulated income. *Sec.11.***[162CTR63,245ITR400,114Taxman69]**

V

Valuable thing **Mad.** **I.Devarajan - Tamil Nadu Farmers Service Co-operative Federation**

The addition of the word thing in section 132(1)(c) of the I.T. Act, 1961,as contrasted with Sections 69A and 69B is significant. What is not comprehended by the expression valuable article which would refer to tangible assets is sought to be brought in by using the expression valuable thing in order to describe or bring in intangible assets also.A chose-in-action would also be a thing and hence money kept in a bank will be a valuable thing so as to be covered by the provisions of section 132. <u>Sec.132</u>.**[13CTR280,131ITR506]**

Valuation date **Mys.** **CWT - Lt. Col.D.C. Basappa**

The liability to pay Income-tax arises not later than the close of the previous year,though the quantification of the amount payable is postponed.A provision for payment of agricultural Income-tax in respect of an accounting year which has expired on or before the valuation date is therefore a debt owed by the assessee on the valuation date within the definition of net wealth contained in section 2(m) of the Wealth-tax Act and is allowable in computing the net wealth of the assessee for purposes of levy of wealth-tax under the Act. <u>Sec.2(m)WT.Act.</u>**[51ITR790]**

2. The valuation date under the Wealth-tax Act, 1957, in relation to any year means the last day of the previous year as defined in section 3 of the Income-tax Act, 1961.Therefore, the year cannot be taken as completed on the valuation date and thus,where a reference is made to any year immediately preceding the valuation date, it is to be taken as a completed year. **[CWT-Karan Thapar, Cal.,** <u>Sec.3.WT.Act.</u>**143CTR185,223ITR531]**

Value so declared **Kar.** **Sanjiv V. Kudva – CIT**

The words consideration declared by the assessee and the words value so declared used in sub-section (2) and in particular the word declared at both places presuppose that the actual value received is the undeclared value which is something different from and higher than the value declared in the sale deed. Any doubt in this regard is removed by the marginal note to the section which reads consideration for transfer in cases of understatement, because it is only in a case where the vendor, having received a higher amount as consideration for the transfer of a capital asset, makes an understatement of consideration in the sale deed to avoid capital gains tax on the difference of amount and to retain it as unaccounted money, the amount mentioned in the sale deed can be regarded as the declared value and the amount actually received as the undeclared value.In a case, where the consideration actually recorded in the sale deed is also the amount actually received by the vendor, merely because the fair market value exceeds by 15 per cent.of the actual sale consideration, it could not be said that it was a case of understatement or that the consideration declared is different from the undeclared consideration which is non-existent. Therefore, the word declared used in section 52(2) attracts a case in which it is established that a vendor has declared a lower consideration in the deed transferring his capital asset having received a higher amount as consideration. *Sec.52(2).***[20CTR1,127ITR354]**

Vocation **Mad. C.Rajagopalachariar - CIT**

A vocation is only a way of living or a sphere of activity for which one has special fitness. It is not necessary that such activity should be one indulged in for earning a livelihood before it can be called vocation. Nor can it be said that a person cannot have more than one vocation. A motive for making a profit is not an essential requisite of a vocation. A vocation does not involve any organised or systematic activity like business. *Sec.10(3)(ii).***[50ITR196]**

2. Vocation is a sphere of activity for which one has special fitness. It is not necessary that such activity should be one indulged in for earning a livelihood before it can be called vocation. A motive for making a profit is not an essential requisite of a vocation. **[CIT - RamParshad,Del.,***Sec.10(3)(ii).***7CTR146, 113 ITR 462]**

Void **Ker. K.R.Narayana Iyer - CIT**

While section 531 of the Companies Act,1956,treats certain transactions as invalid and section 532 treats another category of transfers as void,section 531-A stands in between treating the transfers covered thereby as void against the liquidator. The expression is often used: void as against a person or persons. In strict terminology, a thing cannot be void and valid at the same time.As void denotes a nullity, a thing which is void must be a nullity for all. It is totally non-existent. Therefore, void as against A can mean only that A can treat it as void; or, in other words, A can avoid it.It is, strictly speaking,voidable at the option of A. The fact that a transfer falling within section 531A is void as against the liquidator implies that it is not a nullity in the absolute sense.Since it is void only as against the liquidator,it means that the court will invalidate or ignore the transfer only if the relief is sought by the right person, namely, the liquidator and in appropriate circumstances. For instance, it may be avoided only if it is necessary to satisfy the creditors of the company, or to the extent necessary for that purpose. Therefore, when the liquidator himself does not choose or find it necessary to avoid a particular transfer,or to recover the property concerned, it is not open to a stranger like the Income-tax Officer to ignore it and treat the transferred asset as remaining with the liquidator relying on section 531A of the Act.

*Sec.531.Companies Act .***[202ITR774]**

Voidable Cal. Jaymac Lasetron (P.) Ltd. - CIT

When the word voidable in a particular section of the Act is available it is the duty of the court to decide in a civil suit or proceedings as to whether such act is voidable or not. But when the statute prescribes that a transaction is void there is no scope of ascertainment by way of suit to be instituted by the authority concerned to prove that such a transaction is void. The word void itself gives power to the authority to declare a transaction void. Where a transaction is declared void under section 281 there is no occasion for the authority concerned to go before the civil court for further declaration. It is open to the aggrieved party to take appropriate proceedings for the purpose of nullfying such declaration. There is also a scope of decision of the disputed questions about garnishee by way of trial which is as good as suit under Order 21, rule 46(c),of the Code of Civil Procedure.*Sec.281.***[164CTR366,245 ITR734,116Taxman231]**

Voluntarily AP. CIT - Rajah Dhanrajgirji

The word voluntarily cannot be understood as signifying a meaning opposite to the words by operation of law. If that was the intention of Parliament, it would have used the words by act of parties, which is the expression generally understood as the opposite of the words by operation of law...the expression must be understood as distinct from, and as opposed to, the expression involuntarily.The word involuntarily means, without there being any option, i.e.,under an enforceable obligation.Hence, where a person creates an annual charge to meet an existing, genuine, legal or contractual obligation, it would not be a case of creating a charge voluntarily.*Sec.24(1)(iv).* **[45CTR53,154ITR719,19Taxman52]**

2. The word voluntarily cannot be construed in isolation with reference to the general animus or state of mind of the assessee. From the legal obligation to file a return, no element of fear could be either attributed or inferred.The word voluntarily in the context of section 273A(1), therefore, has to be construed as filing of the return by the assessee without being prompted by the animus to avoid or pre-empt adverse exposure or penal action.The Commissioner, before rejecting the returns as not voluntary, must have material based upon which it is reasonable to infer that, in all probability, but for the filing of the voluntary return, the assessee would have been subjected to penal action or adverse exposure. Fear on the part of the assessee, without anything more, cannot be a ground for not exercising the discretion under section 273A. The fear must be traceable to the imminent or proximate exposure of the assessee to penal action but for the filing of the voluntary return under section 273A and,in order to enquire into this subjective element, there must be in existence objective facts warranting such an inference. **[Sujatha Rubbers-ITO,AP.,** *Sec.273A.* **102 CTR 152,194ITR355,62Taxman13]**

3. The word voluntarily in section 273A of the Income-tax Act, 1961, means out of ree will without any compulsion. Disclosure of concealed income after the Department has seized the incriminating material with regard to the income so disclosed, cannot be voluntary disclosure, because it is made under the constraint of exposure to adverse action by the Department. But it cannot be held as a principle of law that the disclosure of income made after the search/ raid cannot be voluntary. It is a question which has to be decided by the Department in each case on the basis of the material on the record. The criteria for deciding the question is to find out as to whether the Department has any incriminating material with regard to the disclosed income. If there is incriminating material on record with regard to the disclosed income, the disclosure cannot be

voluntary. But if the Department has no incriminating material with regard to the income disclosed, the disclosure is liable to be treated as voluntary having been made without any compulsion or constraint of exposure to adverse action by the Department. In a case where the assessee has disclosed not only the income regarding which the Department has incriminating material, but has also disclosed income with regard to which no incriminating material was seized by the Department, the disclosure of the income with regard to which the Department has no incriminating material, is liable to be treated as voluntary. If an assessee for example,has five accounts and the Department has incriminating material with regard to one of those accounts only, the disclosure of income relating to four accounts with regard to which the Department has no incriminating material, is voluntary, because it was made without any constraint or compulsion, even though the disclosure of the income relating to the account regarding which the Department has incriminating material, is liable to be treated as non-voluntary. **[Bhairav Lal Verma-Union of India,All.,**<u>*Sec273A.*</u>**146CTR16, 230ITR855,97Taxman 489]**

Voluntary **All.** **Hakam Singh - CIT**

The term voluntaryin section 273A of the Income-tax Act, 1961, has been used to indicate an action free of any constraint. A return filed under the constraint of exposure to adverse action by the I.T. department will not be voluntary within the meaning of section 273A. The action of an assessee in filing a return after the books of account had been seized at a raid would be impelled by the compelling circumstances that the assessee was likely to be dealt with under the penal provisions of the I.T. Act.The action of an assessee in filing a return under such a constraint cannot be said to be voluntary. <u>*Sec.273A.*</u>**[124ITR228]**

2. The term voluntary has to be understood as anything done intentionally and without coercion, compulsion or constraint. Coercion may in turn be direct or positive as in cases where physical force is used to compel an act against one's will or it may be implied.That would not however mean that a mere legal obligation to do something should constitute a constraint of the kind which would render any such action involuntary. It follows that the circumstances, conditions or constraints that make a disclosure under the Act involuntary must be constraints other than obligations that arise under the Act, requiring the assessee to take aparticularaction. **[K.L.Swamy-CIT,Kar.,**_Sec.271(1)._**157CTR 489,239ITR386]**

3. The term voluntary means without compulsion; it has to be understood as anything done intentionally and without coercion, compulsion or constraint. A mere legal obligation to do something would not constitute a constraint of the kind as would render any such action involuntary. Therefore, it becomes necessary for the Commissioner before rejecting an application under section 273A on the ground that the return filed by the assessee is not voluntary, to satisfy himself,on the basis of the materials before him, that it is reasonable to infer that, in all probability, the assessee would have been subjected to penal action or adverse exposure.Mere fear on the part of the assessee,without anything more, cannot be a ground for not exercising the discretion : such fear must be traceable to imminent or proximate exposure of the assessee to penal action, but for the filing of the return. **[K.S.N.Murthy- Chairman,CBDT,AP.,** _Sec.273A._**171CTR563,252ITR269,119Taxman310]**

Voluntary disclosure **Cal.** **CIT - Bimal Kumar Damani**

...the voluntary disclosure scheme was applicable to an assessee who, or the first time, disclosed his income,the assessment whereof was neither complete nor pending...... Any disclosure made within the stipulated time before detection would be a voluntary disclosure once it was so claimed by the assessee while disclosing the income.

VDIS **[180CTR452,261ITR87,129Taxman564]**

W

Waiver **Del.** **P.C. Puri. - CIT**

Waiver means abandonment of a right and it may be either express or implied from conduct, but its basic requirement is that it must be an intentional act with knowledge. The Income-tax Officer must direct himself properly to the statutory provisions. He must call to his attention all relevant matters which he is bound to consider.He must exclude from his consideration all irrelevant matters. There must be some indication of the application of mind to these considerations.The Income-tax Officer is given a discretion under the law to reduce or waive the interest. The reduction can only be by an overt act of first determining the interest payable and then reducing it. <u>Sec.215.</u>**[151ITR584]**

Wealth of the assessee **Guj.** **CWT - Mahendrabhai D. Parmar**

Wealth-tax applies in a case where the wealth belongs to a particular person and he has a power of disposition....If part of the income of the assessee goes to a trust,the wealth would be created in favour of the trust and as the assessee would have no power of disposition, transfer alienation in any manner, the property cannot be included in the wealth of the assessee. <u>Sec.2.WT.Act.</u> **[292ITR622]**

Welfare centre **Ker.** **CIT - Upasana Hospital and Nursing Home**

The expression welfare centre is not defined in the Act or Rules. Hence, the meaning of welfare centre should be that as commonly understood. (Item No.(2) of Part I of Appendix I to the Income-tax Rules). <u>Sec.32.</u>**[168CTR348,250ITR78,117Taxman687]**

When a question is raised before the Tribunal **SC** **CIT - Scindia Steam Navigation Co. Ltd.**

When a question is raised before the Tribunal and is dealt with by it,is clearly one arising out of its order. When a question of law is raised before the Tribunal but the Tribunal fails to deal with it, it must be deemed to have been dealt with by it, and is,therefore, one arising out of its order. When a question is not raised before the Tribunal but the Tribunal deals with it, that will also be a question arising out of its order.When a question of law is neither raised before the Tribunal nor considered by it, it will not be a question arising out of its order notwithstanding that it may arise on the findings given by it. A question of law might be a simple one,having its impact at one point,or it may be a complex one, trenching over an area with approaches leading to different points therein. Such a question might involve more than one aspect, requiring to be tackled from different standpoints. ......To import in the meaning of the expression any question of law arising out of such order the concept that the question must have been argued before and dealt with it by the Tribunal in its judgment deciding the appeal, is to impose a fetter upon the jurisdiction of the High Court not warranted by the plain intendment of the statute.A concrete question of law having a direct bearing on the rights and obligations of the parties which may be founded on the decision of the Tribunal is one which arises out of the order of the Tribunal even if it is not raised or argued before the Tribunal at the hearing of the appeal. *Sec.253.* **[42ITR589]**

Where in respect of a particular source of income. .. **MP.** **CIT - Kanchanbai**

The expression whether in respect of a particular source of income, profits and gains, an assessee has once been assessed in section 2 (11)(i)(a) implied that the income,profits and gains of a particular source had been computed in the manner laid down in the Act and included in the total income.The assessment was on the total income,and there could not be an assessment in respect of a particular source of income,profits and gains unless the income, profits and gains there from were included in the total income. Therefore, the assessment spoken of in the proviso to section 2 (11) (i)(a) in respect of a particular source of income,profits and gains meant the computation of the income from the source for the purpose of its inclusion in the total income. *Sec.4.* **[44ITR242]**

Where no such return has been made **P&H.** **CWT -Anil Tayal (HUF) (No. 1)**

The object of section 16A of the Wealth-tax Act, 1957, is to enable the Wealth-tax Officer to refer the issue relating to the value of any asset to a Valuation Officer for the purpose of making assessment. The expression any other case in clause (b) of section 16A(1) is wide enough to include a case where no return has been filed by the assessee.If a narrow view is taken that a reference under section 16A(1) can be made only where the return has been filed,then the expression any other case in clause (b) of section 16A(1) and the expression where no such return has been made in sub-section (4) of section 16A would become redundant. Sec.16(1)WT.Act. **[195CTR420,285ITR243,146Taxman239]**

Where tax has been paid **Bom.** **Navinchandra Mafatlal - CIT**

The expression where tax has been paid in section 23A(4) merely indicates the liability in respect of which tax has been paid. It makes no reference to the mode of payment of the tax or the process of assessment that has to be followed before a liability arises to pay the tax. Sec.109.r.w.s.104.**[27ITR245]**

Where order is the subject of an appeal before **Bom.** **R.H. Muttoo - Kasturbai Walchand**

The phrase where the order is the subject of an appeal before the Appellate Tribunal occurring in proviso (b) to section 25(1) of the Wealth-tax Act, 1957, should be read as meaning that the order is the subject of an effective appeal by the aggrieved party and the reason is that the scheme of the relevant provisions clearly give two remedies to the assessee who may feel aggrieved by the order passed by the authorities subordinate to the Commissioner of Wealth-tax and the scheme also suggests that he should not have recourse to the two remedies concurrently. Implicit in the scheme is the position that the Commissioner shall refrain from exercising his revisional powers when an appeal is pending before the Tribunal and he may exercise his powers in such a way as not to get in conflict with the orders passed by the Tribunal. Sec.25(1)(b)WT.Act. **[166ITR392]**

Who shall ... have powers.. for the imposition of penalty **P&H.** **Dayawanti - CWT**

The words who shall, for the purpose, have all the powers conferred under this section for the imposition of penalty, occurring in section 18(3), indicate that the section does not cast upon the Inspecting Assistant Commissioner an absolute duty of proceeding with the case. All that the section provides is that the Inspecting Assistant Commissioner shall have the powers of imposing penalty which implies that, if at the time when the case comes up for final hearing, the Inspecting Assistant Commissioner is satisfied that the condition precedent for the exercise of his jurisdiction are lacking, it is open to him to decline the reference. A necessary corollary of this is that the AC is under an obligation to entertain and to decide the question whether a penalty of more than Rs. 25,000 is imposable or not before he makes a final adjudication upon the case. If he comes to the conclusion that such a penalty is not imposable, then the very basis of his jurisdiction goes. In that event, he has to decline the reference leaving it to the Wealth-tax Officer to decide the matter.

*Sec.18(3)WT.Act.***[20CTR114,128ITR504]**

Whole and exclusive purpose **MP.** **CIT - Gwalior Sugar Company Ltd.**

Where the whole and exclusive purpose of the expenditure is the purpose of the expender's trade, and the object which the expenditure serves is the same the mere fact that to some extent the expenditure enures to a third party's benefit,cannot,in law, defeat the effect of the finding as to the whole and exclusive purpose. *Sec.37(1).***[42CTR69,150ITR320,72Taxation247]**

Wholly and exclusively **Mad.** **Sree Meenakshi Mills Ltd. -CIT**

The words wholly and exclusively in section 10(2)(xv) pointedly signify that the expenditure should be completely devoted to the business. It need not be essential, necessary or compelling; it may be optional and purely voluntary. But it must be commercially expedient and should have the aim of the continuance and furtherance of the business and an eventual augmentation or stabilisation of profits. *Sec.37(1).***[49ITR156]**

2.The expression wholly and exclusively used in section 10(2)(xv) of the Indian Income-tax Act,1922, does not mean necessarily.

Ordinarily, it is for the assessee to decide whether any expenditure should be incurred in the course of his or its business.Such expenditure may be incurred voluntarily and without any necessity and if it is incurred for promoting the business and to earn profits, the assessee can claim deduction under section 10(2)(xv) of the Act even though there was no compelling necessity to incur such expenditure. The fact that somebody other than the assessee is also benefited by the expenditure should not come in the way of an expenditure being allowed by way of deduction under section 10 (2) (xv) of the Act if it satisfies otherwise the tests laid down by law. **[Sassoon J. David andCo.(P.)Ltd.-CIT,SC,***Sec.37(1).***AIR1979SC 1441,10 CTR383,118ITR261,1Taxman485]**

3. The adverb wholly in the phase laid out or expanded .. refers to the quantum of the expenditure, the adverb exclusively refer to the object or motive of the act behind the expenditure. **[Mysore Kirloskar Ltd.- CIT, Kar.,** *Sec.37(1).* **166 ITR836,84 Taxation50, 30 Taxman467].**

4 The expression wholly and exclusively used in section 37(1) of the Income-tax Act, 1961, does not denote necessarily. It refers to quantum of expenditure. The expression refers to motive, objective or purpose with which the particular expenses have been incurred. Ordinarily,it is for the assessee to decide whether any expenditure should be incurred in the course of its or his business. Such expenses can be incurred voluntarily and without necessity. If they are incurred for promoting the business and to earn the profits, the assessee can claim the deduction. It is also not necessary to show that expenses were not profitable or no benefit was actually derived. The receipt of actual benefit is also not necessary. Any contribution made by an assessee to a public welfare fund which is directly connected or related with the carrying on of the assessee's business or which results in benefit to the assessee's business has to be regarded as an allowable deduction under this section. Such a donation, whether voluntary or at the instance of the authorities concerned,when made to a relief fund or a welfare fund or any other fund for the benefit of the public and with a view to secure benefit to the assessee's business cannot be regarded as payment opposed to public policy. There is no law which prohibits the making of such a donation. The mere fact that making of a donation for a charitable or public cause or in public interest results in the Government giving partronage or benefit can be no ground to deny the assessee a deduction of that amount under this section. **[Addl.CIT - Rajasthan Spinning and Weaving Mills Ltd., Raj.,** *Sec.37(1).***186CTR117,274ITR465,137Taxman367]**

5. By using the words wholly and exclusively, the Legislature has cast a duty on the assessee to establish that the expenditure was to earn income and not for any other purpose. The effect of the words, wholly and exclusively for the purpose must not be diluted.By using three words, i.e. wholly, exclusively and purpose the Legislature had made it mandatory to find out the reason behind investment. For the purpose of earning income through investment the assessee has to take into account the financial prospects of the company concerned. The principle applied to ascertain the intention of the assessee to earn income must be what a man of common prudence will think while expending in a company. If there is no material on record to establish that the expenditure of the assessee is done bona fide to earn income, the deduction under section 57(iii) of the Act will not be available. The dominant purpose of investment made must be to earn profit. The decision taken under the circumstances while making an investment should reveal that there was likelihood to earn profit. Investment or expenditure in a company where there is no hope of earning profit will not be covered by section 57(iii) of the Act (laid out or expended wholly and exclusively for the purpose of making or earning such income). Though it is not unfair to borrow money or take a loan from one concern and invest it in other concern for the purpose of profit or income,while doing so, the assessee must act bona fide with the primary motive to earn profit. **[CIT- Smt. Swapna Roy, All.,**<u>*Sec.57(iii).*</u>**233CTR10,331ITR367, 192Taxman105]**

Wholly and exclusively for the purpose of the business **MP.** **CIT - Shriram Prayagdas and Mahadeo Prasad**

A payment voluntarily made in order to facilitate the carrying on of the business on the ground of commercial expediency may amount to expenditure wholly and exclusively for the purpose of the business. <u>*Sec.37(1).*</u>**[27CTR155,144ITR883]**

Wholly and exclusively for the purposes of such ...... **SC** **CIT - Travancore Sugar & Chemicals**

The expenditure of a capital nature is certainly not an expenditure which is deductible for compulsory profits though it may be an expenditure wholly and exclusively laid out for the purpose of the business. <u>*Sec.37(1).*</u>**[AIR1973SC982,88ITR1]**

Wholly and mainly SC CIT- Baroda Distributors (P.) Ltd.

Clause (i) of Explanation 2 to section 23A of the Indian Income-tax Act, 1922, concerns itself with a company whose business consists wholly or mainly in the dealing in or holding of investments. The word mainly in that clause as well as in the main section 23A must necessarily take its colour from the word wholly preceding that word in those provisions.A company which comes within the scope of these provisions must be one whose primary business must be in the dealing in or holding of investments.If a company engages itself in two or more equally or nearly equally important business activities, then it cannot be said that the company's business consists wholly or mainly in dealing in a particular activity. Even in cases where a company has more than one business activity and one of its activities is more substantial than the others, unless that activity is the primary activity of the company, it cannot be said that the company is engaged wholly or mainly in any one of its business activities. Section 23A applies only to cases where the primary activity of the company is in the dealing in or holding of investments. *Sec.109(ii).*[83ITR377]

Wholly for religious or charitable purpose Mad. Sadayapillai Trust -Agr. ITO

The expression wholly for religious or charitable purpose means that the property under trust is held wholly for religious or charitable purposes. *Sec.11(a)(b).*[AIR1967Mad396,63ITR520]

2. The words wholly for religious or charitable purposes in section 4 (3)(i)show that the income from trust property would only be exempt if all the objects of the trust are of a religious or charitable nature. In case a trust has ten distinct objects and nine of them are of a religious or charitable nature,but the tenth is not and there is nothing to prevent the trustees from applying the property of the trust in carrying out any of the objects of the trust including the object which is not of a religious or charitable nature, the income derived from the property of the trust would not be exempt from Taxation under section 4(3)(i). The reason is that the trustees in such an event can apply the property of the trust exclusively for that object of the trust which is not of a religious or charitable nature. The only relaxation which has been permitted, in such cases,is that if all the primary objects of the trust are of a religious or charitable nature, the existence of an ancillary or secondary object which is not of a religious or charitable nature but which is intended to subserve the religious and charitable objects, would not prevent the grant of the exemption. **[CIT-Jaipur Charitable Trust,Del.,**__Sec.11 (a)(b).__**81ITR1]**

Wholly laid out for the purpose of the business **All.** **Agra Leatheries Ltd. - CIT**

In a case where penalty has been incurred because of the fault of the assessee himself, as for instance by reason of his having carried on his business in an unlawful manner or in contravention of certain rules and regulations, the penalty paid by the assessee for such conduct could not be allowed as wholly laid out for the purpose of the business,because the incurring of the said expenses has not been necessitated by the business but by the conduct of the assessee in trying to carry on the business in an unlawful manner. __Sec.37.__ **[200ITR792]**

Wholly situated **Bom.** **Bhimji R. Naik - CIT**

The control and management mentioned in section 4A(b) of the Indian Income-tax Act must be de facto control and management and not de jure control and management. A liberal meaning is to be given to the words wholly situated and it must be ascertained in every case where in fact the control and management of the business is situated apart from the temporary journeyings of the active partners or the residence of the dormant ones. In the case of a firm the problem ought to be approached from the same angle as in the case of a company. __Sec.6(2).__**[14ITR334]**

Wholly used for the business **Del.** **CIT-Northern India Iron and Steel Co.Ltd.**

It may be that the concept of wholly used is not the same as exclusively used.But it conveys the meaning of exhaustive user of the machinery, as by the assessee. This is possible only when the assessee has control over the user. The term wholly used for the business is referable to the manner of its user. The meaning of the word user cannot be confined to the actual derivation of a financial benefit;it includes a proper control over the machinery in the matter of its utilisation, running, repairing, replacement,etc. __Sec.33.__**[123CTR120,211ITR370,123Taxation451,84Taxman182]**

Wife and husband **SC** **Philip John Plasket Thomas - CIT**

In order to attract the application of section 16(3)(a)(iii) of the Indian Income-tax Act, 1922, the relationship of husband and wife must subsist not only at the time of the accrual of income from the assets but also when the transfer of assets is made. The words wife and husband in section 16(3)(a) must be taken in their primary sense which is clearly indicative of a marital relationship.

*Sec.64.***[49ITR97]**

Wilful **Pat.** **Raghunath Pandey -State of Bihar**

The dictionary meaning of the expression willful is something done intentionally. It is thus a culpable mental state and the necessary presumption of the existence of such mental state can be drawn against the accused under section 278E of the Act,which presumption is, however, rebuttable. *Sec.278E.* **[150CTR145,232ITR 908,148Taxation412,99Taxman371]**

2. It must be intentional, deliberate, calculated and conscious with full knowledge of legal consequences flowing from them.**[Union of India -Bhavecha Machinery,MP.,***Sec.276CC.***320ITR263]**

Wilful default **MP.** **Narayan - Union of India**

Wilful default means that it must be intentional, deliberate, calculated and conscious with full knowledge of the legal consequences flowing from them. *Sec.276CC.***[208ITR82]**

Windfall **Bom.** **Mehboob Productions Pvt. Ltd.-CIT**

When talking of a windfall receipt in connection with the consideration of the question whether such receipt would be income or not,one has to restrict the concept of such a windfall to a case where the unexpectedness of the advantage pertains to the factum of receipt and not to the quantum of receipt. What we are considering as windfall is some unexpected receipt not in the contemplation of the assessee and not directly attributable to or occurring by way of its business profits. On the other hand, where there was clear expectation, though small, of receiving such advantage or profit, then it cannot be properly regarded as windfall merely because the advantage of receipt is much more than could have been reasonably anticipated. *Sec.10(3).*[**106ITR758**]

Winning **Mad.** **CIT - G.R. Karthikeyan**

......the word winning in its plural form was chiefly applied in modern usage to money won by gaming or betting. The race in the instant case involved skill in performance of driving of the vehicle with the least number of penalty points. Accordingly, the ordinary use of the expression winnings would not comprehend the winning of a prize in a case of this kind. *Sec.2(24)(ix).*[**124ITR85**]

Winnings **Cal.** **Indcom - CIT(TDS)**

A plain reading of sections 2(24), 5, 9, 115BBA, 194E, 194J and 201 makes it abundantly clear that if a foreign cricket team, by virtue of agreement among the various teams of different cricket-playing countries, participates in a cricket match played in India and an agreed amount is paid to such a team for participating in such a match in the form of prize money, such prize money, whether for winning or for losing the match, will come within the term winnings and, hence, should be treated to be income within the meaning of section 2(24)(ix) of the Act. *Sec.2(24)r.w.s.115BBA,194E.194J.* **[242CTR337,335ITR485,200Taxman40]**

Within such period not being less than 30 days **Bom.** **CIT- Ekbal and Co.**

Section 22(2) of the Indian Income-tax Act which provides that a notice may be served on an assessee requiring him to furnish a return of his income within such period, not being less than thirty days gives him an interval of thirty clear days from the date of the receipt of the notice to the date on which return is to be furnished. A notice requiring an assessee to furnish a return of his income within thirty days of the receipt thereof is not a valid notice within the meaning of the sub-section. ....within thirty days is within two points of time, one at which the period begins and the other at which it expires, not less than thirty days is outside these two points of time. *Sec.142.*[**13ITR154**]

Without prejudice **SC** **Supdt. (Tech. I),Central Excise, I.D.D. Jabalpur- Pratap Rai**

The implication of the term without prejudice means (i) that the cause of the matter has not been decided on merits, (ii) that fresh proceedings according to law were not barred. Sec.122,128Customs Act.**[114ITR231]**

2. From a bare reading of sub-clause (i) to sub-section (1)(a) of section 143 of the Income-tax Act,1961, it is evident that giving of intimation in terms of the provisions is without prejudice to the provisions of sub-section (2), which means that an intimation sent to the assessee specifying the sum payable by him in terms of that sub-section does not preclude the operation of the provisions of sub-section (2).By force of the expression without prejudice, the jurisdiction of the assessing authority to proceed under sub-section (2) of section 143 is preserved despite intimation under sub-section(1). **[Apogee International Ltd. -Union of India, Del.,** Sec.143. **137CTR93,220ITR248,87Taxation198,133 Taxman414]**

3. The expression without prejudice signifies that the mere filing of a return will not be allowed to be used against the assessee implying its admission.Without prejudice implies future rectification in accordance with law. **[CWT- Apar Ltd., Bom.,** Sec.16A,23(1A).WT.Act.**175CTR312,267ITR705,140Taxman222]**

Without prejudice to any action already taken for the recovery of such sum under.. **Ker.** **CIT - Ettumanoor Motors (P.) Ltd.**

The words without prejudice to any action already taken for the recovery of such sum under the repealed Act appearing in clause (j) of section 297(2) of the Income-tax Act, 1961, does not indicate that action ought to have been taken under the old Act. Those words are only intended to protect recovery proceedings already taken under the repealed Act and not to make such a proceeding a condition precedent to invoke the power under the new Act. Recovery under the new Act will be without prejudice to whatever action in that behalf had already been taken under the old Act. Any other construction would lead to absurdity, for it would then mean that there would be no statute under which recovery is possible even though an assessment has been validly made under the repealed Act. Sec.297.**[165ITR751,31Taxman417]**

Without prejudice to the powers conferred on him by any other provision of this Act **Cal.** **Satish Kapur - CIT**

In section 273A of the Income-tax Act, 1961, the opening words of sub-section (4) without prejudice to the powers conferred on him by any other provision of this Act make it clear that the said sub-section was not a provision over riding the provisions of sub-section (1) and is clearly a provision in addition to sub-section (1). *Sec.273A.* **[187CTR109,265ITR673,136Taxman288]**

Without prejudice to the provisions of section 143(2) **Del.** **CIT - Punjab National Bank**

The expression without prejudice to the provisions of section 143(2) in section 143(1) means that the right of the Assessing Officer to proceed under section 143(2) despite intimation to the assessee of the sum payable as tax or interest is preserved and not taken away. *Sec.143.* **[166CTR340,249ITR763,116Taxman310]**

Without prejudice to the provisions of sub-section (2) **MP.** **CIT- H.E.G. Ltd.**

........expression without prejudice to the provisions of sub-section (2) appearing in the section would mean that once a notice has been issued under sub-section (2),then in that case, the Assessing Officer shall not resort to section 143(1)(a)(i). *Sec.143.* **[171CTR318,255ITR 251,121Taxman475]**

Work **Cal.** **Calcutta Goods Transport Asso. - Union of India**

The word work has been used as a noun in section 194C, not as a verb. The word work may be used in two senses; it may mean either the labour which a man bestows upon a thing,or the thing upon which the labour is bestowed. The fact that two meanings are possible to be given to a word means that there is an ambiguity in the word itself. Even assuming that the decision of the Supreme Court is not restricted to the facts of the case before it, it is clear that without the word any work being defined, the applicability of section 194C to a given situation cannot be determined. The Supreme Court had observed that any work means any work. This does not take the matter any further. The Supreme Court has not gone on to clarify or explain or define the sense in which the word was being used by the Supreme Court. That the word work does not have the widest possible connotation is also clear from the fact that Parliament had sought to bring professional services and other such works in the wider sense within the net of tax deduction at source.

If such work were already covered by section 194C it was wholly unnecessary to introduce separate statutory provisions in this regard. The only conclusion that follows from this is that the word work is to be understood in the limited sense as product or result. The carrying out of work indicates doing something to conduct the work to completion or an operation which produces such result. That being so the mere transportation of goods by a common carrier does not affect the goods carried nor are the goods affected thereby and as such cannot be brought within the scope of section 194C of the Act.Common carriers of goods by road are not liable to deduction of tax at source under section 194C of the Income-tax Act, 1961, and the provisions of the section are not applicable to them. *Sec.194C.*[134CTR132,219ITR486]**

2. The word work in section 194C has to be understood in a limited sense and would extend only to the service contracts specifically included in section 194C by way of Explanation III. Section 194C would apply to payments for carrying out the work such as constructing buildings or dams or laying of roads and air fields or railway lines or erection or installation of plant and machinery, etc. In all these contracts, the execution of the contract by a contractor/ sub-contractor results in production of the desired object or accomplishing the task under the contract. The facilities or amenities made available by a hotel to its customers do not constitute work within the meaning of section 194C. **[East India Hotels Ltd.,Bom., *Sec.194C.*320ITR526,179Taxman7]**

Work of art AP. CWT -SB. Zainab Noorul Sayeeda

Work of art is not defined under the provisions of the Wealth-tax Act, 1957. The expression art treasure is defined under the provisions of the Antiquities and Art Treasures Act, 1972, according to which art treasure means any human work of art. *Sec.5(1)(xii). WT.Act.* **[184CTR596,262ITR306]**

Worked Pun. CIT - Sarveshwar Nath Nigam

The expression worked in the second proviso does not mean worked continuously for the entire year. *Sec.32.***[48ITR853]**

Worker Mad. P.R.CIT - Alagappan

Worker,as commonly understood, is undoubtedly a person who works and according to the definition of worker under the Factories

Act, 1948, a worker may be a person who works for wages or without wages. But such a meaning will not include the person who is the owner of the establishment in which other people are working. The owner himself is never referred to as a worker. Merely because a person can work without wages, it does not mean that every person who works will become a worker for the purpose of a statutory provision because whether a person is a worker within the meaning of the statutory provision will have to be ascertained with reference to the phraseology used in that provision. *Sec.80J.*
[69CTR43,173ITR82,38Taxman203]

Writing off Bom. CIT-General Insurance Corp.of India(No.2)

The expression writing off is a technical term used by the auditors....there are two methods of dealing with a debt which has been written off in the books of account, viz., by giving corresponding credit to the debtor's account or by giving corresponding credit to the bad and doubtful debts account. The first method is only employed where it is desired to close the account of the debtor. The second method is employed where there are some chances of recovery. That, when we talk of writing off, we are not concerned with the credit to be given to an account. That, writing off means raising a debit entry. This can only be to the debit of the profit and loss account. ..... this is the only debit which can be raised as a result of writing off a bad debt. *Sec.36(1)(vii).*
[254ITR204,114Taxman13]

Written down value All. CIT - Tireth Prakash

Written down value, under section 43(6)(b) of the Income-tax Act, 1961, means the actual cost to the assessee less depreciation actually allowed to the assessee and not any notional allowance that might be permissible under the section. *Sec.43(6).* **[81ITR27]**

2. The expression written down value in section 43(6) of the Income-tax Act, 1961, means in the case of assets acquired before the previous year, the actual cost to the assessee less all depreciation actually allowed to him under the various Indian Income-tax Acts. Where ships belonging to a non-resident acquired earlier had not come into Indian waters and income from such ships was not subject to assessment in India and so depreciation was not allowed, there is no question of the assessee enjoying double depreciation once in the computation of its world income and again in the computation of its Indian income. The mere fact that the ships had been acquired earlier will not make any difference because of the clear wording of sections 32, 34(3) and 43 of the Act. The assessee would be entitled to claim depreciation on such ships in spite of the fact that they were acquired more than twenty years ago. If any of the ships are sold,in computing the balancing charge, the depreciation not actually allowed in India, cannot be taken into account in arriving at the written down value of the ships. **[CIT-East Asiatic Co.Ltd.,Cal.***Sec.43(6).***35CTR220,148ITR124,13 Taxman20]**

3. Neither the Act nor the rules specifically lay down the method to be adopted in arriving at such written down value.The ascertainment of written down value being for the purpose of determining the profits and gains from business, it would be more rational to adopt the method discernible from sections 28 to 41 rather than ascertain it by following the method of accounting adopted by the assessee.That the Companies Act envisages under section 205 read with section 350 a straight-line method is no ground for adopting the same method for the purpose of the I.T Act. The written down value of the asset has to be ascertained for the purpose of allowing the deductions towards depreciation and other benefits ....Since the computation of the written down value is for the purpose of the I.T. Act, the method discernible for other provisions of the I.T.Act and the method which presents a more accurate position of income of the assessee should be adopted .The diminishing value method as envisaged by sections 28 to 41 of the I.T.Act has to be employed for calculating the written down value of the assets. **[CIT-Coromandel FertilisersLtd.,AP.,***Sec.80J.r.w.r.19A.***156ITR283, 18Taxman411]**

4. The definition of written down value in section 43(6) of the Income-tax Act, 1961,envisages the computation of the actual cost of each asset, for every assessment year, not only in respect of assets acquired during the previous year but also in respect of assets acquired before the previous year. This naturally has to be done with reference to the factual or legal position that might prevail during the relevant previous year and could be taken into account for the relevant assessment year. The section does not say that the computation of the actual cost of the asset has to be based occasionally on the facts or law as they stood at the time of acquisition of the asset and as could have been taken into account for the assessment year relevant to the previous year of acquisition. Where subsequent information, factual or legal, reveals that the actual cost determined originally was wrong, there can be no doubt that the original figure of actual cost has to be altered, if need be, and, if possible, by reopening the earlier assessments and, if that be not possible, at least for the future. **[Saharanpur Electric Supply Co. Ltd. - CIT, SC,**<u>*Sec.43(6).*</u>**194ITR294,60Taxman412]**

Written off as irrecoverable **Guj.** **Vithaldas H. Dhanjibhai Bardanwala-CIT**

Under section 36 of the Income-tax Act, 1961, it is clear that before any claim for allowance for a bad debt is held established by the Income-tax Officer it must appear that the concerned bad debt was written off as irrecoverable in the account books of the assessee for the relevant previous years. This requirement has become a condition for the grant of claim for bad debt allowance.To that extent, there is a clear departure from the scheme in the earlier Act. Still, so far as the exact requirement of the writing-off of the concerned debt as irrecoverable in the account books of the assessee is concerned, the language used in both the Acts, viz., the Act of 1922 and the Act of 1961, is almost identical. The only requirement of section 36(2)(i)(b) is that the concerned bad debt must have been written off as irrecoverable in the accounts of the assessee. If the debit entries posted by the assessee indicate the said fact the requisite statutory condition has got to be treated as fully complied with. Once the assessee has posted entries in the profit and loss account and corresponding entries are posted in the bad debt reserve account that would be sufficient compliance with the provisions of the statutory requirement for writing off as irrecoverable the concerned debt in the books of the assessee. No further requirement can be spelt out from the express language used by the Legislature. It is not necessary that the assessee must also post corresponding entries in the ledger account of the concerned parties and should close those accounts. <u>*Sec.36(2)*</u>**[21CTR190,130TR95, 6Taxman105]**

ACT NO. XXXII of 1860.

CCXLVIII. (Interpretation of terms)

The following words and expressions in this act shall have the meaning hereby assigned to them, unless there be something in the subject or context repugnant to such meaning :-

2. (India) : The word India shall mean the Territories which are or may become vested in Her Majesty by the statue 21 and 22 vic. 106, entitled an act for the better Government of India.

4. (Local Government) : The words local government shall mean the person immediately administering the executive government in each presidency, lieutenant governorship of province in India.

10. (Land) : The word land shall include and extend to all immovable property, and all hereditaments and tenements whatsoever, whether corporeal or incorporeal of the nature of immovable property, except houses, and all estate or interest therein, whether freehold or chattel, or held by lease, or howsoever otherwise or whether partial or derivative or otherwise and whether in divided or undivided shares.

11. (House) : The word house shall include and extend to all messages and building used for the purpose of habitation and all warehouses, counting houses, factories, and shops and to all out-houses, office, god owns, and building attached to or used with or for the purposes of such massages, buildings warehouses factories or shops.

12.(rack-rent) : The words Rack-rent shall mean the full rent or value at which lands or houses are worth to be let for the year.

13. (Owner): The word Owner as applied to land or houses, shall include any person beneficially entitled in possession to an absolute estate, or to any lesser estate whether freehold or chattel, or partial or derivative, or otherwise, at law or in equity; or any person entitled to any such estate in trust for another person.

14.(Holder) : The word Holder as applied to land or houses, shall include any person in possession or in the receipt of the rents and profits of land or houses under any claim to be entitled to any estate, whether freehold or chattel, partial or derivative or otherwise at law or in equity; and whether on his account or on account of any other person.

15. (Person) : The word Person shall include any corporation.

16. (Representative) : The word Representative shall extend to any person who is legal representative of a person deceased, and shall include in the case of a deceased person subject to the law of England, the heirs of devisees of such a person in regard to real estate, and the executors or administrators or such person in regard to personal estate,and in the case of deceased Mamomedan or Hindoo, the heirs and person legally entitled to succeed to the property of such person it shall also include the successors of a corporations.

17. (Company) : The word Company shall extend to any society , association , fraternity, or partnership of any kind whatever, of or carried on by more than six person.

18. (Trade) : The word Trade shall include any manufacture, and any business, adventure, or concern in the nature of a trade.

19. (Profession) : The word profession shall extend to any employment ,vocation or calling other than trade.

20. (Profit) : The word profit shall include gains of every kind

22.(Oath) : The word oath shall include an affirmation in the case of any person entitled by law to make any affirmation in lieu of any oath or affidavit.

Word	Act/Acts
Company	1869, 1870, 1871, 1872,1886
Defaulter	1869, 1870, 1871, 1872,1886
Firm	1869
Firm	1870,1871,1872
Income	1871,1872,1886
Local authority	1886
Magistrate	1869, 1870, 1871, 1872,1886
Person	1869, 1870, 1871, 1872,1886
Prescribed	1886
Principal officer	1886
Salary	1886

Company means an association carrying on business in British India whose stock or funds is or are divided into shares and transferable, whether such Company be incorporated or not, and whether its principal place of business be situate in British India or not. *[Acts-1870, 1871, 1872, 1886.]*

Defaulter includes a firm making default under this Act.

[Acts-1870,1871,1872.]

2. Defauler includes a company or firm making default under this Act. *[Act-1886]*

Firm includes a Hindu undivided family.

[Acts-1870,1871,1872,1886]

Income means income and profits accruing and arising in British India. *[Acts-1871,1872]*

2. Income means income and profits accruing and arising or received in British India, and includes, in the case of a British subject within the dominions of a Prince or State in India in alliance with Her Majesty, any salary, annuity, pension or gratuity payable to that subject by the Government or by a local authority established in the exercise of the powers of the Governor General in Council in that behalf. *[Act-1886]*

Local authority means any municipal committee, district board, body of port commissioners or other authority legally entitled to, or entrusted by the Government with, the control or management of any municipal or local fund. *[Act-1886]*

Magistrate means any person exercising the powers of a Magistrate, or of a Subordinate Magistrate of the First Class, and includes a Magistrate of Police and a Justice of the Peace.

[Acts 1869,1870,1871]

2. Magistrate means, till the 1st day of September 1872, any person exercising the powers of a Magistrate, or of a Sub-ordinate Magistrate of the First Class, and after the said day, Any Magistrate of the First or Second Class; and it includesa Magistrate of Police and a Justice of the Peace. *[Act-1872]*

3. Magistrate means a Presidency Magistrate or a Magistrate of the first or second class. *[Act-1886]*

Person includes a firm. *[Act-]*

2. Person includes a firm and a Hindu undivied family. *[Act-1886]*

Prescribed means prescribed by the Governor General in Council by notification in the Gazette of India, or by the Governor General in Council or a Local Government by rules made under this Act.

[Act-1886]

Principal officer, used with reference to a local authority or a company or any other public body or assosication not being a local authority or company, means–

(a) The secretary, treasurer, manager or agent of the authority, company, body or association; or

Any person connected with the authority, company, body or association upon whom the Collector has caused a notice to be served of his intention of treating him as the principal officer thereof. *[Act-1886]*

Salary includes allowances, fees, commissions, perquities or profits received, in lieu of or in addition to a fixed salary, in respect of an office or employment of profit; but, subject to any rules which may be prescribed in this behalf, it does not include travelling, tentage, horse or sumptuary allowance, or any other allowance granted to meet specific expenditure. *[Act-1886].*

INDEX AS PER SECTIONS

Sec.	Word	ITR.Pg. No.	Pg.
2(9)	Assessment year	237.202	75
2(12A)*	Books of ac/s.	256.20	99
2(13)	Adventure in the nature of ..	85.220	50
2(13)	Broker, dealer/profession	204.93	101
2(13)*	Business	3.177	104
2(13)	Business	6.1	104
2(13)*	Business	8.41	104
2(13)	Business	74.94	105
2(13)*	Business	83.700	105
2(13)	Business	91.170	105
2(13)	Business	122.203	105
2(13)*	Business	129.295	105
2(13)*	Business	131.554	106
2(13)*	Business	137.346	106
2(13)*	Business	139.999	106
2(13)*	Business	200.131	106
2(13)	Business	212.220	106
2(13)*	Business	240.5	106
2(13)*	Business	247.295	107
2(13)*	Business	259.184	107
2(13)*	Business	260.55	107
2(13) *	Business	276.45	107
2(13)*	Business	298.277	108
2(13)	Business	301.368	108
2(13)*	Business	322.486	108
2(13)	Business	335.527	108
2(13)*	Business and profession	241.172	109
2(13)	Business and vocation	15.263	110
2(13)*	Business/speculation business	131.554	108
2(13)	Trade	70.730	632
2(13)	Trade	107.637	632
2(13)	Trade	224.318	632
2(14)	A controlling interest	329.126	1
2(14)	Agricultural land	94.433	31
2(14)	Agricultural land	123.567	32

Sec.	Word	ITR.Pg. No.	Pg.
2(15)	Object of general public utility	106.752	397
2(18)	Public	98.211	397
2(18)*	Relative	109.353	528
2(18)*	Relative	246.471	529
2(18)	The public	98.291	523
2(22)	Accumulated	62.382	11
2(22)	Accumulated	98.349	12
2(22)	Accumulated profit	121.572	12
2(22)	Accumulated profit of the company	125.118	13
2(22)	Accumulated profits	85.590	12
2(24)	Class of income	17.509	140
2(22)	Deemed dividend	290.433	171
2(22)	Distribution	57.1	188
2(22)	Dividend	33.510	192
2(22)	Dividend	41.275	193
2(22)	Dividend	74.849	193
2(22)	Dividend	237.35	193
2(22)	Payment	109.508	430
2(22)	Share holder	43.352	579
2(22)	Share holder	58.271	579
2(22)	Share holder	83.170	579
2(22)	Shareholder	122.1	580
2(22)	Substantial	41.613	600
2(22)	Substantial part of the business	332.63	601
2(24)	Income	2.121	280
2(24)	Income	3.237	280
2(24)	Income	6.434	280
2(24)	Income	6.534	280
2(24)*	Income	22.13	280
2(24)	Income	26.121	281
2(24)	Income	26.758	280
2(24)	Income	35.33	281
2(24)	Income	86.2	281
2(24)	Income	90.385	281
2(24)	Income	100.97	281

Sec.	Word	ITR.Pg. No.	Pg.
2(31)	Association	33.767	78
2(31)	Association of individuals	4.412	79
2(31)	Association of individuals	5.584	79
2(31)	Association of person	303.102	81
2(31)	Association of persons	88.432	80
2(31)	Association of persons	136.133	80
2(31)	Association of persons	156.273	80
2(31)	Association of persons	171.269	80
2(31)	Association of persons	201.989	80
2(31)	Association of persons	234.12	81
2(31)	Body of individual /an association of	106.111	95
2(31)	Body of individuals	120.564	95
2(31)	Body of individuals	121.985	96
2(31)	Body of individuals	138.19	97
2(31)	Body of individuals	145.702	96
2(31)	Body of individuals	149.131	96
2(31)	Body of individuals	206.647	97
2(31)	Family	97.493	227
2(31)	Individual	7.369	296
2(31)*	Individual	201.989	297
2(31)	Individuals	50.480	296
2(31)	Individuals	74.33	296
2(31)	Person	67.106	437
2(31)*	Person	72.623	437
2(31)	Person	85.28	437
2(31)	Person	109.7	437
2(31)	Person who is substantially	236.981	440
2(35)*	Agent	280.249	28
2(35)*	Principal officer	331.5	472
2(36)	Profession	59.699	484
2(36)	Profession	212.133	484
2(38)	Recognised provident fund	225.201	513
2(40)	Advance tax	328.477	23
2(40)	Assessment	187.98	72
2(40)	Assessment	328.477	74

Sec.	Word	ITR.Pg. No.	Pg.
3	Particular source	62.711	427
3	Previous year	23.82	469
3	Previous year	174.318	470
3	Previous year	255.194	470
3	Previous year	258.56	470
3(1)	Setting up of a business	90.318	569
3(a)	Making up an accounts	68.136	356
3(a)	Making up of an account	243.561	356
4	A Hindu deity	93.348	1
4	Accrues	302.120	10
4	An association of persons	103.688	40
4	Assessed to income- tax	9.284	66
4	Income tax	83.346	293
4	Office	34.92	400
4	Other association of individual	5.246	410
4	Other association of ...	16.412	410
4	Salami	82.464	553
4	Where in respect of a particular so...	44.242	662
4(1)*	Contract becoming frustrated	155.696	155
5*	Accrue and arise	301.207	8
5	Accrues, arises	243.451	10
5	Accruing or arising	3.376	11
5	Accruing or arising	5.216	11
5	Accruing, arising or received	7.48	11
5*	Assessment made under this ..	15.332	74
5*	Chit fund scheme	288.39	139
5	Deemed to accrues	25.50	172
5	Income	238.740	285
5	Income accrues	247.457	286
5	Income assessed in the foreign	113.636	287
5	Income-tax	262.350	293
5	Property	3.246	491
5	Property	32.535	491
5	Property	38.392	491
5	Real income	206.359	505

Sec.	Word	ITR.Pg. No.	Pg.
6(3)	Affairs	287.514	26
9*	Accrue and arise	301.207	8
9	Deemed to accrue or arise in India	117.804	171
9*	Earned	124.391	199
9*	Money lent at interest and brought	184.257	382
9	Operation	18.888	403
9	Operation	23.101	404
9	Property	3.395	491
9	Royalty	172.521	550
9	Royalty	243.459	551
9	Royalty	345.494	552
9	Source	93.44	585
9	Technical services	251.348	617
9	Technical services	267.209	618
9(1)*	Business connection	223.416	112
9(1)*	Business connection	109.158	11
9(1)*	Business connection	3.395	110
9(1)*	Business connection	5.448	110
9(1)*	Business connection	22.241	110
9(1)*	Business connection	58.169	111
9(1)*	Business connection	129.295	112
9(1)*	Business connection	151.286	112
9(1)*	Business connection	164.401	112
9(1)*	Business connection	268.156	113
9(1)*	Business connection	289.438	114
9(1)*	Business connection	120.887	112
9(1)	By such person	332.276	116
9(1)	Contractor	274.501	155
9(1)	Copyright	322.125	158
9(1)*	Effective connection	312.273	200
9(1)	Fees for technical services	251.53	229
9(1) *	Fees for technical services	287.450	229
9(1) *	Fees for technical services	319.139	229
9(1)*	Royalty	332.340	551
9(1)	Royalty	139.806	550

Sec.	Word	ITR.Pg. No.	Pg.
10(22A)	Any income	262.241	50
10(22A)*	Philanthropy	210.299	442
10(23)	Existing solely	327.337	217
10(26)	Harijan	226.696	258
10(26)	Residing	290.481	545
10(26)	Residing in any area specified	320.29	545
10(29)	Authority	94.129	85
10(29)	Marketing of commodity	94.129	372
10(29)	Authority	124.147	85
10(29)	Marketing of commodities	125.136	373
10A	Export turnover	330.175	225
10A*	Full disclosure	343.188	239
10A	Shares are not beneficially held	331.72	580
10AA	Retirement	229.394	546
10B	Production	314.32	482
11	Applied	170.62	60
11*	Charitable object of general public ..	105.546	136
11*	Charitable organization	346.86	136
11*	Charitable purpose	81.417	136
11*	Education	95.279	200
11*	Instrument	128.377	313
11	Property	331.154	493
11	Utilised	245.400	653
11	Wholly for religious or charitable	63.520	667
11	Wholly for religious or charitable	81.1	667
11(1)	Applied	131.497	60
11(1)	In part	131.86	273
11(1)	To the extent to which such income...	345.362	627
11(2)	In the prescribed manner	86.282	277
11(5)*	Debenture	333.248	166
12A	Documents evidencing the creation ..	150.465	194
12A	Religious community	239.528	529
12A	Shall	195.825	574
13	Institution	250.64	313
13(1)*	Debenture	333.248	166

Sec.	Word	ITR.Pg. No.	Pg.
17(1)*	Or otherwise	273.126	405
17(1)	Salary	213.783	555
17(1)	Salary,Profit	243.143	558
17(2)	Concession	286.89	151
17(2)	Perquisite	124.391	435
17(2)	Perquisite	139.937	435
17(2)	Perquisite	152.163	435
17(2)	Perquisite	265.84	436
17(2)	Perquisite	286.89	436
17(2)	Perquisite	345.256	437
17(2)*	Perquisites	220.611	436
17(2)	Salary	201.208	554
17(2)	Salary	345.256	437
17(3)	Mean	151.48	376
18	Due	260.532	195
18*	Income	22.13	280
22	Annual value of property	83.470	44
22	Appurtenance	232.281	61
22	Appurtenant	288.481	61
22	Commercial asset	180.457	141
22	House property	266.685	266
22*	Income from business	207.392	288
22*	Income from business/property	253.732	288
22	Income from house property	146.34	289
22	Income from house property	146.181	290
22*	Income from house property	203.933	290
22	Income from house property	208.645	290
22	Income from house property	263.197	290
22	Income from house property	293.557	290
22	Income from property	102.499	291
22*	Land owner/trader	259.567	339
22*	Lull in business and going out of business	124.561	350
22	Occupy	210.1	399
22	Occupy	271.79	399
22	Owner of a building	2.209	417

Sec.	Word	ITR.Pg. No.	Pg.
24(1)	Repair	169.510	534
24(1)	Voluntarily	154.719	657
24(2)	Receipt	192.495	512
25*	Income	185.6	284
27	House property	331.344	267
28	A separate business	133.85	3
28	Adventure in the nature of trade	118.81	24
28	Adventure in the nature of trade	121.580	25
28	Adventure in the nature of trade	281.354	25
28	Adventure in the nature of trade.	187.316	25
28	Business expenditure / loss	96.568	114
28*	Business income	103.245	114
28*	Business income	221.117	114
28	Business is set up	202.662	115
28*	Capital expenditure	174.689	121
28	Carrying on business	86.402	127
28	Commercial asset	197.1	141
28*	Contract settled	247.79	155
28	Exploitation of a commercial asset	114.769	225
28	Incidental to the carrying on of the	142.800	278
28	Income from business	28.220	288
28	Income from business	207.392	288
28*	Income from business	253.732	288
28	Income from business	263.169	288
28	Income from business	296.661	289
28*	Income from house property	203.933	290
28*	Land owner/trader	259.567	339
28*	Lottery	238.1	349
28*	Payment made on periodical or	207.298	430
28*	Profession / vocation	125.408	485
28*	Professional income	135.146	485
28*	Rental /business income	249.47	533
28	Same business	192.666	562
28	Setting up of business	119.616	570
28	Trading loss	111.263	633

Sec.	Word	ITR.Pg. No.	Pg.
31	Replacement	255.243	537
31*	Revenue/capital expenditure	137.886	548
31(i)	Current repairs	293.201	164
31(1)	Plant	332.235	164
32	Actual cost	118.772	15
32*	Actual cost	127.37	16
32	Actual cost	137.1	16
32	Aircraft, aero engines	230.88	37
32	Assessee	6.124	66
32	Assessee	11.328	67
32	Automotive / Locomotive	260.55	86
32	Being property of the assessee	103.455	90
32	Building	65.377	102
32	Building	68.62	102
32	Building	82.353	102
32	Building	218.511	102
32	Building	233.391	103
32	Building	274.66	103
32	Building owned by the assessee	127.97	104
32	Building owned by the assessee	239.775	104
32*	Chemical works	190.535	138
32	Corrosive chemicals	178.419	158
32	Crane	256.50	160
32	Depreciation	282.132	178
32	Discarded	96.672	185
32	Electrical machinery	204.447	201
32	For the purpose of its business	268.395	234
32	Goodwill	331.192	254
32	Hire	226.914	263
32	Hire	254.445	263
32*	Hire/lease	265.114	263
32	Installed	38.25	312
32	Installed	75.533	312
32	Installed	201.117	312
32*	Jeep	193.483	329

Sec.	Word	ITR.Pg. No.	Pg.
32	Plant	272.383	460
32	Plant	273.154	460
32	Plant and machinery	52.615	462
32	Production	271.331	482
32	Put to use	306.114	500
32	Sold	146.136	584
32	Solely	275.582	585
32	Textiles	241.350	620
32	The and such	8.41	621
32	The number of days on which	227.286	622
32	Thing	225.178	626
32	Transportation	226.914	643
32	Use	278.142	650
32	Used	5.621	650
32	Used	13.415	650
32	Used	221.857	650
32	Used	272.583	651
32	Used	276.625	651
32	Used	298.245	651
32	Used for the purpose of business	5.734	653
32	Used for the purpose of business	278.142	652
32	Used for the purpose of the business	49.177	651
32	Used for the purpose of the business	123.404	652
32	Used for the purposes of business	277.60	653
32	Used for the purposes of the business	328.297	653
32	Welfare centre	250.78	661
32	Worked	48.853	673
32(1)	Actually allowed	254.780	20
32(1)	Amount of scrap value, if any	259.733	38
32(1)*	Became due	86.501	90
32(1)	Building	150.23	102
32(1)	Business and profession	295.376	109
32(1)	Destroyed	195.682	181
32(1)	Employed	201.1036	201
32(1)	For purposes of business of	194.749	233

Sec.	Word	ITR.Pg. No.	Pg.
32A	Plant and machinery	248.622	462
32A	Plant installed	196.667	462
32A	Processing	196.179	478
32A	Processing / production	166.646	480
32A	Production	201.17	481
32A	Production of mineral	221.48	483
32A	Production and manufacturing process	286.470	483
32A*	Road transport vehicle	260.55	549
32A	Road transport vehicle	264.273	549
32A	Small- scale industrial undertaking	229.93	584
32A(5)	Otherwise transferred	263.697	412
32A(5)*	Otherwise transferred	223.324	412
32AB	Shall	254.6	576
33*	Includes	112.288	289
33	Installed	164.633	312
33	Manufacture or production	155.794	368
33*	Office appliance	151.75	400
33*	Plant	118.809	445
33*	Plant	121.996	447
33	Plant	122.288	447
33*	Plant	151.75	449
33	Plant	164.633	451
33	Plant	198.701	454
33*	Profit or loss of a capital nature	130.351	488
33	Residential accommodation	122.296	545
33	Ship	217.609	580
33	Ship	232.709	581
33	Textiles made wholly or mainly of	152.457	620
33*	Transfer	219.478	641
33	Wholly used for the business	211.370	668
33(1)	Iron and steel (metal)	180.155	324
33(1)	Manufacture and produce	186.597	365
33(1)*	Metal	118.39	376
33A	And – or	127.481	43
33A	Plant	116.259	445

Sec.	Word	ITR.Pg. No.	Pg.
37*	Revenue/capital expenditure	137.886	548
37	Taxi	296.627	616
37	Travelling	246.505	644
37	Used for the purpose of business	25.265	651
37	Wholly laid out for the purpose of	200.792	668
37(1)	Commercial expediency	118.379	141
37(1)	Commercial expediency	288.1	142
37(1)	Commercial expediency	324.396	142
37(1)	Expenditure is laid out	117.505	221
37(1)	Expenditure laid out wholly and	208.503	221
37(1)	Expenditure made wholly and ...	116.751	222
37(1)	Expenditure of capital nature	117.505	223
37(1)	For the business of the assessee	135.221	233
37(1)	For the purpose of business	81.664	234
37(1)	For the purpose of business	288.1	234
37(1)	For the purpose of the business	101.817	235
37(1)	In the nature of	314.329	274
37(1)	Not a business loss	120.789	392
37(1)	Opposed to public policy	101.817	404
37(1)	Penalty for an infraction of the	140.782	431
37(1)	Revenue expenditure	131.650	548
37(1)	Revenue expenditure.	71.752	548
37(1)	Whole and exclusive purpose	150.320	664
37(1)	Wholly and exclusively	49.156	664
37(1)	Wholly and exclusively	118.261	664
37(1)	Wholly and exclusively	166.836	665
37(1)	Wholly and exclusively	274.465	665
37(1)	Wholly and exclusively for the	88.1	666
37(1)	Wholly and exclusively for the	144.883	666
37(2A)	Entertainment	122.972	205
37(2A)	Entertainment	215.165	206
37(2A)	Entertainment	240.297	206
37(2A)	Entertainment expenditure	145.37	207
37(2A)	Expenditure in the nature of ...	120.140	220
37(2A)	Expenditure in the nature of ...	126.430	221

Sec.	Word	ITR.Pg. No.	Pg.
40(A)	Deposit	227.480	176
40(A)	Deposit	231.294	176
40(A)	Deposit	240.758	176
40(A)	Deposit	251.661	176
40(A)	Deposits	227.473	177
40(A)	Deposits	273.432	178
40(A)	Financial company	281.316	230
40(A)	Investment company	281.316	323
40(A)	Perquisites	220.611	436
40(A)	Stock	190.61	597
40A(2)	The A.O.is of the opinion	303.271	621
40A(3)	Expenditure	109.1	218
40A(3)	Expenditure	160.504	219
40A(3)	Expenditure	169.12	219
40A(3)	Expenditure	190.61	219
40A(3)	Expenditure	191.667	219
40A(3)	Expenditure	193.646	219
40A(3)	Practicable	112.134	464
40A(3)	Sum	121.680	609
40A(7)*	Provision made by the assessee	156.585	496
40A(8)	General deposits	200.252	244
41	Books of the assessee	77.338	100
41	Capital receipt	81.777	123
41	Cessation of liability	140.286	133
41	Cessation of the liability	330.435	133
41	Obtained	103.312	397
41	Received	46.771	513
41	Remission of the liability	114.853	**530**
41(1)	Cessation or remission of the	158.78	134
41(1)	Obtained	237.821	398
41(2)	Sale and barter	66.692	560
41(2)	Due	200.1	195
41(2)	Money	237.24	382
41(2)	Demolished or destroyed	265.13	174
41(2)	Due	266.135	195

Sec.	Word	ITR.Pg. No.	Pg.
43(3)	Plant	110.281	445
43(3)	Plant	116.231	445
43(3)	Plant	117.15	446
43(3)*	Plant	118.809	446
43(3)*	Plant	121.996	446
43(3)	Plant	126.196	448
43(3)	Plant	129.728	448
43(3)	Plant	132.401	448
43(3)	Plant	137.110	449
43(3)*	Plant	144.585	449
43(3)	Plant	156.283	450
43(3)	Plant	157.86	450
43(3)	Plant	170.431	451
43(3)	Plant	173.100	451
43(3)	Plant	173.374	452
43(3)	Plant	191.460	454
43(3)	Plant	192.20	454
43(3)	Plant	207.494	455
43(3)	Plant	217.622	456
43(3)	Plant	225.154	456
43(3)	Plant	229.776	457
43(3)	Plant	232.270	457
43(3)	Plant	236.615	458
43(3)	Plant	240.1	458
43(3)	Plant	240.816	458
43(3)	Plant	247.25	459
43(3)	Plant	253.520	459
43(3)	Plant	260.55	459
43(3)	Plant	274.529	461
43(3)*	Road transport	281.297	549
43(5)	A breach of the contracts/speculative...	107.845	589
43(5)	A settlement of the contract	144.57	3
43(5)	A speculative transaction / A breach	154.756	4
43(5)	Any speculation business	322.43	57
43(5)*	Contract settled	247.79	155

Sec.	Word	ITR.Pg. No.	Pg.
43B	Actually Paid	290.282	21
44*	Borne by	277.97	100
44	Expenditure	106.969	218
44	More reliable data	13.141	384
44	More reliable data	17.173	384
44	Seller	202.535	566
44	Spending	240.139	595
44A	Expenditure	134.68	218
44A	Similar association	171.504	582
44A*	Mutuality	171.504	386
44AB*	Audit	175.254	85
44AC	Right to receive	248.554	549
44AC*	Income	185.375	284
44BB	Aircraft	249.162	36
44BB	Services	320.268	568
44C	Profits whichever is greater	26.256	490
44D*	Permanent establishment	272.99	434
45	Capital loss	183.493	121
45	Cost of acquisition	221.810	158
45	Exchange	95.656	215
45	Exchange	177.198	215
45*	Full value of the consideration ..of	66.622	240
45*	Gift	122.313	247
45*	Gift	127.354	247
45*	Hold	207.148	264
45*	Improve	196.939	270
45*	Jewellery held for personal use	154.573	331
45*	Or the extinguishment of any rights	119.399	405
45	Relinquishment	95.656	530
45	Relinquishment	177.198	530
45	Sale or Transfer	24.33	560
45*	Transfer	131.42	637
45*	Transfer	163.125	638
45*	Transfer	165.787	639
45*	Transfer of a capital asset	273.1	643

Sec.	Word	ITR.Pg. No.	Pg.
80HH*	Manufacture	203.542	358
80HH*	Manufacture	221.1	360
80HH*	Manufacture	223.776	360
80HH	Manufacture	226.245	360
80HH*	Manufacture	241.319	360
80HH	Manufacture	284.222	363
80HH	Manufacture	292.444	363
80HH*	Manufacturing process	272.181	369
80HH*	Production	222.418	481
80HH*	Production	308.98	483
80HH*	Production	314.309	482
80HH	Production or produce	239.375	483
80HHA*	Industrial undertaking	251.806	304
80HHA*	Manufacture	251.806	304
80HHB	Rendering technical services	195.81	531
80HHC	Processing,manufacture &production	213.851	479
80HHC	Agricultural primary commodities	280.62	33
80HHC	Any other person	272.314	52
80HHC	Attributable	295.454	82
80HHC	Derived	297.17	179
80HHC	Derived from	262.669	180
80HHC	Derived from /attributable	293.214	181
80HHC	Direct costs	329.426	184
80HHC	Goods	291.8	253
80HHC	Goods or merchandise	267.488	254
80HHC*	Income from other sources	330.199	291
80HHC	Merchandise	313.149	376
80HHC	Minerals	237.131	377
80HHC	Premium	262.88	468
80HHC	Profit	266.521	487
80HHC	Profit	272.29	487
80HHC	Profit	294.1	487
80HHC	Profit	326.533	488
80HHC	Profits	280.1	489
80HHC	Reasons beyond his control	273.8	511

Sec.	Word	ITR.Pg. No.	Pg.
80IA	Manufacture	320.665	364
80IA*	Manufacture or production	320.79	368
80IB	Derived from	317.218	180
80IB	Manufacture	329.348	363
80IB	Manufacture	333.192	364
80IB	Processing	318.455	479
80IB	Production	295.148	482
80IC	Manufacturing	309.344	369
80J	Articles	93.548	62
80J	Articles	208.541	62
80J	Begins to manufacture	281.255	90
80J	Capital	133.365	117
80J	Capital employed	113.109	118
80J	Capital employed	152.308	119
80J	Capital employed	181.214	119
80J	Catching fish	214.239	131
80J	Derived	215.64	179
80J*	Derived	261.737	179
80J	Four assessment years immediately	236.130	238
80J	Machinery and plant used in any	110.103	353
80J	Manufacture	68.325	357
80J*	Manufacture	192.128	358
80J*	Manufacture	203.542	358
80J	Manufacture	206.367	359
80J*	Manufacture	221.1	360
80J*	Manufacture	223.776	360
80J*	Manufacture	241.319	360
80J	Manufacture	245.538	361
80J	Manufacture	281.255	362
80J	Manufacture and produce	248.744	365
80J	Manufacture or produce	217.849	367
80J	Manufacture or production of ...	203.833	369
80J*	Manufacturing process	272.181	369
80J	Per annum	152.471	433
80J	Previously used for any purpose	156.463	471

Sec.	Word	ITR.Pg. No.	Pg.
80P	Gross total Income	299.444	255
80P	Marketing	253.659	372
80P	Marketing	323.666	372
80P	Purchase	156.422	499
80P(2)	Letting	212.124	340
80P(2)	Marketing	115.709	370
80P(2)	Marketing	182.53	371
80P(2)	Marketing	231.814	371
80P(2)	Of its members	231.814	400
80P(2)	Or	300.24	405
80P(2)	Provided credit facility	114.70	494
80P(2)	Providing credit facilities	122.913	494
80P(2)	Providing credit facilities	135.355	494
80P(2)	Providing credit facilities	204.713	494
80P(2)	Providing credit facility	170.455	495
80P(2)	The whole of the amount of profits	322.283	625
80P(2)	Urban Consumer Co. op. Society	75.445	649
80RRA*	Employer	170.137	202
80TT	Lottery	123.187	349
80TT*	Lottery	189.698	349
80TT	Lottery	248.718	349
84	Galvanisation and manufacturing	128.472	244
84*	Investment	136.205	320
84	Manufacturing	126.333	369
84	Profits	198.282	488
84	Reconstruction	137.851	513
84	Six per cent. per annum	122.283	583
84	Splitting up of the business already	137.851	596
84(1)	Capital employed	122.203	119
84(2)	Transfer	134.483	638
94(3)	Exceptional and not systematic	265.1	215
104*	Assessable income	28.78	63
104	Assessable income	46.1135	63
104	Company dealing in or holding ..	106.45	145
104	Company whose business consists ..	215.817	145

Sec.	Word	ITR.Pg. No.	Pg.
127	Case	192.387	129
127	Case	211.256	129
127	Case	294.363	130
127	Power	268.465	463
131(3)	Impound any document or thing	322.381	269
132	Any	315.137	48
132	Articles or things	93.244	62
132	Information	101.112	307
132	Other person	315.137	411
132	Reason to believe	139.1043	506
132	Requisition	287.172	538
132	Search	93.450	565
132	Search initiated	323.467	565
132	Seizure	93.244	566
132	Seizure	98.581	566
132	Valuable thing	131.506	655
132(1)	Search	101.854	565
132(3)*	Mala fide	217.83	350
132(5)	Person concerned	98.581	439
132A	Detection	261.87	182
132A	Taken into custody	244.845	611
137	Appropriate authority	170.62	61
137	Public servant	10.429	499
137	Purchase	254.22	500
139	At any time	79.278	81
139	Defective return	213.862	173
139*	Due time	266.1	196
139	Invalid return	213.862	318
139*	Omission	30.57	401
139*	Total income	131.680	629
139(1)	His income	125.624	263
139(2)*	Mistake apparent from the record	125.630	64
139(8)	Month	155.422	384
139(8)	Reduce	145.266	521
142	Within such period not being less	13.154	670

Sec.	Word	ITR.Pg. No.	Pg.
147*	And also	331.236	43
147	Any person	79.242	53
147	Any person	137.213	53
147	Assessment	80.759	71
147	Definite information	15.224	174
147	Definite information	14.431	173
147	Discover	12.8	168
147	Discovers	14.431	167
147	Discovers	32.418	167
147	Escape	3.171	210
147	Escaped	1.143	210
147	Escaped	3.438	211
147	Escaped assessment	6.51	211
147	Escaped assessment	32.289	211
147	Escaped assessment	236.832	212
147	Escaped assessment	247.192	213
147	Escaping assessment	40.307	213
147	Failure	30.57	227
147	Failure on the part of the...facts	268.203	227
147*	Full disclosure	343.188	239
147	Fully and trully	339.595	241
147	Has escaped assessment	2.71	258
147	Has reason to believe	257.512	259
147	In any year	2.30	271
147	Income profit or gains which have ..	40.407	292
147	Information	32.418	304
147	Information	35.1	305
147	Information	40.163	301
147	Information	48.837	305
147	Information	51.479	305
147	Information	52.1	305
147	Information	66.322	306
147	Information	67.11	306
147	Information	80.188	306
147	Information	84.584	306

Sec.	Word	ITR.Pg. No.	Pg.
147	Regular assessment	249.456	527
147	To assess	34.601	627
147	To inform	64.218	627
147	Too low a rate	66.604	628
148*	Hybrid systems of accounting	208.930	627
148	Issue	44.325	326
148	Issue and serve	46.28	326
148	Issue and serve	53.100	328
148	Issue and serve	101.106	236
148	Issued and served	105.479	328
148	Service of notice	96.141	568
149	For any year and that year	44.437	232
149*	Issue and serve	157.13	326
153	Direction	69.500	184
153	Finding	47.16	230
153	Finding	59.652	230
153	Finding and direction	52.335	231
153	Finding and Direction	195.438	231
153(1)*	Forward	225.963	227
153(3)	An order of any court	155.739	242
153(3)	Direction	137.213	185
153(3)	Directions and findings	62.729	185
153A	Abatement	290.114	5
153A	Abatement	346.130	5
153A	Pending	346.130	432
154	A mistake apparent on ...	82.50	2
154	A mistake on the face of the record	262.209	2
154	Apparent	248.647	59
154	Apparent from record	32.350	60
154	Assessment or reassessment	125.730	74
154	Debatable issue	284.42	165
154	Error apparent on the face of the	253.507	210
154	Mistake apparent from the record	100.347	379
154	Mistake apparent from the record	102.213	379
154	Mistake apparent from the record	122.863	380

Sec.	Word	ITR.Pg. No.	Pg.
214	Regular assessment	106.38	522
214	Regular assessment	110.289	522
214	Regular assessment	130.928	523
214	Regular assessment	131.239	523
214	Regular assessment	151.389	523
214	Regular assessment	155.548	523
214	Regular assessment	171.224	524
214	Regular assessment	179.580	524
214	Regular assessment	186.217	524
214*	Regular assessment	197.563	525
214	Regular assessment	199.832	526
214	Regular assessment	268.244	527
215	Assessed tax	125.429	64
215	Assessed tax	149.703	65
215	Assessed tax	193.134	65
215	Assessed tax	235.433	66
215*	Justified	151.542	333
215*	Reduced by the amount of tax	235.433	521
215	Regular assessments	146.452	528
215	Waiver	151.584	661
216	Estimate and underestimate	123.712	213
220	Debt	45.622	166
220(1)*	Detriment	276.278	182
220(2)	Hardship	280.70	258
220(A)	Genuine hardship	280.70	246
220(A)	Genuine hardship	306.196	246
221	Advance tax	101.531	22
221	Tax	101.531	612
222	Assessee in default	121.868	69
222	Payable	122.227	429
226	Assets	47.311	77
226	May become due	52.321	375
226(3)	Garnishee	259.586	244
230A	Any person	182.342	54
230A	Such person	150.351	607

Sec.	Word	ITR.Pg. No.	Pg.
254(2)	An error apparent on the face of the	320.12	42
254(2)	Apparent	250.113	59
254(2)	Mistake apparent from the record	140.361	380
254(2)	Mistake apparent from the record	296.595	380
254(2)	Mistake apparent on the face of the	305.227	381
255	Assessing Officer	298.390	70
256	Accompanied by a fee of Rs.100/-	26.741	6
256	An application made	46.272	39
256	Arising out of the order passed by it	54.777	61
256	Assessed	87.13	63
256	Point of law	140.418	462
256	Question of law arising out of such	78.474	502
260A	A substantial question of law	248.152	5
260A	Perverse decision	260.341	442
260A	Perversity	330.291	442
260A	Substantial question of law	245.297	602
260A	Substantial question of law	245.730	602
260A	Substantial question of law	248.155	603
260A	Substantial question of law	250.291	603
260A	Substantial question of law	262.692	604
263	An incorrect assumption of facts	243.83	42
263	Any order	330.568	51
263	Erroneous	99.375	207
263	Erroneous	220.456	207
263	Erroneous	251.707	207
263	Erroneous and prejudicial to the ...	88.323	208
263	Erroneous and prejudicial to the ...	243.83	208
263	Erroneous Judgment	290.395	209
263	Erroneous order	289.226	209
263	Erroneous Order	332.167	210
263	Examining the record	259.386	214
263	If he considers	96.310	268
263*	Income tax Officer	252.888	294
263	Mistake	33.290	378
263	Mistake apparent from the record	46.609	379

Sec.	Word	ITR.Pg. No.	Pg.
271(1)	Concealment	131.680	149
271(1)	Concealment	243.818	150
271(1)	Concealment	265.562	150
271(1)	Concealment	298.109	150
271(1)	Concealment of income	291.519	150
271(1)	Fails to furnish the return	99.448	226
271(1)	Fraud or wilful default	282.321	238
271(1)	Furnishing inaccurate particulars	322.158	243
271(1)	Furnishing of inaccurate particulars	91.467	243
271(1)*	Good faith	190.93	251
271(1)*	Good faith	239.386	251
271(1)	Inaccurate	291.519	278
271(1)	Inaccurate	292.11	278
271(1)	Income as returned by such person	57.270	287
271(1)	Month	97.285	383
271(1)	Month	106.846	383
271(1)	Month	124.485	383
271(1)	Month	274.485	384
271(1)*	On which a penalty has been	88.110	403
271(1)	Particulars	322.158	427
271(1)	Particulars	325.378	427
271(1)	Payable	6.616	429
271(1)	Penal	87.163	431
271(1)	Reasonable cause	154.220	510
271(1)	Satisfied	278.3	563
271(1)	Tax so assessed	126.466	615
271(1)	The amount of tax, if any, payable by	88.192	621
271(1)	The person liable to penalty	130.955	623
271(1)	The tax , if any,payable by him	86.657	624
271(1)	To account	256.20	626
271(1)*	Total income	131.680	629
271(1)	Voluntary	239.386	659
271(3)	No income liable to tax	140.428	390
273A	Cooperation	129.53	158
273A	Full and true disclosure	198.54	238

Sec.	Word	ITR.Pg. No.	Pg.
281B	For the purpose of protecting..	252.642	235
281B	Opinion	246.707	404
281B	Transfer	278.187	642
282	A proper notice	88.374	2
282	An authorised agent of the assessee	24.381	42
282	Deemed to have been effected	301.69	172
282	Deemed to have been served	154.556	172
282	Proper service of notice	136.108	491
282	Serve	30.381	567
282	Served with notice	2.484	567
283(2)	Immediately before	90.412	269
288	Thenceforward	19.616	625
288(2)	Authorised representative	141.82	85
288(4)	On which a penalty has been imposed	88.110	403
292B	Invalid return	310.71	318
293	Any assessment made under this Act	50.93	49
293	Good faith	115.61	250
293	Good faith	288.595	252
293*	Mala fide	217.83	356
293	Under the Act	62.367	647
293A*	Mines and Mining	236.652	377
297	Assessment	66.680	71
297	Assessment	109.629	71
297	Assessment	179.30	72
297	Procedure for the assessment	66.680	473
297	Without prejudice to any action	165.751	671
297(2)	All the provisions of this Act shall ...	167.794	37
297(2)	Investment	59.31	320
297(2)	Issued	75.479	327
297(2)	May	116.516	374
297(2)	Proceedings for the assessment	62.544	474
Rules*	Hire/lease	265.114	263
Rule5*	Chemical works	190.535	138
Rule-5	Electrical machinery	242.636	201
Rule-5	Goods carriage	256.50	253

GIFT TAX ACT

Sec.	Word	ITR No.	Pg.
2	Person	120.126	438
2(viii)	Donee	255.529	194
2(xii)	Disposition	76.675	188
2(xii)*	Gift	94.433	246
2(xii)	Gift	122.313	247
2(xii)	Gift	127.354	247
2(xii)*	Gift	133.108	248
2(xii)*	Gift	172.632	248
2(xii)	Consideration	215.165	151
2(xii)	Gift	226.598	248
2(xii)	Gift	240.539	248
2(xii)	Gift	245.831	249
2(xii)	Gift	255.529	249
2(xxiv)	Transaction	67.612	634
2(xxiv)	Transaction	70.812	634
2(xxiv)	Transaction entered into	76.675	635
2(xxiv)	Transaction	82.599	634
2(xxiv)	Gift	114.90	246
2(xxiv)	Transaction	169.708	634
2(xxiv)	Family settlement	221.842	227
2(xxiv)	Transaction	246.609	634
2(xxiv)	Allotment of shares	307.31	37
3,21.	Body of individual whether ..	120.126	95
3(b)	Maintenance/Gift	248.103	355
5(1)	In the course of marketing of	98.504	274
4	Adequate consideration	106.390	21
4	Adequate consideration	245.347	22
5(1)	On the occasion of marriage	164.369	402
5(1)	On the occasion of the ...	201.741	402
16	Reason to believe	293.321	509
16	Satisfaction	293.321	563

WEALTH TAX ACT

Sec.	Word	ITR No.	Pg.
2*	Assets	156.525	77
2	Wealth of the assessee	292.622	661
2(e)	Agricultural land	56.608	30
2(e)	Agricultural land	59.428	31
2(e)	Agricultural land	105.133	31
2(e)	Agricultural land	149.588	32
2(e)	Agriculture	75.318	35
2(e)	An asset in the hands .....on the	133.800	40
2(e)	Annuity	83.117	45
2(e)	Annuity	177.339	45
2(e)	Annuity	192.318	46
2(e)	Annuity	215.454	47
2(e)	Annuity	228.190	47
2(e)	Annuity and property	76.471	47
2(e)	Asset	180.289	76
2(e)	Asset	186.421	76
2(e)	Asset	215.30	76
2(e)	Assets	196.8	77
2(e)	Assets	220.78	78
2(e)	Assets	274.257	78
2(e)	Assets	277.344	78
2(e)	Commutation of the value of an	129.642	144
2(e)	Is	184.34	325
2(e)	Net wealth	127.731	389
2(e)	Property	93.332	492
2(e)	Property	215.330	492
2(e)	Property	274.553	493
2(e)*	Out house	76.584	413
2(ea)	Building	304.235	103

Sec.	Word	ITR No.	Pg.
4	Such assets	115.160	605
4(1)	Adequate consideration	91.179	21
4(1)	Assessee	111.384	67
5	Non-resident	156.373	392
5*	Assets	156.525	72
5*	Industrial undertaking	235.5	303
5(1)	An article ... of personal use	168.363	39
5(1)	Any policy of insurance	156.525	55
5(1)	Belonging	197.250	91
5(1)	Belonging to	262.403	91
5(1)	Exclusive use	245.800	215
5(1)	Growing Crops	221.39	256
5(1)	House	130.393	264
5(1)	House	219.733	264
5(1)	House	233.460	265
5(1)	House	237.152	265
5(1)	House	264.482	265
5(1)	House	273.500	265
5(1)	House	313.28	265
5(1)	Industrial undertaking	208.552	301
5(1)	Industrial undertaking	209.368	30
5(1)	Industrial undertaking	216.566	302
5(1)	Industrial undertaking	236.885	303
5(1)	Investment	257.275	321
5(1)	Jewellery	102.105	329
5(1)	Jewellery	140.374	329
5(1)	Jewellery	169.245	330
5(1)	Jewellery	177.297	330
5(1)	Jewellery	211.811	330
5(1)	Jewellery	214.721	331
5(1)	Land appurtenant thereto	151.320	336
5(1)	Manufacture	236.885	360
5(1)	Manufacture or process	240.636	366
5(1)	One house used for residential	107.104	403

Sec.	Word	ITR No.	Pg.
18(2A)	In good faith made full	99.414	273
18(2A)	In good faith made full	134.636	273
18(3)	Who shall ..have powers ....for t	128.504	664
18B	Good faith	121.722	250
18B	Good faith	197.269	251
21	Capital of that concern	239.448	122
21(1)*	Property	173.479	492
21(2)	In the like manner..	136.162	274
23(1A)*	Without prejudice	267.705	671
25	Erroneous order	261.611	209
25	Proceedings are dropped	214.130	474
25	Record	208.26	516
25	Record	244.574	519
25	Record	252.586	519
25	Record	276.623	520
25	The subject of an appeal	50.578	524
25(1)	Where the order is the subject of	166.392	553
34AB	Prescribed	188.741	469
35	Mistake apparent from the	105.115	379
39*	Reheard	211.472	528
45(d)	Company carrying on an	60.715	144
45(d)	Engaged	74.465	204
Rule-1	Business premises	195.367	115
Sch.	Business	243 .271	106
Sch.	Quoted share	261.763	502

CONSTITUTION OF INDIA